Physics of Nerves and
Excitatory Membranes

Physics of Nerves and Excitatory Membranes

Thomas Heimburg

WILEY VCH

Author

Thomas Heimburg
Niels Bohr Institute
Blegdamsvej 17
University of Copenhagen
Copenhagen
DK, 2100

Cover Design: Wiley
Cover Image: © Thomas Heimburg

Typesetting Lumina Datamatics

Library of Congress Card No.: applied for

British Library Cataloguing-in-Publication Data
A catalogue record for this book is available from the British Library.

Bibliographic information published by the Deutsche Nationalbibliothek
The Deutsche Nationalbibliothek lists this publication in the Deutsche Nationalbibliografie; detailed bibliographic data are available on the Internet at http://dnb.d-nb.de.

Print ISBN 9783527331802
ePDF ISBN 9783527690626
ePub ISBN 9783527690619
oBook ISBN 9783527690596

Contents

Preface

The first scientist to study the thermodynamics and the velocity of nerve impulses was Hermann von Helmholtz around 1850. Helmholtz made numerous contributions not only to important problems in physiology but also to theoretical physics. He also wrote a popular book, *On the Sensations of Tone as a Physiological Basis for the Theory of Music* [1] – seemingly unrelated to nerves. It expresses a fascination with the effect that music has on us and attempts to relate it to the physical construction of the ear. The question of how the ear is connected to the perception of sound in the brain is still not well understood. It may be that sound is only perceived in the ear and then converted into signals of a different nature. However, many people use analogies to music to describe brain function and perception: "good vibrations," "ringing a bell," "being in harmony," "in tune," and "being on the same wavelength" [2] – or a "cognitive dissonance." "That sounds good" expresses something that makes sense.

How is it that the brain can recognize the doubling of a frequency (an octave) and reconstruct the 12 tones of the musical scale by the circle of fifths? Why does the brain experience "harmony" when listening to music? In neural models based on electrical phenomena, there is no good reason why this should be so. Are the aforementioned terms just useful analogies? Then why can mechanical changes be measured during the nerve pulse, why is the nerve mechanically excitable, why is the pulse velocity close to the speed of sound in membranes, and why can the brain be stimulated by ultrasound?

In this book, I will present a new concept of the function of excitatory membranes, nerves, and the brain that is based on thermodynamics and the theory of adiabatic processes, of which sound is one aspect. It has important early proponents such as Hermann von Helmholtz, Archibald V. Hill, J.V. Howarth, J.M. Ritchie, and most importantly, I. Tasaki of the NIH. In a thermodynamic treatment of nerves, all variables must be considered if one wishes to understand the properties of cell membranes.

For this reason, a nerve impulse has electrical as well as mechanical, thermal, and chemical components. A thermodynamic view of excitatory membranes differs significantly from the electrophysiological description of nerves as developed by Julius Bernstein, Alan V. Hodgkin, Andrew F. Huxley, Bernard Katz, Erwin Neher, and Bert Sakmann, among others. The latter concepts are based exclusively on electrochemistry and have no explicit mechanical or thermal aspects, except for the idea that there may be macromolecules that are sensitive to these variables, such as mechanoreceptors, which influence the electrical properties. The electrophysiological approach is best represented by the Hodgkin–Huxley model of 1952, which won the Nobel Prize in 1963. It is described in Part IV of this book. The Hodgkin–Huxley model is inherently based on dissipative phenomena that increase entropy.

In contrast, the thermodynamic soliton theory at the heart of this book is based on adiabatic phenomena, i.e., on processes that conserve entropy. The two approaches are therefore mutually exclusive.

I have been thinking about the thermodynamics of excitable membranes for more than two decades. My interest actually goes back even further, to my PhD training at the Max Planck Institute, where I grew up in a field of tension between physical chemists and molecular biologists, who did not seem to understand each other very well. The very different approaches of the two fields were embodied by two Nobel laureates. The first is Manfred Eigen, a physicist with a deep interest in biology. He won his prize for the experimental and theoretical description of fast reaction kinetics and went on to develop a thermodynamic theory of biological evolution. The other scientist is Erwin Neher, who was honored with Bert Sakmann for developing the patch-clamp technique to study the function of protein ion channels. In their field, thinking tends to focus on the function of individual molecules. Whereas physicists tend to seek beauty in finding simple explanations for seemingly complicated processes, molecular biologists marvel at the complexity of biological systems, which they see as the result of 4 billion years of evolution. There are compelling arguments to be made for both approaches.

Because of my physics training, I tend to take a physics or physical chemistry approach to biology. The term *physiology* was coined by Emil du Bois Raymond in the first half of the nineteenth century to describe a physical approach to biology. I have great sympathy with Erwin Schrödinger's influential book *What Is Life?* [3]. Schrödinger believed that biology was based solely on physical and especially thermodynamic processes. I believe that many processes only seem complicated because one has not yet found the most appropriate physical level of description for a particular problem.

In particular, many problems turn out to be much simpler if you try to understand them not as the sum of the properties of individual molecules (as is typically done in molecular biology), but rather at the level of properties that emerge at larger scales. This approach is common in physics and can lead to simple laws even when the atomic or molecular structure of the matter under consideration is complex. An example is the emergence of waves in continuous matter, which depend on elastic moduli.

Waves can be understood using a simple differential equation that requires no knowledge of the molecular structure of the medium. This is because the wavelength is typically many orders of magnitude larger than atoms and molecules. Waves cannot be understood at the level of the elementary components of a system (e.g., atoms or molecules), with the possible exception of crystals. This means that the concept of waves cannot be deduced from the structure of molecules. Waves in biology are outside the scope of molecular biology. We will show here that such considerations apply to nerves.

My own expertise is in the field of membrane thermodynamics. In this book, I will show that phenomena, such as the nerve impulse, can be convincingly understood by applying a thermodynamic formalism to nerves and excitatory membranes. The propagation of the pulse, its temporal length, the occurrence of ion channel-like events in nerve membranes, and the action of anesthetics are all rooted in thermodynamic couplings and are contained in one of the deepest thermodynamic concepts: the fluctuation-dissipation theorem. It originated in a seminal paper on thermodynamics by Albert Einstein in 1910 dealing with the phenomenon of critical opalescence [4] – one of Einstein's many important works that defined a whole new field. This theorem describes the couplings between the fluctuations of thermodynamic variables, the susceptibilities, and the time scales of processes. The emergence of lipid channels in melting transitions, as described in Part VIII, is reminiscent of critical opalescence.

Thermodynamics in itself is not really a theory for any particular problem because it is absolutely generic. It works at the atomic level as well as at astrophysical scales, in solids as well as in soft or biological matter. Thermodynamics is more like the language of physics. It has words and a grammar. Terms such as energy, temperature, pressure, and entropy form the words, and the laws of thermodynamics form the grammar of this language. There are strong couplings between different variables, expressed by Maxwell's relations. For example, $(\partial S/\partial p)_T = (\partial V/\partial T)_p$ relates the pressure dependence of entropy and the volume expansion coefficient, a coupling that is not immediately intuitive.

As Albert Einstein noted [5]:

> *"Therefore the deep impression which classical thermodynamics made upon me. It is the only physical theory of universal content concerning which I am convinced that, within the framework of the applicability of its basic concepts, it will never be overthrown."*

Any violation of a thermodynamic principle falsifies a theory, regardless of its field of application. A correct description of biological processes can never be in conflict with any thermodynamic law. I hope that the power of a thermodynamic approach to biology will be evident in the description of nerves and excitatory membranes presented in this book.

There are also other important aspects of scientific investigation. In his 1637 essay "Discourse on the Method," René Descartes described his scientific method and formulated four steps:

1. *"The first was never to accept anything for true which I did not clearly know to be such; that is to say, carefully to avoid precipitancy and prejudice, and to comprise nothing more in my judgment than what was presented to my mind so clearly and distinctly as to exclude all ground of doubt.*
2. *The second, to divide each of the difficulties under examination into as many parts as possible, and as might be necessary for its adequate solution.*
3. *The third, to conduct my thoughts in such order that, by commencing with objects the simplest and easiest to know, I might ascend by little and little, and, as it were, step by step, to the knowledge of the more complex; assigning in thought a certain order even to those objects which in their own nature do not stand in a relation of antecedence and sequence.*
4. *At the last, in every case to make enumerations so complete, and reviews so general, that I might be assured that nothing was omitted."*

For me, the first is the most important: to rid ourselves of any prejudice and to base an argument only on what we ourselves know beyond a shadow of a doubt, without relying on the authority of others. This may seem difficult because it means that we have to check every single step of our argument and convince ourselves that it is true beyond any reasonable doubt. It is particularly demanding in biology because of the sheer volume of experimental data and the immense number of molecular components. The latter leads to an obsession with molecular detail that obscures the vision of macroscopic simplicity. Hence, the strong emphasis on thermodynamics in this book. As long as arguments are based on thermodynamics, they must be correct. Phenomena derived from thermodynamics are therefore not just hypotheses or models.

In this book, we show that thermodynamics can be used to explain a surprisingly wide range of phenomena, including nerve impulses, the appearance of channels, and the action of anesthetics.

That the properties of excitatory membranes should follow thermodynamic laws has been pointed out by various scientists, starting with Helmholtz. I learned a lot from Konrad Kaufmann's self-published writings on the thermodynamics of nerves (e.g., [6, 7]) and from extensive and often heated discussions – how else should one discuss thermodynamics? – which made me want to learn non-equilibrium thermodynamics and Einstein's theory of fluctuations [4]. Another thing I learned from Konrad is to read authors rather than theories. He believes that the quickest way to understand physics is to find a handful of trusted authors and read everything they have written, rather than reading everything on a subject regardless of the author. I thought that was very wise advice. Authors I trust very much include Hermann von Helmholtz, Albert Einstein, Archibald Vivian Hill, Alan Lloyd Hodgkin, and Ichiji Tasaki.

I am grateful to the many students and postdocs in my research group who enthusiastically carried out experiments and theory and who always reinforced my belief that we were onto something special and important. You can find them as co-authors of our group's publications.

Finally, and most importantly, I owe much to my friend and collaborator Andrew Dumont Jackson for forcing me into rigor, precision, and simplicity, and the importance of subtle and precise language. Although I will probably never achieve his linguistic and mathematical finesse, I will always strive for his perfection and musicality. Andrew is an expert in nonlinear physics (among countless other things), and he introduced me to the theory of solitons. Without working with him, the soliton theory of nerve impulses would not exist. It is the wonderful result of a casual discussion about the music of Johann Sebastian Bach in the corridors of the Niels Bohr Institute. Our work is a perfect example of the fact that science and collaboration cannot be planned. Fruitful collaborations are the result of perfect moments.

Thomas Heimburg
Copenhagen
18 February 2025

Part I

Introduction

In this section, we describe the historical development of our current understanding of nerves since the time of Galenos of Pergamon, Descartes, Swammerdam, Galvani, and Volta. The term physiology was originally introduced in the 19th century to describe a physical view of biological phenomena, especially nerves. On the one hand, this has led to our current electrophysiological picture of nerve impulses. However, we also show that, in parallel, there has always been research that took a more general thermodynamic view of nerve impulses and was interested not only in the electrical properties but also in the heat generation and mechanical properties of nerves. A hundred years ago, some prominent researchers were already of the opinion that it had been proven beyond doubt that the nerve pulse is a reversible wave-like (electromechanical) phenomenon. This perspective will play an important role in this book.

I.1

Early Nerve Studies

Over the millennia, philosophers and scientists have pondered life and its connection to the soul. Aristotle believed that the source of sensations such as pain was the heart. The brain did not play a major role. Such considerations were not based on experimental evidence. Dissections of dead bodies were not performed. This changed during the Roman period, when in the third century AD, Galenos of Pergamon performed anatomical studies. There are several lines of thought, starting from Galenos and René Descartes to Luigi Galvani, Alessandro Volta, and Julius Bernstein, which eventually led to the current electrophysiological view of nerve action. Today, nerve action is mostly understood as an electrical phenomenon related to the flow of charged ions and the charging of cell membranes, which act as capacitors. Over the centuries, however, there have been other views and arguments that attribute nerve action also to mechanical or thermodynamic phenomena. These views are not always easy to distinguish, since electrostatics is part of thermodynamics, and the Nernst potentials in electrophysiological theories resemble a Boltzmann distribution of ions in an electrostatic potential. Furthermore, several experiments on nerves are not included in the current understanding of nerve action. In this introduction, we briefly review some of the history of the understanding of the brain and nerve function.

I.1.1 Galenos of Pergamon and the Animal Spirits

While the early Greek understanding of nerves and the brain was not based on anatomical studies, this changed during the Roman Empire. Unlike his predecessors, the Greek anatomist Galenos of Pergamon (129–216 AD) (Figure I.1.1) dissected the bodies and brains of animals (probably never humans). His findings are summarized in a 14-volume book called *Methodus medendi*.

It served as the basis for medical education until the seventeenth century. For Galenos, the human being was a synthesis of soul and body subject to both spiritual and material influences. According to Galenos, there are various spirits that control the interaction of the different parts of the body: the *spiritus animals* (animal spirit), which connects the brain to the nerves, the *spiritus vitalis* (vital spirit), which connects the heart to the arteries, and the *spiritus naturalis* (natural spirit), which connects the liver to the veins. One must imagine these spirits as a kind of light matter flowing in nerves, arteries, or veins.

Physics of Nerves and Excitatory Membranes, First Edition. Thomas Heimburg.

Figure I.1.1 Galenos of Pergamon (129–216 AD). *Source*: National Library of Medicine/Wikimedia Commons/Public Domain.

As for the brain, Galenos concluded that it controls the function of the muscles through nerves. He noted that special nerves control well-defined muscles through the animal spirits. Galenos' work was considered so complete that many of these studies were not repeated until the Renaissance, and deviations from Galenos' writings found in later experiments were attributed to errors or degeneracies.

I.1.2 Descartes and the Hydraulic Nerve

René Descartes was born on 31 March 1596 in La Haye en Touraine, France, and died on 11 February 1650 in Stockholm. For many reasons, Descartes is one of the most important figures in the history of science. In his famous monograph *Discours de la méthode* (Discourse on the Method), published anonymously in Leiden in 1637, he described the analytical method of scientific investigation. This method outlined an approach to understanding the world based on one's own common sense rather than reference to authority and belief. His advice is never to use anything in a line of argument that you have not checked beyond doubt to be true. His method is analytical, working from the simple picture to the more complex.

Descartes was very interested in anatomy and apparently carried out many anatomical studies himself. In his book *Traité de l'homme*, he supported the idea of Galenos that a human being has two parts, the soul and the body, which communicate with each other through animal spirits. While the concept of an animal spirit is somewhat mysterious in the ancient writings, in Descartes' view it is akin to a liquid. For Descartes, the human body was a hydraulic machine similar to those existing in the royal gardens in Saint-Germain-en-Lay. In these gardens, various figures and statues were moved by the hydraulic pressure of a system of water pipes.[1] He argued that the blood is pumped by the heart, filled with oxygen in the lungs, and that it flows subsequently to the different body parts. The blood then flows back to the heart in veins. The most animated parts of the blood (with the highest pressure) flow into the brain, where the pressure fills the nerves with the animal spirit. The nerves control the muscles via a hydraulic mechanism in the same way as the water pressure controls the motion of the statues in the Royal Gardens. Small aliquots of the animal spirits in the nerves are transferred into the muscles. Via this mechanism, the muscle expands in diameter and shortens in length. A drawing by Descartes of the muscles and nerves surrounding the eyeball is shown in Figure I.1.2 (center). The soul plays the role of the engineer that opens and closes the valves of the pipe system. An obvious consequence of this view is that a muscle must change its volume when the *animal spirits* are injected.

Descartes was a friend and correspondent of Galileo Galilei. When Galileo was condemned by the Catholic Church in 1633, Descartes did not dare to publish “*Traité de l'homme*,” which was written in the same year. It only appeared after Descartes' death in

Figure I.1.2 Left: René Descartes (1596–1650). Jan Baptist Weenix/Wikimedia Commons/Public Domain. Center: Drawing of the eye ball surrounded by muscles and nerves from “*Traité de l'homme.*” Right: Jan Swammerdam (1637–1680). DALIBRI/Wikimedia Commons/Public Domain.

1 “... visitors described an automaton Neptune with a streaming blue beard, brandishing his trident, naked astride a chariot pulled by seahorses, accompanied by three round-bellied, horn-playing tritons. Farriers, “their faces black with filth and sweat,” hammered iron on an anvil and - “that which is most pleasant and seems made to provoke laughter” - drenched their eager audiences with surprise sprays of water. Mercury posed by a window with one foot carelessly propped, “loudly intoning a trumpet.” Elsewhere, Orpheus played his lyre for an audience of animals and trees who, including the trees, stretched and craned toward him. A towering Perseus descended upon a mighty dragon arising from beneath the waves. Perseus swung his sword to behead the fearsome beast, sending it, slain, back down into the watery depths; whereupon farther back in the grotto, Andromeda promptly lost her chains. Meanwhile busy figures of artisans - blacksmiths, weavers, millers, carpenters, knife-grinders, fishermen - went about their sundry tasks.” Quote from [1].

Latin translation ("*De homine*"). Parts of it were summarized in "*Discours de la méthode.*" Descartes was nevertheless subject to prosecution. The University of Utrecht condemned him in 1643, and he fled to Den Haag in order to avoid the public burning of his books. In the later 1640s, he was also condemned by the University of Leiden. This might be the reason he moved to Stockholm in the last year of his life.

The hydraulic theory of Descartes was disproven by the Dutch microscopist Jan Swammerdam (1637–1680), who showed that the muscle does not change its volume during contraction. Thus, it does not act as a hydraulic piston [2]. He also showed that a frog muscle disconnected from the frog's brain still contracts when the nerve is irritated. Thus, the contraction of the muscles did not require the brain as the source of an *animal spirit*.

> *"From these experiments, therefore, it may, I think, be fairly concluded, that a simple and natural motion or irritation of the nerve alone is necessary to produce muscular motion, whether it has its origin in the brain, or in the marrow, or elsewhere [3]."*

Swammerdam rather argued that the action of nerves can be likened to a vibration.

> *"I would have it seriously considered, that it cannot be demonstrated by any experiments, that any matter of sensible or comprehensible bulk flows through the nerves into the muscles. Nor does any thing else pass through the nerves to the muscles: all is a very quick kind of motion, which is indeed so rapid, that it may be properly called instantaneous. ... Therefore the spirit, as it is called, or that subtile [sic] matter, which flies in an instant through the nerves into the muscles, may with the greatest propriety be compared to that most swift motion, which, when one extremity of a long beam or board is struck with the finger, runs with such velocity along the wood, that it is perceived almost at the same instant at the other end [3]."*

I.1.3 Gassendi, Willis, and Newton

The hydraulic theory of Descartes was very popular in his days. However, it left open the question of how the soul interacts with the nerves and muscles, i.e., how the hydraulic machine can be subject to willful decisions. It is fair to say that this question has remained unanswered until today.[2]

The French philosopher Pierre Gassendi (1592–1655) (Figure I.1.3) (left) is another important figure who critically discussed both Aristotelian philosophy and the teachings of Descartes. He was seen as the main adversary of Descartes in his days. His view on nerves and muscles was mechanical, without an explicit notion of a soul. In his opinion, the hydraulic theory of Descartes could not work because it required solid hollow pipes. However, Gassendi noted that the nerves are made of a soft and tender material, which would not be suitable to transmit a force sufficient to make a muscle contract. He remarked:

---

2 For the following paragraphs, we follow a historical review of Wallace [4] and use quotes cited therein.

Figure I.1.3 Left: Pierre Gassendi (1592–1655). Œuvres complètes de Voltaire/Wikimedia Commons/Public Domain. Center: Thomas Willis (1621–1675). Bodleian Libraries/Public domain. Right: Isaac Newton (1643–1727). James Thronill/Wikimedia Commons/Public Domain.

> *"For if the nerves, as is sometimes said, issued from a firm and solid origin, the attraction which occurs between the part that is moved and the origin of movement could truly be attributed to the nerves; but since the nerves issue from the marrow, which is a soft, tender and slack material, and since the nerves themselves at their origin are both very soft and very tender, consequently they cannot be fit for attracting the [moving] parts, nor for being the true and physical organs of movement." ... "What then can we say comes to the muscle from the brain, by the intermediary of the nerve, without which the muscle is not capable of moving? Certainly this seems to be nothing other than the commandment to move, which is somehow signified to the muscle by the arrival of the spirits transmitted by the nerve, such that without this front of spirits , the muscle remains sleeping, but if it has been excited thereby, as if awoken, it acts."*

Gassendi also noticed that the muscle can contract by itself when stimulated from the exterior. As a consequence, Gassendi believed that the nerve does not transmit a force but rather a signal that makes the muscle contract. The muscle possesses its own machinery that allows it to contract, and the nerves just transmit the order.

> *"Nothing is transmitted from the brain to the muscle by the intermediary of the nerve, except this order, and this commandment, which is like the Will or the Appetite. And the Understanding, or Phantasy as mistress and directress, signifies to the tendon or to the muscle as to a slave, so that it should make such and such a movement, and that it should move such and such a body part."*

In summary, Gassendi's view was that the nerve mediates muscular contraction. However, it transmits a signal and not a force. Further, the signal resembles light or fire because it is too fast to be transmitted by the motion of particles or fluids.

William Croone (1633–1684) was another philosopher who noted that the spirits in the nerve are not sufficient to inflate a muscle. He rather proposed that the nerves inject tiny amounts of a spirit into the blood of a muscle. The consequent contraction of the muscle is caused by a chemical reaction induced by those spirits. In the tradition of Gassendi, the English doctor Thomas Willis (1621–1675), (Figure I.1.3) (center) who made the first detailed anatomic descriptions of the nervous system and the brain, published his book *Cerebri Anatome* in 1664, and two more books called *De motu musculari* (1670) and *De anima brutorum* (1670). In contrast to Descartes, he made use of microscopes, just like Antonie Philips van Leeuwenhoek (1632–1723), who is often called the father of microbiology. Just like van Leeuwenhoek, he could not find any hollow structures in nerves. Therefore, he proposed the so-called solid nerve, i.e., a fiber that is not a hollow hydraulic pipe but rather a solid tissue that transports stimuli of a much more subtle nature than liquids:

> *"The passages of the nerves are not hollowed out like those of the arteries and veins, because their substance is not only impervious to any stylus, but the use of a lens or microscope confirms that there is no cavity present in them. ... The nerves are clearly formed of a compact and firm substance so that the subtle humor, which is the vehicle of the spirits, may pass through their structures not otherwise than spirits of wine through the tensed cords of a lute, only by slowly creeping through. Hence it may be argued that the animal spirits require no manifest cavity within the nerves for their expansion."*

He believed that the spirits in the neuron are rather a kind of waves:

> *"For just as in a river, various kinds of waves are thrown up by the winds or by things cast into it, so the animal spirits, when awoken by objects, go forth to carry out the functions of sense and motion, stretching hither and thither within the nervous kind, and being agitated there in other ways." ... "But since it is not trivial to distinguish spirits and waters by their motions and consistency, perhaps it will better illustrate the matter if the spirits ... are compared to the beaming forth of different rays of light. ... The animal spirits irradiate and swiftly cross all parts of the nervous system, both primary and secondary, so that light is scarcely carried faster through a diaphanous medium, than the communication of spirits is made from one end of the nervous system to the other."*

Willis also referred to the nervous impulse as a vibration as in a stringed instrument. This is clearly a view that is different from that of Descartes.

The aforementioned views became most popular through the writings of Issac Newton (Figure I.1.3 (right)). He added short but important sections on nerves to the second edition of his *Principia Mathematica* from 1713. Newton also used the term "solid nerve" and explained the signal of the nerve in the following manner:

> *"And now we might add something concerning a certain most subtle spirit which pervades and lies hid in all gross bodies; by the force and action of which spirit the particles of bodies attract one another at near distances, and cohere, if contiguous; and electric bodies operate to greater distances, as well repelling as attracting the neighboring corpuscles; and light is emitted, reflected, refracted, inflected, and heats bodies; and all sensation is excited, and the members of animal bodies move at the command of the will, namely, by the vibrations of this spirit, mutually propagated along the solid filaments of the nerves, from the outward organs of sense to the brain, and from the brain into the muscles."*

Thus, Newton adopted views from Gassendi and Willis and favored an electrical or light-like mechanism for the transmission of a signal (the spirits). Like Gassendi, Croone, and Willis, he proposed that muscle contraction is the consequence of chemical reactions induced by the influx of spirits from the nerves.

For more details, we refer to the very detailed review of Wallace [4].

I.1.4 Galvani, Volta, and the Electrical Nerve

The controversy between Luigi Galvani (1737–1798) (Figure I.1.4) (left) and Alessandro Volta (1745–1827) (Figure I.1.4 (right)) is famous. It led to the present understanding that the nerve pulse is predominantly an electrical phenomenon.

Luigi Galvani was an Italian scientist from Bologna. He made three important experiments that established that the nerve pulse is an electrical phenomenon. He showed that the legs of a frog contracted when electrically excited by either a Leyden jar (a capacitor), a machine producing a charge, or by touching the sciatic nerve with a charged scalpel. He noted:[3]

> *"I dissected and prepared a frog and placed it on a table, on which was an electrical machine... widely removed from its conductor and separated by no brief interval [distance]. When by chance one of those who were assisting me gently touched the point of a scalpel to the medial crural nerve of this frog, immediately all the muscles of the limbs seemed to be so contracted that they appeared to have fallen into violent tonic convulsions. But another of the assistants, who was on hand when I did electrical experiments, seemed to observe that the same thing occurred whenever a spark was discharged from the conductor of the machine..."*

After finding that the muscles of frogs contracted when electrical sparks occurred in the proximity of the preparation, or the nerve was touched in the presence of such electrical

Figure I.1.4 Left: Luigi Galvani (1737–1798). Unknown author/Wikimedia Commons/CC BY-SA 3.0. Right: Alessandro Volta (1745–1827). Wolfgang Peter/Wikimedia Commons/Public Domain.

3 This section follows a review by Geddes and Hoff [5] and the English translations of Galvani's writings by Green [6].

machines, Galvani also wondered whether muscle action could be induced by atmospheric electricity, i.e., during thunderstorms:

> *"Having discovered the effects of artificial electricity on muscular contractions ..., there was nothing we would sooner do than to investigate whether atmospheric electricity, ..., would afford the same phenomena, ... Therefore we erected, in the fresh air, in a lofty part of the house, a long and suitable conductor, namely an iron wire, and insulated it and to it, when a storm arose in the sky, attached by their nerves either prepared frogs, or prepared legs of warm animals .... Also we attached another conductor, namely another iron wire, to the feet of the same, and this as long as possible, that it might extend as far as the waters of the well indicated in the figure. Moreover, the thing went according to our desire, just as in artificial electricity; for as often as the lightning broke out, at the same moment of time all the muscles fell into violent and multiple contractions, so that, just as the splendor and flash of the lightning are wont, so the muscular motions and contractions of those animals preceded the thunders, and, as it were, warned of them; ..."* [5]

Since the effect of lightning could be observed prior to thunder, but at the same time as the lightning, it was shown that the lightning was the cause of the muscle contraction. Thus, he demonstrated that nerve contraction is an electrical phenomenon.

Galvani made a second experiment of a different nature, which is related to the contact of nerves with metals. He describes this experiment as follows:

> *"Wherefore, since I had sometimes seen prepared frogs placed on iron gratings which surrounded a certain hanging garden of my house, equipped also with bronze [some accounts refer to copper or brass] hooks in their spinal cord, fall into the customary contractions, not only when the sky was lightning but also sometimes when it was quiet and serene, I thought these contractions derived their origin from the changes which sometimes occur in atmospheric electricity. Hence, not without hope, I began diligently to investigate the effects of these changes in these muscular motions in various ways. Wherefore at different hours, and for many days, I inspected animals, appropriately adjusted therefor; but there was scarcely any motion in their muscles. Finally, weary with vain expectation I began to press the bronze hooks, whereby their spinal cords were fixed, against the iron gratings, to see whether by this kind of device they excited muscular contractions, and in various states of the atmosphere, and of electricity whatever variety and mutation they presented; not infrequently, indeed, I observed contractions, but bearing no relation to varied state of atmosphere or of electricity. " ... "Nevertheless, since I had not inspected these contractions except in the fresh air, for I had not yet experimented in other places, I was on the point of seeking such contractions from electricity of the atmosphere, which had crept into the animal and accumulated in him and gone out rapidly from him in contact of the hook with the iron grating; for it is easy in experimentation to be deceived, and to think one has seen and discovered what we desire to see and discover."* [5]

Galvani found that the frog muscles connected to the metal hooks contracted even in the absence of lightning.

The important requirement for Galvani's second experiment was that the hook to which the nerve was connected consisted of two different metals. Galvani believed that the nerve served as a kind of capacitor (Leyden jar) that was charged with *animal electricity*, and that the hook represented a conductor that discharged the capacitor. According to Galvani, the muscle contracts upon discharging the nerve.

Galvani's experiments quickly became very popular. They were demonstrated in public, and many scientists and laymen reproduced them.

Alessandro Volta (1745–1827) was born and lived in Como (Italy). He is known as the inventor of the battery (Voltaic cells). Basically, the Voltaic cell consists of two different metals separated by an electrolyte. Volta arranged such elementary batteries in series and noted that the effect of the battery is similar to that of a Leyden jar. He writes:

> *"Thirty, forty or more pieces of copper, or rather silver, applied each to a piece of tin, or zinc, which is much better, and as many strata of water or any other liquid which may be a better conductor, such as salt water, ley [lye], etc., or pieces of pasteboard skin etc. well soaked in the liquids; such strata are interposed between every pair or combination of two different metals in an alternate series, and always in the same order of these three kinds of conductors are all that is necessary for constituting my new instrument, which as I have said, imitates the effect of the Leyden flask, or of electric batteries by communicating the same shock as these do."* [5]

Thus, according to Volta, in Galvani's second experiment, the metal hook actually formed a battery, and the nerve liquids acted as electrolytes. Therefore, it was actually Volta who paved the way for the emergence of a bioelectric potential from chemical sources built up in electrolytic solutions. The reinterpretation of Galvani's experiments and the rejection of the concept of *animal electricity* is known as the Galvani–Volta controversy and holds an important place in the history of science.

As a reaction to Volta's experiments, Galvani performed a third experiment, in which he bent the leg of the frog such that it touched the injured nerve and showed that the muscles contracted. This shows that the biological tissue itself can act as a battery. The voltage created in this arrangement is called the injury potential, due to its origin from the injured sciatic nerve.

Galvani's late experiments were confirmed and refined by Carlo Matteucci (1811–1868) by employing a galvanometer. He showed that the injured nerve in contact with a muscle creates an electrical current, and that the contraction of the muscle is correlated with a transient current and the disappearance of the bioelectric potential, which later came to be known as the *action potential*.

I.2

The Early Period of Electrophysiology

Three important scientists in the early phase of neuroscience were Emil du Bois-Reymond (1818–1896), Hermann von Helmholtz (1921–1894), and Julius Bernstein (1839–1917) (Figure I.2.1). They can be considered the fathers of a physical approach toward the understanding of nerve activity.

I.2.1 Emil du Bois-Reymond

Du Bois-Reymond (Figure I.2.1) (left) was a German scientist whose family originally came from Switzerland. He developed and improved a number of electrical instruments. Galvanometers are sensitive devices that depend on the magnetic field induced by an electrical current. This effect was first described by Hans Christian Ørsted, who demonstrated the effect of a current on a compass needle in 1820. In the same year, Johann Schweigger from Halle developed the first galvanometer, named after Luigi Galvani. Du Bois-Reymond improved the galvanometers in order to conduct the first quantitative experiments on electrical phenomena in nerves. He published a book entitled *Untersuchungen über tierische Elektrizität (Investigations of Animal Electricity)* over the span of nearly 40 years (1848–1884). He developed the first purely physical theory of nerves, called the *Molekeltheorie (molecular theory)*, postulating a polarization of molecules in the nerve during a nerve pulse that is absent in the resting nerve. With this theory, he followed his credo, formulated as a young researcher in 1842 [7]:

> *"Brücke (his colleague Ernst von Brücke) and I are dedicated to promote the view that no forces exist in organisms except those of general physical and chemical nature. Where those are presently not sufficient for an explanation we have to search for applications in the concrete case following the physical-mathematical procedure, or we have to assume new forces that are of same dignity as the physical-chemical methods, and which are only based on material properties of attractive or repulsive nature."*[1]

---

1 Original quote in German: *"Brükke und ich, wir haben uns verschworen, die Wahrheit geltend zu machen, daß im Organismus keine anderen Kräfte wirksam sind, als die gemeinen physikalisch-chemischen; daß, wo diese bislang nicht zur Erklärung ausreichten, mittels der physikalisch-mathematischen Methode entweder nach ihrer Art und Weise der Wirksamkeit im konkreten Fall gesucht werden muß, oder daß neue Kräfte angenommen werden müssen, welche, von gleicher Dignität mit den physikalisch-chemischen, der Materie inhärent, stets auf nur abstoß ende oder anziehende Componenten zurückzuführen sind."*

Physics of Nerves and Excitatory Membranes, First Edition. Thomas Heimburg.

Figure I.2.1 Left: Emil du Bois-Reymond (1818–1896). Smithsonian Institution/Wikimedia Commons/Public Domain. Center: Hermann Helmholtz (1821–1894). Armin Kübelbeck/Wikimedia Commons/Public Domain. Right: Julius Bernstein (1839–1917). Biographical Encyclopedia/Wikimedia Commons/Public Domain.

With this statement, he distanced himself from the vitalistic view of Luigi Galvani, which is based on an animal spirit inherent and unique to living organisms. Together with Ernst Wilhelm von Brücke, Carl Ludwig, and Hermann von Helmholtz, he established *Physiology* as a scientific discipline based on physical principles. In 1958, du Bois-Reymond became a full professor of physiology at the University of Berlin. From 1875 to 1877, he published *Gesammelten Abhandlungen zur allgemeinen Muskel- und Nervenphysik (Collected Works on the General Physics of Muscles and Nerves).* In his time, Du Bois-Reymond was one of the most renowned scientists with an international reputation, and his work formed the foundation for most later studies that followed a physical approach. He also became one of the founding fathers of the German Physical Society. It was du Bois-Reymond who coined the term *action potential* for the electrical impulse.

I.2.2 Hermann von Helmholtz

Hermann von Helmholtz (Figure I.2.1) (center) was a physiologist by training, but later became the first chair of theoretical physics at the University of Berlin. He is a prominent example of a scientist breaching several disciplines who, as a consequence, made important contributions to the progress of understanding. Helmholtz was a friend of du Bois-Reymond, who considered him a member of his *Organic Physics* Club.

In his PhD thesis, he measured for the first time the velocity of a nerve pulse, which he found to be about 30 m/s in the sciatic nerve of a frog [8, 9]. He immediately noticed that this velocity is much slower than that expected for electrical currents (which basically travel with a velocity of the order of the speed of light). Von Helmholtz was also the first scientist who tried to measure the heat production in nerves [10]. He found that the heat production could not be detected and therefore had to be very small. This interesting finding will be discussed in more detail in Chapter I.4. Interestingly, von Helmholtz also formulated the first law of thermodynamics (Über die Erhaltung der Kraft [11]), which had been independently stated some years earlier by Julius von Mayer (1842). Helmholtz based this law on metabolism and the heat production of living organisms and thereby positioned himself (like du Bois-Reymond) against the vitalism of Galvani. Later, von Helmholtz introduced the term *free energy* into thermodynamics and made many other important contributions to electrochemistry and physics in general. The thermodynamic foundation of electrochemistry formulated by Hermann von Helmholtz, Walther Nernst (1864–1941), and Wilhelm Ostwald (1852–1932) is the basis of all later approaches to nerve physiology, particularly those developed by Julius Bernstein. Helmholtz also investigated the physiology of seeing and hearing. He proposed that seeing is related to three receptors for red, green, and blue, which are connected to nerves [12]. Furthermore, he wrote a famous book on the *sensation of tone* [13], in which he postulated that the cochlea of the inner ear contains elements similar to the strings of a piano that resonate with perceived sound.

In the context of this book, we will be mostly involved with Helmholtz's measurements of the speed of the action potential and his recordings of heat production in nerves, described in Chapter I.4.

I.2.3 Julius Bernstein

Julius Bernstein (1839–1917) (Figure I.2.1 (right)) was a student of du Bois-Reymond and later an assistant of Helmholtz. He became a professor of physiology at the University of Halle in 1872. His background gave him the qualifications for his approach to electrophysiology (a term that was actually coined by Bernstein). Bernstein improved the quality of the galvanometers such that he could describe the shape of the action potential for the first time.

Figure I.2.2 shows the so-called *negative variation* of the action potential [14]. If a neuron is stimulated, two action potentials travel in opposite directions (a fact that was already described by du Bois-Reymond). If the action potential propagates along the neuron, one can measure a current between two positions, 1 and 2, along the neuron (Figure I.2.3). The direction of the current depends on the temporal position of the pulse.

Bernstein's famous book entitled *Electrobiology* from 1912 [15] had a great influence on later researchers (including Alan L. Hodgkin and Andrew F. Huxley). This book outlined the properties of semipermeable walls, electrochemical potentials, and resting potentials of cells and nerves, which were influenced by contemporary investigations of Nernst and Ostwald. In this view, the nerve membrane is semipermeable to positive ions, and Nernst potentials build up as a consequence. Figure I.2.4 (left) shows the ion distribution of an intact neuron at rest. No currents flow (from Bernstein 1912 [15]). If the nerve is cut, ions can flow freely, and the so-called injury current can be measured (Figure I.2.4 (right)).

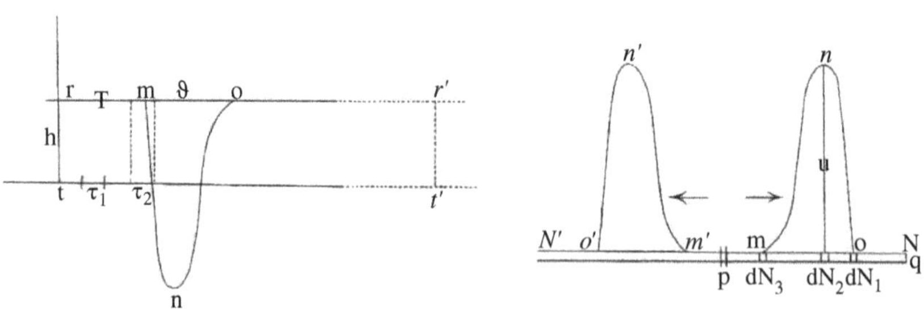

Figure I.2.2 The action potential as determined by Bernstein. Left: The "negative variation" of the potential. Right: The emergence of two pulses traveling in two directions from the site of stimulation. From [14]/with permission of Springer Nature.

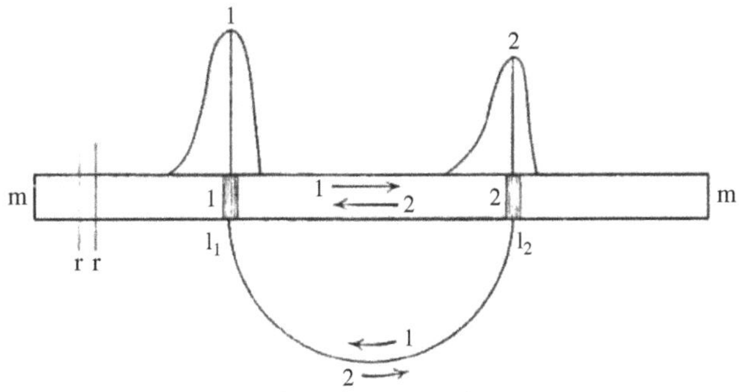

Figure I.2.3 The action potential propagating from positions 1 to 2. Indicated are the currents from position 1 → 2 and from 2 → 1 that depend on the position of the pulse. From [15]/with permission of Springer Nature.

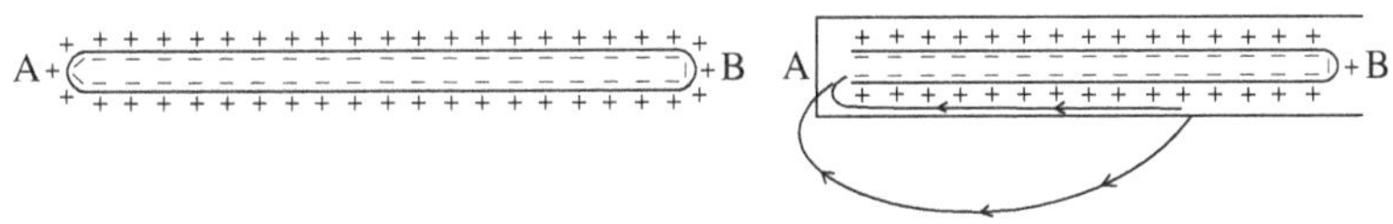

Figure I.2.4 Bernstein's membrane theory. Left: The intact nerve fiber is surrounded by a semipermeable membrane across which a potential difference is found. Right: If the nerve fiber is cut or injured, one finds a current from the outer layer to the inner layer of the neuron (injury current). From [15]/with permission of Springer Nature.

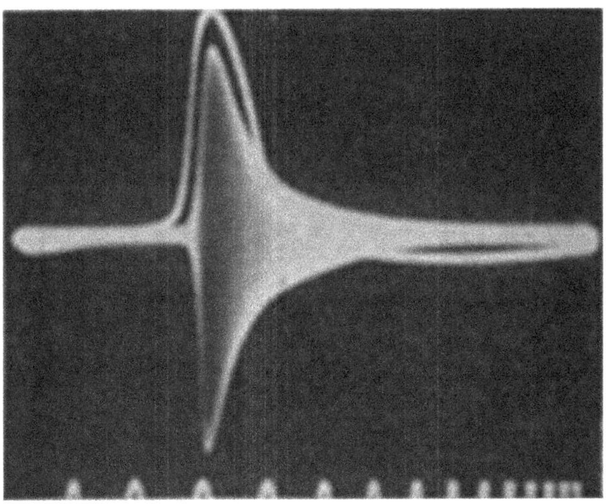

Figure I.2.5 The change of the conductance during the action potential as measured by Cole and Curtis (1939). [17]/with permission of The Journal of General Physiology.

Bernstein's most important achievement is the *Julius-Bernstein hypothesis* or *Membrane Theory* from 1902 [16]. It is based on the assumption that membranes become transiently permeable to positive ions (in particular to potassium) during the action potential. The transient change in membrane conductance was confirmed later by Cole and Curtis in 1939 [17], who used an oscilloscope to visualize the action potential (Figure I.2.5). The use of oscilloscopes had already been proposed in *Electrobiology* in the last paragraph of the book [15, p. 212].

Bernstein's model is the first modern theory of nerve conduction. It is the seminal precursor of the later theory by Hodgkin and Huxley that involved separated currents of sodium and potassium through the nerve membrane. It is important to note that the action potentials described by Bernstein are purely negative without hyperpolarization and overshoot. This is different from Hodgkin and Huxley's model, which is described in the next chapter.

I.3

The Hodgkin–Huxley Model and Beyond

I.3.1 The Model of Hodgkin and Huxley

The presently most popular model for the nerve pulse is the Hodgkin–Huxley model by Alan L. Hodgkin and Andrew F. Huxley (Figure I.3.1). The basic idea of the model is similar to that of Bernstein discussed in the previous chapter. Bernstein's model assumed that during the nerve pulse, the electrical resistance of the membrane to ions breaks down. It was postulated that this would lead to a change in ion concentrations on both sides and consequently to a change in the transmembrane Nernst potential.

The Hodgkin–Huxley model assumes instead that the resistance breaks down for sodium and potassium ions separately, with distinctly different time and voltage dependences. Hodgkin and Huxley's considerations were based on voltage clamp experiments on giant squid axons. These axons have the advantage of being thick enough ($\approx$1 mm) to allow for the insertion of electrodes over the complete length of the axon and to apply a constant voltage difference across the nerve membrane. This allowed for measuring currents and conductivity across the membrane in the absence of a propagating pulse. In 1952, they found that the currents in voltage clamp experiments showed a distinct voltage dependence [18, 19]. Surprisingly, after an increase in voltage, there were two phases of the response in time. During the first phase, they found a current against the applied voltage, and during the second phase, currents flowed along the potential gradient. The finding was different in seawater with free sodium and choline seawater without free sodium (which was replaced by choline, which is too big to pass through membranes). This is shown for a voltage jump of 42 mV in Figure I.3.2 (left). The two phases also displayed a different dependence on the presence of the neurotoxins tetrodotoxin (TTX) and tetraethylammonium (TEA).The negative phase was blocked by TTX, while the positive phase was blocked by TEA. The dependence of the different phases of the currents was interpreted as evidence for two conduction processes selective for specific ions.

The conclusion was that there exist two different currents in squid axons: an inward sodium current and an outward potassium current, both along their concentration gradients. Since these currents depend on the magnitude of the voltage jump, leading to a different time-dependence, Hodgkin and Huxley introduced voltage- and time-dependent sodium and potassium conductances. Their dependence on voltage and time was parametrized from the voltage clamp experiments and inserted into electrical cable theory, which was developed over the previous century (detailed in [20]). This leads to a

Figure I.3.1 Left: Alan Lloyd Hodgkin (1914–1998). Right: Andrew Fielding Huxley (1917–2012). Nobel Prize Outreach/Wikimedia Commons/Public Domain.

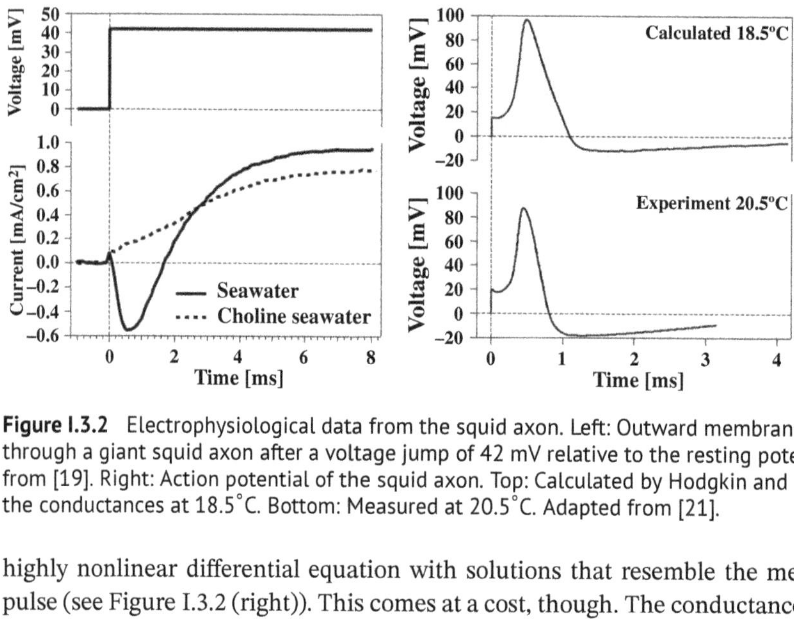

Figure I.3.2 Electrophysiological data from the squid axon. Left: Outward membrane current through a giant squid axon after a voltage jump of 42 mV relative to the resting potential. Adapted from [19]. Right: Action potential of the squid axon. Top: Calculated by Hodgkin and Huxley from the conductances at 18.5°C. Bottom: Measured at 20.5°C. Adapted from [21].

highly nonlinear differential equation with solutions that resemble the measured nerve pulse (see Figure I.3.2 (right)). This comes at a cost, though. The conductance of the membrane was not derived from first principles, but was fitted to experiment. This leads to about 23 parameters for the sodium and potassium conductances alone. For this reason, the Hodgkin–Huxley model is rather a description of the nerve pulse than an explanation. However, it made all kinds of valuable points, including the existence of a refractory period, the maximum amplitude of the action potential, and the stimulation threshold. Figure I.3.3 illustrates the difference between Bernstein's model and the Hodgkin-Huxley model. Further, it suggested that the conductances of the ions originate from molecular carrier objects that are now believed to be ion channel proteins.

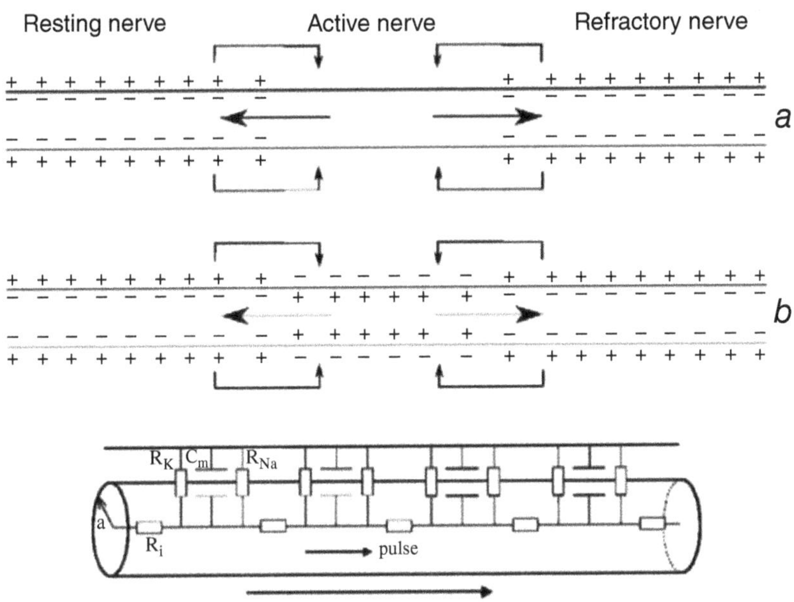

Figure I.3.3 The action potential. Top: In the view of Julius Bernstein, the nerve pulse is generated by a local breakdown in membrane resistance, leading to a local depolarization of the membrane to zero. Center: In contrast, the Hodgkin–Huxley model involves a reversal of the potential because the resistance of the membrane breaks down transiently but separately for sodium and potassium. Bottom: In the view of Hodgkin and Huxley, the nerve is a membrane cylinder that forms a capacitor. Embedded into the membrane are two different kinds of resistors for sodium and potassium, respectively. From [22]/with permission of Elsevier.

The Hodgkin–Huxley model connects voltage clamp currents with a propagating action potential. This strongly suggests that the transmembrane currents contain information on the physical state of the membrane that is necessary to understand the propagation process. Hodgkin and Huxley [21] note that

> *"... the equations and constants were derived entirely from 'voltage clamp' records, without any adjustment to make them fit the phenomenon to which they were were subsequently applied."*

Since the Hodgkin–Huxley model relies on the parametrization of the voltage- and time dependences of ionic currents, any model that results in a parametrization of the net currents would lead to a successful description, even if not related to separate sodium and potassium currents. Hodgkin and Huxley were aware of this. They noted in the discussion of their seminal paper:

> *"The agreement must not be taken as evidence that our equations are anything more than an empirical description of the time-course of the changes in permeability to sodium and potassium. An equally satisfactory description of the voltage clamp data could no doubt have been achieved with equations of very different form, which would probably have been equally successful in predicting the electrical behaviour of the membrane."*

Anything that models the time-course of inward and outward currents would do the job, even if the currents had a completely different origin and were not related to sodium and potassium currents. However, Hodgkin and Huxley concluded that

> *"It is concluded that the response of an isolated giant axon of Loligo to electrical stimuli are due to reversible alterations in sodium and potassium permeability arising from changes in the membrane potential."*

The observation of separate sodium and potassium currents is a finding that is not necessary to yield the action potential. However, experiments with radioactive tracers confirmed the concept of ion currents correlated with the action potential [23–26]. While this is easy to demonstrate for potassium, sodium influx during electrical activity was not always demonstrated. In desheathed rabbit vagus nerve, Keynes and Ritchie [27] found that

> *"... ^{22}Na and ^{45}Ca gave negative results in that the exchange rates were not significantly altered by electrical activity or treatment with various drugs, but a study of the movements of radioactive potassium proved more rewarding."*

I.3.2 Ion Channel Proteins as Pores

While the fitting of the nerve pulse does not require separate sodium and potassium currents (see previous section), the postulation of separate conductances for sodium and potassium was quite influential. The dependence of the inward current on TTX effective at very low concentrations suggests that there exists a molecular origin for the sodium current. Likewise, the sensitivity of the potassium current to TEA and other drugs suggests a molecular origin of potassium currents.

It was debated whether proteins facilitate conduction by being transporters or by the formation of pores. Bertil Hille recalled the emergence of the concept of ion channel proteins ref. [28]:

> *"In our first papers we brought a clear list of 'molecular' assumptions to the table (Figure I.3.4). They included the following ideas: ions are passing through aqueous pores that we called channels, ion channels are proteins, the channels for Na$^+$ and K$^+$ are different, they have swinging gates that open and close them, we can study their architecture by using electric currents to measure gating, permeation and block, and channel blockers are molecules that enter the pores and physically plug them."*

Together with W. Schwarz, Hille proposed in 1978 that the potassium channel is a water-filled pore fitting several ions [29], a concept that is in agreement with the crystal structure of a potassium channel by MacKinnon and collaborators (see Figure I.3.5 [30]). The structure of the KcsA potassium channel shows a fourfold symmetry consistent with the four gating particles proposed by Hodgkin and Huxley to model the parametric potassium currents [21].

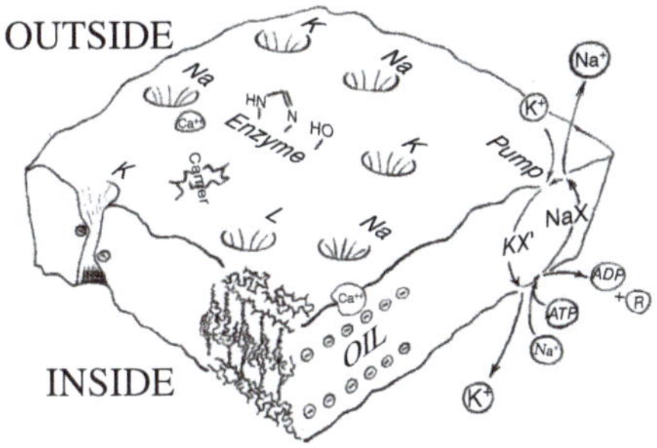

Figure I.3.4 Schematic drawing of membrane pores. From [28]/with permission of Springer Nature.

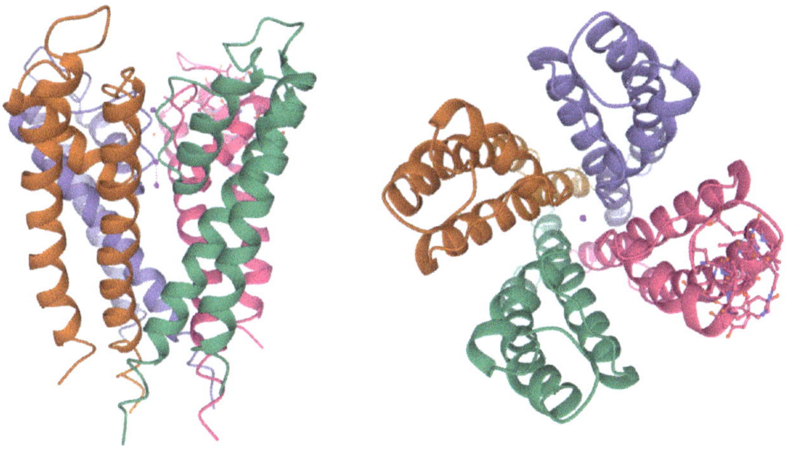

Figure I.3.5 The crystal structure of the transmembrane part of the KcsA potassium channel from MacKinnon and collaborators [30] seen from the side (left panel) and from the top (right panel). The dots in the center indicate potassium ions penetrating the protein pore. Adapted from [30].

I.3.2.1 The Finding of Quantized Ion Currents by Neher and Sakmann

The line of argument leading to the concept of pores in ion channel proteins was mostly indirect in so far as it relied on the interpretation of macroscopic ion currents. However, in 1976, Erwin Neher and Bert Sakmann (Figure I.3.6) found in patch clamp experiments that one finds discrete steps in transmembrane currents (see Figure I.3.7, [31]). In patch clamp experiments, a small glass pipette is placed on a cell surface. Slight suction generates a so-called giga seal, indicating a resistance of more than a giga-ohm (GΩ). Most currents that can be measured are assumed to originate from currents through the membrane patch. One finds tiny currents that seemingly correspond to the opening and closing of a gate, with current amplitudes on the order of picoamperes (pA). These tiny currents required the development of very good noise-free amplifiers. The technique is described in detail in a highly cited publication from 1981 [32].

If one assumes a pore filled with electrolytes, one arrives at a pore diameter of less than a nanometer for currents of a few pA. If one attributes the currents to the opening and closing of gates, then it is compelling to attribute them to small nano-sized pores filled with water

Figure I.3.6 Erwin Neher (*1944), who invented the patch clamp technique together with Bert Sakmann (*1942). *Source*: Molgen/Wikimedia Commons/Public Domain.

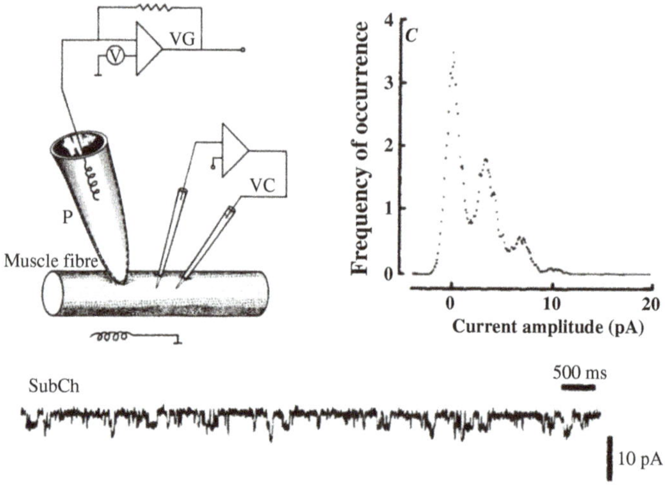

Figure I.3.7 The patch clamp setup, a current trace and the current histogram of the seminal paper of Neher and Sakmann from 1976. [31]/with permission of Springer Nature.

within a macromolecule. In Neher and Sakmann's seminal paper, for instance, they investigated the acetylcholine receptor in the presence of various drugs, including acetylcholine [31]. They found channel open-times of the order of a few tens of milliseconds.

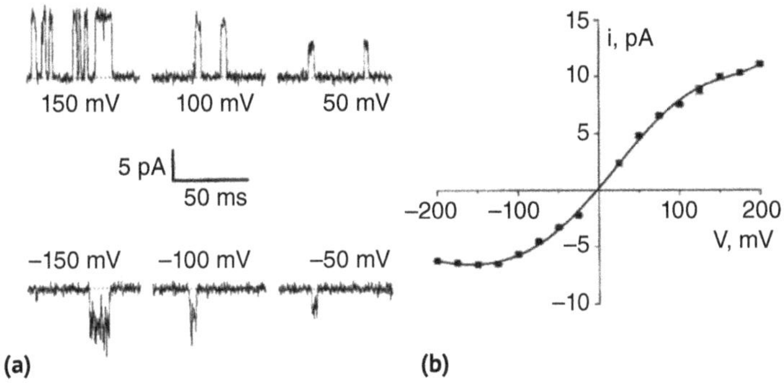

Figure I.3.8 Voltage gating for the KcsA potassium channel. (a) Recordings are shown of single KcsA channels in symmetrical 100 mM KCl at the voltages indicated. The closed state is indicated by dashed lines. (b) Open-channel current–voltage relation. From [33]/with permission of Rockefeller University Press.

I.3.2.2 Gating

The channel events recorded in membranes containing channel proteins often depend on voltage (or other intensive thermodynamic variables). This is shown for the KcsA potassium channel in Figure I.3.8a. The frequency of open-channel events increases with both negative and positive voltage. This leads to a nonlinear current–voltage profile, shown in Figure I.3.8b. This phenomenon is known as voltage gating.

Similarly, it is believed that there exist proteins gated by

- drugs such as neurotransmitters
- voltage
- ions such as calcium
- mechanical tension (mechanosensitive channels)
- temperature

which are discussed in more detail in Chapter IV.5. In fact, it seems that the dogma is that for each thermodynamic variable there exists a particular channel or receptor that responds to changes in these variables. This is an attempt to relate the thermodynamics of biological membranes to single molecules, which will be challenged in this book.

Summarizing, the Hodgkin–Huxley model together with the finding of proteins that affect selective conductance of the nerve membrane for different ions leads to an electrical view of the nerve membrane that is dominated by single molecules and their interaction with drugs and by changes in the thermodynamic variables affecting receptor molecules. For this reason, the nerve pulse is seen as a chain reaction of chemical reactions involving channel proteins that ultimately leads to a propagating electrical pulse called an *action potential.*

I.3.3 Biochemical Pathways

The association of individual macromolecules with conduction events in the patch clamp experiments by Neher and Sakmann was probably the first single-molecule measurement in nanoscience. It had a huge impact on the biological sciences. The picture that emerged

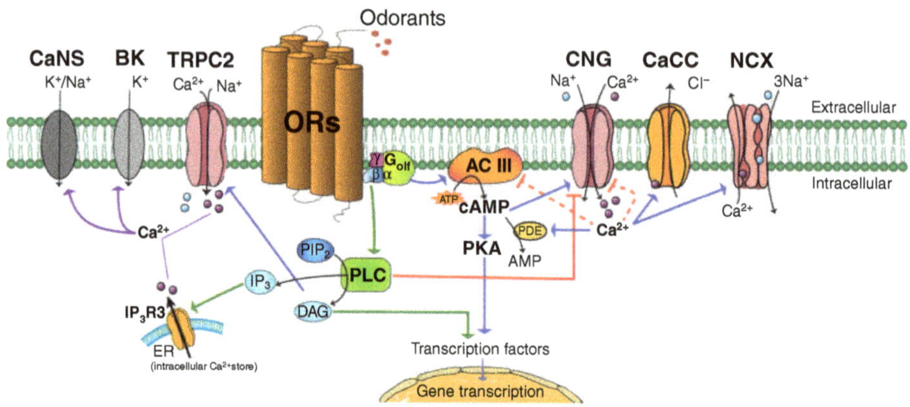

Figure I.3.9 Olfactory pathway modeled as a sequence of binary interactions between macromolecules and their ligands. From [34]/with permission of Elsevier.

was that individual channels are regulated by drugs and other small components and affect the function of the channel. This leads to changes in the electrostatic potential of the membranes. The interactions between receptors and drugs are mostly binary and local.

Combining many binary interactions between molecules into reaction networks is a concept called *pathways*. A simple example of such a pathway diagram is shown in Figure I.3.9 for an olfactory pathway taken from [34]. In this diagram, various ion channels are controlled by drugs and lead to signaling processes.

These pathways may be complex. Probably the most complete pathways are the Roche Biochemical Pathways. These are two very detailed maps showing the metabolic pathways and the cellular and molecular processes. They can be found on the Roche webpages. Specific macromolecules are connected to other proteins and drugs by arrows. The arrows are associated with rate constants that control the reaction rate. It is implicitly assumed that the individual reactions can be separated, and their thermodynamics can be described separately from the other reactions. The underlying hypothesis is that one can describe life by adding up all the paths. However, this is not possible in thermodynamics, which is generally a theory of the whole system under consideration, which cannot be broken down into subprocesses without losing information. The pathways are a nice visualization of what is connected to what in biochemistry, but they don't allow you to describe the thermodynamics and thus the function of a cell, which is often assumed to be possible. The description of processes in the cell at the molecular level does not include the emergent phenomena that are central to this book.

I.4

Another Line of Thought

Since the early beginnings of electrophysiology, i.e., the time of Emil du Bois-Reymond, there has existed another line of thought linking nervous impulses to thermodynamic phenomena. This included conservation of heat, mechanical changes during the action potential, and the propagation of electromechanical waves.

I.4.1 Schrödinger and Physics in Biology

In 1944, Erwin Schrödinger (Figure I.4.1) wrote a remarkable essay called *What is Life?* [35], which was based on the assumption that life is exclusively based on the laws of physics. He is therefore in line with the views of du Bois-Reymond and von Helmholtz (see Section I.2.1). This view does not give any room for vitalism or non-material spirits. Schrödinger states that life is based on the laws of thermodynamics and the increase of entropy on its path toward equilibrium. In particular, life is based on the second law of thermodynamics, which describes how a biological system evolves spontaneously.

Schrödinger's insightful consideration led to a prediction about the nature of the molecular origin of hereditary information, which had a seminal influence on the postulation of the double-helical structure of the DNA by James D. Watson and Francis Crick [36].

The dogma of molecular biology, formulated by Francis Crick, states that the information for living processes is stored in the genes of the DNA, which encodes proteins, while the proteins are responsible for function. The biochemical pathways are built on the assumption that cellular function is a property of single molecules that interact via association constants. Schrödinger's view is not entirely consistent with the dogma of molecular biology because entropy is not a feature of individual molecules but rather of an ensemble (e.g., a cell) as a whole. At larger scales, one may find phenomena that cannot be explained by interactions between single molecules. Such phenomena are called emergent phenomena. Examples include (see Figure I.4.2):

- waves such as surface water waves or sound waves, as well as waves on membranes.
- phase transitions such as the cooperative melting of ice or other solids, i.e., order transitions from crystals to liquids
- boiling of water
- freezing point depression, i.e., the change in melting point due to the entropy of dissolution

Physics of Nerves and Excitatory Membranes, First Edition. Thomas Heimburg.

Figure I.4.1 Erwin Schrödinger (1887–1971). Image is public domain. Nobel foundation/Wikimedia Commons/Public Domain.

Figure I.4.2 Cooperative phenomena in water that cannot be understood on molecular scale. Left: Water waves. Center: Melting of ice to water. Right: Melting of snow induced by salt crystals.

In fact, the entire fields of hydrodynamics, wave mechanics, and thermodynamics cannot be thought of at local scales.

I.4.2 Early Studies on Mechanical Stimulation

If one reads modern textbooks, one finds a narrative that nearly exclusively focuses on the electrical properties of nerves. However, there have been plenty of studies that investigated other phenomena, namely of mechanical, optical, and thermal nature. It is important to know these studies because they imply that the nervous impulse is a phenomenon of much broader nature than just being electrical. These phenomena are described in their historical context in the following sections, and in much more detail in Part III.

Nerves can be stimulated mechanically by compression, stretching, and by cutting the nerve. Additionally, one finds that the nerve pulse itself has a propagating mechanical component, which can be observed both by mechanical means (e.g., in atomic force experiments) and by optical means that monitor changes in absorption, scattering, birefringence, and polarization changes of fluorescence markers caused by changes in the geometry and state of the nerve and the nerve membrane. The history of early mechanical nerve studies is nicely summarized by Tigerstedt in 1880 [37] (Figure I.4.3). Robert Tigerstedt was a Finnish/Swedish scientist living in Helsinki and Stockholm from 1853 to 1923. Tigerstedt wrote in his monograph *Studien über die mechanische Nervenreizung (Studies on Mechanical Stimulation of Nerves):*

> *"The scientists who have worked on the stimulation of nerves and muscles nearly exclusively made use of electricity, or to be more precise of electrical currents as stimulus, whereas stimulation by chemical, thermal or mechanical stimuli ... only were used as an exception."* (translated from German)

Since originally electrical methods were not well developed, many early studies focused on exciting the nerve mechanically and recording the contraction of a muscle, e.g., the stimulation of a frog sciatic nerve and the contraction of muscles in the frog leg. As early as 1766, Haller [38] described that nerves can be stimulated mechanically by cutting or compressing the nerve. He called this a phenomenon widely known from ancient times. Fontana [39] found that one can stimulate nerves by fast compression, fast stretching, or cutting the nerve. In contrast, slow compression or slow stretching rather inhibits nerve pulse propagation. He also found that the fast cutting of a nerve with a very sharp knife does not stimulate the nerve. This indicates that it is not the electrical shortcut that generates the nerve pulse but rather the mechanical perturbation of the nerve. It also shows that the stimulus must be fast, i.e., it is rather the change of the force in time rather than a force that is important. This is also the case for electrical stimulation.

Figure I.4.3 Robert Tigerstedt (1953–1923). Finna/Public Domain.

Mechanical stimulation of nerves by fast repetition of hits (mechanical tetanus) was also reported by du Bois-Reymond in 1849 [40, pp. 52–54]. Du Bois-Reymond could further show that the nerve pulse could be measured electrically after mechanical stimulation, thus proving that the signal created after mechanical stimulation is the same action potential as after electrical stimulation, and that the action potential displays both an electrical and a mechanical component. It is therefore obvious that du Bois-Reymond – who was the first one to describe the action potential – did not think about it as a purely electrical phenomenon. In [40], he also described mechanical, thermal, and chemical stimulation. The mechanical means were diverse – from repetitive hits of the nerve with the back of a knife to rotating gears. Often the problem in these experiments was that the stimulation damaged or even destroyed the nerve. To create a reliable mechanical tetanus, one has to be well able to control the force and frequency of the beats. Heidenhain [41, 42] constructed a device which was called a *tetanomotor*, which first contained an electromechanical stimulator and later gears. Julius Bernstein studied mechanical stimulation by a tetanus administered by repetitive beats with the back of a knife [43]. He was interested in the fatigue of the nerves after repetitive mechanical stimulation. He noticed that the amplitude of muscle contraction decayed after repetitive stimulation, but recovered after about one half of an hour. While the aforementioned studies mostly considered a mechanical tetanus as stimulus, there also have been attempts to measure stimulation after one single hit by a falling weight [44, 45]. The experiments by Wundt are described in detail in [37]. As Tigerstedt noted, most of these stimuli were too large and damaged or destroyed the nerve. In order to avoid this, Tigerstedt constructed an electromechanical device in which a small weight was lifted by an electromagnet and then dropped [37]. Tigerstedt used weights of the order of 0.5 g that were dropped from a few millimeters of height. By these means, he could reproduce and improve the experiments by Bernstein [43].

I.4.2.1 The Mechanical Action Potential

In 1912, E. Wilke reconsidered the unexplained fact that nerves can be stimulated mechanically. He hypothesized that the nerve resembles a viscoelastic hydrogel in which deformation pulses can propagate [46]. As pointed out by Wilke, all materials can carry mechanical waves. However, one has to explain how a mechanical wave can give rise to the electrical signature of an action potential. Wilke pointed out that hydrogels possess piezoelectric properties. For instance, if a piece of an agar-agar gel is compressed on one side, one can measure a potential difference on the order of 10 mV relative to the uncompressed part of the gel. For this reason, a deformation wave in such a gel would give rise to a propagating electrical signal. The magnitude of the electrical signal would be different for different kinds of hydrogels. Together with his colleague E. Atzler, he measured mechanical deformations both in hydrogels and in frog nerves [47]. They reckoned that the thickness change might be too small to be measured. However, due to conservation of volume, a contraction of the nerve was expected. They further assumed that the nerve pulse may only last for a very short period of time. In order to measure it nevertheless, they fixed the nerve on one end and connected it to a vibrating quartz thread on the other end. The idea was that the thread would vibrate for a longer time than the duration of the nerve pulse (see Figure I.4.4). After electrical stimulation, they could in fact see a contraction of the nerve as recorded by

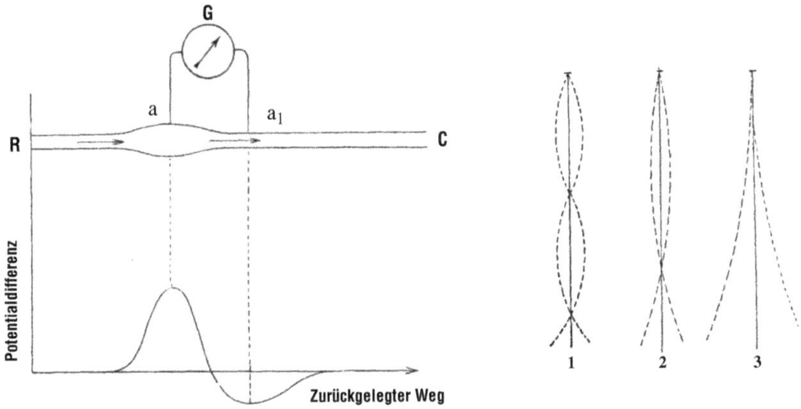

Figure I.4.4 The experiment by Wilke and Atzler. Left: Schematic drawing of the propagation of a peristaltic pulse in a nerve compared to the distance traveled. From [46]/with permission of Springer Nature. Right: The motion of thin glass fibers attached to the end of the nerve indicates a contraction of the excited nerve under the influence of an action potential. From [47]/Elsevier/CC BY 4.0.

Figure I.4.5 Ishiji Tasaki (1910–2009) was a Japanese-American scientist working most of his career at the NIH in Bethesda. Eugene M. Izhikevich/Scholarpedia/CC BY-SA 3.0.

the vibration of the thread. To our knowledge, this is the first recording that demonstrated a mechanical displacement during the nerve pulse.

Investigations on mechanical (or dimensional) changes in nerves were taken up again by the great Ichiji Tasaki (Figure I.4.5). Tasaki was a famous biochemist who made significant contributions to the understanding of nerves, including the proposal of saltatory conduction in myelinated nerves and the role of calcium during nerve pulses. His contributions also include mechanical, thermal, and optical measurements during the action potential – all measurements that had rarely been performed by other people prior to him.

In 1980, he and his colleague Kuni Iwasa made the first measurements of thickness changes during the nerve pulse [48, 49] (Figure I.4.6 (left)). They designed a simple device that measured forces and dislocations during the action potential. They found transient changes in a squid axon of the order of 1 nm. While this sounds like a small change, it is about 20% of the thickness of a nerve membrane. These findings were supported in further

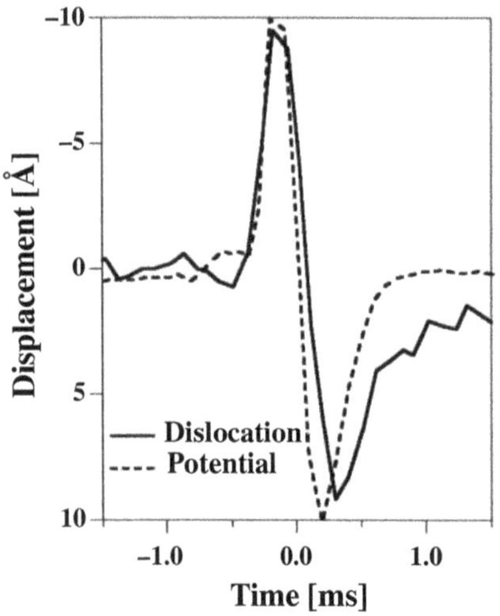

Figure I.4.6 The thickness changes of a squid axon during the action potential. One can see that thickness and voltage are in phase. Thickness changes are about 1 nm, corresponding to about 20% of the membrane thickness. Adapted from [48].

studies [50–56]. Tasaki and collaborators also found that axons contract in length [55, 57], as proposed by Wilke [46, 47].

More recent experimental findings will be discussed in Chapter III.4.

I.4.3 From Helmholtz to Hill: The Absence of Heat Production in Nerves

One of the most striking properties of the nerve pulse is its heat signature. The interest in the heat production of nerves dates back to Helmholtz in 1848. The history of related findings will be described as follows.

In 1845, Emil du Bois-Reymond was the first to measure electrical currents and voltage changes in stimulated muscles and nerves [58], which were found to be of similar magnitude. In 1847, Hermann von Helmholtz proposed the first law of thermodynamics [11]. It states that the change of energy in a system is equal to the sum of the heat absorbed or released and the work performed.[1] Du Bois-Reymond's nerve and muscle experiments and his own considerations aroused Helmholtz's interest in the heat generation of muscles and nerves. Given the similarity of their currents, he expected that the chemical reactions and heat generation in muscles and nerves would also be similar. In experiments on frog muscles, he was able to measure temperature changes of the order of 0.035°C after repeated muscle contractions [10]. Using the same experimental setup, he found that the temperature changes in the nerves leading to this muscle were unmeasurable (i.e., less than 0.002°C, which was the sensitivity limit of his experiment). This finding indicates that the mechanisms of heat production in muscles and nerves must be very different. In particular, the lack of measurable heat production suggests that no metabolism takes place during

1 This chapter borrows from [59], where the heat phenomena in nerves were discussed in more depth.

Figure I.4.7 Archibald Vivian Hill (1886–1977). Unknown/Wikimedia Commons/Public Domain.

the nerve impulse and that the process under consideration could be of a reversible physical nature. Helmholtz found this negative result interesting enough to devote a separate section to heat production in the nerves [10].

Stimulated by von Helmholtz's intriguing finding, subsequent research performed similar experiments while increasing the sensitivity of the experimental setup. In 1868, Heidenreich [60] improved the sensitivity of heat recordings by about a factor of 10 using a thermopile. Like Helmholtz, he found no heat release during the nerve impulse. In 1890, Rolleston [61] came to a similar conclusion using a platinum resistance thermometer. In 1891, Stewart [62] repeated the experiments on the mammalian nerve using a similar setup. He also found no heat production. In 1897, Cremer [63] also failed to find any heat production in the olfactory nerves of pike, carp, and barbel. In 1912, Archibald Vivian Hill [64] (Figure I.4.7) performed experiments on the sciatic nerves of frogs using a very sensitive thermopile consisting of 30 thermocouples in series. He summarized:

> *"By a thermo-electric method it is shown that tetanus, up to 25 secs., of a live nerve, does not cause a change of temperature (other than at the seat of excitation) of more than about* $+6 \cdot 10^{-6}$ *°C. There is no evidence of any change at all, but the method does not allow conclusions beyond this limit. For every single propagated disturbance the change of temperature therefore cannot exceed about* 10^{-8} *°C., a hundred million*th *of a degree. This corresponds to an oxidative process, in which only one molecule of oxygen is used in a space of visible size, viz. a 3.7 μcube. This suggests very strongly, though of course it does not finally prove, that the propagated nervous impulse is not a wave of irreversible chemical breakdown, but a reversible change of a purely physical nature."*

Later in his life [65], he commented on this early phase of heat measurements:

> *"Why did people go on trying to measure the heat production of nerve, in spite of repeated failure? Chiefly, I suppose, in order to settle the question of whether the nerve impulse is the sort of physical wave in which the whole of the energy for transmission is impressed on the system at the start."*

W.M. Bayliss [66] summarized the situation in his textbook *Principles of General Physiology*:

> *"The result makes it impossible to suppose that any chemical process resulting in an irreversible loss of energy can be involved in the transmission of a nerve impulse, and indicates that a reversible physicochemical one of some kind is to be looked for."*

Thus, the conclusion to be drawn from the pre-World War I experiments was that, unlike a contracting muscle, the nerve impulse did not generate any measurable heat. Therefore, it was considered most likely that the nerve impulse was a reversible physical phenomenon similar to a mechanical wave. While Hill's paper was very influential after its publication, it is almost forgotten today.

I.4.3.1 Archibald V. Hill and His School

In the following decades, Hill became the leading authority on the measurement of heat production in muscles and nerves. In 1922, he was awarded the Nobel Prize for Physiology and Medicine. The problem of the undetectable heat production of the nerve pulse continued to bother Hill. In the second half of the 1920s, he and his collaborators Downing and Gerard managed to improve the sensitivity of the temperature measurements [67]. Between 1926 and 1935, he continued studying the problem of heat generation in nerves [68–76].

In a seminal publication in 1958, Abbott, Hill, and Howarth [77] (see also [78]) first resolved the initial heat response during the action potential in nerves from the legs of spider crabs. They called their article "The positive and negative heat production associated with a nerve impulse." It was found that, during the rising phase of the action potential, heat is released into the environment of the nerve. This heat is mostly reabsorbed (to at least 70%) in a second phase of the action potential. The existence of a reabsorption of heat is clearly surprising.

Hodgkin acknowledged the problem arising from Abbott et al.'s result in his monograph "The Nervous Impulse" from 1964 [79, p. 70]:

> *"In thinking about the physical basis of the action potential perhaps the most important thing to do at the present moment is to consider whether there are any unexplained observations which have been neglected in an attempt to make the experiments fit into a tidy pattern. ... perhaps the most puzzling observation is one made by A. V. Hill and his collaborators Abbott and Howarth (1958). ... On reinvestigating the initial heat of crab nerve with better time resolution, Hill and his colleagues found that it was diphasic and that an initial phase of heat liberation was followed by one of heat absorption. ... a net cooling on open-circuit was totally unexpected and has so far received no satisfactory explanation."*

During the next 30 years, Hill's studies were continued and refined by close collaborators, in particular Howarth, Abbott, and Ritchie [80–87]. In all of these papers, the authors came to the conclusion that the charging and discharging of a membrane capacitor (as expected in the electrical models for nerve conduction [21]) cannot explain the observed changes in heat. The authors propose that most of the reversible heat production is due to entropy changes within the membrane during the action potential.

Details of the experimental findings will be discussed in Chapter III.6.

I.4.4 Thermodynamic Properties of Nerves

As shown in the recent sections, the nerve pulse is accompanied not only by electrical changes but also by mechanical changes and transient changes in temperature (heat production). Thus, it seems that the nerve pulse is a thermodynamic rather than purely electrical event. There is also the remarkable fact that the velocity of the nerve pulse is very similar to the speed of sound in the membrane (~30 m/s in the sciatic nerve of a frog), which was already mentioned by Helmholtz [8, 9]. He noted that the nerve pulse does not propagate with the speed of light (as one might expect for electrical phenomena) but rather with a velocity that is similar to that of a sound wave in organic media. Helmholtz also noted that the propagation velocity is very temperature-dependent. It is slower at lower temperatures.

So, why has the electrical part gained so much attention while the other contributions are rarely considered? This is, in my opinion, merely a consequence of that the voltage changes measured during the action potential are large compared to the temperature and dimensional changes. However, what is it that we mean by "large"? The voltage change of about 100 mV measured during the action potential is easily measured with modern instrumentation. A thickness change of 1 nm, however, requires more sophisticated experiments, and a temperature change of 10 μK during a time period of 1 ms is on the edge of what is possible to measure even today. Therefore, one is tempted to conclude that voltage changes are big, while thickness and temperature changes are unimportant side effects. However, the only thing that we can truly state is that our instrumentation has different sensitivities. The units of the different observables are different, and it is hard or impossible to compare the numbers. This, however, becomes feasible when one relates the numbers to the units of energy. The electrical energy of the nerve membrane is the capacitive energy, $E_{\rm el} = \frac{1}{2}C_m V^2$, of the membrane integrated over the pulse length [59]. The thermal energy of the nerve would be $E_{\rm therm} = c_V \cdot \Delta T$ integrated over the pulse. It turns out that the thermal energy is considerably larger than the electrical energy [81, 84, 88], as reviewed in [59]. The temperature changes therefore do not represent an unimportant side phenomenon. They rather seem to represent the dominant signal.

This peculiar conclusion led to increased interest in the non-electrical signals in the recent one-and-a-half decades (2005–2020). El Hady and Machta attempted to model the mechanical changes by electrostriction, i.e., a mechanism in which the voltage difference across the nerve membrane causes an electrical force that leads to mechanical changes [89]. Thus, they assumed that the mechanical changes are an epiphenomenon caused by voltage changes. They did not provide a theory for the voltage changes. However, they suggested that the mechanical changes might well be in agreement with the Hodgkin–Huxley model. This is probably an incomplete conclusion because it is explicitly assumed in the

Hodgkin–Huxley model that the capacitance of the membranes is constant. Changing this assumption would lead to including piezoelectric effects into the nerve pulse, which implies different physics involving elastic constants of the membrane and its thermodynamics.

However, the change in capacitance plays an important role in a model that represents a true alternative to the Hodgkin–Huxley model. This is the soliton theory that likens the nerve pulse to a localized sound pulse [90]. Intrinsically contains mechanical and thermal contributions to the action potential.

I.4.4.1 The Nerve Pulse as a Density Pulse or Soliton

As described earlier, there are numerous facts that speak in favor of a thermodynamic description of nerve pulses:

- The velocity of the nerve pulse is of the order of the speed of sound in membranes.
- There exists a transient release and re-uptake of heat of the nerves with the environment that is larger than the heat of charging a capacitor, and the total heat exchange is zero.
- Mechanical changes in nerve dimensions and associated changes in membrane viscosity have been measured during the action potential.
- The size of action potentials scales from millimeter to centimeter size (see Section III.3.5, [91]).

Various authors have proposed that the nerve pulse is of mechanical nature, among them Hill [46, 47, 64, 66] but also Kaufmann [92], who pointed out that the nerve pulse responds to the thermodynamics of the membrane. Tasaki and collaborators suggested that a phase transition is involved in the nerve pulse [93, 94].

In this spirit, Heimburg and Jackson proposed that the action potential is a propagating sound pulse involving an order–disorder phase transition in the membrane [90] depicted in Figure I.4.8. Such pulses are called solitons – but in the present context, they are just a special form of sound propagation in membranes that happens to lead to localized pulses due to the thermodynamic features of biological membranes close to transitions. The theory of nerve solitons is described in Part VII.

The soliton picture relies on the finding that the elastic constants and the state of a membrane change drastically in melting transitions. For this reason, a compressional wave

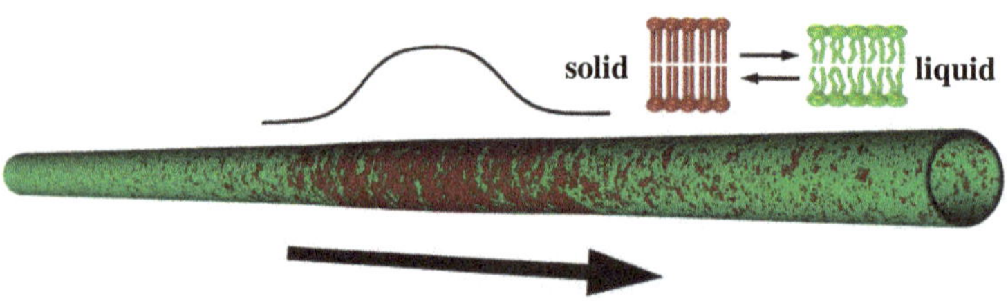

Figure I.4.8 Schematic picture of a propagating mechanical pulses in a membrane cylinder that involves a change in membrane state from liquid to solid. From [95]/with permission of Elsevier.

or pulse will lead to a large reversible exchange of heat, changes in membrane thickness, and changes in membrane permeability. All these properties have been found in nerves. Since membranes in transitions display large thermal fluctuations, one also finds channel-like conduction events in the membranes that resemble protein ion channels (see Section VIII.1.3). Significant parts of this book are dedicated to the thermodynamic description of the nerve pulse.

I.5

Scope of This Book

In the previous chapter, we suggested that biological membranes possess many thermodynamic properties of biological relevance, which are macroscopic in nature and not necessarily purely electrical. These properties give rise to the possibility of electromechanical pulses (solitons) that resemble the nervous impulse. They are correlated with thermodynamics fluctuations that have the appearance of ion channel proteins. They yield a control of these properties by the thermodynamic variables, in particular voltage, temperature, pressure, and the chemical potential of certain drugs and membrane components, including proteins. In this book, we focus on the phenomena that are outlined in Chapter I.4 called *Another Line of Thought*. We show that surprisingly many properties of nerves and other excitable membranes naturally arise from thermodynamics. Since thermodynamics is a complete theory, one can make many testable predictions on the properties of biomembranes and understand, e.g., the pressure reversal of anesthesia.

This book is structured in the following manner:

- In Part II, we give a brief introduction into thermodynamics. It provides the technical background to many derivation later in the book.
- In Part III, the details of the physical changes observed during nerve activity are described.
- In Part IV, we introduce the basic principles of electrophysiology and discuss the advantages and shortcomings of such a description.
- In Part V, we describe the physics of synthetic and biological membranes. This includes the thermodynamics of melting transitions, elastic constants, capacitance, and polarization, and how these properties are influenced by the intensive thermodynamic variables.
- In Part VI, we show how the thermodynamics of nerve membranes leads to fluctuations and maxima in both the elastic constants, the permeability, the capacitance, and the fluctuation lifetimes. These phenomena enable the nerve membrane to carry electromechanical pulses with a given lifetime, and facilitate the formation of channel-like phenomena.
- In Part VII, we derive the hydrodynamics of biological membranes, and how that leads to hydrodynamic pulses and solitons. These solitons are shown to describe many features of nerve pulses.
- In Part VIII, we describe the properties of spontaneous pore formation, leading to lipid ion channels, and how they relate to protein channels.

Physics of Nerves and Excitatory Membranes, First Edition. Thomas Heimburg.

- In Part IX, we discuss in depth the properties of anesthetics and how a thermodynamic description of the nerve pulse includes a natural explanation of the poorly understood phenomenon of anesthesia. We discuss other findings of medical relevance, including nerve stretching, fever, the role of lithium in bipolar disorder, and essential tremor.

While many of these topics have been published in articles, there has never been a compilation in the form of a monograph, in which the wealth of biological thermodynamics has been outlined in its wonderful beauty. We hope that this book will be a helpful contribution toward achieving a complete picture of biological membranes, and that it demonstrates the power of a thermodynamic treatment for the understanding of life.

Part II

Thermodynamics

Thermodynamics is the basic pillar of physics. It is not a theory applied to particular systems, but rather the language of physics. It defines the meaning of terms such as temperature and energy, which are used in all fields of physics, and is valid on all scales. It is a language of exceptional beauty. In his autobiographical notes, Albert Einstein wrote [1]:

> *"A theory is the more impressive the greater the simplicity of its premises is, the mere different kinds of things it relates, and the more extended is its area of applicability. Therefore the deep impression which classical thermodynamics made upon me."*

In this book, we make extensive use of thermodynamics. For this reason, we provide a brief introduction to thermodynamics that contains everything that is necessary to understand the derivations in this book. Skip this part if you feel familiar with the basic concepts, or use this part as a reference.

II.1

Fundamental Laws in Thermodynamics

II.1.1 Observations

The beginning of a systematic development of thermodynamics is intrinsically linked to the investigation of ideal gases. Robert Boyle (Figure II.1.1) (left) and Edme Mariotte independently found that the product of volume and pressure in a gas is constant, i.e., the smaller the volume, the larger the pressure:

$$pV = \text{const.} \quad \text{(Boyle, 1662, and Mariotte, 1679)} \tag{II.1.1}$$

More than 100 years later, Joseph Louis Gay-Lussac (Figure II.1.1) (right) found that the constant in this law is proportional to temperature [2]:

$$pV \propto T \quad \text{(Gay-Lussac, 1802).} \tag{II.1.2}$$

It is obvious that the constant must be proportional to the amount of gas in a container. If the volume is twice as large, two times the amount of gas fits into the container while the pressure remains the same. This leads to the ideal gas law:

$$pV = N \cdot kT = n \cdot RT \tag{II.1.3}$$

N is the number of particles, and n is the number of moles in the system under consideration. One mole of a substance is the weight in grams corresponding to the mass number (e.g., 18 g H_2O, 2 g H_2, 44 g CO_2,...) and always contains $N_A = 6.0223 \cdot 10^{23}$ mol^{-1} molecules (Avogadro's number). The gas constant $R = 8.314$ J/mol· K is given in units per mole. Boltzmann's constant is $k = R/N_A = 1.3807 \cdot 10^{-23}$ J/K and is given per molecule. For this reason, nRT and NkT have the same units.

II.1.2 Variables

The ideal gas law creates a relation between four variables of the gas: pressure p, volume V, number of moles of particles n, and temperature T. Two of these variables, pressure and temperature, are called intensive variables, i.e., variables that are independent of system

Physics of Nerves and Excitatory Membranes, First Edition. Thomas Heimburg.

Figure II.1.1 Left: Robert Boyle (1627–1691) Rijksmuseum / Wikimedia Commons / Public domain. Right: Joseph Louis Gay-Lussac (1778–1850). Unknown Source / Wikimedia Commons / Public domain.

size. The other two variables, number of moles n and volume V, are called extensive variables, which are proportional to system size. In the example of an ideal gas, if two gas containers of equal size and identical pressure and temperature are connected, the pressure and temperature remain unchanged while the number of particles and the volume double.

Examples of intensive variables used in this book include:

- temperature, T
- pressure, p
- lateral pressure or surface tension, Π
- force in a spring, F
- electrostatic potential, Ψ
- chemical potential of particle species i, μ_i

Examples for extensive variables used in this book include:

- energy, E
- volume, V
- area, A
- extension of a spring, x
- charge, q
- number of moles of particle species i, n_i

II.1.3 Conjugated Variables

Energy displays units of joules, J. With respect to energy, intensive and extensive variables come in conjugated pairs, where their products also have units of joules, e.g., $p \cdot V$, $\Pi \cdot A$, $F \cdot x$, $\Psi \cdot q$, and $\mu_i \cdot n_i$.

II.1.4 Functions of State

An important requirement in the physical sciences is that an experiment shall be reproducible under identical conditions. No physical law shall depend on the history of a sample, e.g., which person performed the experiment or what happened to an experimental system the previous day. This implies that the experiment should be independent of time, location, and the particular order of changes made to the system. If such conditions cannot be defined, one cannot express fundamental laws.

This leads to the concept of a function of state, Z. The integral over dZ from state a to state b must be independent of the path and must only depend on the values of the variables.

$$\int_a^b dZ \quad \text{independent of integration path.} \tag{II.1.4}$$

This implies that the integral over a closed path returning to the initial state must be zero, independent of the integration path:

$$\oint dZ = 0. \tag{II.1.5}$$

One also says that dZ is a perfect differential. It does not matter which path one goes to state with a different p, V, and T.

From this, it follows that pV in the ideal gas law is a function of state because this product only depends on the fixed number of particles and temperature. At the same temperature and volume, one will always find the same pressure independent of experimental history. This means that

$$\oint d(pV) = \oint pdV + \oint Vdp = 0, \tag{II.1.6}$$

and $\oint pdV = -\oint Vdp$. However, this does not imply that either $\oint pdV$ or $\oint Vdp$ is zero. These integrals do not represent functions of state.

II.1.5 The First Law of Thermodynamics

The first law of thermodynamics was first expressed by Julius Robert Mayer in 1842 [3] and by Hermann von Helmholtz [4]. Let us introduce the internal energy, which shall be a function of state. The first law implies that energy may be transformed from one form into another but is conserved as a whole. The energy of a thermodynamic system may change by adding heat to or performing work on the system:

$$dE = dQ + dW, \tag{II.1.7}$$

where dQ denotes heat added or removed from the system, and dW denotes the work performed on that system. In a closed system, the energy is conserved, and the only changes that can happen are conversions of energy from one state to another. As demonstrated in the following, this implies a change in entropy.

II.1.5.1 Work

The work is given as

$$dW = \underbrace{-pdV}_{\substack{\text{work on}\\\text{volume}}} \underbrace{-\Pi dA}_{\substack{\text{work on}\\\text{area}}} \underbrace{-Fdx}_{\substack{\text{work on}\\\text{spring}}} \underbrace{+\Psi dq}_{\substack{\text{electr.}\\\text{work}}} \underbrace{+ \ldots + \sum_i \mu_i dn_i}_{\substack{\text{chem.}\\\text{work}}} . \tag{II.1.8}$$

Since $\oint dE = 0$, but $\oint dW \neq 0$, it follows that $\oint dQ \neq 0$. Both dQ and dW are imperfect differentials (they depend on the integration path).

II.1.5.2 Carnot Cycle

The Carnot cycle is a theoretical construct used to determine the efficiency of a heat engine and also serves to define the concept of entropy and absolute temperature. It is named after the French engineer Sadi Carnot (1796–1832) [5] (Figure II.1.2 (left)). It consists of a closed sequence of alternating isothermal and adiabatic expansions and compressions of an ideal gas, which is schematically shown in Figure II.1.2 (right).
We define:

- isothermal process: $dT = 0\,(\text{T} = \text{const})$
- adiabatic process: $dQ = 0 \Longrightarrow dE = dW$

Let us now consider a gas cylinder with a piston containing n moles of particles in a volume V_1. The cylinder is in contact with a water bath at temperature T_a. The ideal gas law shall apply. Further, we will assume that the energy of the ideal gas does not depend on

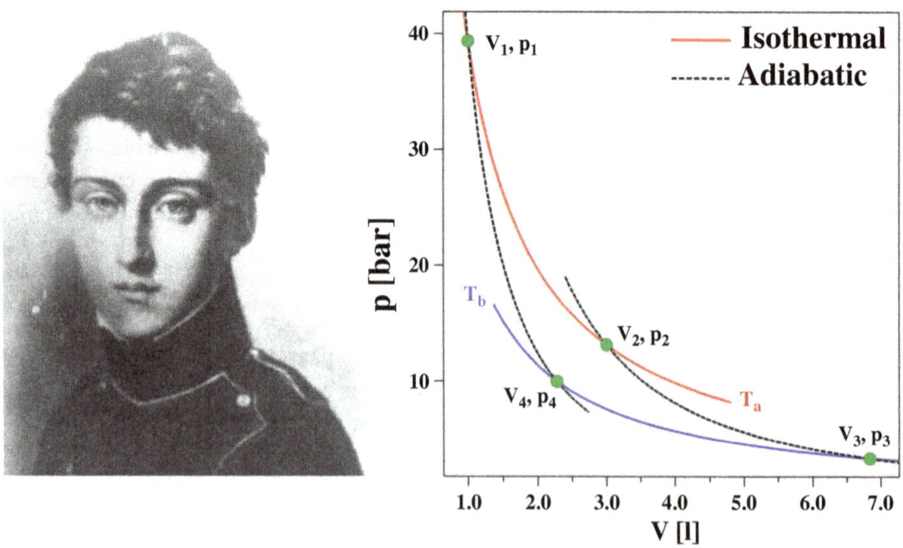

Figure II.1.2 Left: Nicolas Leonard Sadi Carnot (1796–1832). Unknown author / Wikimedia Commons / Public Domain. Right: Carnot cycle calculated for $n = 1$ mol of gas atoms, $T_a = 200\,°\text{C}$ and $T_b = 0\,°\text{C}$. $V_1 = 1$ L and $V_2 = 3$ L.

either pressure or volume because the gas particles do not interact. It only depends on the number of particles and the temperature such that

$$E = \frac{3}{2}nRT. \tag{II.1.9}$$

This result from statistical mechanics attributes an energy of $\frac{1}{2}nRT$ to each degree of freedom. An ideal gas has three degrees of freedom in the three spatial directions. Adiabatic expansions of ideal gases are governed by the equation of state

$$T^{3/2}V = \text{const.} \tag{II.1.10}$$

We derive this in Section II.1.10.

In the following, we describe a Carnot cycle with four changes in state: In the first step, the gas volume is slowly expanded isothermally from volume V_1 to V_2. In this expansion process, the energy stays constant because the temperature of the gas does not change. Therefore, $\Delta W_{12} = -\Delta Q_{12}$. An amount of heat ΔQ_{12} is absorbed from the water bath at T_a, and a work ΔW_{12} is performed on the outside:

$$\text{Process 1: } \Delta W_{12} = nRT_a \ln\left(\frac{V_2}{V_1}\right) = -\Delta Q_{12};\ \Delta E_{12} = 0. \tag{II.1.11}$$

In the second step, the gas cylinder is taken out of the heat bath and adiabatically expanded to volume V_3. During the adiabatic expansion, $\Delta Q_{23} = 0$, and therefore $\Delta W_{23} = \Delta E_{23}$. The temperature decreases due to the equation of state of the adiabatic expansion. We chose the volume V_3 such that it corresponds to temperature T_b:

$$\text{Process 2: } \Delta W_{12} = \Delta E_{23} = nR(T_b - T_a);\ \Delta Q_{23} = 0. \tag{II.1.12}$$

In the third step, we place the gas cylinder in a second water bath at temperature T_b. Now, we isothermally compress the gas until volume V_4 is obtained. An amount of heat ΔQ_{34} is released into the water bath at temperature T_b.

$$\text{Process 3: } \Delta W_{34} = nRT_a \ln\left(\frac{V_4}{V_3}\right) = -\Delta Q_{34};\ \Delta E_{34} = 0. \tag{II.1.13}$$

In the fourth step, the gas with volume V_4 and temperature T_b is adiabatically compressed until one returns to the state with T_a and V_1:

$$\text{Process 4: } \Delta W_{41} = \Delta E_{41} = nR(T_a - T_b);\ \Delta Q_{41} = 0. \tag{II.1.14}$$

Now we have a close cycle. Using the equation of state for the adiabatic expansion, we can show that $V_1/V_2 = V_4/V_3$. The total energy and work change in the cycle are given by:

$$\Delta E_{\text{tot}} = \Delta E_{12} + \Delta E_{23} + \Delta E_{34} + \Delta E_{41} = 0$$
$$\Delta W_{\text{tot}} = \Delta W_{12} + \Delta W_{23} + \Delta W_{34} + \Delta W_{41} = nR(T_a - T_b)\ln \frac{V_2}{V_1}\Delta Q_{\text{tot}} = -\Delta W_{\text{tot}}$$

This shows that the internal energy E is a function of state with $\oint dE = 0$, while heat Q and work W are not functions of state, because $\oint dW = -\oint dQ \neq 0$. Heat and work do not describe the state of the system but rather an interaction with the environment.

They depend on the particular way in which the thermodynamic system interacts with the outside world.

During the Carnot cycle, an amount of heat ΔQ_{12} is taken from the water bath at temperature T_a. In this cyclic process, some of that heat is transferred to the water bath at temperature T_b, and some of it is converted into work. In the example shown in Figure II.1.2 (right), with $V_1 = 1$ l, $V_2 = 3$ l, $T_1 = 300$ K, and $T_2 = 200$ K, a heat of 913 J is transferred from water bath 1 to water bath 2, and an equivalent amount of work is performed.

II.1.6 The Introduction of Entropy

It can be shown with the help of the aforementioned Carnot cycle that

$$\oint \frac{dQ_{12}}{T_a} + \frac{dQ_{34}}{T_b} = 0 \tag{II.1.15}$$

Since one can construct Carnot cycles with an arbitrary number of alternating isothermal and adiabatic processes using many different water baths, one can more generally state that

$$\oint \frac{dQ}{T} = 0 \tag{II.1.16}$$

Thus, dQ/T describes a function of state. We define

$$\frac{dQ}{T} \equiv dS_r \qquad \text{or} \qquad dQ = TdS_r \tag{II.1.17}$$

The function of state S_r was called **entropy** by Rudolf Clausius (Figure II.1.3 (left)). Here, we add an index "r" to denote that all processes in the Carnot cycle are reversible as long as they are performed very slowly. S_r is a function related to heat and temperature. The aforementioned law was used by William Thomson (Lord Kelvin) (Figure II.1.3 (right)) to define an absolute temperature scale, which was named after him.

Figure II.1.3 Left: Rudolf Clausius (1822–1888). Unknown / Wikimedia Commons / Public Domain. Right: William Thompson (Lord Kelvin, 1824–1907) Messrs. Dickinson / Wikimedia Commons / Public Domain.

II.1.7 The First Law and the Second Law

II.1.7.1 The First Law

Together with the previous two sections, the differential of the internal energy can be written as

$$dE = TdS - pdV - \Pi dA - Fdx + \Psi dq + \cdots + \sum_i \mu_i dn_i, \tag{II.1.18}$$

or

$$dS = \frac{1}{T}dE + \frac{p}{T}dV + \frac{\Pi}{T}dA + \frac{F}{T}dx - \frac{\Psi}{T}dq - \cdots + \sum_i \frac{\mu_i}{T}dn_i. \tag{II.1.19}$$

At this point, one can introduce the variables that are conjugated with respect to the entropy: $1/T$ is conjugated to the energy E, p/T is conjugated to the volume V, Π/T is conjugated to the area A, F/T is conjugated to the position x, $-\Psi/T$ is conjugated to the charge q, and μ_i/T is conjugated to the number of moles n_i of species i. The products of the conjugate pairs, E/T, pV/T, …, all carry the units of entropy.

II.1.7.2 The Second Law

Carnot's cycle describes an ideal heat engine during which all processes are under the external control of the experimenter. For real processes, the efficiency is always smaller compared to the ideal process because not everything in real processes can be controlled from the outside. Two sources of non-ideality are internal friction, dissipation due to imperfect shielding of the adiabatic processes, or the finite time scale of the expansion, which may be faster than the relaxation processes in the gas cylinder. That can be written as

$$dS \geq \frac{dQ}{T}. \tag{II.1.20}$$

Let us divide the entropy into a reversible part consisting of $dS_r = dQ/T$, which describes the exchange of heat of a system with its environment that can perfectly be controlled from the outside, and an irreversible part dS_i that is not controlled by the experimenter and can therefore not be converted into work. While dS_r can be used to perform work on the external world, dS_i describes the spontaneous changes inside a closed system that cannot be used to perform work. All processes in the ideal Carnot cycle are reversible, i.e., $dS = dS_r = dQ/T$. The second law can be written as:

$$dS = dS_r + dS_i \quad \text{where} \quad dS_r = \frac{dQ}{T} \quad \text{and} \quad dS_i \geq 0. \tag{II.1.21}$$

This is called the second law of thermodynamics. It states that the entropy change dS_i in a closed system (no heat exchange and no work performed) is always greater than or equal to zero. The analysis of the irreversible part of the entropy, dS_i, and the associated entropy production is central to the discipline of nonequilibrium thermodynamics and the theory of fluctuations described in Chapter II.4.

II.1.8 Enthalpy and Free Energy

Functions of state are functions that only depend on the values of the variables. The following functions are functions of state: pV, T, E, and S. Therefore, the following functions are also functions of state:

$$\begin{aligned} H &= E + pV && \text{enthalpy} \\ F &= E - TS && \text{Helmholtz free energy} \\ G &= H - TS && \text{Gibbs free energy } \textit{or free enthalpy}. \end{aligned} \tag{II.1.22}$$

The differentials of these functions are

$$\begin{aligned} dE &= TdS - pdV + \cdots \\ dH &= dE + d(pV) + \cdots = TdS + Vdp + \cdots \\ dF &= dE - d(TS) + \cdots = -SdT - pdV + \cdots \\ dG &= dH - d(TS) + \cdots = -SdT + Vdp + \cdots \end{aligned} \tag{II.1.23}$$

The $+\ldots$ describe the other terms in the expression for the work, such as $-\Pi dA$ or $+\Psi dq$. If the respective extensive variables A and q are kept constant, we can omit the dots.

For the differential of the Helmholtz free energy, we find that

$$dF = 0 \qquad \text{if } V = \text{const. and } T = \text{const.} \tag{II.1.24}$$

Thus, at constant temperature and volume, the free energy displays an extremum. Since the entropy approaches a maximum, the free energy at equilibrium is at a minimum. The Helmholtz free energy is also called the work function, because at constant temperature it describes the amount of work that can be performed in an equilibration process by a spontaneous process without interference from the outside.

For the differential of the Gibbs free energy, we find that

$$dG = 0 \qquad \text{if } p = \text{const. and } T = \text{const.} \tag{II.1.25}$$

Thus, at constant temperature and pressure, the Gibbs free energy displays an extremum. Since the entropy approaches a maximum, the free energy at equilibrium is at a minimum. Since most experiments in biology are performed under conditions of constant pressure and temperature, the Gibbs free energy is the function that describes equilibration processes.

In a similar manner, one can determine functions of state that describe the equilibration at conditions of constant area A or lateral pressure Π: $H = E + \Pi A$, $F = E - TS$, $G = H - TS$, and $dE = SdT - \Pi dA$, $dH = SdT + Ad\Pi$, $dF = -SdT - \Pi dA$, and $dG = -SdT + Ad\Pi$. For electrostatic problems, one obtains minima for constant electrostatic charge, q, or field, Ψ: $dE = SdT + \Psi dq$, $dH = SdT - qd\Psi$, $dF = -SdT + \Psi dq$ and $dG = -SdT - qd\Psi$, which approach minima at constant temperature and constant charge or field, respectively.

One can adapt the definitions of the free energies accordingly when it is known which variables are kept constant during the experiment, i.e., which variables are controlled by the experimenter.

II.1.9 The Heat Capacity

Let us consider a sample to which heat is added from the outside under conditions of constant volume. Then

$$c_v \equiv \left(\frac{dQ}{dT}\right)_V = \left(\frac{dE}{dT}\right)_V, \tag{II.1.26}$$

because $dE = TdS - pdV$, $TdS = dQ$, and $-pdV = 0$ at constant volume. In a similar manner, one can derive the heat capacity at constant pressure:

$$c_p \equiv \left(\frac{dQ}{dT}\right)_p = \left(\frac{dH}{dT}\right)_p, \tag{II.1.27}$$

because $dH = TdS + Vdp$, $TdS = dQ$, and $Vdp = 0$ at constant pressure. Heat capacities can be measured with a differential scanning calorimeter. For biological samples, usually the heat capacity at constant pressure is recorded.

The integrals of the heat capacity over temperature can be used to determine changes of internal energy and enthalpy:

$$\Delta E = \int_{T_1}^{T_2} c_V dT \quad \text{and} \quad \Delta H = \int_{T_1}^{T_2} c_p dT. \tag{II.1.28}$$

Since $dQ = TdS$ and $dS = dQ/T$, entropy changes can also be determined from the heat capacity profiles:

$$\Delta S = \int_{T_1}^{T_2} \frac{c_V}{T} dT \quad \text{at constant volume} \tag{II.1.29}$$

and

$$\Delta S = \int_{T_1}^{T_2} \frac{c_p}{T} dT \quad \text{at constant pressure} \tag{II.1.30}$$

II.1.10 Adiabatic Compression

An adiabatic compression describes a case where $TdS = 0$ and consequently, $dS = 0$ during the change in state. Considering the compression of an ideal gas in which temperature and volume may change, we can write the differential of the entropy as

$$dS = \left(\frac{\partial S}{\partial T}\right)_V dT + \left(\frac{\partial S}{\partial V}\right)_T dV = 0. \tag{II.1.31}$$

Since $(TdS/dT)_V = c_v$, $(dS/dV)_T = p/T$ (see eq. (II.1.19)) and $p = nRT/V$ (ideal gas law), we find that

$$dS = \frac{c_V}{T} dT + \frac{nR}{V} dV = 0. \tag{II.1.32}$$

Integration leads to

$$\Delta S = \int_{T_0}^{T} \frac{c_V}{T} dT + \int_{V_0}^{V} = \frac{nR}{V} dV = c_V \ln \frac{T}{T_0} + nR \ln \frac{V}{V_0} = 0. \tag{II.1.33}$$

The heat capacity at constant volume is given by $(dE/dT)_V$, which for the ideal gas is $c_V = \frac{3}{2} nR$. Therefore, we obtain the equation of state

$$T^{3/2} V = T_0^{3/2} V_0 = \text{const.} \tag{II.1.34}$$

This is the equation of state for the adiabatic compression of an ideal gas that we already used in the introduction of the Carnot cycle.

II.1.11 Summary: Key Ideas of Chapter II.1

- The ideal gas law is given by $pV = nRT$.
- A function of state is a function that only depends on the value of thermodynamic variables in a given state, but not on the history of how this state was achieved. Functions of state are pV, the internal energy E, the entropy S, the enthalpy H, the Helmholtz free energy F, and the Gibbs free energy G. A closed integral yields zero, $\oint dZ = 0$.
- The Carnot cycle is a consideration about heat engines consisting of alternating reversible isothermal and adiabatic compressions and expansions. It was used by Rudolf Clausius to introduce the entropy S, with $dS = dQ/T$. S is a function of state.
- Heat, Q, and work, W, are not functions of state. They describe an exchange with the outside world. A closed integral is not generally equal to zero, $\oint dQ \neq 0$ and $\oint dW \neq 0$. The magnitude of these integrals depends on the history of the system, or on the machine used to transform heat into work. For this reason, heat and work are not quantities that describe the state of a system.
- The Helmholtz free energy F assumes a minimum at equilibrium, if during an experiment volume V and temperature T are kept constant. F is also called the *work function* because at constant temperature it describes the work a system can perform during a (spontaneous) equilibration process.
- The Gibbs free energy assumes a minimum at equilibrium, if during an experiment pressure p and temperature T are kept constant. These conditions are present in a typical biological lab experiment.
- The first law of thermodynamics describes the conservation of internal energy. All changes in the internal energy of a system are either due to work performed on the system, or heat added to the system.
- The second law states that in a closed system, the entropy does not spontaneously decrease. A system that is not at equilibrium will, on average, increase its entropy as a function of time.

- The heat capacity is the amount of heat absorbed by a system when temperature is increased. It is different for different systems. One distinguishes the heat capacity recorded at constant volume, c_V, and recorded at constant pressure, c_p. Since the typical lab experiment is done at atmospheric or constant pressure, we typically refer to c_p.

II.2

Some Statistical Thermodynamics

One of the most fundamental breakthroughs in physics is the statistical interpretation of entropy developed by Ludwig Boltzmann (Figure II.2.2). It relates the entropy maximum to the most likely state of a system. In order to develop his concepts, Boltzmann had to introduce two ideas into physics that were not widely accepted:

1. The concept of random chance
2. The concept of atoms, i.e., small elementary units that can be counted

While the elements were long known in chemistry, the concept of atoms was long disputed in physics. The famous physicist and epistemologist Ernst Mach asked: "Atoms? Have you ever seen one?" In fact, there existed a widespread rejection of the concept of atoms. It is difficult to "see" atoms due to their smallness and the quantum laws that rule on small scales. However, we see traces of them in particle detectors or in X-ray and neutron scattering experiments. The concept of random chance was considered ugly, since most scientists marveled the laws of physics as deterministic laws that allow us to predict the future.

Only by adopting the concept of atoms and elementary particles could Einstein explain diffusion and the photoelectric effect in two papers published in 1905 that made him famous [6, 7]. In several further papers in the first decade of the twentieth century, Einstein applied the statistical ideas of Boltzmann, which allowed him to predict the magnitude of Avogadro's number. After Einstein, the existence of elementary particles was widely accepted.

II.2.1 The Canonical Ensemble

If one wants to develop a microscopic picture for a macroscopic theory, one has to take into account that the microscopic theory for large systems must be in agreement with macroscopic thermodynamics.

Let us consider an ensemble of N identical microscopic systems with constant volume. The total ensemble is thermally isolated ($dQ = 0$) and it does not perform work on the outside world ($dW = 0$). Due to the first law of thermodynamics, this implies that the total energy is constant. The N sub-volumes shall be able to exchange energy in the form of small amounts of heat. Each volume may have different energies, with the constraint that the

Physics of Nerves and Excitatory Membranes, First Edition. Thomas Heimburg.

total energy of all systems is conserved. There is no exchange of particles. This system is called the "canonical ensemble."

II.2.2 The Boltzmann Distribution

Assumption: Let us assume a system that is divided into N identical subsystems, as shown in Figure II.2.1. Of these, n_i systems have the same internal energy E_i. A given distribution of energy states shall be abbreviated as (n).[1]

Due to the conservation of the number of subsystems and the total internal energy, we obtain

$$\sum_i n_i = N$$
$$\sum_i n_i E_i = E_{\text{tot}}. \tag{II.2.1}$$

For a given set of n_i and E_i, the number of ways to distribute these energies over N subvolumes is given by a multinomial distribution

$$\Omega(n) \equiv \Omega(n_1, n_2, n_3, \ldots) = \frac{N!}{n_1! \cdot n_2! \cdot n_3! \cdot \ldots n_n! \cdot}. \tag{II.2.2}$$

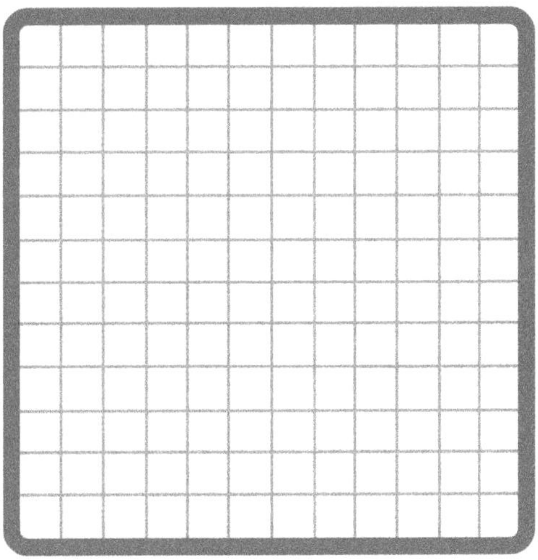

Figure II.2.1 The canonical ensemble consists of a thermally insulated box with constant volume. It is subdivided into N small identical containers that may have different energy contents. The question posed is: What is the most likely distribution of the total energy over all these containers?

1 For the derivations in this chapter, see also [8].

Figure II.2.2 Ludwig Boltzmann (1844–1906). Unknown author/Wikimedia Commons/Public Domain.

Method of Lagrangian Multipliers:

We search for the maximum of the function $F(x_1, x_2, \ldots)$, provided that two boundary conditions are fulfilled: $G(x_1, x_2, \ldots) = \text{const.}$ and $H(x_1, x_2, \ldots) = \text{const.}$ The conditional maximum can be found by solving the following set of differential equations with Lagrangian multipliers α and β that have to be determined:

$$\frac{\partial}{\partial x_i}(F - \alpha G - \beta H) = 0$$

for all coefficients i.

We search for the conditional maximum of $\Omega(n)$ provided that the boundary conditions $\sum n_i = N = \text{const.}$ and $\sum n_i E_i = E_{tot} = \text{const.}$ are fulfilled. Since $\ln \Omega(n)$ is a monotonously growing function of $\Omega(n)$, the position of the maximum of $\Omega(n)$ is also the maximum of $\ln \Omega(n)$. We can obtain the most likely distribution while taking the boundary conditions into account by using the method of Lagrangian multipliers (see the box):

$$\frac{\partial}{\partial n_j}\left(\ln \Omega(n) - \alpha \sum_i n_i - \beta \sum_i n_i E_i\right) = 0 \qquad j = 1, 2, 3, \ldots \tag{II.2.3}$$

The logarithm of the factorial of a number n can be approximated with the help of Stirling's formula: $ln(x!) \approx x \ln x - x$ (this is very accurate for large x). Therefore, it follows from eq. (II.2.2) that

$$\ln \Omega(n_1, n_2, \ldots) \approx \left(\sum_i n_i\right) \ln\left(\sum_i n_i\right) - \sum_i n_i \ln n_i \tag{II.2.4}$$

Evaluation of eq. (II.2.3) using expression (II.2.4) yields:

$$\ln \underbrace{\left(\sum_i n_i\right)}_{N} - \ln n_j - \alpha - \beta E_j = 0 \qquad j = 1, 2, 3, \ldots \tag{II.2.5}$$

which leads to

$$n_j = Ne^{-\alpha}e^{-\beta E_j}. \tag{II.2.6}$$

With the boundary condition eq. (II.2.1), $\sum n_j/N = 1$, one obtains

$$e^{\alpha} = \sum_j e^{-\beta E_j}. \tag{II.2.7}$$

Therefore, the probability distribution is given by

$$P_j = \frac{n_j}{N} = \frac{e^{-\beta E_j}}{\sum_i e^{-\beta E_i}} \qquad j = 1, 2, 3, \ldots \tag{II.2.8}$$

which contains the Lagrangian multiplier β. The ratio of the probabilities of two energy states is

$$\boxed{\frac{P_j}{P_i} = e^{-\beta(E_j - E_i)} \qquad \text{Boltzmann distribution}} \tag{II.2.9}$$

The partition function, Z, is defined as

$$Z \equiv \sum_i \exp(-\beta E_i) \tag{II.2.10}$$

and therefore

$$P_i = \frac{\exp(-\beta E_i)}{Z} \quad \text{probability of state } i \quad \rightarrow \quad E_i = -\frac{1}{\beta}(\ln\ P_i + \ln\ Z). \tag{II.2.11}$$

One can perform the same calculation for a canonical ensemble where the pressure (instead of volume) is kept constant. This will then lead to a similar probability distribution:

$$P_i = \frac{\exp(-\beta H_i)}{Z} \qquad \text{with} \qquad Z \equiv \sum_i \exp(-\beta H_i), \tag{II.2.12}$$

where H_i is the enthalpy of a microstate. This is shown in [8].

II.2.3 Statistical Interpretation of Entropy

In the following, we will find a statistical thermodynamics interpretation of the entropy S and determine a value for the Lagrangian multiplier β. For this, we assume that the statistical thermodynamics interpretation of thermodynamics must be consistent with classical thermodynamics. The mean energy (which is the macroscopic observable) is given by

$$\langle E \rangle = \sum_i E_i P_i \tag{II.2.13}$$

and its differential is

$$\begin{aligned} d\langle E \rangle &= \sum_i E_i dP_i + \sum_i P_i dE_i \\ &= \sum_i -\frac{1}{\beta}(\ln\ P_i + \ln\ Z)\, dP_i + \sum_i P_i \left(\frac{\partial E_i}{\partial V}\right)_N dV. \end{aligned} \tag{II.2.14}$$

Here, the E_i are taken from eq. (II.2.11): $E_i = -\ln(P_i \cdot Z)/\beta$. Let us define a pressure p_i in substate i and an average pressure $\langle p \rangle$

$$p_i = -\left(\frac{\partial E_i}{\partial V}\right)_N \quad ; \quad \langle p \rangle = \sum_I P_I p_I. \tag{II.2.15}$$

Using $\sum dP_i = 0$ and $d\sum(P_i \ln P_i) = \sum(\ln P_i dP_i)$, it follows from eq. (II.2.14) that

$$\begin{aligned} d\langle E \rangle &= -\frac{1}{\beta} d\left(\sum_i P_i \ln P_i\right) - \langle p \rangle dV && \text{statistical thermodynamics} \\ dE &= TdS - pdV && \text{classical thermodynamics} \end{aligned} \tag{II.2.16}$$

By assuming a correspondence principle, we obtain:

$$TdS = -\frac{1}{\beta} d\left(\sum_i P_i \ln P_i\right) \qquad \text{correspondence principle} \tag{II.2.17}$$

The meaning of the constant β remains to be determined.

The constant β can be determined using the fact that S is a function of state. If one has two systems, A and B, which are only thermally coupled but otherwise separated, then the entropy of the coupled system is the sum of the entropy of system A and system B:

$$S_{AB} = S_A + S_B. \tag{II.2.18}$$

The total probability P_{ij}^{AB} to find subsystem A in state i and subsystem B in state j is equal to the product of the probabilities P_i^A and P_j^B:

$$P_{ij} = P_i^A \cdot P_j^B \quad \longrightarrow \quad -\sum_{ij} P_{ij} \ln P_{ij} = -\sum_i P_i^A \ln P_i^A - \sum_j P_j^B \ln P_j^B. \tag{II.2.19}$$

The term $P \ln P$ is therefore additive, similar to entropy. Therefore, $\sum P \ln P$ is proportional to the entropy S. Thus,

$$dS \propto -d\left(\sum_i P_i \ln P_i\right) \quad \text{or} \quad dS = -k\, d\left(\sum_i P_i \ln P_i\right), \tag{II.2.20}$$

where we introduced the constant k. By comparison with eq. (II.2.17), we finally obtain

$$\beta = \frac{1}{kT}, \tag{II.2.21}$$

where $k = 1.38065 \cdot 10^{-23}$ J/K is the Boltzmann constant. The entropy is now finally given by

$$\boxed{S = -k \sum_i P_i \ln P_i + \text{const.}}, \tag{II.2.22}$$

where P_i is the probability of a single microstate.

II.2.4 Thermodynamic Averaging

We now know that $\beta = 1/kT$. Therefore, the probability of state i in the Boltzmann distribution is now given by

$$P_i = \frac{e^{-\frac{E_i}{kT}}}{\sum_i e^{-\frac{E_i}{kT}}} \quad \text{for } V = \text{const., and}$$
$$P_i = \frac{e^{-\frac{H_i}{kT}}}{\sum_i e^{-\frac{H_i}{kT}}} \quad \text{for } p = \text{const.} \tag{II.2.23}$$

Thus, it is slightly different for ensembles with constant volume and with constant pressure. The average of an intensive variable A is now given by

$$\langle A \rangle = \sum_i P_i A_i, \tag{II.2.24}$$

where $\langle ... \rangle$ denotes the thermodynamic average. We have already used this relation in eq. (II.2.13) in order to find the average internal energy. The variable A could, for instance, also correspond to the enthalpy, H, the volume, V, or the charge, q.

II.2.5 Microcanonical Ensemble

If one considers a subsystem in which all states have the same energy, the probability of each microstate is the same even if the state might be distinguishable from another state. If one defines the degeneracy $\Omega = 1/P$ and takes into account that $\sum P_i = 1$ (where all P_i are identical), eq. (II.2.22) turns into

$$S = k \ln \Omega + \text{const.} \qquad \text{or} \qquad \Omega = \exp\left(\frac{S}{k}\right). \tag{II.2.25}$$

The distribution of a system over many different configurations with identical energy E is called the "microcanonical ensemble."

II.2.6 Summary: Key Ideas of Chapter II.2

- Statistical thermodynamics introduces the concept of atoms and of random chance.
- The canonical ensemble consists of a thermally insulated box with constant volume, containing N identical subsystems, which may exchange heat with each other and possess different energies.
- The distribution of internal energy among these subsystems follows the Boltzmann distribution:

$$P_i = \frac{\exp\left(-\frac{E_i}{kT}\right)}{\sum \exp\left(-\frac{E_i}{kT}\right)} \tag{II.2.26}$$

This represents the most probable distribution of energies in the canonical ensemble.

- The statistical representation of entropy is

$$S = -k\sum P_i \ln P_i + \text{const.} \tag{II.2.27}$$

- The mean energy is given by

$$\langle E\rangle = \sum P_i E_i \tag{II.2.28}$$

where P_i are the probabilities of the subsystems, and E_i are their respective energies.

- Similarly, one can determine averages of other variables. In systems with constant pressure, one can determine the mean volume $\langle V\rangle = \sum P_i V_i$ or the mean area $\langle A\rangle = \sum P_i A_i$.
- If the energies of all subsystems i are the same, one finds the microcanonical ensemble, for which

$$S = k\sum \Omega + \text{const.} \tag{II.2.29}$$

II.3

Nonequilibrium

II.3.1 Einstein's Considerations About Boltzmann's Law

The probability theory and entropy concept discussed earlier require knowledge of the energy levels and their degeneracies. Without this information, Boltzmann's concept is not useful for calculating anything in the real world because it assumes that one has a molecular theory of the system under consideration.

This was noted by Albert Einstein in a landmark publication from 1910 [9]. Considering Boltzmann's expression

$$S = k \ln \Omega, \tag{II.3.1}$$

he remarked:

> "*Ω is commonly equated with the number of different possible ways (complexions) in which the state considered - which is incompletely defined in the sense of a molecular theory by observable parameters of a system - can conceivably be realized. In order to be able to calculate Ω, one needs a complete theory (perhaps a complete molecular-mechanical theory) of the system under consideration. Given this kind of approach, it therefore seems questionable whether Boltzmann's principle by itself has any meaning whatsoever, i.e., without a complete molecular-mechanical or other theory that completely represents the elementary processes (elementary theory). If not supplemented by an elementary theory or - to put it differently - considered from a phenomenological point of view, equation (II.3.1) appears devoid of content.*"

This means that, seemingly, Boltzmann's principle is not useful if one does not have a theory for the system. However, how do we then deal with systems that we do not know in all detail, e.g., complex biological systems?

Einstein noted, however, that Boltzmann's principle is useful even in the absence of a molecular theory. He states:

> "*... However, Boltzmann's principle does acquire some content independent of any elementary theory if one assumes and generalizes from molecular kinetics the proposition that the irreversibility of physical processes is only apparent.*"

This is an important statement, which Einstein applied to describing the critical opalescence of some liquid mixtures that was described earlier by Smoluchowski.

II.3.2 The Entropy as a Potential

The Boltzmann distribution describes a distribution of states with different values for the extensive variables around a most likely state. Boltzmann's principle can be written as

$$P = P_0 \exp\left(\frac{\Delta S}{k}\right), \tag{II.3.2}$$

where $P \propto \Omega$ is a probability, and $\Delta S = S - S_0$. Einstein [9] proposed to include into the considerations that the entropy S displays a maximum at an equilibrium value of the variables $\xi_1, \xi_2, \ldots$, denoted by $\xi_{1,0}, \xi_{2,0}, \ldots$. S can be Taylor-expanded around the maximum, which yields for the entropy $S(\xi_1, \xi_2, \ldots)$

$$\begin{aligned} S &= S_0 + \underbrace{\sum_i \left(\frac{\partial S}{\partial \xi_i}\right)_0 (\xi_i - \xi_{i,0})}_{=0} + \frac{1}{2}\sum_{ij} \underbrace{\left(\frac{\partial^2 S}{\partial \xi_i \partial \xi_j}\right)_0}_{\equiv -g_{ij}} (\xi_i - \xi_{i,0})(\xi_j - \xi_{j,0}) + \cdots \\ &\equiv S_0 - \frac{1}{2}\sum_{ij} g_{ij}\alpha_i\alpha_j + \cdots, \end{aligned} \tag{II.3.3}$$

or

$$\Delta S \approx -\frac{1}{2}\sum_{ij} g_{ij}\alpha_i\alpha_j,$$

where $\alpha_i = \xi_i - \xi_{i,0}$ is the deviation of the thermodynamic variable α from equilibrium, $g_{ij} \equiv -(\partial^2 S/\partial \xi_i \partial \xi_j)_0$ are the elements of a positive definite matrix, and $\Delta S = S - S_0$. This is shown for two variables ξ_1 and ξ_2 in Figure II.3.1. The solid black line indicates a possible trajectory from a nonequilibrium state back to equilibrium. $S(\xi_1, \xi_2)$ now turns into a potential.

The probability P to find a state with $S < S_0$ can be written as

$$P \approx P_0 \exp\left(-\sum \frac{g_{ij}\alpha_i\alpha_j}{2k}\right) \tag{II.3.4}$$

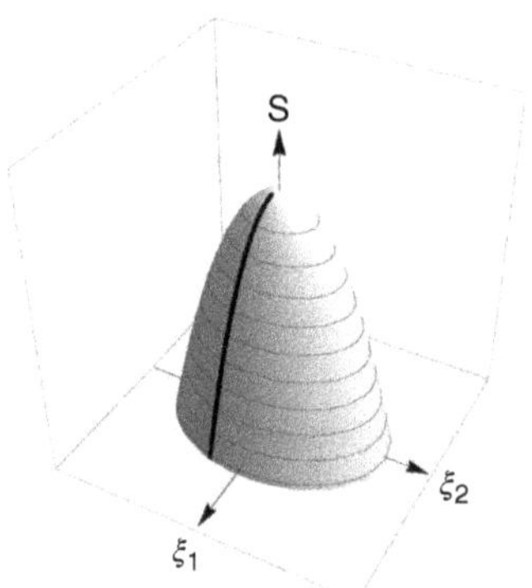

Figure II.3.1 The entropy as a harmonic potential depending on two variables ξ_1 and ξ_2.

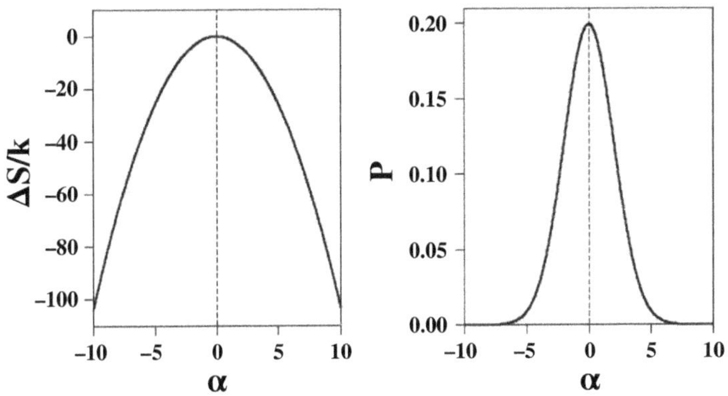

Figure II.3.2 Example for an entropy potential and the corresponding Gaussian distribution.

II.3.3 The Gaussian distribution

If there is only one single variable ξ corresponding to a fluctuation $\alpha = (\xi - \xi_0)$, this reduces to

$$P = P_0 \exp\left(\frac{\Delta S}{k}\right) = P_0 \exp\left(-\frac{g\alpha^2}{2k}\right). \tag{II.3.5}$$

This is equivalent to the Gaussian distribution

$$P(\alpha) = \sqrt{\frac{1}{2\pi\sigma^2}} \exp\left(-\frac{\alpha^2}{2\sigma^2}\right), \tag{II.3.6}$$

where σ^2 is the variance of the distribution.

By comparison, the normalized probability distribution from eq. (II.3.5) for the fluctuations of α is given by

$$P(\alpha) = \sqrt{\frac{g}{2\pi k}} \exp\left(-\frac{g\alpha^2}{2k}\right), \tag{II.3.7}$$

with $\sigma^2 = k/g$. The harmonic approximation of the entropy around equilibrium and the mathematical form of the Gaussian distribution are equivalent. This is shown in Figure II.3.2. The variance σ^2 corresponds to the mean square fluctuations discussed in the next chapter.

II.3.4 Thermodynamic Forces and Fluxes

Let us assume that the system is not equilibrated, and $S \neq S_0$. The system will then evolve back to the maximum entropy state with a rate

$$\frac{dS}{dt} = \sum_{ij} \underbrace{\frac{\partial S}{\partial \xi_i}}_{\equiv X_i} \underbrace{\frac{\partial \xi_i}{\partial t}}_{\equiv J_i}, \tag{II.3.8}$$

where the thermodynamic forces X_i are defined as

$$X_i = \frac{\partial S}{\partial \xi_i}. \tag{II.3.9}$$

They correspond to the forces in mechanics, which are derivatives of energy. The thermodynamic fluxes J_i are defined as

$$J_i \equiv \frac{\partial \xi_i}{\partial t} \tag{II.3.10}$$

The force as a gradient of the entropy potential is shown on the right side in Figure II.3.3. The function $\sigma = dS/dt$ is called the entropy production.

By using eq. (II.3.4), the forces can be determined to be

$$X_i = \left(\frac{\partial S}{\partial \xi_i}\right) = -\sum_j g_{ij}\alpha_j. \tag{II.3.11}$$

According to eq. (II.1.19), the entropy change in a reversible process can be written as

$$dS = \frac{1}{T}dE + \frac{p}{T}dV + \frac{\Pi}{T}dA + \frac{F}{T}dx - \frac{\Psi}{T}dq - \cdots + \sum_i \frac{\mu_i}{T}dn_i.$$

This is valid for equilibrated systems. It the system is not equilibrated, the forces correspond to the deviations of the intensive variables from their equilibrium value:

$$\begin{aligned}
X_E &= \Delta\left(\frac{1}{T}\right) && \text{force leading to a flux in energy, } E\text{, or heat}\\
X_V &= \Delta\left(\frac{p}{T}\right) && \text{force leading to a change in volume, } V\\
X_A &= \Delta\left(\frac{\Pi}{T}\right) && \text{force leading to a change in area, } A\\
X_x &= \Delta\left(\frac{F}{T}\right) && \text{force leading to a change in position, } x\\
X_q &= -\Delta\left(\frac{\Psi}{T}\right) && \text{force leading to a flux in charge, } q \text{ (electrical current)}\\
&\vdots\\
X_i &= -\Delta\left(\frac{\mu_i}{T}\right) && \text{forces leading to a flux of particles, } n_i
\end{aligned} \tag{II.3.12}$$

In equilibrium, all forces are zero.

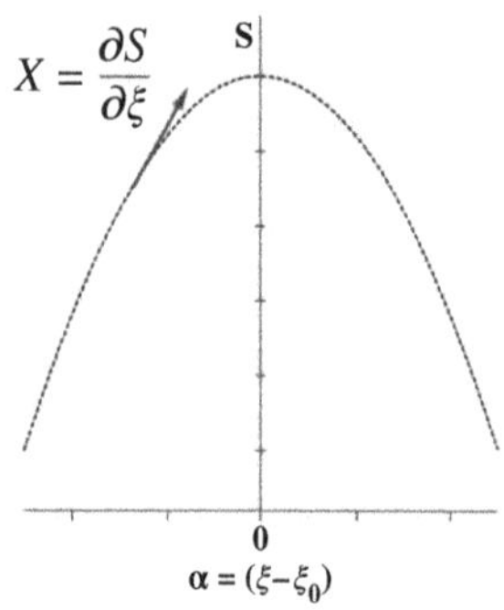

Figure II.3.3 The introduction of a thermodynamic force as the slope of the potential.

II.3.5 Onsager's Phenomenological Equations

Equation (II.3.11) shows that the forces are a linear combination of the variables α_i. In viscous media, the change of a variable (the flux) is proportional to a force. This leads to Lars Onsager's phenomenological equations [10, 11], which are simple linear relations for the coupling of fluxes and forces:

$$\begin{aligned}\frac{d\xi_1}{dt} &\equiv J_1 = L_{11}X_1 + L_{12}X_2 + \cdots + L_{1n}X_n \\ \frac{d\xi_2}{dt} &\equiv J_2 = L_{21}X_1 + L_{22}X_2 + \cdots + L_{2n}X_n \\ &\vdots \\ \frac{d\xi_n}{dt} &\equiv J_n = L_{n1}X_1 + L_{n2}X_2 + \cdots + L_{nn}X_n\end{aligned} \qquad \text{(II.3.13)}$$

where the L_{ij} are coefficients of a positive definite symmetric matrix for which it can be shown that

$$L_{ij} = L_{ji} \qquad \text{reciprocal relations} \qquad \text{(II.3.14)}$$

These are the famous reciprocal relations for which Onsager was rewarded the Nobel Prize in 1968. In the case that one considers a problem with only one single variable α, one can write this as a single phenomenological relation

$$\frac{d\xi}{dt} \equiv J = LX, \qquad \text{(II.3.15)}$$

which implies that the magnitude of the flux is proportional to the conjugated force. We know of many relations in physics that are of this nature, e.g., Stoke's law, in which the velocity of a ball dragged through a viscous fluid is proportional to a force, or Ohm's law providing a linear relation between voltage and current. In the first case, the coefficient L is related to the viscosity, in the second case, it is related to the electrical resistance.

II.3.6 Summary: Key Ideas of Chapter II.3

- Boltzmann's law, $\Delta S = k \ln \frac{\Omega}{\Omega_0}$ is only useful if there is a molecular theory for the system under consideration. Without a theory, it is (according to Einstein) seemingly devoid of content.
- The probability of a state is given by

  $$\frac{P}{P_0} = \frac{\Omega}{\Omega_0} = \exp\left(\frac{\Delta S}{k}\right).$$

- By taking into account that the entropy is at a maximum in equilibrium, one can expand the entropy in a Taylor series in which the leading term is a quadratic function

  $$\Delta S \approx -\frac{1}{2}\sum_{ij} g_{ij}\alpha_i\alpha_j.$$

where $\alpha_i = \xi_i - \xi_{i,0}$, and the ξ are variables of the system. Thus, the entropy resembles a harmonic potential just as the potential V of a spring in analytical mechanics.

- The thermodynamic forces are given as

$$X_i = \frac{\partial S}{\partial \xi_i}.$$

They lead to conjugated fluxes in the extensive variables defined as

$$J_i = \frac{\xi_i}{dt}.$$

For instance, the electrical current is driven by $X_q = -\Psi/T$, which is related to a voltage. These forces drive the system back to equilibrium. At equilibrium, the forces are zero.

- In overdamped viscous systems, one can relate the fluxes and the forces with Onsager's phenomenological equations

$$\frac{d\xi_1}{dt} \equiv J_1 = L_{11}X_1 + L_{12}X_2 + \cdots + L_{1n}X_n$$

$$\frac{d\xi_2}{dt} \equiv J_2 = L_{21}X_1 + L_{22}X_2 + \cdots + L_{2n}X_n$$

$$\vdots$$

$$\frac{d\xi_n}{dt} \equiv J_n = L_{n1}X_1 + L_{n2}X_2 + \cdots + L_{nn}X_n$$

for which the coefficient $L_{ij} = L_{ji}$ forms a positive definite symmetric matrix.

II.4

The Fluctuation Relations

II.4.1 Forces and Fluctuations

Let us consider the entropy as a quadratic function of the extensive thermodynamic variables ξ (e.g., energy E, volume V, area A, charge q) as introduced in the previous chapter, and define the deviations from the equilibrium value as the fluctuations, $\alpha_i = \xi_i - \xi_{i,0}$. As shown in Section II.3.2, we can approximate the entropy as

$$\Delta S \approx -\frac{1}{2}\sum_{ij} g_{ij}\alpha_i\alpha_j, \tag{II.4.1}$$

where $\Delta S \leq 0$, and the g_{ij} are the elements of a positive definite matrix.

The expansion of the entropy into a quadratic function of the extensive thermodynamic variables forms the basis of the entire discipline of linear nonequilibrium thermodynamics as developed by Onsager and Prigogine. Let us consider the fluctuations of one variable, α_i. The probability that a particular fluctuation α_i occurs is given by (see Section II.3.2):

$$P(\alpha_i) = \sqrt{\frac{g_{ii}}{2\pi\ k}}\exp\left(-\frac{g_{ii}\alpha_i^2}{2k}\right), \tag{II.4.2}$$

which corresponds to a Gaussian distribution. The mean square fluctuation, α_i^2, is given by

$$\begin{aligned}\langle\alpha_i^2\rangle &= \int \alpha_i^2 P(\alpha_i) d\alpha_1 d\alpha_2 d\alpha_3 ... d\alpha_n \\ &= \int \alpha_i^2 \sqrt{\frac{g_{ij}}{2\pi\ k}}\exp\left(-\frac{g_{ij}\alpha_i^2}{2k}\right) d\alpha_1 d\alpha_2 d\alpha_3 ... d\alpha_n\end{aligned}, \tag{II.4.3}$$

where $\langle ... \rangle$ denotes the statistical mean. The definite integral $\int_{-\infty}^{\infty} x^2 \exp(-a^2x^2)dx = \sqrt{\pi}/2a^3$. Therefore,

$$\langle\alpha_i^2\rangle = \frac{k}{g_{ii}}. \tag{II.4.4}$$

Equation (II.4.4) corresponds to the mean square amplitude of the fluctuations or the width of the Gaussian distribution. This implies that the half-width of the distribution is related to the curvature of the entropy potential, a fact that we will make use of below in order to derive the relaxation times of fluctuations.

Physics of Nerves and Excitatory Membranes, First Edition. Thomas Heimburg.

The thermodynamic forces were defined as the derivatives of the entropy with respect to the extensive variable (see eq. (II.3.11)).

$$X_i = \frac{\partial S}{\partial \alpha_i} = -\sum_j g_{ij}\alpha_j. \tag{II.4.5}$$

This is completely in analogy to the definition of forces in classical mechanics (as the derivative of a potential), but more general, since it includes extensive variables such as energy and charge, as well as particle numbers. Further, the potential under consideration here is the entropy.

Some of the thermodynamic forces (Section II.3.4) are given in eq. (II.3.12).

From eq. (II.4.5), it follows that $\partial X_i/\partial\alpha_i = -g_{ii}$. As a consequence, eq. (II.4.4) leads to

$$\langle \alpha_i^2 \rangle = -k\left(\frac{\partial X_i}{\partial \alpha_i}\right)^{-1}, \tag{II.4.6}$$

meaning that the fluctuations are related both to the curvature of the entropy potential and to the change of the forces upon the change of a variable. This is consistent with eq. (II.4.4) because, according to eq. (II.4.5), $\partial X_i/\partial\alpha_i = -g_{ii}$.

II.4.2 Susceptibilities

The susceptibilities or response functions are functions like the heat capacity, compressibility, or capacitance. They are given by the derivative of an extensive variable with respect to an intensive variable.

II.4.2.1 Heat Capacity and Enthalpy Fluctuations

The heat capacity tells us how much heat is absorbed by a sample upon an infinitesimal increase in temperature. The heat capacity at constant volume is given by

$$c_V = \left(\frac{\partial Q}{\partial T}\right)_V = \left(\frac{\partial E}{\partial T}\right)_V, \tag{II.4.7}$$

where Q is the heat, the internal energy E is an extensive variable, and the temperature T is an intensive variable.

Let us now consider the fluctuations in internal energy, $\alpha_E \equiv E - \langle E\rangle$, where $\langle E\rangle$ is the equilibrium position. The differential of the temperature can be rewritten as $\partial T = -T^2\partial(1/T)$. It is a function of the conjugated thermodynamic force, $X_E = 1/T$. In particular,

$$c_V = -\frac{1}{T^2}\left(\frac{\partial \alpha_E}{\partial X_E}\right)_V = -\frac{1}{T^2}\left(\frac{\partial X_E}{\partial \alpha_E}\right)_V^{-1} \overset{\text{eq.II.4.6}}{=} +\frac{\langle \alpha_E^2\rangle}{kT^2}. \tag{II.4.8}$$

Since $\left\langle (E-\langle E\rangle)^2\right\rangle = \langle E^2\rangle - \langle E\rangle^2$, another way of writing this is

$$c_V = \frac{\langle E^2\rangle - \langle E\rangle^2}{kT^2}. \tag{II.4.9}$$

This implies that the heat capacity is proportional to the fluctuations in internal energy. This relationship could also be derived by differentiating the statistical mean of the internal energy that is given by $\langle E\rangle = \sum_i E_i \exp(-E_i/kT)/\sum_i \exp(-E_i/kT)$. Thus, the fluctuation relations would be strictly true even if the assumption of a harmonic entropy potential is not made.

A similar relation can be derived for the heat capacity at constant pressure:

$$c_p = \frac{\langle H^2\rangle - \langle H\rangle^2}{kT^2} \tag{II.4.10}$$

II.4.2.2 Compressibility

II.4.2.2.1 Isothermal Volume Compressibility and Volume Fluctuations

The isothermal volume compressibility describes how much the volume changes upon an infinitesimal change in pressure. It is defined as

$$\kappa_T^V = -\frac{1}{\langle V\rangle}\left(\frac{dV}{dp}\right)_T. \tag{II.4.11}$$

Here, V is the intensive variable, and p is the intensive variable. We define $\alpha_V \equiv V - \langle V\rangle$, where $\langle V\rangle$ is the equilibrium position of the volume. Let us consider the force $X_V = p/T$. The differential of the force is given by $dX_V = \frac{1}{T}dp - \frac{p}{T^2}dT$. At constant temperature, this reduces to $d(\frac{p}{T}) = \frac{1}{T}dp$, or $dp = TdX_V$. Therefore

$$\kappa_T^V = -\frac{1}{\langle V\rangle T}\cdot\left(\frac{\partial\alpha_V}{\partial X_V}\right)_T = -\frac{1}{\langle V\rangle T}\left(\frac{\partial X_V}{\partial\alpha_V}\right)_T^{-1} \overset{\text{eq.II.4.6}}{=} +\frac{\langle\alpha_V^2\rangle}{\langle V\rangle kT} \tag{II.4.12}$$

or

$$\kappa_T^V = -\frac{1}{V}\left(\frac{\partial V}{\partial p}\right)_T = \frac{\langle V^2\rangle - \langle V\rangle^2}{\langle V\rangle\, kT} \tag{II.4.13}$$

II.4.2.2.2 Isothermal Area Compressibility and Area Fluctuations

Following the derivations for isothermal volume compressibility, the isothermal area compressibility is given by

$$\kappa_T^A = -\frac{1}{\langle A\rangle}\left(\frac{\partial A}{\partial\Pi}\right)_T = \frac{\langle A^2\rangle - \langle A\rangle^2}{\langle A\rangle\, kT}, \tag{II.4.14}$$

where Π is the lateral pressure.

II.4.2.3 Bending Elasticity and Curvature Fluctuations

The curvature of a membrane is the inverse of the radius of curvature, $c = 1/R$. The membrane possesses two principal curvatures because it is two-dimensional. They are defined by

$$c_1 = \frac{1}{R_1} \quad \text{and} \quad c_2 = \frac{1}{R_2}. \tag{II.4.15}$$

The curvature is an extensive variable.

The work W_c to bend a membrane was given by Helfrich [12]:

$$W_c = \frac{1}{2}K_B(c_1 + c_2 - c_0)^2 + K_G c_1 c_2, \tag{II.4.16}$$

where c_1 and c_2 are the two principal curvatures of a membrane, and c_0 is a spontaneous curvature corresponding to the curvature that the membrane assumes at equilibrium in the absence of forces. K is the bending modulus, and K_G the Gaussian modulus, i.e., two constants that describe the elastic features of the membrane. The bending elasticity κ_B is defined as the inverse of the bending modulus:

$$\kappa_B = \frac{1}{K_B}. \tag{II.4.17}$$

Let us consider a simple symmetric membrane for which the spontaneous curvature is zero. We wish to bend the membrane along one principal axes, e.g., we introduce a curvature $c_1 \equiv c$. The curvature c_2 shall be zero. Then, the Helfrich expression simplifies to

$$W_c = \frac{1}{2\kappa_B}c^2. \tag{II.4.18}$$

The differential of the entropy is given by

$$dS = \frac{1}{T}dE - \frac{1}{T}dW_c, \tag{II.4.19}$$

where

$$dW_c = \frac{c}{\kappa_B}dc. \tag{II.4.20}$$

Let us define the fluctuation of the curvature as $\alpha_c = (c - \langle c \rangle)$. From eq. (II.4.20), it follows that the thermodynamic force X_c is given by

$$X_c = \frac{\partial S}{\partial \alpha_c} = -\frac{1}{T}\frac{dW_c}{d\alpha_c} = -\frac{c}{\kappa_B T}. \tag{II.4.21}$$

Therefore, $dX_c = \frac{1}{T^2}\frac{c}{\kappa_c}dT - \frac{1}{\kappa_c T}dc$. At constant temperature, $dX_c = -\frac{1}{\kappa_c T}dc$. Equation (II.4.6) leads to

$$\frac{\partial \alpha_B}{\partial X_B} = -\kappa_B T \overset{\text{eq.(II.4.6)}}{=} -\frac{\langle \alpha_i^2 \rangle}{k}. \tag{II.4.22}$$

This can also be written as

$$\kappa_B = \frac{\langle c^2 \rangle - \langle c \rangle^2}{kT}. \tag{II.4.23}$$

This is the fluctuation relation for the bending elasticity.

II.4.2.4 Capacitance and Charge Fluctuations

The capacitance of a membrane is introduced via the relation

$$q = C_m V, \tag{II.4.24}$$

where q is the charge on the capacitor, and $V = -\Delta\Psi$ is the voltage. Let us introduce the capacitive susceptibility,

$$C_m = \frac{dq}{dV} = -\frac{dq}{d\Psi}. \tag{II.4.25}$$

This is also a susceptibility because q is an extensive variable and Ψ is an intensive variable. The work for charging a capacitor is given by $W_q = \frac{1}{2}C_m q^2$. The expression for the differential of the entropy is therefore

$$dS = \frac{1}{T}dE - \frac{1}{T}dW_q, \tag{II.4.26}$$

where

$$dW_q = C_m q dq. \tag{II.4.27}$$

Let us define the fluctuation of the charge as $\alpha_q = (q - \langle q \rangle)$. From eq. (II.4.20), it follows that the thermodynamic force X_q is given by

$$X_q = \frac{\partial S}{\partial \alpha_q} = -\frac{1}{T}\frac{dW_q}{d\alpha_q} = -\frac{q}{C_m T}. \tag{II.4.28}$$

Therefore, $dX_q = \frac{1}{T^2}\frac{q}{C_m}dT - \frac{1}{C_m T}dq$. At constant temperature, $dX_q = -\frac{1}{C_m T}dq$. Equation (II.4.6) leads to

$$\frac{\partial \alpha_q}{\partial X_q} = -C_m T \overset{\text{eq.(II.4.6)}}{=} -\frac{\langle \alpha_q^2 \rangle}{k}. \tag{II.4.29}$$

This can also be written as

$$C_m = \frac{\langle q^2 \rangle - \langle q \rangle^2}{kT}. \tag{II.4.30}$$

This is the fluctuation relation for the capacitive susceptibility.

In a similar manner, one can write the susceptibilities for all other pairs of extensive and intensive variables. The coupling of the susceptibilities to the fluctuations is important and will play a crucial role in determining the magnitude of the elastic constants from the heat capacity.

II.4.3 Time Scales

As shown in the previous chapter, fluctuations are a measure of the width of the probability distribution and do not intrinsically display a connection with time. Therefore, relating the term "fluctuation" to something dynamic is misleading. However, fluctuations display lifetimes. Let us consider the most simple Onsager relation for the case that there exists only one force and one flux related to the variable α (see eq. (II.3.15)):

$$J_\alpha = \frac{d\alpha}{dt} = LX_\alpha, \tag{II.4.31}$$

where L is an Onsager coefficient. According to eq. (II.3.11), the force X_α is given by

$$X_\alpha = -g\alpha, \tag{II.4.32}$$

where g is the curvature of the entropy potential, which in the present case depends only on one single variable. Therefore, we obtain the simple differential equation

$$\frac{d\alpha}{dt} = -Lg \cdot \alpha, \tag{II.4.33}$$

with the solution

$$\alpha(t) = \alpha_0 \exp(-Lgt) \equiv \alpha_0 \exp\left(-\frac{t}{\tau}\right), \tag{II.4.34}$$

where we introduce a relaxation time

$$\tau = \frac{1}{Lg}. \tag{II.4.35}$$

Since g represents the curvature of the entropy potential, the relaxation time is related to the shape of the potential. If there are more than one variable, and the variables are coupled with each other, the relaxation process may be multi-exponential.

We will later use the relaxation timescale to determine the lifetimes of membrane pores.

II.4.4 Summary: Key Ideas of Chapter II.4

The susceptibilities are derivatives of an extensive variable with respect to a conjugated intensive variable. The susceptibilities are always proportional to the mean square fluctuations of the intensive variable.

- The heat capacities are proportional to the fluctuations in energy or enthalpy

$$c_V = \frac{\langle E^2\rangle - \langle E\rangle^2}{kT^2} \quad \text{and} \quad c_p = \frac{\langle H^2\rangle - \langle H\rangle^2}{kT^2}.$$

- Volume and area compressibility are proportional to the fluctuations in volume and area:

$$\kappa_T^V = \frac{\langle V^2\rangle - \langle V\rangle^2}{\langle V\rangle \quad kT} \quad \text{and} \quad \kappa_T^A = \frac{\langle A^2\rangle - \langle A\rangle^2}{\langle A\rangle \quad kT}.$$

- The bending elasticity is proportional to the fluctuations in curvature:

$$\kappa_B = \frac{\langle c^2\rangle - \langle c\rangle^2}{kT}.$$

- The capacitance (or capacitive susceptibility) is proportional to the fluctuation of charge on a capacitor

$$C_m = \frac{\langle q^2\rangle - \langle q\rangle^2}{kT}.$$

- The lifetime of a relaxation process is equal to the fluctuation lifetime. If there is only one variable in the system, Onsager's phenomenological equations lead to a single exponential relaxation behavior described by

$$\alpha(t) = \alpha_0 \exp(-Lgt) \equiv \alpha_0 \exp\left(-\frac{t}{\tau}\right)$$

with a relaxation time of

$$\tau = \frac{1}{Lg}$$

Part III

Properties of Nerves

In many respects, nerve cells are just like other cells, i.e., they possess a plasma membrane, a cell body, and they contain organelles and a nucleus. However, nerve cells also display dendrites and axons, which are quasi one-dimensional extensions of the soma. This special geometry permits the cells to transmit signals, which are known as action potentials. Most of the literature considers the electrical aspects of nerve pulse transmission [1–5]. This book rather focuses on the less well-known thermodynamic aspects of signal transmission, Therefore, while here we also introduce the electrical properties, axon myelination, and the theoretical modeling of the action potential (see Part IV), we refer the reader to the extensive literature on electrophysiology for a more profound understanding of the electrical properties of nerves. Beyond them, here we describe mechanical, thermal, and magnetic pulses, as well as myelination and its consequences for pulse velocity.

III.1

Structure of Nerves

Neurons display a unique geometry with axons and dendrites. Some cell types can be rather large, with axons exceeding 1 m in humans, while other cells are quite small, with sizes much less than a millimeter. Many cell shapes were first described by Santiago Ramon y Cajal in the late nineteenth century. He is considered one of the fathers of neurohistology and has published several dozen papers on the shapes of neurons. He used a dying technique invented by Camillo Golgi. One of his most important contributions was to realize that the nerves in the brain do not represent a continuous network as proposed by his competitor Golgi. Instead, the brain consists of many individual cells that are connected via synapses. He was awarded the Nobel Prize for Physiology and Medicine in 1906 together with Golgi. For a historical account of the importance of Ramon y Cajal's work, see [6].

III.1.1 Types and Sizes of Nerves and Neurons

For a long time, Ramon y Cajal's drawings were the best representations of nerves available (see Figure III.1.1). However, with the advance of fluorescence labeling of proteins, more details became available. Figure III.1.2 shows various neurons from rat brains. It can be seen that, while the neurons possess dendrites, they often are still smaller than 1 mm. In many cases, this extension is smaller than the length of the nerve pulse, as shown in Sections III.3.1 and III.3.5. Figure III.1.2 shows four different neurons: a brainstem motor neuron (maximal length 1.4 mm), a stellate cell (230 μm), a hippocampal pyramidal cell (850 μm), and a Purkinje cell (230 μm) – all of them significantly smaller than a typical single action potential (cell images from [8]). Figure III.1.3 (left) shows a spinal motoneuron motor from a cat with a diameter of about 1.5 mm [9]. Figure III.1.3 (right) shows human pyramidal neuron with a diameter of about 225 μm [10], including its synapses. The physical length of nerve pulses compared to cell sizes suggests the need for a macroscopic thermodynamic description (see Part VII).

When visualizing nerves and measuring their dimensions, longer structures are often hidden. An interesting approach has been taken by Deisseroth and collaborators, who fixated a mouse brain in a polymer and removed the lipid membranes. This made the brain transparent such that axons and complete neuron structures labeled with fluorescence markers could be followed in wide-field microscopy experiments without dissection. The authors named their technique "CLARITY" [11, 12]. Some parts of an intact mouse hippocampus are shown in Figure III.1.4. It is possible to see the three-dimensional structure of neurons and arrays of neurons. It is even possible to follow axons over longer distances.

Physics of Nerves and Excitatory Membranes, First Edition. Thomas Heimburg.

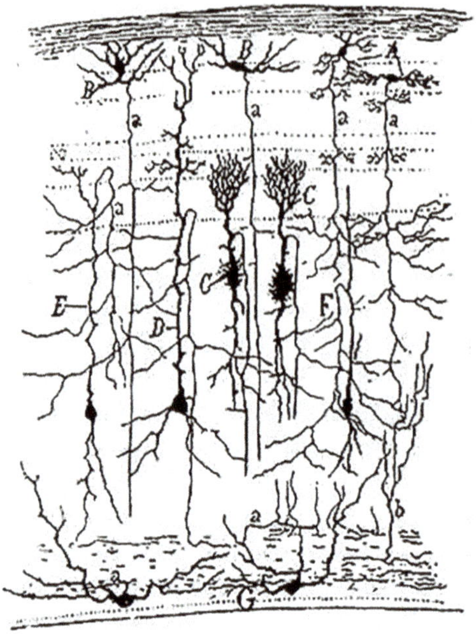

Figure III.1.1 Drawing of Santiago Ramon y Cajal of the nerve cells from sparrow tectum from 1888. Adapted from [7].

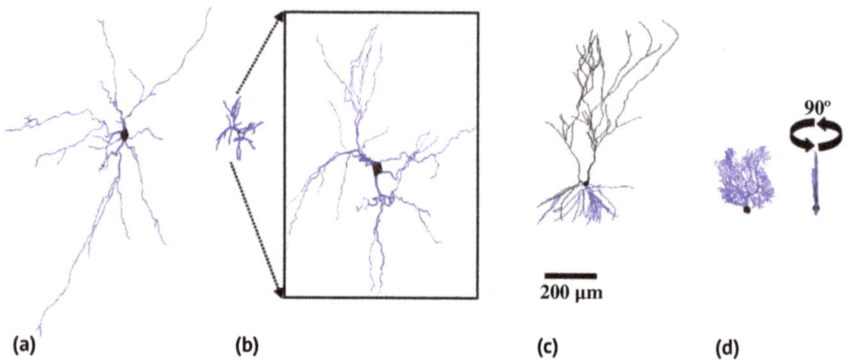

Figure III.1.2 Size and shape of neurons from rat brains. (a) Brainstem motoneuron with a diameter of 1.4 mm. (b) Stellate cell with a diameter of 230 µm. The inset box is the same cell amplified 4 times. (c) Hippocampal CA3 pyramidal cell with a diameter of 850 µm. (d) Purkinje cell with a diameter of 230 µm. The scale bar is 200 µm. [8]/with permission of Elsevier.

III.1.2 Myelin

Nerves display conduction velocities between 1 and 100 m/s. Myelinated nerves conduct faster than non-myelinated nerves. The myelin consists of many membrane layers surrounding the nerve axon (Figure III.1.5). These membranes are derived from Schwann cells and wrap around the axons. In between regions with a myelin sheath, one finds nodes of Ranvier, which are short unmyelinated segments of the axon between Schwann cells. In Figure III.1.6 (top, left), the nodes in teased fibers are indicated by arrows. The average

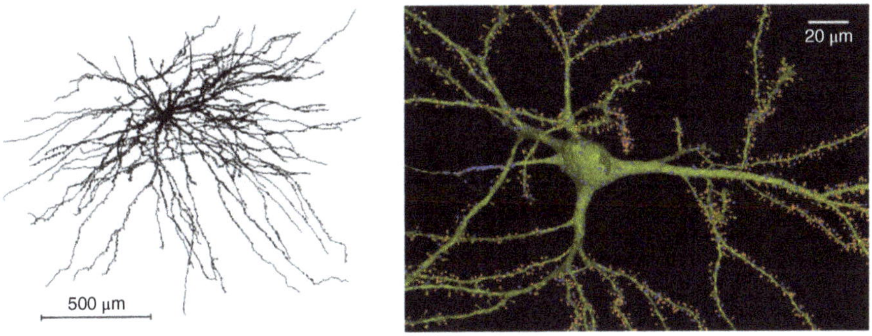

Figure III.1.3 Left: Spinal motoneuron from a cat. The dots represent synaptic buttons. The diameter from left to right is about 1.5 mm. From [9]/with permission of John Wiley & Sons. Right: A human pyramidal neuron with a total horizontal extension of 225 μm. [10]/with permission of American Association for the Advancement of Science – AAAS.

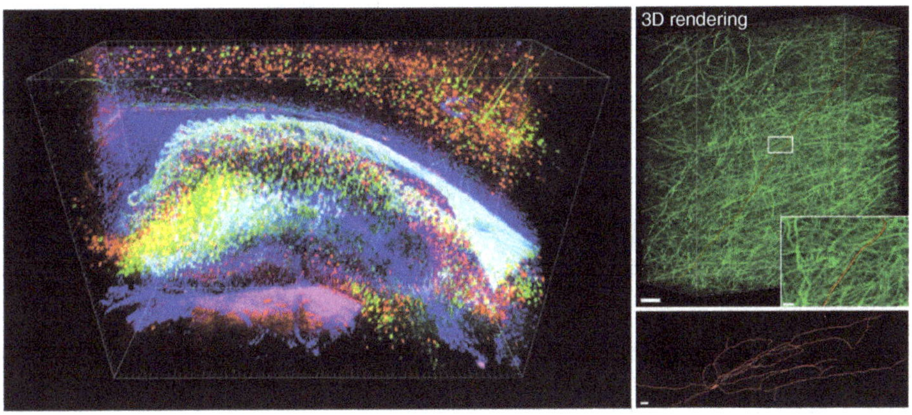

Figure III.1.4 Fluorescence imaging of mouse hippocampus using the CLARITY technique. Left: Intact mouse hippocampus. Right, top: Neurofilament-positive axonal fibers. A traced axon across volume is shown in red. Scale bar: 500 μm. Right, bottom: Three-dimensional reconstruction of a neuron. [12]/with permission of Springer Nature.

internode distance is 285 μm. In motor neurons, the distance between nodes of Ranvier is about 1 mm [4].

The myelin membranes wrap around the axon in many layers (see the dark region in a cross-section along an axon in Figure III.1.6 (top, right)). In the bottom panels of Figure III.1.6 (top, left), one can see cross-sections vertical to axons. In panels B, D, and F, one counts 30–40 layers. Myelinated nerves display high conduction velocities of up to 100 m/s [13, 14].

In the Hodgkin–Huxley model for nerve pulse conduction (see Part IV), the role of the myelin serves is interpreted as an electrical insulation. It changes the ratio of currents along the neuron and across the membrane, resulting in increased conduction velocity. The nodes of Ranvier are non-insulated, and the electrical signal is postulated to jump from node to node. This is called "saltatory conduction," which was first described by Tasaki [15] and later by [16]. In the soliton theory (see Part VII), the myelin layer is rather a mechanical

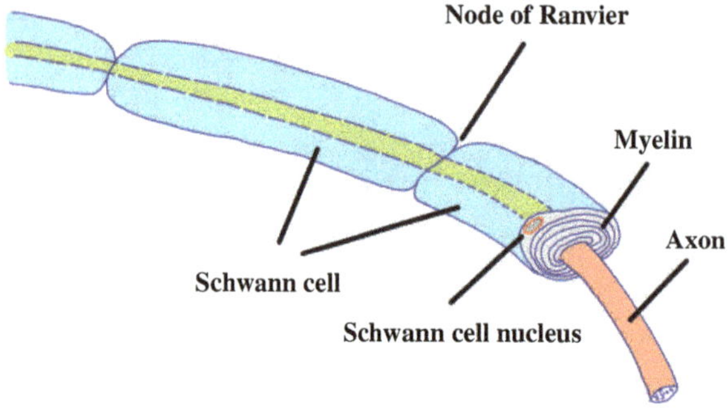

Figure III.1.5 Schematic drawing of an axon surrounded by a myelin sheath created by Schwann cells. The myelin layer serves both as a geometric barrier for membrane undulations. It is also an electrical insulation, which prevents long-distance coupling of the axon membrane with water.

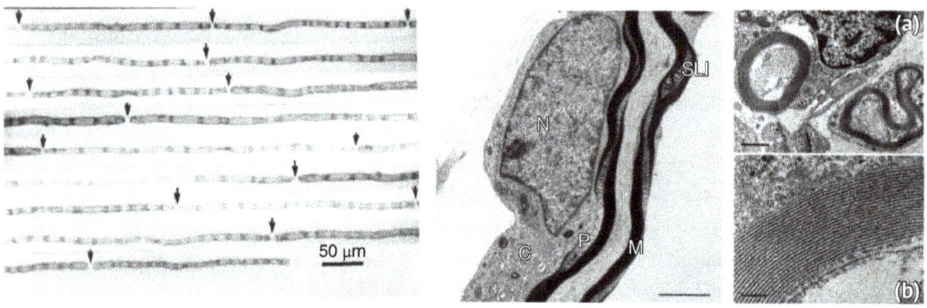

Figure III.1.6 Left: Consecutive segments of a teased fiber show regular spacing of nodes of Ranvier. The average internode distance in these fibers is 285 μm. [17]/with permission of Elsevier. Center: Electron microscopy of myelin-forming rat Schwann cell/mouse dorsal root ganglia neuronal coculture. The Schwann cell nucleus is marked by N, its cytoplasm by C, and compact myelin of a myelinated fiber by M. Scale bar: 2 μm. [18]/with permission of Springer Nature. Right: Electron microscopy of myelin-forming rat Schwann cell/dorsal root ganglia neuron coculture. In panel (b), details of myelin structure are shown. Scale bar in (a) is 1 μm and (b) 0.14 μm. [18]/with permission of Springer Nature.

insulation that prevents the nerve from having transverse mechanical modes. It also prevents the axon from long-range coupling to interfacial water. Both effects render the nerve pulse faster.

III.1.3 Phase Transitions in Nerve Membranes

Nearly all cells – from nerve cells and cancer cells, the surfactant covering the surface of the alveoli in the lung, and bacterial cell membranes – display broad lipid melting transitions about 8–15 °C below body or physiological temperature [19]. An example is given in Figure III.1.7. These melting transitions are discussed in detail in Section V.2.4. During these transitions, the membrane becomes softer, more compressible, and also more

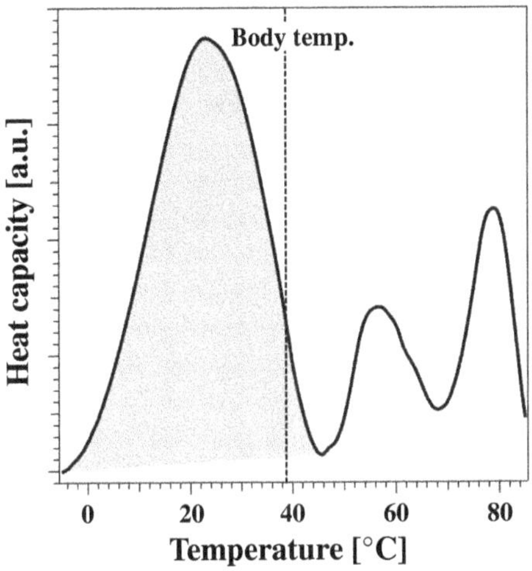

Figure III.1.7 Melting of nerve membranes from pig spine. The gray-shaded region is the melting of the lipids with a maximum at 23 °C. The other two peaks represent protein unfolding. Adapted from [19].

permeable to ions and small molecules. This will play a central role in Part VII, which discusses the emergence of electromechanical pulses in nerve axons, and in Part VIII, which covers the emergence of channel-like phenomena in membranes.

III.1.4 Summary: Key Ideas of Chapter III.1

Nerve cells display many shapes and sizes.

- Many neurons display sizes of around 1 mm or smaller.
- Motor neurons have axons that can be a meter long.
- Motor neurons are myelinated. Their myelin sheath consists of Schwann cell membranes wrapping around the nerve axon.
- The myelin sheath can both be considered as an electrical insulator and a mechanical confinement for the axon.
- The myelination is interrupted between adjacent Schwann cells. These locations are called *nodes of Ranvier.*
- The distance between the two nodes is between a few hundred micrometers and a millimeter.
- The propagation velocity of a nerve pulse in myelinated nerves is much faster.
- The membranes of nerves display melting transitions of their lipids 8–15 °C below physiological temperature.

III.2

Electrical Properties of Nerves

Most neuroscience textbooks focus on the electrical properties of nerves. These are easy to measure because the typical timescale of a nerve pulse is 1 ms and the typical amplitude is on the order of 100 mV. Early recordings have been available since the adaptation of galvanometers to such recordings by Emil du Bois-Reymond in the first half of the nineteenth century. Since much attention has been given to the electrical properties of neurons, we refer to standard textbooks for detail [3, 4]. The biophysical modeling of the nerve transmembrane voltage and the propagation of the action potential is described in Part IV.

III.2.1 Voltage

The electrical changes occurring during the nervous impulse can be measured by both intracellular and extracellular electrodes.

III.2.1.1 Intracellular Recordings

The nerve pulse consists of a segment of the axon that displays transient variations in transmembrane voltage, which are called action potentials (a term created by Emil du Bois-Reymond). The most studied neuron is probably the giant squid axon (see Figure IV.1.1). It displays a diameter of up to 1 mm, which makes it easy to insert electrodes into the axon and measure the voltage across the axonal membrane. Such a recording is called an intracellular recording. It records the voltage difference between the inside of the axon and a reference electrode placed outside the cell.

The action potential displays a variation in voltage with an amplitude of approximately 100 mV (see Figure III.2.1) relative to a baseline called the *resting potential* (see Part IV). The resting potential is usually negative, e.g., around −70 mV in squid axons. In electrophysiology, the nerve membrane is considered a capacitor. An amplitude on the order of 100 mV changes the sign of the transmembrane voltage, reaching a maximum of about +30 mV in squid axons. Most action potentials last a few milliseconds, but it is dependent on temperature and other variables. The conduction velocity is between 1 and 100 m/s, depending on the type of nerve. The timescale, velocity, and spatial dimensions of the nerve pulse are discussed in more detail in Section III.3.1.

The action potential displays a first phase during which the voltage decreases in magnitude relative to the resting potential. This is called the depolarization phase. In the second

Physics of Nerves and Excitatory Membranes, First Edition. Thomas Heimburg.

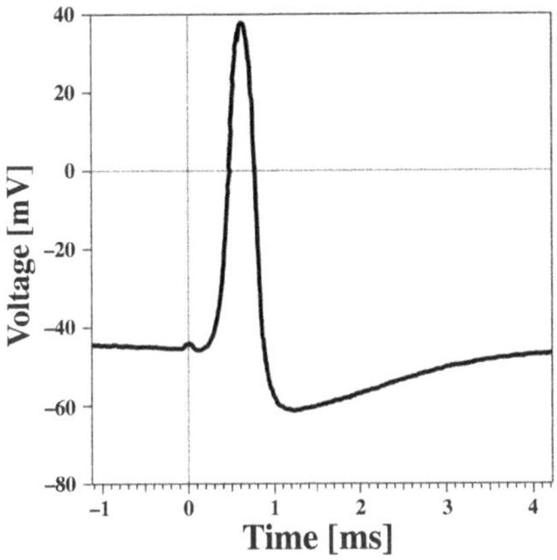

Figure III.2.1 Transmembrane recording of the voltage difference between the inside and outside of a squid axon during an action potential. One finds an amplitude on the order of 100 mV. The first phase of the action potential is called depolarization, and the second one is called hyperpolarization. Adapted from [20].

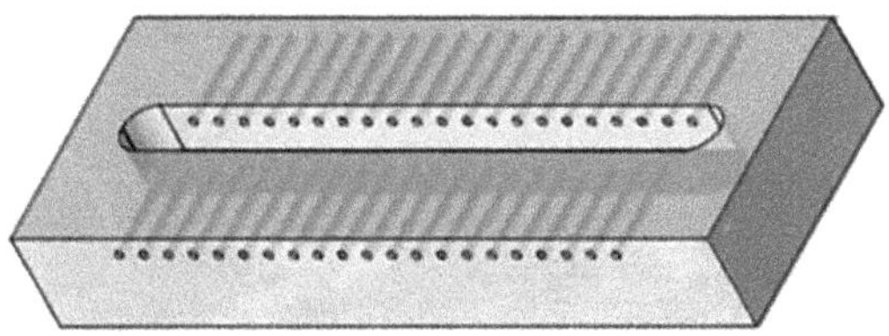

Figure III.2.2 A convenient manner to measure the action potential is by placing the nerve on a plexiglass chamber containing pairs of needles or pins that serve as stimulation of recording electrodes. The nerve is placed on these needles, which can both be used to stimulate the nerve and to perform an extracellular recording of the action potential. From [21]/American Physical Society/CC BY 3.0.

phase, the voltage returns to the resting potential (repolarization). For some nerves, e.g., the squid giant axon, one finds a hyperpolarization phase during which the voltage is more negative than the resting potential. While the hyperpolarization phase is pronounced in the squid axon, it is not visible in the action potentials of many other neurons (e.g., the action potential of giant lobster axons in Figure III.4.7).

III.2.1.2 Extracellular Recordings

The intracellular recording of action potentials becomes more difficult if neurons are very thin or if one investigates nerve bundles. A convenient way to record action potentials is an extracellular recording. It is done by placing two electrodes on the surface of the nerve and measuring the potential difference. Figure III.2.2 shows a simple chamber for such a measurement. A plexiglass container is filled with an electrolyte solution. On the surface, an array of needles is placed, and the nerve is placed on top of them. The needles can be used as electrodes, both for stimulating the nerve and for measuring the action potential at various positions. The time delay between stimulation and recording at a distance Δx can be used to determine the pulse velocity.

Let us assume electrical recordings are performed at the surface of the nerve membrane. When the action potential is measured at two positions separated by distance Δx, one measures the difference in voltage at two positions, $\Delta V(x,t) = V(x+\Delta x,t) - V(x,t)$. If the electrodes are close to each other (i.e., closer than the width of the pulse), one can approximate

$$\Delta V(x,t) \approx \frac{\partial V(x,t)}{\partial x}\Delta x. \tag{III.2.1}$$

If the pulse travels with constant shape and constant velocity v, this yields

$$\Delta V(x,t) \approx \frac{1}{v}\frac{\partial V(x,t)}{\partial t}\Delta x. \tag{III.2.2}$$

We find, therefore, that the extracellular signal recorded in time displays a shape corresponding to the first derivative of the true shape. Its amplitude is proportional to the distance between the electrodes, and it is inversely proportional to the pulse velocity, v. If the distance between the two electrodes is larger than the pulse shape, then both electrodes detect the pulse separately, and one observes a positive pulse followed by a negative pulse. This is shown in Figure III.2.3. The left-hand panels schematically show the voltage at the two electrodes, with varying distances Δx. The center panels show the voltage difference ΔV between the two signals. The top traces show the signal when the two electrodes are further

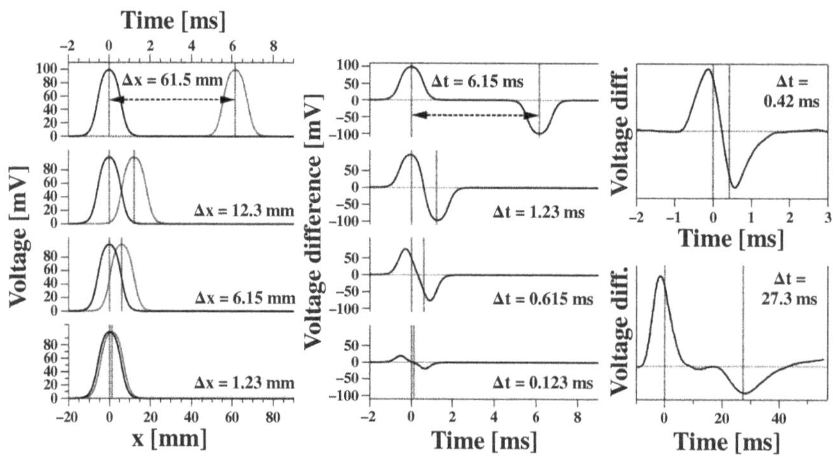

Figure III.2.3 Calculated appearance of intracellular and extracellular voltage recordings for different positions of two electrodes. We assume a propagation velocity of 10 m/s and a pulse duration of about 1 ms, corresponding to a pulse width of 10 mm. Left panels: The two electrodes placed at different positions record the voltage change at different time. Distances between the two electrodes are 61.5, 12.3, 6.15, and 1.23 mm (from top), respectively. Center panels: Theoretical voltage recordings as a function of time from the difference between two electrodes positioned on the nerve (extracellular recording). When the electrodes are much closer to each other than the width of the pulse, the voltage signal resembles the first derivative of the true signal. If the electrodes are farther apart than the width of the pulse, one finds a pulse from each electrode resembling the true voltage signal (top). Right, top: Extracellular recoding on a single axon from lobster connectives. The electrodes have a distance of 2.5 mm, while the width of the pulse is about 6 mm. The difference signal resembles the first derivative of the true signal. Adapted from [22]. Right, bottom: Extracellular recording on a garfish olfactory nerve. The distance of the electrodes is 6 mm, and the velocity of the pulse is 0.22 m/s (corresponding to a time interval of 27.3 ms). Each electrode records the voltage pulse separately. Adapted from [23].

apart than the width of the pulse. The other three traces show that for close electrodes, the shape of the signal ΔV corresponds to the first derivative of the signal on the left-hand side, and that the amplitude of the signal is proportional to the distance between the electrodes. The two traces in the panels on the right-hand side show experimental examples recorded with such an array. The top trace shows the recording of an action potential from lobster connectives (taken from [22]) with a pulse width of 6 mm and a distance between the electrodes of 2.5 mm (i.e., closer than the width of the pulse). The bottom recording shows an action potential from a garfish olfactory nerve. The distance between the electrodes is 6 mm, while the pulse width is about 2 mm (i.e., the electrodes are farther apart than the width of the pulse).

We learn from these considerations that extracellular recordings lead to a different pulse shape when compared with intracellular recordings, and that the amplitude of the pulse depends on the electrode arrangement. In real experiments, the electrodes may not always have perfect contact with the nerve membrane, resulting in deviations from the aforementioned idealized calculation. However, one often finds that the extracellular recordings have the shape of first derivatives, and that the amplitude of the signals can be much smaller than those from intracellular recordings. One should therefore handle quantitative information from extracellular recordings with caution.

III.2.2 Stimulation Threshold

The stimulation of a nerve pulse is a nonlinear process. It can be excited in many different ways, e.g., by voltage or a local current, mechanically, and by local cooling. In order to stimulate an action potential, one must exceed a minimum threshold. Figure III.2.4 shows the stimulus–response profiles of three different nerves. Shown is the maximum amplitude of the action potential measured by different methods as a function of stimulation voltage or current for the ventral cord of earthworm, lobster connectives [24], and the human median

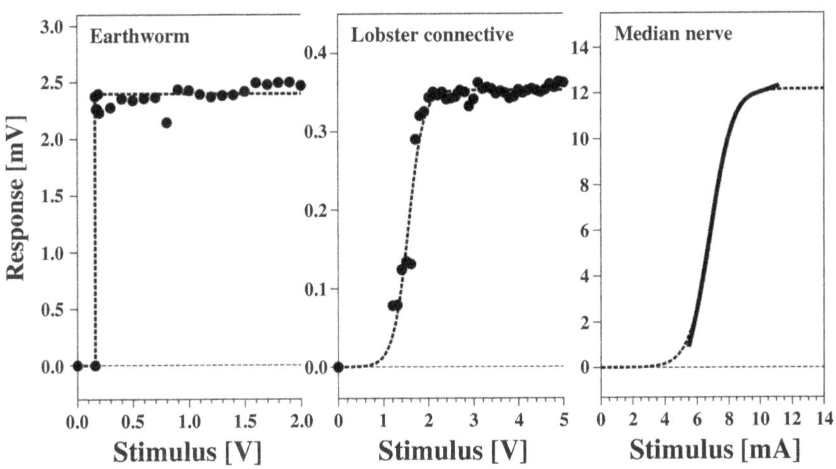

Figure III.2.4 Stimulus–response profiles. Left: Ventral cord of earthworm. Center: Lobster connective. Right: Human median nerve. Data adapted from [24] and [25].

nerve [24, 25]. All three nerves contain several axons. The earthworm nerve shows an all-or-none behavior [26], i.e., no action potential is found below a stimulus voltage of 160 mV, and action potentials of maximum amplitude are found for stimulation voltages above 160 mV. The lobster connective (the nerve bundle from the lobster brain to the first ganglion with three giant axons and some small fibers) displays a rather sigmoidal shape, as does the human median nerve that consists of a few thousand nerve fibers [27]. The action potentials in nerve bundles are called compound action potentials. They represent the sum of the action potentials of individual fibers. Even if the stimulation of a single fiber is all-or-nothing, the compound action potential may display a continuous sigmoidal shape if the fibers display a distribution of slightly different threshold values due to different size and distance from the stimulation electrodes. In such cases, a lower amplitude of the compound action potential reflects that only a fraction of the neurons carry an action potential, rather than all fibers displaying an amplitude that is smaller by a constant factor.

The stimulation threshold finds a different explanation in the Hodgkin–Huxley model (see Part IV) and the soliton theory (see Part VII). In the Hodgkin–Huxley model, a nerve pulse arises when there is a net current from the outside to the inside carried by sodium and potassium ions combined. For small stimulation voltages, the sodium current does not suffice to create a net current toward the inside and there is no spike. In the soliton theory, the threshold is the consequence of a phase transition in the nerve membrane. For a stimulus that is strong enough to move the membrane through its melting transition, a signal emerges.

III.2.3 Refractory Period

The refractory period is a time interval between two stimuli during which the nerve cannot be excited a second time or where the amplitude of the second signal is smaller than the first signal. A history of the investigation of the refractory period was given by Tasaki [28]. Gotch and Birch found in 1899 that a second stimulation directly after a first one does not generate an additional action potential [29]. It was also found that the refractory period displays a significant temperature dependence. At lower temperatures, the refractory period is much longer [29, 30]. The refractory period is basically as long as the nerve pulse itself [15]. It is the time interval the nerve requires to return to its steady state in a relaxation process of either proteins or the nerve membrane.

Figure III.2.5 shows schematically how the refractory period can be measured (adapted from [31]). The figure shows a stimulus and an action potential that arrives after some time at some distance from the stimulation site. After a stimulation interval, a second pulse is excited and measured at the same site. If the stimulation interval is shorter or of similar order to the width of the nerve pulse, the second pulse cannot be excited at all, or its amplitude is lower.

Figure III.2.6 (left) shows a realization of the experiment in Figure III.2.5 for the human median nerve (adapted from [31]). The stimulation interval was changed from 0.75 to 5 ms. The width of an action potential is approximately 1.5 ms. For shorter stimulation intervals, the second pulse displays significantly smaller amplitudes, and it cannot be excited at standard stimulation strength for time intervals shorter than 0.75 ms, which corresponds

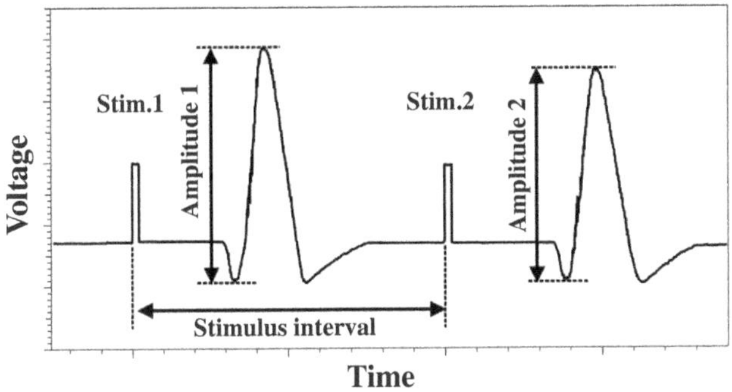

Figure III.2.5 Schematic drawing of an experiment with two stimulations at the same site and the resulting response at the recording electrodes at a different site after the first and the second stimulus as a function of time. Adapted from [31].

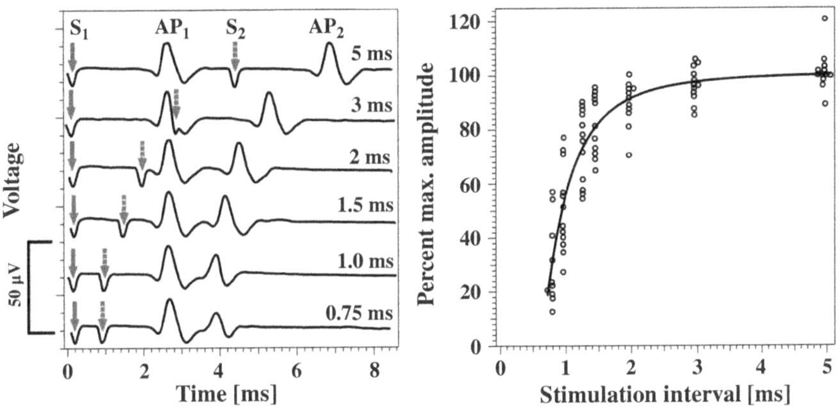

Figure III.2.6 Realization of the experiment in Figure III.2.5 on a human median nerve while varying the stimulus interval. Adapted from [31].

approximately to the positive phase of the action potential. Similar results were reported by [32]. The human median nerve is a nerve bundle. It seems plausible that the continuous change of the action potential amplitude upon shortening of the stimulation interval is the result of a reduction of the fraction of single neurons that carry an action potential. Borg [33, 34] showed that in single myelinated fibers of human motor nerves, the refractory period may be an all-or-nothing phenomenon. They gave an example where a stimulus interval of 2.2 ms resulted in two full-amplitude action potentials, whereas for a stimulation interval of 2.0 ms, no second action potential was detected. Thus, it seems that each nerve fiber has an all-or-none refractory period during which no second pulse can be excited with constant stimulus. The continuous change in amplitude of the compound action potential shown in Figure III.2.6 (right) will then be the result of a distribution of refractory periods of the many fibers of nerve bundles. Thus, it seems plausible to assume that the refractory period and the stimulation threshold in the previous section are related.

III.2.4 Summary: Key Ideas of Chapter III.2

The nerve pulse displays an electrical signature.

- One finds voltage changes by about 100 mV between the inside and outside of the nerve axon, i.e., in intracellular recordings.
- Many nerve signals are recorded by extracellular recordings. When two close electrodes touch the outer membrane, the signal is the difference between the potentials at the two electrodes, and the signal appears as a first derivative in time.
- There exists a stimulation threshold for action potentials. In small nerves or single neurons, it can take the shape of an all-or-none process.
- The action potentials in a nerve bundle are called compound action potentials. One typically finds a sigmoidal stimulus–response profile, which can be a consequence of a distribution of single nerve fibers with slightly different thresholds. It can also be the consequence of the finite width of transitions in the nerve membrane.
- If a nerve is stimulated two times, no second pulse arises when the stimulus interval is smaller than the width of the action potential in time. This is called the refractory period.

III.3

The Dimensions of the Nerve Pulse

The nerve pulse has extensions in time and in space. Here, we show that the spatial extension of a nerve pulse is much larger than generally believed.

III.3.1 Velocity, Timescale, and Dimension of Action Potentials

The first to measure the velocity of a nerve pulse was Hermann von Helmholtz [35, 36]. He found a velocity of 27 m/s in the sciatic nerve of a frog. He further described the temperature dependence. He found considerably lower velocities at lower temperatures. The most important conclusion Helmholtz drew from this result was that this velocity is of the order of the speed of sound in a hydrocarbon layer rather than of the order of the speed of light, as one would expect for electrical phenomena.

He wrote (translated from the German original) [35]:

> *"At present we know from the investigations on the electromotive properties of the nerves by 'du Bois-Reymond' that the activity of the nerves, by which the transmission of a stimulus is mediated, is at least closely connected with a changed arrangement of their material molecules, and is perhaps even essentially conditioned by them. According to this, conduction in the nerve would belong to the series of propagating molecular effects of ponderable bodies, to which, for example, sound conduction in the air and in elastic substances ... is to be counted."*

It is indeed surprising that the conduction velocity in a nerve is on the scale of 1–100 m/s, depending on myelination and size. For instance, the action potential in the squid giant axon has a velocity of about 21.2 m/s at room temperature [37], the velocity in hippocampal neurons from rat brain is of the order of 2 m/s [38], and the action potential in the giant neurons from lobster connectives is 3.1 m/s [22] (see Figure III.3.1). The compound action potential in the human median nerve displays a velocity of 57 m/s [13], the compound sensory action potential (CSAP) of the median nerve is 69 m/s [13], and the compound action potential of the femoral nerve of grasshoppers is 1 m/s [39] (see Figure III.3.2). Motor neurons may display velocities up to 100 m/s. In Part VII, we suggest that the nerve pulse can be described by a sound pulse (soliton) with pulse propagation velocities of about 60% of the sound velocity. The velocity of an electromechanical soliton in the fluid biomembrane

Physics of Nerves and Excitatory Membranes, First Edition. Thomas Heimburg.

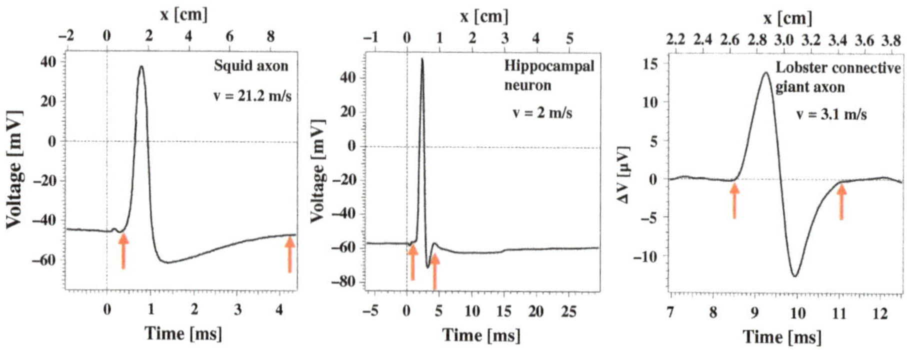

Figure III.3.1 Timescale and length scale of action potentials in single neurons. The spatial and temporal length of the pulse is assumed to be the distance between the arrows. Left: Action potential of a squid axon at 20°C. The velocity is 21.2 m/s and the length of the pulse is $\Delta x = 8.2$ cm. Adapted from [20]. Center: A fast-firing medial-septum-diagonal band hippocampal neurons from rat brain. The velocity of the pulse is 2 m/s and the length of the pulse is $\Delta x = 0.7$ cm. Adapted from [38]. Right: Action potential of a giant axon from the lobster connectives. The velocity is 3.1 m/s and the length of the pulse is $\Delta x = 0.8$ cm. Adapted from [22].

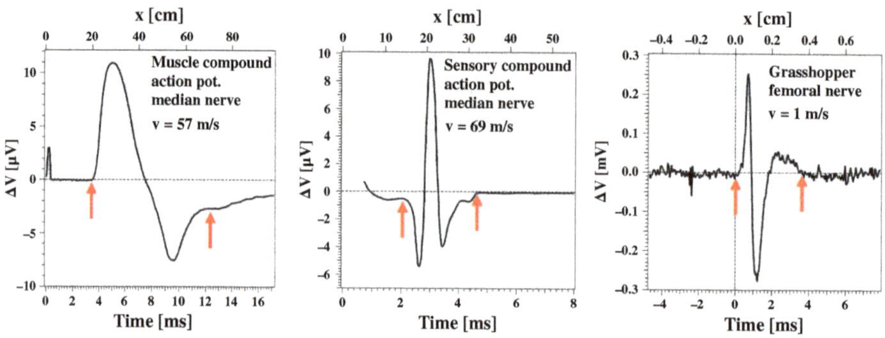

Figure III.3.2 Timescale and length scale of compound action potentials in nerves. Left: Compound muscle action potential (CMAP) from the median nerve in humans. The velocity is 57 m/s and the length of the pulse is $\Delta x = 51$ cm. Adapted from [13]. Center: CSAP of the median nerve. The velocity is 69 m/s and the length of the pulse is $\Delta x = 18$ cm. Adapted from [13]. Right: Compound action potential of the femoral nerve in grasshopper (locust) legs. The velocity is 1 m/s and the length of the pulse is $\Delta x = 0.37$ cm. Adapted from [39].

is about 100 m/s when viscoelastic coupling to the aqueous medium is not considered, and a velocity down to 1 m/s when the viscoelastic coupling is considered (see Section VII.2.2). Thus, the range of velocities of a nerve pulse is similar to the range of the sound velocity as Helmholtz suggested.

Some nerve pulses can display extremely slow propagation velocities. The action potential in the pike olfactory nerve displays a velocity of about 5 cm/s (see Figure III.3.3). However, the duration of the pulse is very long. As a result, the physical length of the pulse of about 3 cm is on a similar scale to that of other nerves described earlier.

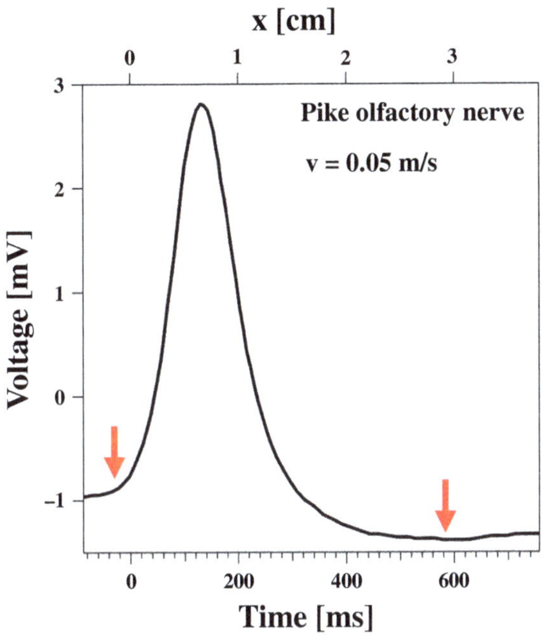

Figure III.3.3 Pike olfactory nerve displays a very long action potential and a very small propagation velocity. Adapted from [40].

III.3.2 Dependence of the Nerve Pulse Velocity on Axon Diameter

There has been significant uncertainty about the velocity of nerve pulses as a function of diameter. The Hodgkin–Huxley model described in Part IV is based on cable theory. It predicts that $v \propto \sqrt{d}$. In 1938, Pumphrey and Young analyzed the conduction velocity in cephalopods (*Loligo* and *Sepia*) as a function of axon diameter (Figure III.3.4 (right)). They plotted the diameter versus the velocity on a double-logarithmic plot with the intention of finding the exponent of a power law that best describes the data. The authors found that the data are best described by

$$V \propto d^{0.614}. \tag{III.3.1}$$

Since the data display considerable uncertainty, especially for the very thin fibers whose diameters were approximated by assumptions, laws like $v \propto d$ (Figure III.3.4 (left)) and $v \propto d^{0.5}$ also describe the data reasonably well on a double-logarithmic scale. Matsumoto and Tasaki [41] interpreted the data as evidence for a$v \propto \sqrt{d}$ law. The fitted line in Figure III.3.4 (right) uses a power law with an exponent of 0.38. It seems clear that the fit also depends largely on whether one fits the raw data in a linear representation or in a double-logarithmic plot. Further, the data display significant variations, and one may find different exponents if one removes the data points for very small diameters that could not be accurately measured. When it comes to myelinated nerves, the data are somewhat clearer. Hurst [42] found that the velocity of myelinated nerves excised from cats displays a linear relation to the nerve diameter, ranging from $v = 10$ to 120 m/s for fibers between 2 and 20 μm.

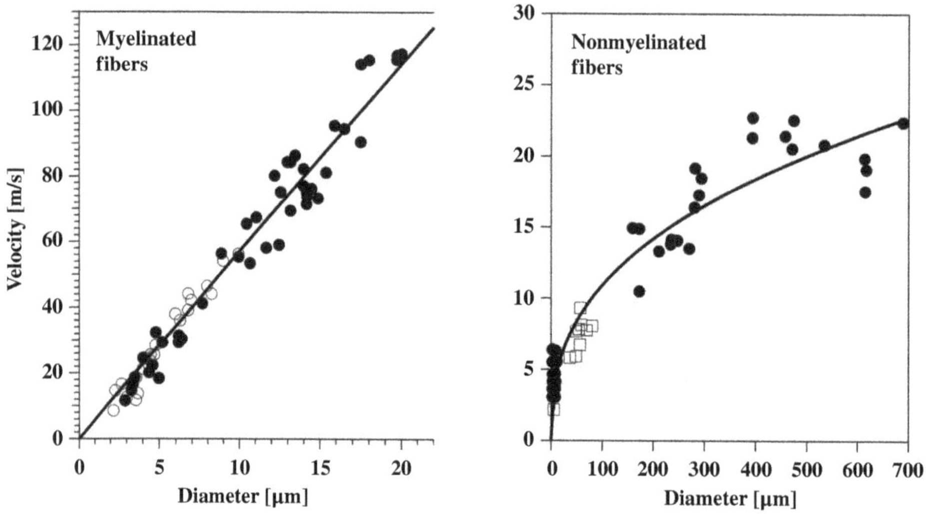

Figure III.3.4 Velocity v versus diameter d. Left: The conduction velocity v of myelinated nerve fibers from adult cats (filled circles) and kittens (open circles) is well described by $v \propto d$. Adapted from [42]. Right: The conduction velocity in nonmyelinated fibers from *Loligo* squid (filled circles) and *Sepia* squid (open squares) is best described by $v \propto d^{0.38}$. Adapted from [43].

In the aforementioned experiments, all experiments were done on different fibers for each data point. In contrast, Goldman [44] used a single nerve from a worm and changed its diameter by stretching it from a length L_0 to a length L (Figure III.3.5 (center)). He found three regimes: in the first regime with $1 < L/L/0 < 2$, the velocity increased linearly with length. In the second regime with $2 < L/L0 < 4.3$, the velocity remained constant and was independent of length. Above $4.3 < L/L/0$, the velocity slightly decreased. These findings were interpreted as follows: it was suggested that in the first regime, the nerve was simply straightened out without changing its length or the diameter of the axon, such that $L/L_0 = 2$ actually represents the true length of the nerve. For $L/L_0 > 4.3$, the nerve was ruptured and did not remain intact. Only the regime of $2 < L/L_0 < 4.3$ represented true stretching of the nerve. These findings were confirmed by [45] (Figure III.3.5 (top)). Goldman further found that the diameter was proportional to the length such that the diameter became smaller upon stretching. In the constant velocity regime, the diameter was reduced by about 40% upon stretching from $L/L_0 = 2$ to $L/L_0 = 4$ (Figure III.3.5 (bottom)). The conclusion from this experiment is that the conduction velocity is independent of diameter in these nerves.

We are left with an unclear result. There exists no clear relation between diameter and velocity that would be valid for myelinated, nonmyelinated, and worm nerves.

III.3.3 Temperature Dependence of the Conduction Velocity

The conduction velocity of the nerve pulse is quite temperature-dependent. Figure III.3.6 shows the conduction velocity in the squid giant axon [46]. At 25°C, the conduction velocity is about three times higher than at 1.8°C. The temporal length of the action potential is

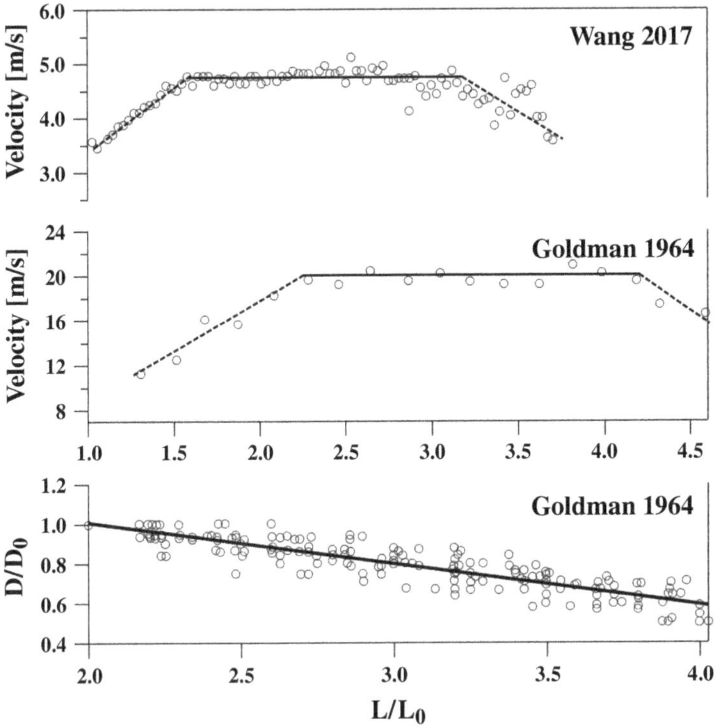

Figure III.3.5 Top and Center: The conduction velocity in earthworm axons as a function of nerve stretching. L is the length of the nerve compared to the length L_0 of the unstretched nerve. In most parts of the profile, one finds that the conduction velocity is independent of stretching. Bottom: One finds a linear relation between the length and diameter. Thus, the data show that for worm nerves, the conduction velocity is independent of diameter. Data adapted from Wang [45] (top) and Goldman [44]/with permission of John Wiley & Sons.

quite different at different temperatures, which is partially a consequence of the different velocities. Ritchie and Keynes [47] used desheathed rabbit vagus nerve and pike olfactory nerve at temperatures around 0°C (i.e., far from physiological temperature), with timescales of 100–250 ms. This means that action potentials can, in fact, be long under nonphysiological conditions.

III.3.4 The Timescale of a Nerve Pulse

While the velocity of nerve pulses largely varies from 1 to 100 m/s, the timescale of the pulse at physiological temperature is surprisingly uniform. The start and end of the action potentials in Figures III.3.1 and III.3.2 are somewhat arbitrarily marked by red arrows. The action potentials in these figures all display a timescale of 2.6–9 ms (see Table III.3.1), i.e., they are of similar order. In contrast, the action potentials in garfish olfactory nerve, as shown in Figure III.6.3, can last 100 ms or longer.

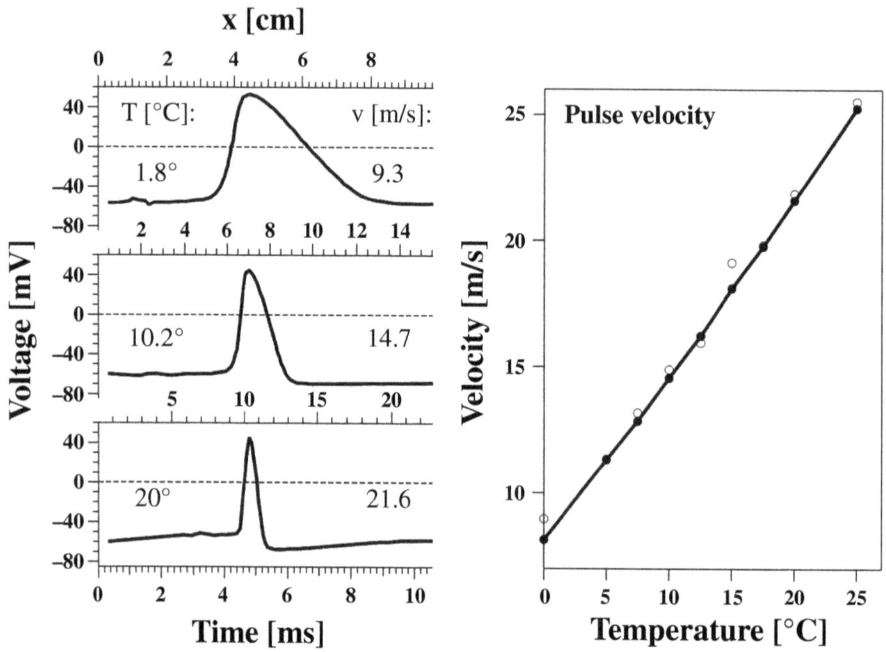

Figure III.3.6 The temperature dependence of the pulse velocity in squid giant axons. Adapted from [46].

Table III.3.1 The velocity, timescale, and physical length of action potentials in neurons and nerves determined from Figures III.3.1 and III.3.2, and the extremely slow action potential from pike olfactory nerve at 0°C.

System	Velocity (m/s)	Δt (ms)	Δx (cm)	Source
Single neurons:				
Squid axon	21.2	3.9	8.2	[20]
Hippocampal fast-firing neuron	2	3.5	0.7	[38]
Giant axon lobster connective	3.1	2.6	0.81	[38]
Nerves:				
Median nerve CMAP	57	9.0	51.4	[13]
Median nerve CSAP	69	2.6	17.9	[13]
Locust femoral	1	3.65	0.37	[39]
Pike olfactory	0.042	270	1.13	[40]

III.3.5 The Spatial Dimensions of a Nerve Pulse

How extended is the nerve pulse? On the webpages of many neuroscience groups or in popular articles, one frequently finds images as the one in Figure III.3.7 that suggest that action potentials display a size of similar order to the diameter of the dendrites of neurons, i.e., of the order of micrometers. It is very easy to see that this is incorrect if one calculates the length of a nerve pulse by multiplying the timescale of the nerve pulse Δt (=2–10 ms, see Table III.3.1) by its velocity v (1–100 m/s) using the simple equation [48]:

$$\Delta x = v \cdot \Delta t, \tag{III.3.2}$$

where Δx is the spatial length of the nerve pulse. Figures III.3.1 and III.3.2 show the extension of various nerve pulses in space (in cm) and in time. One finds a range of possible sizes of nerve pulses ranging from 2 mm to 50 cm depending on the nerve. Pulses in long motor nerves are longer because they display large conduction velocities. Therefore, nerve pulses are in fact very large – larger than some cells. As shown in Section III.1.1, pyramidal cells have a size of 100–200 μm, spinal and brainstem motoneurons have a size on the order of 1 mm, and hippocampal pyramidal cells also display a size of about 1 mm. This implies that the dimensions of the action potential in neural tissue are often larger than a complete cell and orders of magnitude larger than synapses. The axons of motor nerves, such as the median nerve, have a length of about a meter. In such nerves, the action potential can extend up to 50 cm. Since nerve pulses are quite long, it is likely that the molecular details of the

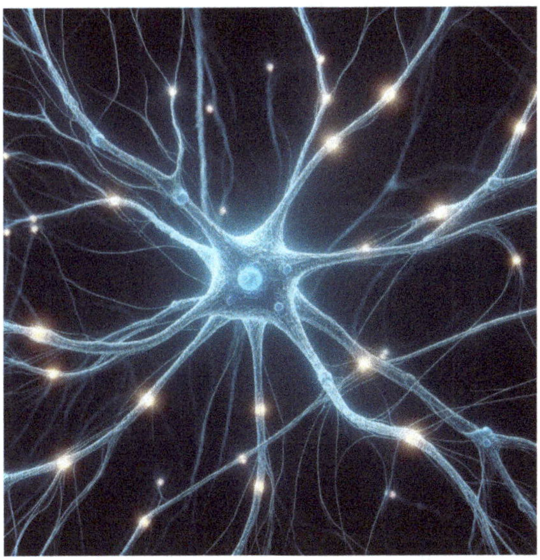

Figure III.3.7 Networks of neurons are commonly depicted as shown here. Here, action potentials are indicated as bright spots on the dendrites. This suggests that action potentials have dimensions on the order of micrometers (i.e., similar to the diameter of a dendrite). Such images are incorrect because they hugely underrepresent the spatial length of the action potential. This image was AI-generated with Microsoft Image Creator based on DALL E3. However, this image is representative of many artistic representations of neurons in the brain used by neuroscience groups on the internet and in popular publications. [14]/Elsevier/CC BY 4.0.

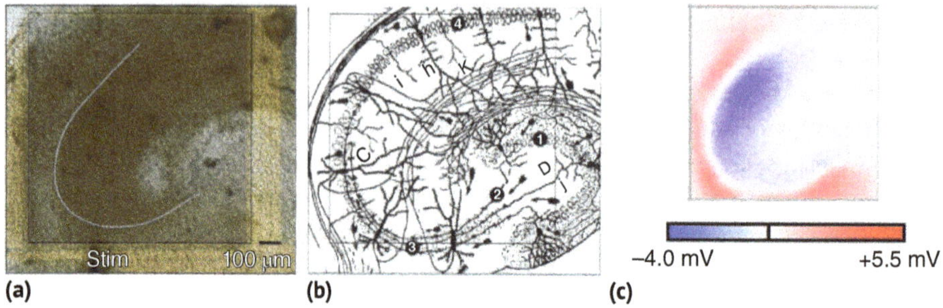

Figure III.3.8 (a) Micrograph of cultured slice on a silicon chip (two-dimensional multitransistor array) of 1 × 1 mm. (b) Classical drawing of a Golgi-stained section of rat hippocampus (Ramon y Cajal, 1911), matched in orientation and size. (c) Image of field potential. Recording area: 1 × 1 mm. From [49]/with permission of The American Physiological Society.

membranes do not matter for the mechanism of the action potential. The pulse just sees an average quantity, i.e., average mass density, average charge density, average conductance, average capacitance, and average membrane compressibility. The dimensions of pulses in myelinated motor neurons are significantly larger than the distance between nodes of Ranvier (1–1.5 mm). This is inconsistent with a saltatory conduction model in which pulses jump from node to node because the electrical properties of the myelinated nerve are averaged over the length of an action potential.

The action potential in some cases may span complete brain regions. Figure III.3.8 shows a two-dimensional slice of hippocampal tissue on a multi-transistor array (chip) of about 1 × 1 mm (from [49]). The right panel shows an electrical signal traveling across this region, which is of much larger dimensions than the individual pyramidal cells. The image shows that the brain slice excitations are macroscopic, i.e., much larger than the cells.

The macroscopic nature of nerve pulses and action potentials is important for a thermodynamic theory of nerve pulses, which is described in Part VII.

III.3.6 Summary: Key Ideas of Chapter III.3

In this chapter, we described the dimensions of the nerve pulse. We found that:

- Nerve pulses at physiological conditions display durations of 2–10 ms.
- The propagation velocity of action potentials ranges between 1 and 100 m/s.
- Myelinated nerves display a linear relationship between velocity and axon diameter. Squid axons display a relation of velocity to diameter that is consistent with $v \propto d^{0.5}$, but other exponents of a power law are also possible. Worm nerves display a conduction velocity that is independent of nerve diameter. Thus, the relationship between diameter and velocity is unclear.
- The conduction velocity is a function of temperature. It becomes slower with decreasing temperature.
- The spatial length of an action potential ranges between 2 mm (in single unmyelinated neurons) and 50 cm (in myelinated motor neurons).

- The dimensions of action potentials are mostly larger than single neurons (e.g., pyramidal cells with a size of 200 μm) and can span over several neurons or entire brain regions.
- The length of the action potential in motoneurons can reach up to 50 cm, which is 500 times larger than the typical distance between nodes of Ranvier (internodes) in myelinated nerves (about 1 mm).
- This makes it unlikely that pulse propagation works by saltatory conduction, i.e., pulses that jump from node to node. Over the length of a nerve pulse, physical properties assume average values.
- The propagation velocity of action potentials is of the same order as the speed of sound in biological membranes.

The macroscopic nature of the nerve pulse suggests that a proper explanation of nerve activity lies at the macroscopic scale. A thermodynamic theory of the nerve pulse will be presented in Part VII.

III.4

Mechanical Properties of the Nerve Pulse

The action potential is usually considered a purely electrical phenomenon. However, during the nerve pulse, one also finds changes in other variables, e.g., the thickness and the length of the neuron, and the molecular order of the lipids and macromolecules in the nerve membrane. Such changes are reviewed in this chapter. As already shown in the introduction (Section I.4.2.1), it was long known that nerves can be excited mechanically [50] (see Section I.4.2), and that they contract during nerve activity [51, 52]. In this chapter, we report other, more recent findings about the mechanical properties of the action potential.

III.4.1 Mechanical Recordings of the Action Potential: Thickness Changes

Mechanical changes in nerve thickness can be measured by atomic force microscope (AFM) cantilevers or optically by light scattering (see next section). Figure III.4.1 (left) shows a well-known experiment by Iwasa and Tasaki [53]. They measured thickness changes in squid axons using a cantilever method described in [54] that is reminiscent of an AFM. They found a deflection of the cantilever by about 1 nm, which corresponds to about 20% of the thickness of a biological membrane. With the same setup, one can also measure a force exerted on the cantilever, which for the squid axon can be up to 700 nN (see two recordings in Figure III.4.1 (right)). Mechanical and electrical recordings are in phase.

Gonzalez-Perez et al. found similar changes in membrane thickness by AFM [22]. They used single giant axons from lobster connectives and found a vertical dislocation of up to 1 nm at constant force, which was in phase with and roughly of the same shape as the electrical recording (Figure III.4.2 (left)). Figure III.4.2 (right) shows six recordings on different neurons. The magnitude of the signal varies due to the quality of the contact between the cantilever and the nerve surface. The upper bound of the recordings was always around 1 nm, similar to the magnitude of the changes observed by Iwasa and Tasaki in Figure III.4.1 (left).

Another very useful technique is interferometry (high-speed quantitative phase imaging), as described in Ling et al. [55]. The advantage of this technique is that one can follow thickness changes of membranes in a neural culture in a manner that is simultaneously time- and space-resolved. In [55], this is demonstrated in several movies. Figure III.4.3 shows the height profile in a complete neuron at one given point in time. One can recognize changes

Physics of Nerves and Excitatory Membranes, First Edition. Thomas Heimburg.

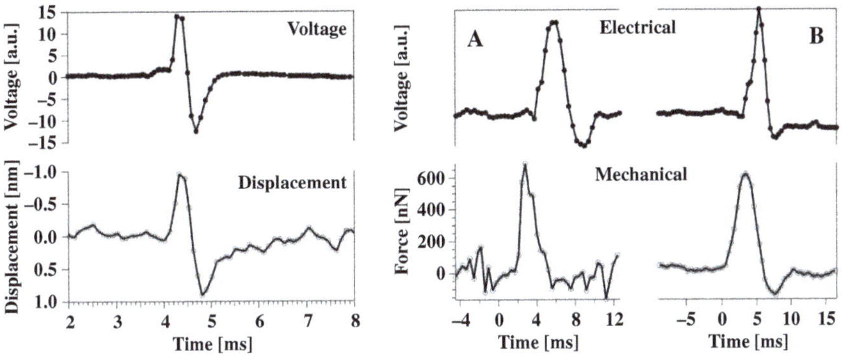

Figure III.4.1 Left: Thickness change (vertical dislocation) of a squid axon (bottom) under the influence of the action potential (top) as measured by a mechanical cantilever. The magnitude of the thickness change is about 1 nm. Adapted from [53]. Right: The force of crab nerve fibers excreted on a cantilever. Two different experiments are shown. Adapted from [54].

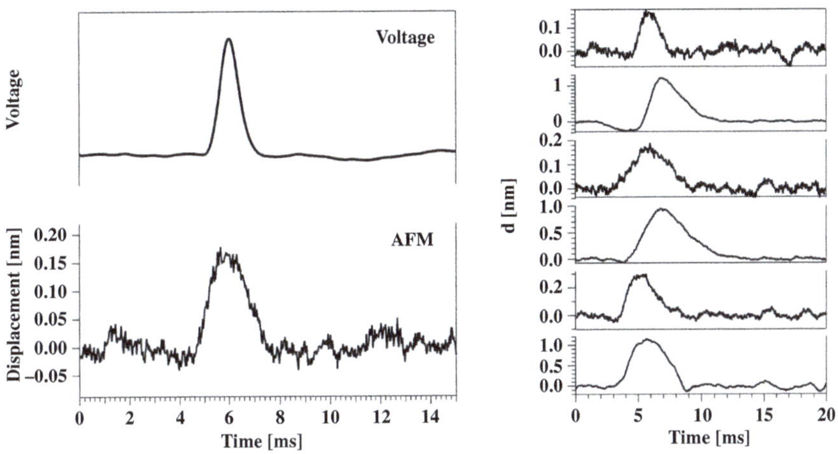

Figure III.4.2 Thickness change (bottom) of giant axons from lobster connectives during the action potential (top) measured by an AFM. Left: In this experiment, the magnitude of the vertical dislocation is 0.15 nm. Right: The vertical dislocation in similar experiments on other giant axons ranges up to 1 nm, consistent with the data of Iwasa and Tasaki in Figure III.4.1 (left). Adapted from [22].

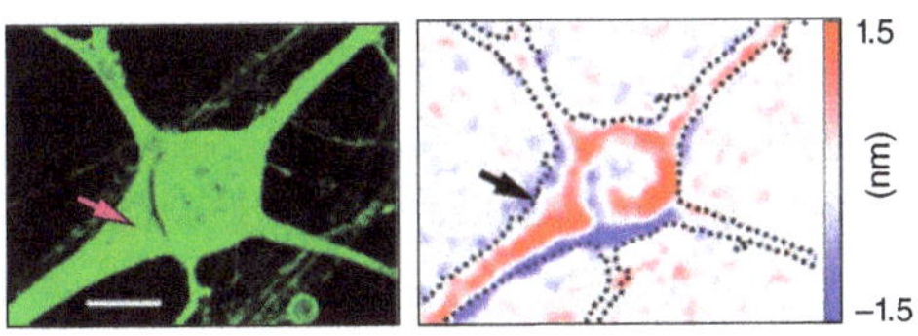

Figure III.4.3 Nanometer-scale membrane deformations from neurons from rat cortical tissues measured by high-speed interferometric imaging show a thickness change of 1–2 nm. Left: Confocal microscopy image. Right: Interferometric image of the spike-induced deformation on the same nerve [55]/with permission of National Academy of Sciences.

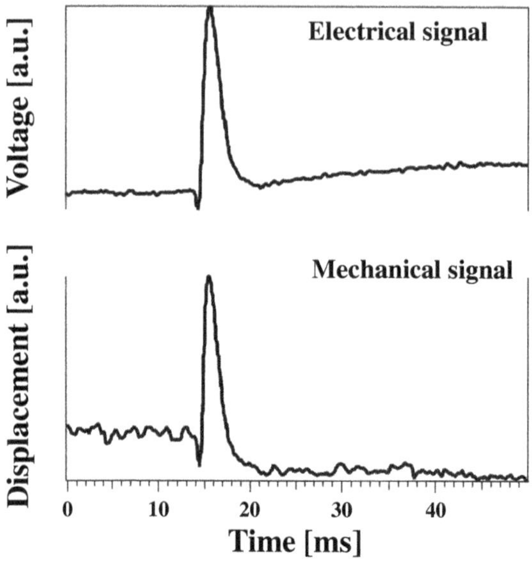

Figure III.4.4 Voltage and thickness change of nerve terminals of the mammalian neurohypophysis. The mechanical signal was measured by an AFM and the electrical signal by a fluorescence marker [57]/with permission of Elsevier.

of the order of ±1.5 nm. Thus, in all of the aforementioned recordings, thickness changes of a neuron of the order of 1 nm were found. It should be pointed out that this change roughly corresponds to the change in thickness when one or two membranes go through a melting transition (thickness change of about 17% [22, 56] of a membrane thickness of about 5 nm). Thickness changes were also observed in synapses. Figure III.4.4 shows the voltage change and the thickness change of nerve terminals in the neurohypophysis.

In all recordings, electrical and mechanical changes were in phase and qualitatively nearly identical. This suggests that the data represent different projections of the same phenomenon and should not be seen as independent events.

III.4.2 Contraction of Nerves

To our knowledge, the first direct experiments demonstrating mechanical changes during nerve pulse propagation were performed by Wilke and Atzler [51, 52] who studied the contraction of frog nerves. The starting hypothesis of their experiment was the possibility of mechanical deformation waves in the nerve axons. They considered a peristaltic wave traveling along a nerve cylinder containing an incompressible fluid. Since they expected that thickness changes in the axons were very small, they rather focused on the contraction of the nerve because it integrates the mechanical changes over the whole length of the axon. They attached a very thin glass fiber to the end of a nerve and found that the fiber vibrated after stimulation of the nerve (see Figure I.4.4). They concluded that the nerve pulse is associated with mechanical changes, and that nerve contraction can indeed be demonstrated.

Since the action potential travels along the length of the axon, the time span of contraction is not equal to the timescale of the action potential but rather proportional to the dwell time of the action potential in the total axon. The duration of the contraction can be many times longer than the timescale of the action potential at any given position. Tasaki and Iwasa [58] used a cantilever to measure changes in length and the force acting on the cantilever in a crab nerve (in units of a counterweight where 1 μg ≡ 9.81 nN). They found both a force of about 177 nN and a shortening of the axon of about 6 nm on a timescale much longer than that of the action potential. This is shown in Figure III.4.5. The magnitude of these changes probably has to be taken with caution because nerves are very soft, they are under the load of their weight, and they possess inertia. Therefore, the magnitude of length changes is probably underestimated.

Figure III.4.6 (left) shows that under repetitive stimulation of a desheathed crab nerve, the shortening of the nerve lasts as long as the pulse train is present [59]. A total length change of about 80 nm was observed, i.e., the contraction is much larger than the thickness change of the neurons described in the previous section. Olfactory nerves show very slow propagation velocities (see Table III.3.1), and several pulses can be present in the axons at the same time. During repetitive stimulus, the length changes of each action potential accumulates in the axon. Figure III.4.6 (right) shows the accumulating increase in force (measured in μg in units of a counterweight) in a garfish olfactory nerve [23]. A maximum force of 10 μg ≡ 98 nN was found. Both force and contraction should be proportional to the number of pulses coexisting in the axon.

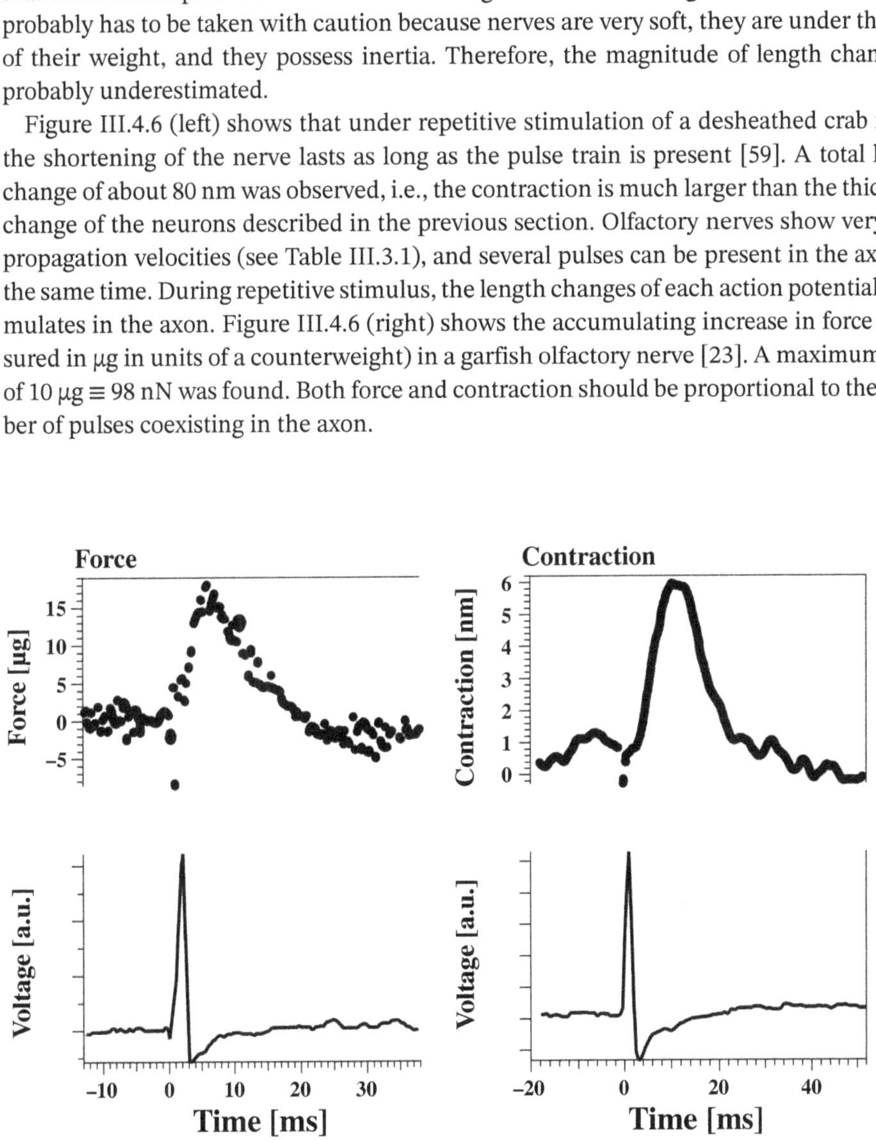

Figure III.4.5 Shortening of a crab nerve under the influence of a single action potential. Left: Force along the axon in units of a counterweight. At maximum a force of 18 μg ≡ 177 nN was found. Right: The shortening of the nerve. Adapted from [58].

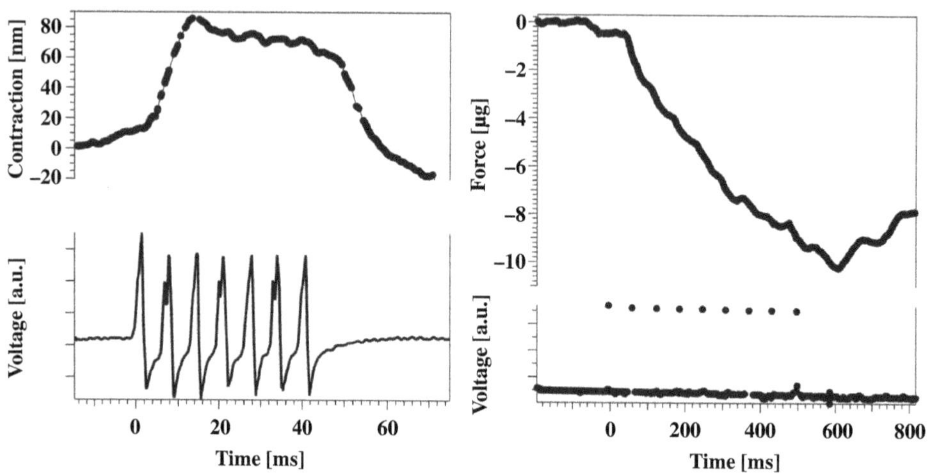

Figure III.4.6 Contraction of nerves during repetitive stimulation. Left: Records of tetanic contraction of a 20-mm long desheathed crab nerve. The contraction lasts as long as there are action potentials in the nerve. Adapted from [59]. Right: Shortening of a garfish olfactory nerve associated with the generation of a train of nerve pulses. The top trace shows downward movements corresponding to the development of a force tending to pull the tip of a piezoelectric bender downwards. The maximum force is approximately 10 µg ≡ 98 nN. The lower traces (9 dots) indicate the nerve stimulations. Adapted from [23].

III.4.3 Mechanical Stimulation of Nerve Pulses

In 1880, Robert Tigerstedt published his book *Studien über mechanische Nervenreizung* (*Studies of Mechanical Nerve Stimulation*) [50]. He noted that the possibility to excite nerves mechanically was known since ancient times. This book reviews in much detail all attempts to stimulate nerves by mechanical perturbations and stretch known at the time. He begins with (translated from the German original):

> *"The researchers who have so far worked with nerve and muscle irritability have almost exclusively used electricity, or, to put it more precisely, electrical currents as stimuli, whereas chemical, thermal and mechanical stimuli, not to speak of other types of electrical stimuli, have only been used in exceptional cases. ... Electrical stimuli are particularly suitable for neurophysiological examinations in many ways. They can be modified within a wide range, can be repeated as often as necessary and are easy to apply to the organs to be examined; moreover, they have less of a destructive effect on the irritable organs themselves than any other stimulus available to us."*

Thus, Tigerstedt believed that the choice to stimulate a nerve electrically is more a matter of convenience because it is easy to repeat and modify. However, he clearly states that nerves can also be excited by other means. He discusses chemical and thermal stimulation but notes that these stimuli are more difficult to apply, repeat, control, and modify than electrical stimuli. Thus, he concludes, the only other convenient manner to stimulate a nerve is by mechanical force. One well-known example of mechanical stimulation is the patellar

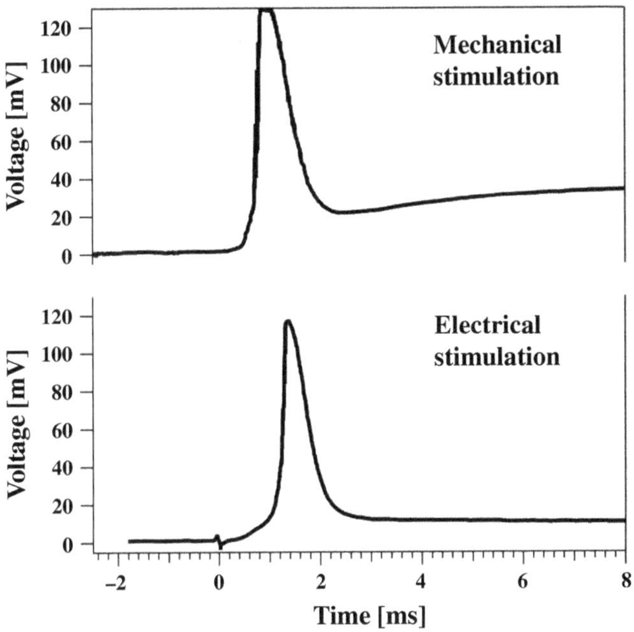

Figure III.4.7 Action potentials in giant axons of lobster connectives. Top: After mechanical stimulation. Bottom: After electrical stimulation. The two signals are very similar. Adapted from [61].

reflex, where the tendon in the knee is mildly hit with a hammer, leading to a nerve signal. There have been various studies focusing on mechanical stimuli, applied either by hitting a nerve or by sudden stretch. Gray and Ritchie reported that fast local stretching of single motor nerve fibers isolated from the nerve supplying the gastrocnemius of frogs by a piezoelectric device was able to stimulate an electrical response [60]. Julian and Goldman stimulated giant axons from lobster connectives both mechanically and electrically, and found very similar action potentials recorded by electrodes (see Figure III.4.7, [61]). Terakawa and Watanabe stimulated squid axons by a mechanical stimulus exerted by a sudden, local, reversible injection of electrolyte into the axon [62].

In Figure III.4.8, we show the stimulation of a nerve from the walking leg of a lobster. The stimulation was done by locally hitting the nerve with the cantilever of a reversed pickup. The timescale of compression was about 1 ms. The panel on the lower right shows the voltage driving the cantilever. About 2 ms after maximum force, one can record an action potential with extracellular electrodes.

III.4.4 Stimulation by Ultrasound

As shown in the previous section, nerves can be excited by mechanical stimuli. Therefore, it may not be surprising that the brain can be excited by sound. In focused ultrasound neurostimulation, ultrasound is focused by a parabolic array of ultrasound transducers on a small region in the brain. For instance, Kamimura and collaborators showed that one can cause the contraction of the legs of a mouse by selectively stimulating specific brain

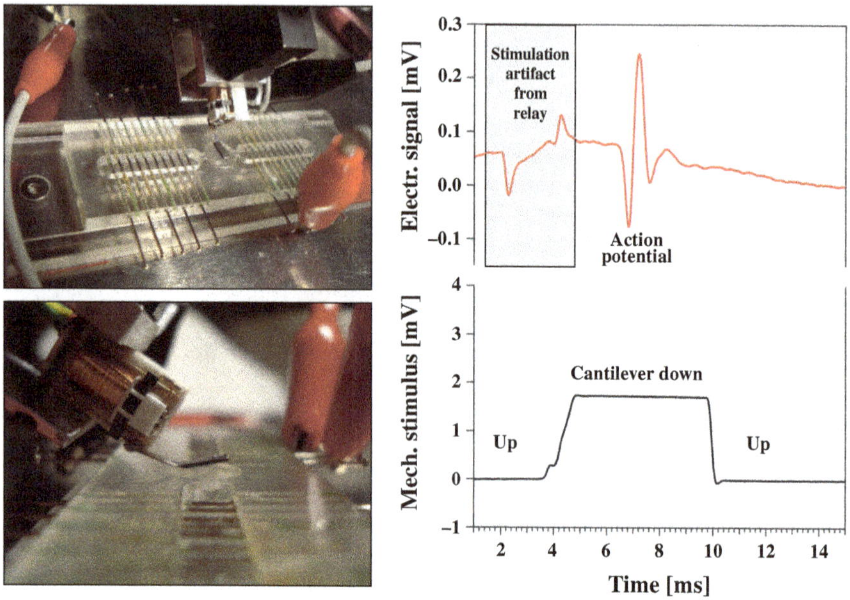

Figure III.4.8 Mechanical stimulation of a nerve from the walking leg of a lobster with a cantilever. Left: Images of the inverted pickup. Right, bottom: The voltage for driving the cantilever exerts a mechanical force. Maximum force is achieved within 1 ms. Right, top: Within 2 ms, one finds an action potential recorded extracellularly with two electrodes. Unpublished data, courtesy of Vardan Azizian, NBI 2015.

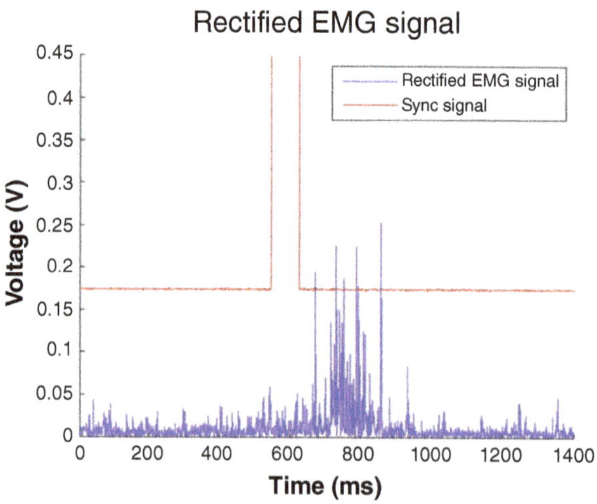

Figure III.4.9 Electromyography signal recorded after ultrasound neurostimulation of a mouse brain. From [64]/with permission of Elsevier.

regions with sound [63]. This experiment is described in more detail in Chapter IX.5. Figure III.4.9 shows the response of the brain to an ultrasound stimulus in an electromyography recording.

However, it is still not well understood why ultrasound can stimulate nerves. It seems obvious that nerves must be susceptible to mechanical perturbations. Several authors have proposed that the soliton theory could provide a way for explaining the phenomenon [65, 66, 67]. The soliton theory is described in Part VII. Typical frequencies of ultrasound are 500 kHz–5 MHz, corresponding to wavelengths of 0.3–3 mm in water. This is of a similar length to the width of an action potential in nonmyelinated neurons (Table III.3.1).

III.4.5 Summary: Key Ideas of Chapter III.4

In this chapter, we presented various mechanical features of nerves. We showed that:

- Nerves display thickness changes on the order of 1 nm during the action potential, which corresponds to about 20% of the thickness of the nerve membrane.
- Electrical and mechanical phenomena are in phase.
- Nerves contract when an action potential is traveling through the neuron. In experiments, this change is on the order of 10–100 nm.
- Nerves can be excited mechanically, for example, by a sudden compression.
- Nerves can be excited by ultrasound.

These results suggest that the nerve pulse is not a purely electrical event, but rather that mechanical and electrical signatures are just different aspects of a more general phenomenon. Both the electrical and mechanical properties of the action potential imply that the nerve can be stimulated electrically and mechanically. We will see in Chapter III.6 that there are also thermal phenomena that accompany the nerve pulse, and that the nerve pulse can be excited thermally.

III.5

Optical Changes During the Action Potential

It has long been known that geometrical changes in nerves can be observed by light scattering. Figure III.5.1 (left) shows a recording of white light scattering at a 45° angle during the action potential in a squid axon [68]. Light scattering monitors changes in geometry, refractive index, or both. The time course is the same as the electrical recording (upper trace). Further evidence for light scattering changes during the action potential was reported by Tasaki and collaborators [69] for spider crab nerves, squid nerves, and nerves from lobster legs. Light scattering changes were also reported for neurons of aplysia (a large sea slug) [70]. The authors concluded: *"An analysis of the data indicates that the radial component of the index of refraction of the membrane increases and the tangential components decrease concomitant with an increase in membrane potential. This is suggestive of a rapid reorientation of dipoles in the membrane during an action potential."* This is consistent with the hypothesis of [71] on the coupling of a dipolar reorientation of membrane molecules and reversible heat release in nerves. One possible explanation for such dipolar reorientation is a change in the phase state of the lipid membrane, as it is known that the dipole potential of lipid monolayers differs between the gel and fluid states of the membrane [72–75].

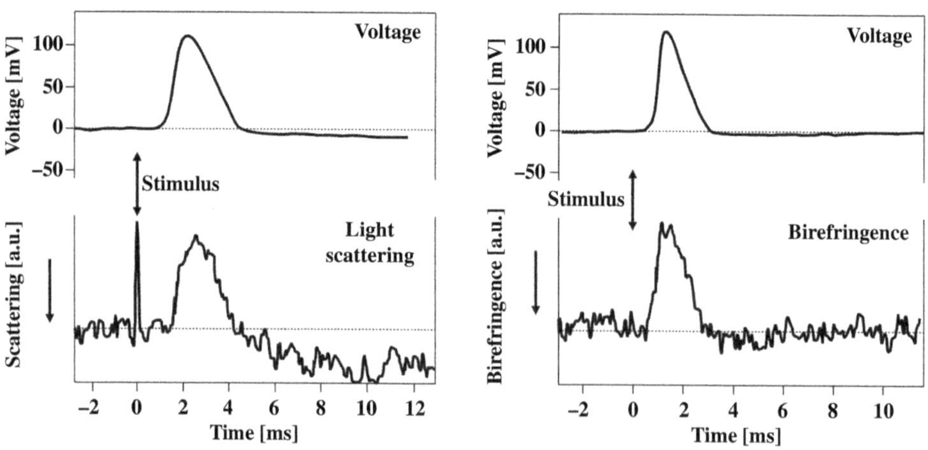

Figure III.5.1 Optical recordings on the squid axon. Left: Light scattering at 45° compared to voltage change. Right: Birefringence compared to voltage changes. Adapted from [68].

Physics of Nerves and Excitatory Membranes, First Edition. Thomas Heimburg.

The right-hand panel of Figure III.5.1 shows birefringence changes during the action potential. Birefringence is a material property indicating that the refractive index differs for different polarizations of light. During a birefringence recording, the relative changes in the intensity of light scattering with different polarizations are recorded. Figure III.5.1 (right) shows that one finds birefringence changes that are proportional to the action potential. The authors attributed these changes partially to a reorientation of elongated proteins that are tangential to the membrane surface and the reorientation of lipids of the Schwann cells and the axon. The authors also reported similar birefringence changes in crab nerves. Changes in birefringence in crab spider nerves and squid axon were also reported by [69]. These changes were attributed to retardation, i.e., a relative change in the intensities of different polarizations leading to a rotation in the optical axis [76]. As mentioned, such optical activity could originate from the rotation of proteins and lipids and is most likely a property of the nerve membrane. Changes in retardation during action potentials in the pike olfactory nerve were reported by [40] (see Figure III.5.2 (right)). Kobatake and collaborators [77] showed that there are changes in fluorescence polarization of a lipid dye during the action potential (see Figure III.5.2 (left)), indicating that the lipid membrane is at least partially involved in the changes in optical properties.

One can summarize the optical properties as follows: The optical changes are in phase with the action potential and the mechanical changes. They must be due to changes in geometry and/or in refractive index and molecular orientations. Thus, they indicate structural changes in the excited part of the neurons at both macroscopic and molecular levels. The changes in birefringence and retardation indicate alterations in the molecular orientation of molecules in the membrane. In particular, the changes in orientation of molecules from tangential to normal to the membrane are consistent with changes in the head group

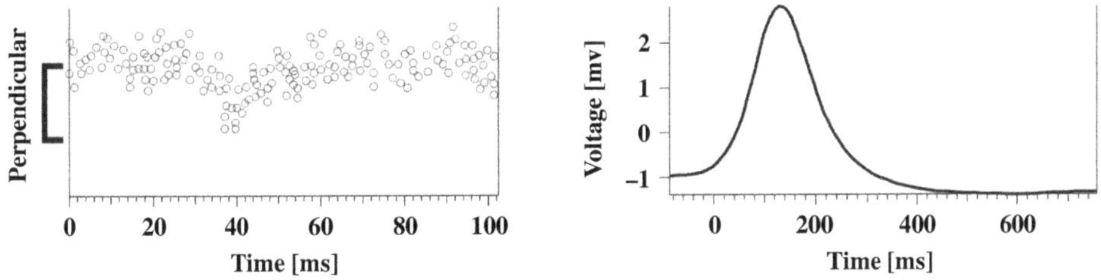

Figure III.5.2 Left: Changes in fluorescence polarization during excitation in a crab nerve stained with Pyronin B. The bottom record was taken under the same experimental conditions except that the analyzer was rotated by 90°. Adapted from [77]. Right: Birefringence changes during action in the pike olfactory nerve. Upper curve: retardation changes. Lower curve: action potential. Adapted from [40].

orientation of lipids. Changes in lipid order are also found when measuring the fluorescence polarization of lipid dyes. These alterations suggest changes in the physical state (for instance, a fluid–gel transition) of the membrane molecules, including the membrane lipids that are close to a melting transition during the action potential.

Summarizing, one can conclude that the electrical, mechanical, and optical changes in the nerve membrane are in phase and most likely represent the same phenomenon.

III.5.1 Summary: Key Ideas of Chapter III.5

Nerve signals are visible in optical measurements. One finds:

- Light is scattered from axons at different angles in a manner consistent with the propagation of the action potential, indicating changes in refractive index, changes in nerve geometry (i.e., mechanical changes), or both.
- Different polarizations of the incoming light beam are scattered to varying degrees. This effect is called birefringence or retardation.
- Birefringence changes indicate the rearrangement of macromolecules and/or changes in the head group orientation of membrane lipids: this implies that the action potential is accompanied by changes in molecular structure in the membrane, i.e., in the lipids or associated macromolecules.
- The fluorescence anisotropy of lipid markers changes under the influence of the action potential, indicating that structural changes in the lipid membrane contribute to the nerve pulse.
- Changes in optical activity are in phase with and qualitatively similar to mechanical and electrical changes.

III.6

Heat Production and Temperature Changes During the Nerve Pulse

In the previous chapter, we showed that during the action potential, one finds mechanical changes. Here we show that nerve pulses are also accompanied by reversible changes in temperature. The historical developments that led to this finding are discussed in Section I.4.3.

III.6.1 Reversible Heat Production of Nerve Pulse Propagation

In a seminal publication in 1958, Abbott, Hill, and Howarth [78] first resolved the initial heat response during the action potential in nerves from the legs of spider crabs. They titled their paper *"The positive and negative heat production associated with a nerve impulse."* It was found that during the rising phase of the action potential, heat is released into the environment of the nerve. This heat is mostly reabsorbed (at least 70%) in a second phase of the action potential. Some experiments from [78] are shown in Figure III.6.1. The existence of heat reabsorption is clearly surprising, and Hodgkin [1] found this to be inconsistent with his electrophysiological theory of the action potential (see Section I.4.3).

At that time, it was not clear whether the heat was completely reabsorbed or whether some heat was dissipated. Furthermore, there was no satisfactory explanation for both the qualitative and quantitative behaviors of the heat production. It should be noted, however, that the shape of the heating profile resembles the reversible adiabatic heating during a mechanical wave [79].

In investigating the origin of the reversible heat, Howarth et al. [80] considered whether the reversible heat could originate from the discharging and recharging of the membrane capacitor during the action potential. They called this concept the *condenser model*. They found that this could not account for most of the heat because the energy of the capacitor was too small to be explained by this mechanism. Instead, they proposed that the reversible heat indicates a reversible change in the entropy of the nerve membrane during the action potential. In fact, this is the core idea of the soliton theory of the nerve impulse discussed in Part VII. In this theory, the heat release is attributed to the large entropy change in a phase transition within the nerve membrane during the action potential.

The main results of the heat experiments are summarized in a 1985 review by Ritchie and Keynes [47] (Figure III.6.2). They ruled out the possibility that the heat reabsorption phase of the action potential was due to heat diffusion into the bulk

Physics of Nerves and Excitatory Membranes, First Edition. Thomas Heimburg.

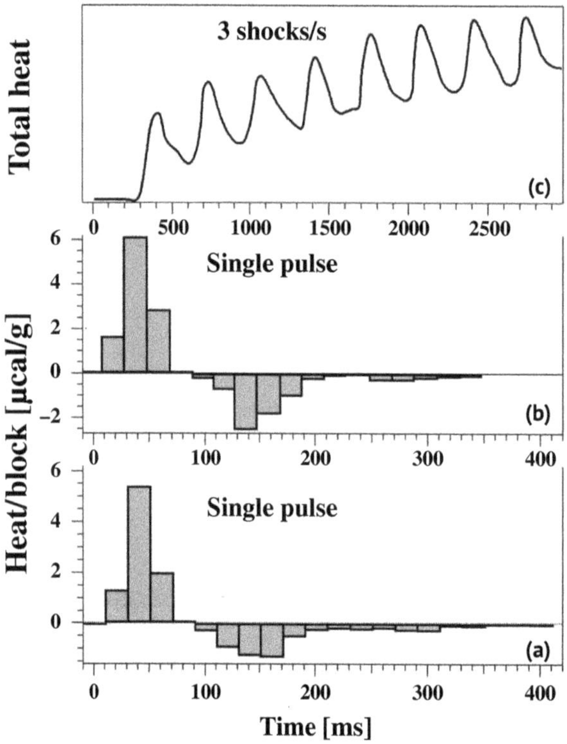

Figure III.6.1 The initial heat in nerves from spider crab. Adapted from [78]. Both phases can be clearly distinguished. (a, b) Heat block analysis for two different nerves after a single stimulation [78] (average of five responses). The data represent the rate of heat release per block of 20 ms. It can be seen that the initial heat consists of a phase of heat release followed by a phase of heat reabsorption. The sum over all heat blocks is the net heat released. (c) Total heat release (the integral of traces as in (a) and (b)) during repetitive stimulation with 3 shocks/second. Each maximum corresponds to heat release and reabsorption of heat. The slow drift represents a difference between heat release and absorption after a pulse (shown in Figure III.6.1c), which might be due to constant metabolism happening in the background.

medium. Local heating of a nerve showed that heat diffusion away from the nerve is much slower than the heat reabsorption measured during the action potential. This control experiment is shown in Figure III.6.2b [47]. They also found that the action potential is exactly in phase and approximately proportional to the square of the transmembrane voltage (Figure III.6.3). Thus, the heat release is proportional to the energy of charging and discharging a capacitor. After some initial doubts in earlier papers about whether the heat reabsorption is complete, they also found that all the heat is reabsorbed within experimental accuracy. In fact, they were able to show that even for action potentials with a hyperpolarization phase, the total heat followed the voltage (Figure III.6.3) [47].

The aforementioned experiments were limited to long action potentials of 100–200 ms. However, most nerve impulses have a duration of the order of 1 ms, which is about two orders of magnitude faster. Measuring heat release in such nerves requires different

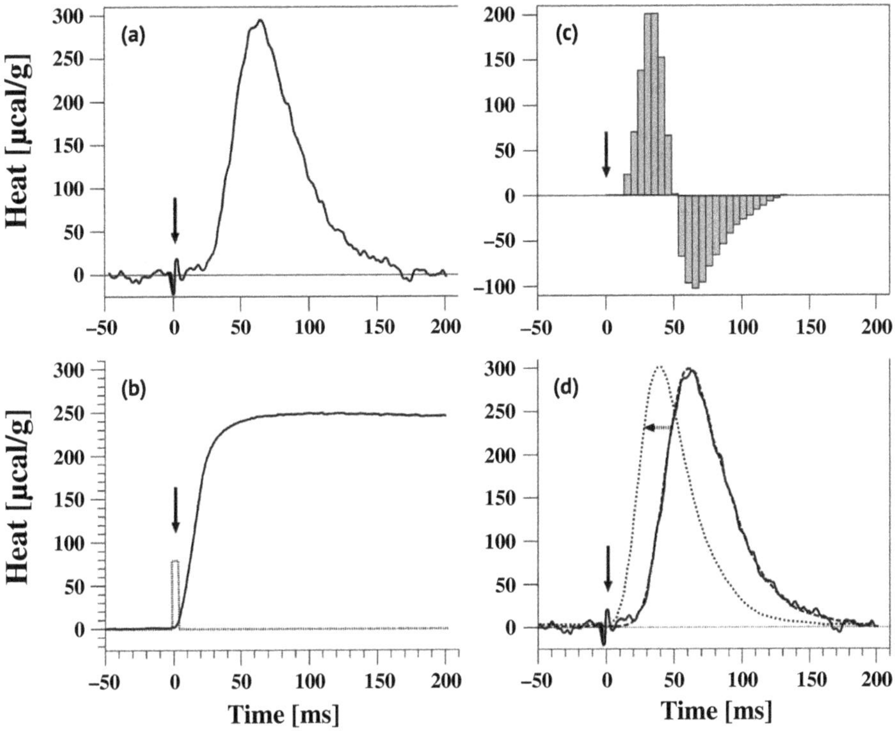

Figure III.6.2 Heat production in non-myelinated fibers of garfish olfactory nerve. Adapted from [47]. (a) Thermal response after a stimulation indicated by the stimulation artifact. (b) Control experiment after applying a local heated pulse on the membrane. (c) Heat flux out of and into the nerves obtained by heat block analysis. (d) Deconvolution of the measured signal using the response function of the instrument, leading to the true signal (dashed line). The derivative of this signal corresponds to the true heat flux in (c).

methods. Tasaki and Iwasa obtained a faster response time using a heat-sensitive polyvinylidene fluoride film [81]. They recorded an action potential of 25 ms length in lobster nerve and later in garfish olfactory nerve and myelinated nerve fibers from bullfrog, with time scales of 10–20 ms [23]. As in the earlier measurements by Abbott, Howarth, and Ritchie, Tasaki and coworkers found the integrated heat to be very small and consistent with almost complete reabsorption of any released heat. The results are shown in Figure III.6.4. The authors concluded that the heat production is in phase with the nerve impulse, but that the total heat is unlikely to be caused by the charging and discharging of the membrane capacitor. The time courses of voltage and heat were not always identical (Figure III.6.4 (left a and b)). They proposed that the heat production might be due to ion exchange in the superficial layer of the axoplasm [77, 23], an idea generally favored by Tasaki. He proposed that the nerve impulse is accompanied by a phase transition in the proteins on the surface of the nerve [77, 82].

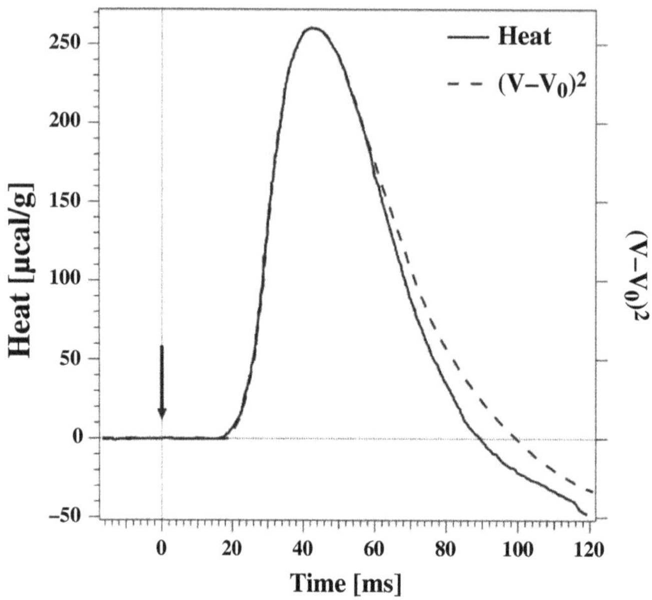

Figure III.6.3 Deconvoluted heat response of a garfish olfactory nerve (solid line) compared to the voltage of the pulses ($V^2 - V_0^2$) in arbitrary units. The heat signal is in phase with the voltage changes and proportional to the energy of the charged capacitor. Adapted from [47].

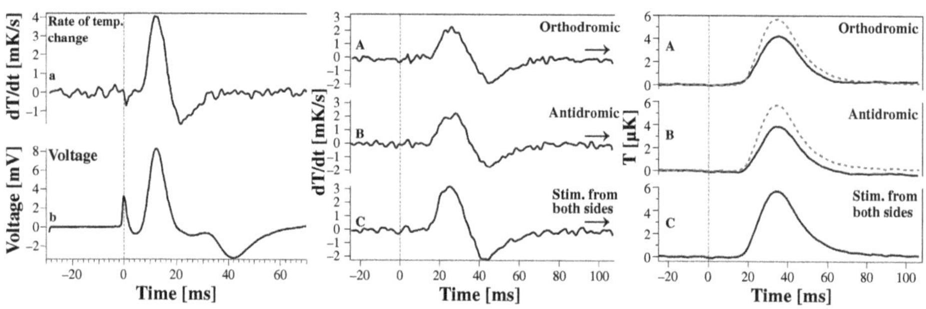

Figure III.6.4 Thermal response of garfish olfactory nerve. Adapted from [23]. Stimulation occurs at a time of 0 ms. Left: rate of temperature change at 20 °C (trace a) and the accompanying action potential (trace b). The temperature change and the voltage are nearly in phase. Center: rate of temperature change for an orthodromic (A), antidromic (B), and combined (C) nerve propagation experiment at 10 °C. Right: the integrated rate yields the absolute temperature change, which is of the order of 40 µK and nearly completely reversible. The stimulation from both sides yields a larger temperature change than stimulation from one end. This is indicated by the gray dashed line in the top and middle panel on the right.

III.6.2 Excitation of an Action Potential by Sudden Changes in Temperature

There are indications that sudden drops in temperature can initiate an action potential [83]. This was also claimed by Kobatake, Tasaki, and Watanabe [77], who wrote:

"It has long been known that an action potential can be evoked by sudden cooling in various excitable tissues. This implies that the process in the axon membrane underlying depolarization (excitation) is exothermic, and that depolarization is endothermic. Since an endothermic reaction is encouraged by heating, a rise in temperature should lead to repolarization of the excited membrane. This was tested for a prolonged action potential by applying a brief heat pulse [83]. It was found that the action potential could be terminated by a heat pulse in an all-or-none manner. This observation supports the view that the termination of the action potential (and, hence, depolarization) is a cooperative and endothermic reaction. Recent studies on temperature change during action potential also support this view [80]."

Thus, excitability is in agreement with the temperature changes described in the previous section. Both Spyropoulos [83] and Kobatake et al. [77] also indicate that action potentials can be excited chemically. In 2012, Shapiro et al. reported that nerves can be stimulated by infrared light – seemingly as a consequence of local heating [84]. Thus, the fact that nerves possess a heat signature during pulse propagation also gives rise to the fact that they can be stimulated thermally.

III.6.3 Summary: Key Ideas of Chapter III.6

The action potential is accompanied by thermal changes. It was found that:

- There is no measurable heat production during the action potential, indicating that the nerve pulse as a whole is isentropic.
- The depolarization phase of the action potential is accompanied by a heat release from the nerve membrane, while that depolarization phase of the action potential is accompanied by an equivalent heat reabsorption.
- The reversible release and reabsorption of heat is too large to be explainable by reversible charging of the axonal membrane capacitor during the action potential.
- Therefore, the thermal energy of an action potential is larger than the electrical energy and represents the dominant contribution.
- The magnitude of the heat changes is rather consistent with significant entropy changes in the membrane, such as during a phase transition.
- A phase transition in the axonal membrane during the action potential was also suggested by experimental results by Spyropoulos and Tasaki and collaborators [83, 77].
- The thermal signature of the nerve pulse resembles that of an adiabatic wave, e.g., a sound or density wave.
- As shown in previous chapters, mechanical changes indeed exist during the action potential.

The physics of the temperature changes in nerves is discussed in much more detail in [79].

III.7

Magnetic Fields Generated During the Action Potential

A noninvasive method to stimulate the brain is transcranial magnetic stimulation (TMS). This is schematically shown in Figure III.7.1. In this technique, a magnetic coil is placed on the skull. Magnetic pulses induce nerve activity in a noninvasive manner [85, 86]. The nerve pulse has an electrical component with voltage changes along the nerve, which leads to currents along the nerve axon. Thus, magnetic pulses induce currents in axons due to electromagnetic induction. Vice versa, a current in the axon leads to a magnetic field surrounding the nerve axons.

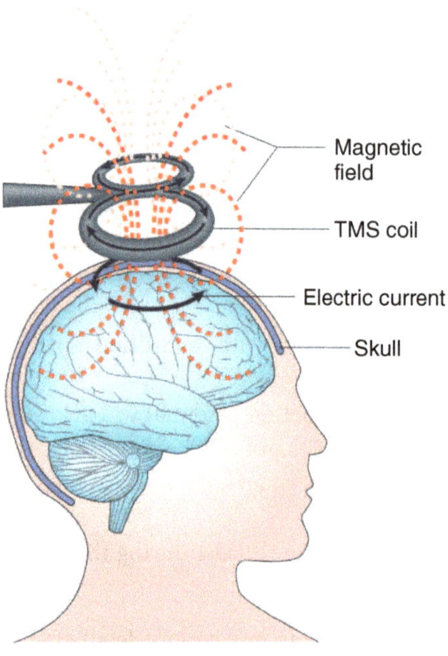

Figure III.7.1 Schematic view of TMS. From [86]/with permission of Springer Nature.

Physics of Nerves and Excitatory Membranes, First Edition. Thomas Heimburg.

III.7.1 Magnetic Fields of Nerves

Magnetic recordings on single nerves were pioneered by Wikswo and collaborators in the 1980s [87–90]. They used a SQUID (superconducting quantum interference device) magnetometer, which consists of superconducting rings and some Josephson junctions. A magnetic field induces currents in this ring. For more detailed descriptions of SQUIDs and their application in the biomedical field, see [91, 92].

Figure III.7.2 (left) shows the principle of a magnetic measurement on a crayfish nerve. The nerve passes through the ring, which measures the magnetic field generated by currents along the axons. The magnetic and electric measurements on the nerve are shown in the center panel. The amplitude of the field measured by the SQUID is about 300 picoTesla. From the electrical voltage recordings, one can calculate the currents expected longitudinally along the axon and calculate a magnetic field, and vice versa [88]. Figure III.7.2 (right) shows that the electrical signal calculated from the magnetic measurement and the voltage recording agree nearly perfectly. Thus, it can be concluded that magnetic phenomena are caused by electrical currents along the axon.

The SQUID technique is well-suited to measure the fields originating from a nerve that can be placed such that it passes through the coil. However, it is not well-suited for non-invasive measurement. Jensen and collaborators have designed an alternative method to measure tiny magnetic fields [93]. It is based on the magnetic field-dependent Zeeman splitting of the energy levels of atoms in a cesium gas. With this technique, one can measure sub-picoTesla magnetic fields. For a measurement, a millimeter-sized container containing the gas is placed next to a nerve. Figure III.7.3 (left) shows the electrical signal of a frog sciatic nerve at 0, 0.2, 0.8, and 1.2 V stimulation. Only the latter two are above the stimulation

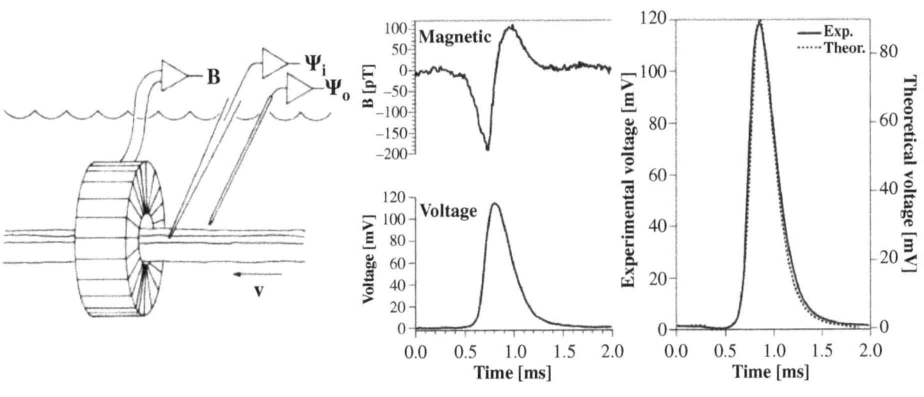

Figure III.7.2 Magnetic recording of the action potential. Left: Schematic drawing of the experiment on a giant axon from a crayfish. A toroidal coil is placed around the nerve. It records the magnetic field *B* of the action potential. Electrical properties of the action potential are measured with two electrodes measuring the potential inside (Ψ_i) and outside (Ψ_o) of the giant axon. Center, top: The magnetic field *B* of the action potential measured by the magnetometer. Center, bottom: The transmembrane voltage of the action potential measured by electrodes. Right: The voltage trace of the action potential compared with the voltage calculated from the magnetic field. They are in close agreement. The calculation is based on assuming a current along the nerve in a voltage gradient that generates a magnetic field [90]/with permission of Elsevier.

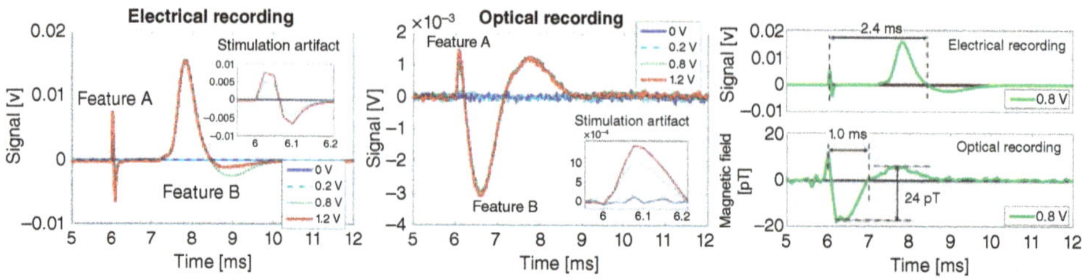

Figure III.7.3 Electrical and optical measurements of the nerve impulse of the sciatic nerve of a frog for different stimulation voltages. Left: Electrical measurement. Center: Optical measurements. Right: Electrical signal for 0.8 V stimulation and magnetic field calculated from the optical recording. From [93]/Springer Nature/CC BY 4.0.

threshold. The center panel shows the magnetic signal (measured in V) from the magneto-optical recording. It is very similar to the shape of the signal of the crayfish in Figure III.7.2 (center). In the right-hand panel, the optical recording has been translated into units of magnetic field. Its amplitude is about 24 picoTesla. A further method to detect the magnetic fields generated by nerves is by electronic spins in diamond [94, 95].

We show here that magnetic fields are generated by the electrical currents caused by the action potential. Seemingly, the magnetic recording leads to identical results as electrical recordings. However, Roth and Wikswo showed that there exist experimental conditions where the magnetic recording can yield information that cannot be accessed by electrical means [96].

III.7.2 Summary: Key Ideas of Chapter III.7

Nerve signals can be observed with ultra-sensitive magnetometers.

- Using superconducting quantum interference devices (SQUIDs), one finds that the magnetic fields generated during the action potential are consistent with the electrical currents along the axons or nerves.
- One also finds magnetic fields by using quantum-optical measurements, i.e., the influence of the magnetic fields on the Zeeman splitting of an atomic cesium gas or the effect of the fields on electronic spins in a diamond.
- The action potential is both an electromagnetic and a mechanocaloric phenomenon.

III.8

Collisions of Nerve Pulses

In electrophysiology, it is generally assumed that colliding nerve pulses annihilate. This is the theoretical result of the Hodgkin–Huxley model that is described in detail in Part IV. In this model, the annihilation of nerve pulses is a consequence of the finite relaxation time of open and closed states of channel proteins, which results in a refractory period during which a nerve cannot be excited after a previous excitation. However, we show in this chapter that nerve pulses can, in fact, pass through each other.

III.8.1 Penetration of Action Potentials

An action potential can travel along a neuron in both directions. The direction from the soma to the end of the axon is called the orthodromic pulse. The direction from the end of the dendrites toward the soma is called the antidromic pulse. Thus, it is in principle possible that two action potentials collide if a neuron is excited at two opposite ends. For the understanding of the nervous impulse, it is important to understand how two pulses interact.

The first paper investigating collisions was by Ichiji Tasaki in 1949 [97]. Tasaki showed that nerve pulses can annihilate upon collision in motor nerve fibers from toads. In Part IV, it is described how the Hodgkin–Huxley model for the action potential in nerves relies on the conformational changes of membrane proteins that have lifetimes of several milliseconds. A consequence is that, in order for a nerve to be excited a second time, the proteins have to relax back into their original state. This is called "refractoriness" or "refractory period" (see Section III.2.3). Within the relaxation period, nerves cannot be excited a second time, or the second pulse is smaller than the first one. When pulses collide in an axon, each pulse would propagate into a regime of a perturbed membrane, and it seems natural to assume that the refractory period is responsible for the annihilation of the two pulses. However, this is not the explanation that Tasaki gave. He remarked:

> *"By collision, transmission of impulses is blocked, not on account of the refractoriness left behind by the impulses, but through lack of internal stimulating current by which the normal transmission is affected."* and *"Some physiologists seem to believe that, when two impulses collide at one spot on a nerve fibre, the transmission of each of these impulses is blocked by refractoriness left behind by the other impulse. But this is an utterly mistaken*

Physics of Nerves and Excitatory Membranes, First Edition. Thomas Heimburg.

idea. We have seen above that, at the moment when the last node of Range in the fiber is involved in action, the current in- and outside the fiber becomes insignificantly small. In other words, there is in this stage no internal stimulus through which the transmission is effected. It is a property of the bioelectric process to subside by itself."

Thus, Tasaki believed that at the location where two pulses meet, the currents along the axon add up to zero, and that this is the reason why the action potentials do not propagate. There are not many other historical papers that show directly that nerve pulses annihilate. However, a common technique is the so-called collision block experiment, which is discussed in the next section.

Gonzalez-Perez and collaborators revisited this problem [21]. They conducted experiments on intact worm axons, lobster leg nerves, and giant axons from lobster connectives. The setup of the experiment is shown in Figure III.8.1. The intention of this figure is to

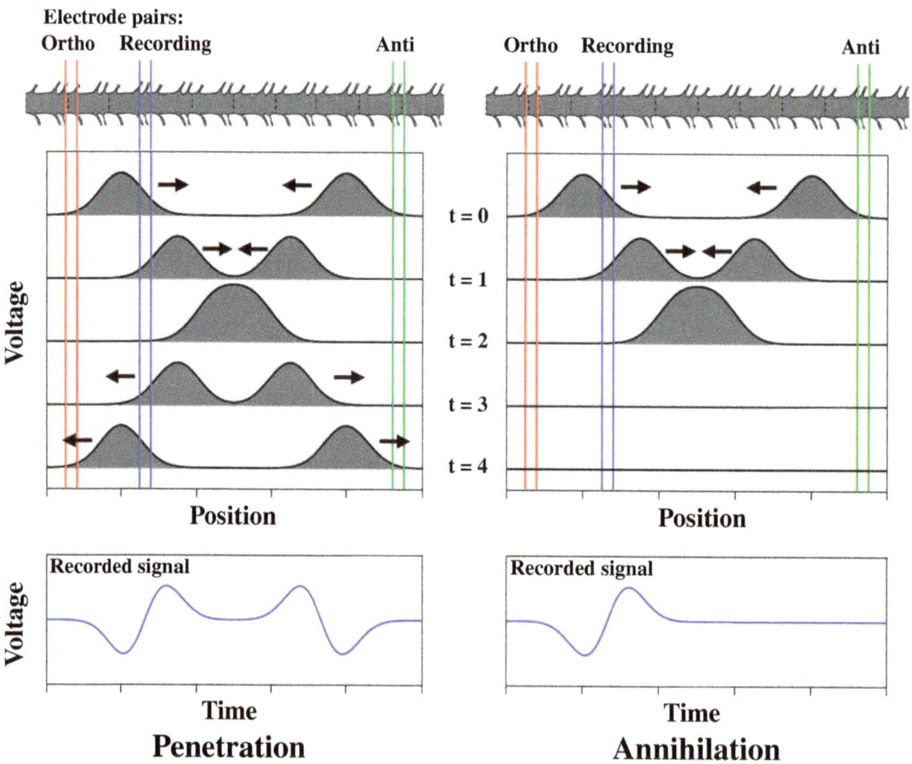

Figure III.8.1 Schematic image of the hypothetical voltage signals expected in a collision experiment. Action potentials are excited with two electrodes at the left (red, orthodromic) and at the right (green, antidromic) side. Action potentials are recorded with the blue electrodes. Left: Case 1: Pulses penetrate. If two pulses are stimulated at both ends, the orthodromic pulse arrives at the recording electrodes before the antidromic pulses. Since pulses travel in opposite direction, the antidromic signal is inverted relative to the orthodromic signal. Right: The same experiment in the case of pulse annihilation. The orthodromic signal arrives at the recording electrodes but the antidromic signal cannot be measured. From [21]/American Physical Society/CC BY 3.0.

illustrate what the outcome of a collision experiment would be in the case of pulse penetration and in the case of pulse annihilation. Shown is an experiment with two pairs of electrodes at the left end (for stimulation of an orthodromic pulse) and the right end (for stimulation of an antidromic pulse) of a neuron or nerve (shown in red and green). A third pair of electrodes is placed closer to the left end such that an orthodromic pulse passes by the recording site prior to a collision, and the antidromic pulse after a collision. In such a setup, the recording electrodes will observe the antidromic pulse only when the pulses pass through each other. When pulses annihilate, only the orthodromic pulse can be observed. Figure III.8.1 (bottom panel) shows the expected outcome of these two cases. The shape of the pulses reflects the fact that two extracellular electrodes are used (see Chapter III.2 and Figure III.2.3).

Figure III.8.2 shows the experimental results of an experiment on an earthworm ventral cord from [21]. The top trace shows the pulse shape when only the orthodromic side is stimulated. The center trace shows the result when only the antidromic side is stimulated. The bottom trace shows the experimental result when both sides are stimulated simultaneously (solid line). The outcome is similar to the predicted result in Figure III.8.1 (left). For comparison, the superposition of the orthodromic and antidromic pulses is shown (dotted line). They are nearly identical, which indicates that the orthodromic and antidromic pulses passed through each other unperturbed.

The ventral cord of the earthworm contains several giant axons and a few small axons. One might argue that the outcome of the collision experiment reflects that orthodromic and

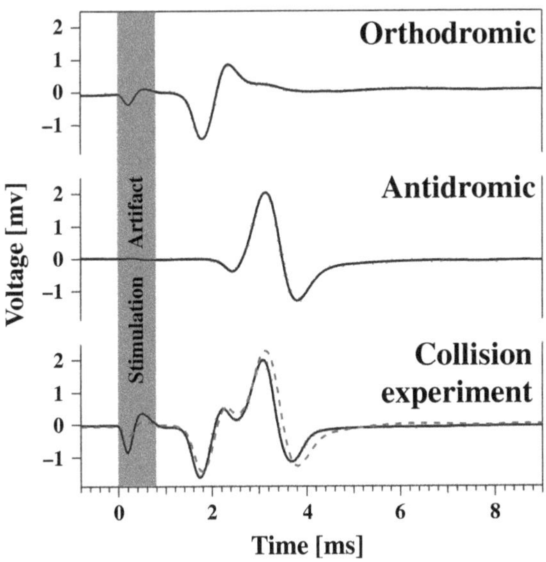

Figure III.8.2 Collision experiment on the ventral cord of an earthworm. Top: Stimulation at the orthodromic end only. Center: Stimulation at the antidromic end only. Bottom: Stimulation at both ends. The solid line is the experimental result. It resembles the hypothetical trace in the left-hand panel of Figure III.8.1. The dashed line is the superposition of the top and center traces. This experiment shows that the orthodromic and antidromic pulses have passed through each other unperturbed and did not annihilate. From [21]/American Physical Society/CC BY 3.0.

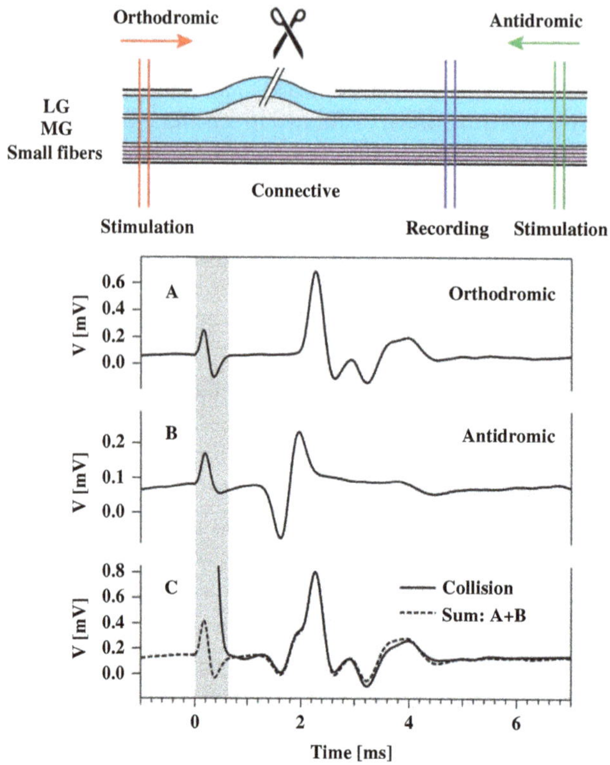

Figure III.8.3 Same experiment as in Figure III.8.2 with the a single giant axon in the ventral cord of a lobster. Other giant axons were cut with scissors (Top). Bottom panel, top trace: Stimulation at the orthodromic end only. Bottom panel, center trace: Stimulation at the antidromic end only. Bottom panel, lower trace: Stimulation at both ends. The result is similar to the experiment in the ventral cord of earthworms, indicating the orthodromic and antidromic pulses in a single axon have passed through each other unperturbed and did not annihilate. From [22]/with permission of Elsevier.

antidromic pulses traveled in different axons. This can be ruled out by increasing stimulation power on both ends such that all fibers are excited at the same time. Such an experiment is shown in Figure III.8.5 [22]. Another way to ensure that orthodromic and antidromic pulses traveled in the same giant axon is to cut the other giant axons. Figure III.8.3 shows an experiment on a single giant axon of lobster connectives. This experiment shows that the recording in a single axon is nearly identical to the superposition of individual pulses, which confirms that penetration of orthodromic and antidromic pulses is found in single giant axons.

III.8.2 Collision Block

When a nerve is excited locally, two pulses emerge that travel in opposite directions. Let us consider the situation depicted in Figure III.8.4a. If the nerve is stimulated at position S2, two pulses are created: an antidromic pulse (A_2) traveling to the left and an orthodromic pulse (O_2) traveling to the right. Likewise, at position S_1, an orthodromic pulse O_1 is created that travels to the right. The corresponding antidromic signal is not shown. The electrical signals shall be recorded at two positions, R_1 and R_2. When the pulse O_1 and the pulse A_2 meet and annihilate, one only measures pulse O_2 at the recording electrodes R_2. This is the idea of a collision block experiment, which is shown in Figure III.8.4b. Likewise, at position R_1, one would only measure pulse O_1 but not pulse A_2. However, if pulses pass through each other, one would measure both pulse O_1 and A_2 at position R_1, and one would measure pulses O_2 and O_1 at position R_2 (Figure III.8.4c).

Berg and collaborators [98] performed a collision block experiment on intact worm ventral cords and measured at position R_2 only. They found that one can observe pulse O_2 but not pulse O_1 and concluded that nerve pulses annihilate. However, this interpretation makes assumptions about what pulses O_1 and A_2 do when they collide that are not tested directly.

Wang and collaborators [45] performed a similar experiment but recorded the signal at both positions R_1 and R_2 simultaneously (Figure III.8.5), i.e., they performed a collision experiment as shown in Section III.8.1 and a collision block experiment as in [98] at the same time on the same nerve. They found that in a direct annihilation experiment at position R_1 one detects both the orthodromic pulse O_1 and the antidromic pulse A_2. This suggests that the pulses pass through each other. At the same time, one finds only pulse O_2 at position R_2. This experiment confirms the experimental finding by [98] but challenges its

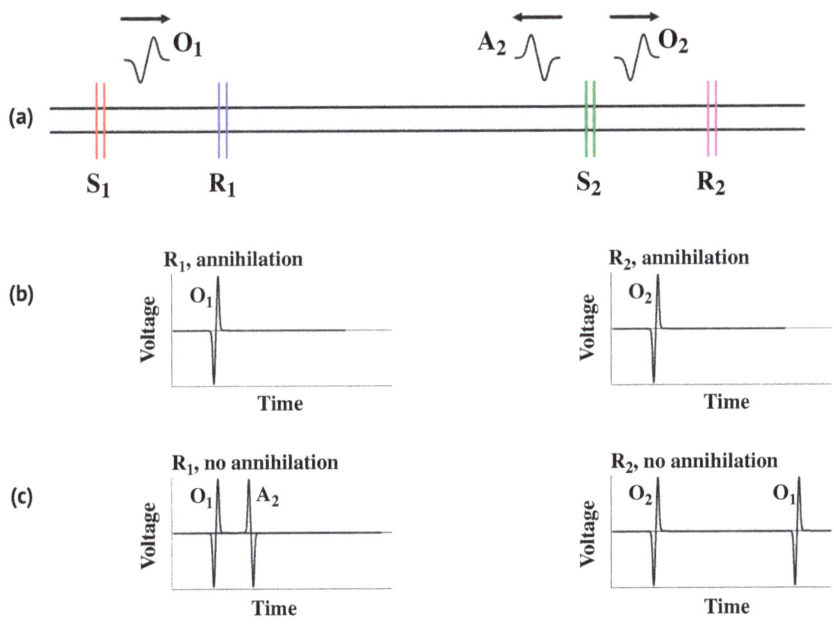

Figure III.8.4 (a) Schematic drawing of a collision (with recording at electrode pair R_1) and collision block experiment (with recording at electrode pair R_2). During stimulation at electrode pairs S_1 and S_2, one creates an orthodromic pulse O_1 at S_1, an orthodromic pulse O_2 at S_2 and an antidromic pulse at S_2. (b) When pulses O_1 and A_2 collide and annihilate, on expects to observe pulse O_1 at R_1, and pulse O_2 at R_2. (c) If O_1 and A_2 pass through each other, one observes both pulse O_1 and A_2 at R_1, and both O_2 and O_1 at position R_2.

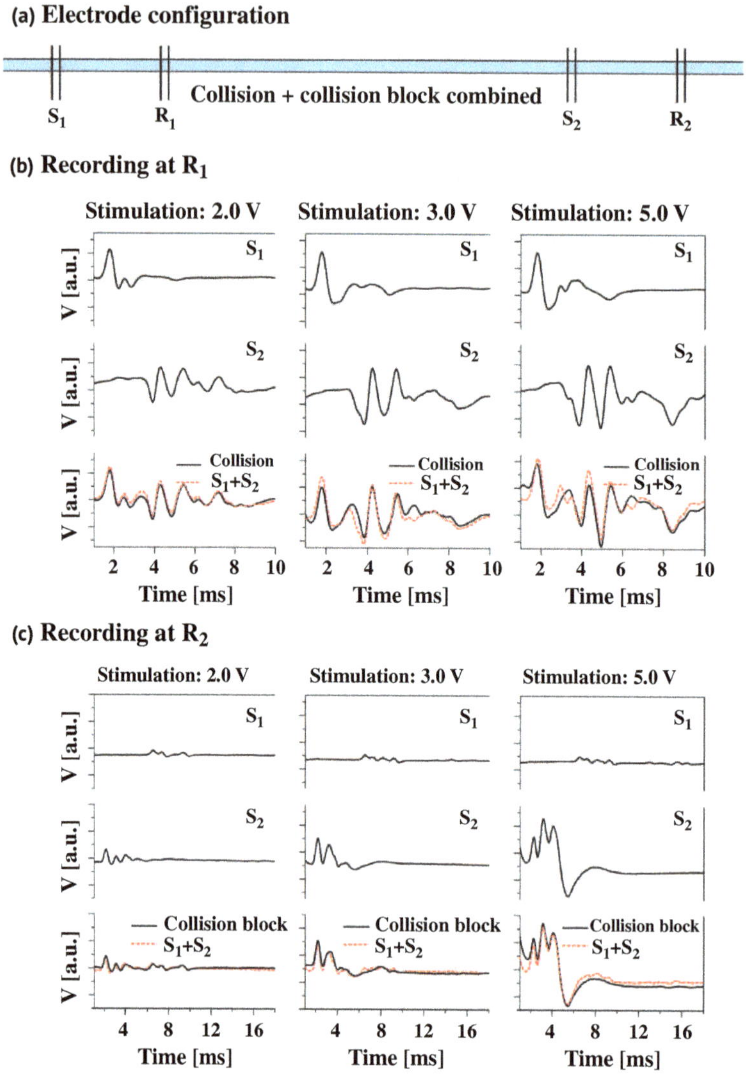

Figure III.8.5 (a) Schematic drawing of the experiment. (b) Collision experiment on a walking nerve of a lobster at electrode pair R_1 at three stimulation voltages. The top traces show the orthodromic pulse only, the center trace the antidromic pulse only. The bottom traces show the result when stimulation occurred both at S_1 and S_2. The recording shows that, within error, the superposition of antidromic and orthodromic signals is identical to the result after stimulation at both sites, independent of stimulation voltage. This proves that the signals have penetrated. (c) Collision block experiment on the same nerve as in panel (b) shows that only the orthodromic signal from the stimulation at S_2 can be observed.

interpretation. The absence of pulse O_1 in a collision block experiment does not automatically imply that pulses O_1 and A_2 annihilated. The tentative explanation for the findings in Figure III.8.5 was that the high stimulation voltages (2–4 V) in such experiments make the stimulation sites impenetrable rather than proving pulse annihilation. The recording at position R_1 is the more direct experiment because it does not involve the transmission of any nerve pulse through the stimulation sites.

III.8.3 Summary: Key Ideas of Chapter III.8

The collision of two action potentials in the same neuron can either lead to pulse annihilation, as expected in the electrophysiological view of the action potential (the Hodgkin–Huxley model) described in Part IV, or to pulse penetration, as expected by the soliton theory described in Part VII. In experiments, one finds that:

- In neurons, action potentials can travel in both directions. The pulse from the soma to the dendrites is called the orthodromic pulse. The pulse from the dendrites to the soma is called the antidromic pulse.
- In worm nerves, the walking leg of the lobster, and in single giant lobster axons, one finds pulse penetration, i.e., orthodromic and antidromic pulses pass through each other.
- This finding is not undisputed and there are authors expressing different views (e.g., [98]), partially based on the results of collision block experiments.
- Direct collision experiments and collision block experiments do not lead to the same result and seem to measure different phenomena.
- Collision experiments may be an important tool to investigate the physical nature of the action potential to help discriminate between thermodynamic and electrophysiological approaches to explain nerve pulses.

Part IV

Basic Principles of Electrophysiology

This book focuses on the macroscopic thermodynamic properties of nerves and nerve membranes. We will derive an electromechanical picture of a nerve impulse that can be controlled by changes in the intensive variables. However, the currently used electrophysiological description of the nerve impulse is the Hodgkin–Huxley model, in which control of the properties of the nerve impulse is mainly due to ion channel proteins. Here, we introduce the basic principles of electrophysiology. This part will serve as a reference for comparison with the thermodynamic nerve theory outlined in Parts VII and VIII.

IV.1

Some Historical Considerations

In his membrane theory, Julius Bernstein proposed that during the action potential, the resistance of the nerve membrane breaks down and ions can pass through the membrane [1, 2]. The conductance of the membrane was unspecific for all cations without any particular selectivity. An ion flux was thought to affect the electrostatics of the membrane and lead to transient changes in the transmembrane voltage. Bernstein's theory was reconsidered during the 1930s to 1950s by scientists in the United States and Great Britain, in particular by Kenneth S. Cole, Alan Lloyd Hodgkin, Andrew Fielding Huxley, Bernard Katz, Richard Darwin Keynes, and others.

Cole is considered the father of modern biophysics and neuroscience. His goal was to study biological phenomena from a physical perspective – similar in spirit to the "*organic physics club*" of du Bois-Reymond, Helmholtz, and Brücke. Figure IV.1.1 shows a *Loligo Forbesi* squid and the giant axon action potential. He studied the electrical properties of biological membranes, in particular the axons of the squid. Squid axons are a great advantage for electrophysiological studies because they have large diameters of up to 1 mm. They were characterized by J.Z. Young and coworkers [5, 6]. Cole

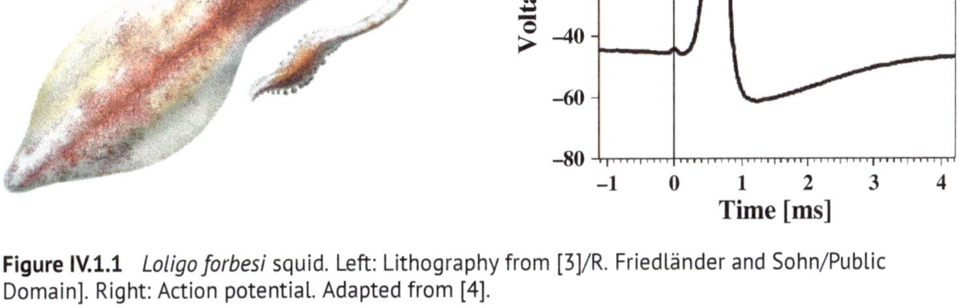

Figure IV.1.1 *Loligo forbesi* squid. Left: Lithography from [3]/R. Friedländer and Sohn/Public Domain]. Right: Action potential. Adapted from [4].

focused on impedance spectroscopy, which showed that squid axons had strange nonlinear electrical features such as rectification and inductive elements. His work was summarized in his 1968 monograph "*Membranes, Ions, and Impulses*" [7]. It led to a reinvestigation of Bernstein's theory, as it became apparent that the electrical properties of nerve membranes were nonlinear and possessed the properties of resistance, capacitance, and even induction. This led to many studies investigating the permeability of nerve membranes to various ions, and the propagation of electrical perturbations in cylindrical cables. The presence of an inductive (i.e., an inertial) term in the impedance spectra of nerve membranes poses a particular challenge, as we will discuss later. This research culminated in the famous Hodgkin–Huxley model. In 1952, Alan Lloyd Hodgkin and Andrew Fielding Huxley proposed a model for the action potential in the nerve that would become one of the pillars of modern electrophysiology. They were awarded the Nobel Prize in Physiology and Medicine in 1963. It is based on cable theory (described in Chapter IV.2), which consists of an application of Ohm's law and Kirchhoff's law to cylindrical cables, combined with the experimental observation that the conductance of the nerve membrane to ions is not constant during an action potential. This was already the case in Bernstein's theory of 1902, which proposed a transient breakdown of the resistance of the nerve membrane during the voltage pulse. Cable theory, as described in Chapter IV.2, makes use of capacitance and resistance to ions that are uniform along the length of the cable and are constant in time. The Hodgkin–Huxley model is an attempt to modify cable theory to account for the dependence of the conductances on voltage and time.

Hodgkin and Huxley studied the conduction of sodium and potassium ions across the membranes of squid axons and concluded that the conductance of the two ions is the result of two different mechanisms. Based on the parameterization of the nonlinear ion conductance as a function of voltage and time, they derived a differential equation with pulse-like solutions that look very similar to the action potential in the squid axon. This model is based on so-called equivalent circuits, in which membrane components are compared to electrical elements such as resistors and capacitors.

IV.2

Cable Theory

Cable theory is the attempt to understand the electrical properties of a neuron in analogy with electrical circuits. Its application to nerves has been proposed, e.g., in [8] (see also the excellent textbook by Johnston and Wu [9]). The nerve membrane is considered a resistor with capacitive properties, and the ionic currents are treated as electrical currents flowing through resistors. Cable theory is important because it forms the basis for the Hodgkin–Huxley model.

IV.2.1 Currents in a Cylindrical Axon

In the following, we will consider the axon of a nerve as an infinite hollow cylinder with a constant radius, surrounded by a membrane. Both the cytosol within the cable and the membrane shall display homogeneous and constant properties (see Figure IV.2.1). The coordinate along the cable is given by the x-axis. At position $x = 0$, a current I_0 is injected. This mimics the sudden in- or out-flux of ions at a synapse, or a sudden change in the local permeability of the membrane.

From the position of the injection, the following currents of interest flow away from the site of the injection: (i) the current i_i along the internal liquid within the cable at position x, and (ii) the current i_m through the membrane at position x.

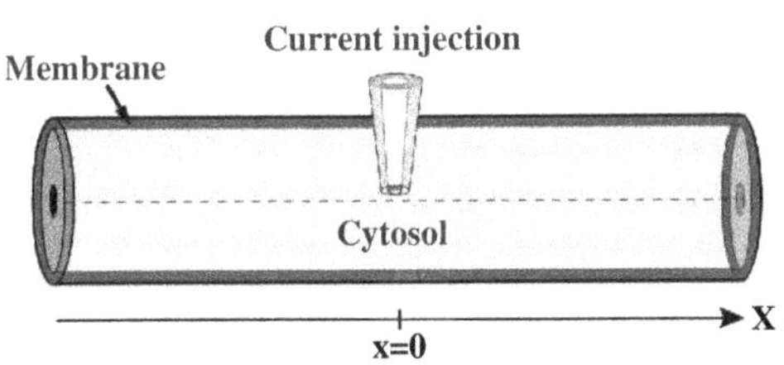

Figure IV.2.1 The neuronal axon is considered a cylinder filled by a cytosol. This is called a "cable." It is surrounded by a membrane. Both cytosol and membrane display a resistance for ionic currents. In the center, a current is injected, either through a small pipette or by a local stimulation, e.g., at the site of a synapse.

Physics of Nerves and Excitatory Membranes, First Edition. Thomas Heimburg.

IV.2.1.1 The Current Along the Cable

We assume that the electrical potential on the outside is constant, but the electrical potential inside the cable might change due to the flux of the electrical current that is injected at $x = 0$. This creates a transmembrane voltage V_m that depends on the position x. The voltage difference along the cable is equal to the difference in the transmembrane voltage. Figure IV.2.2 shows the currents i_i and i_m along the cable and across the membrane, and the change in the transmembrane potential V_m as a function of position. The radius of the cylinder is given by a, and the total length of an element of this cable shall be l. Ohm's law is given by $V = R \cdot I$. Ohm's law along the cable can thus be written as

$$\Delta V_m = -\mathcal{R}_i \cdot i_i, \tag{IV.2.1}$$

where ΔV_m (unit [V]) is the transmembrane voltage difference between the two ends of the cable. The negative sign results from the convention of how the direction of the transmembrane voltage is defined. The total internal cable-specific current along the cable is i_i (unit [A]), and $\mathcal{R}_i$ (unit [Ω]) is the total resistance over the length Δx (unit [m]) of the cable. This resistance is proportional to the length of the cable and inversely proportional to its cross-sectional area. Therefore, we can rewrite eq. (IV.2.1) as:

$$\Delta V_m = -\underbrace{\frac{R_i}{\pi a^2}}_{\equiv r_i} \cdot \Delta x \cdot i_i = -r_i \cdot \Delta x \cdot i_i, \tag{IV.2.2}$$

where R_i is a specific resistance that carries the units [Ωm], and r_i is a cable-specific resistance with units [Ω/m]. R_i is independent of the geometry of the cable, but r_i depends on its radius.

Let us now consider a small segment dx along the cable. The voltage change across that segment shall be given by dV_m. Thus, Ohm's law in eq. (IV.2.2) can be written as

$$dV_m = -r_i \cdot dx \cdot i_i \quad \text{or} \quad \frac{\partial V_m}{\partial x} = -r_i \cdot i_i. \tag{IV.2.3}$$

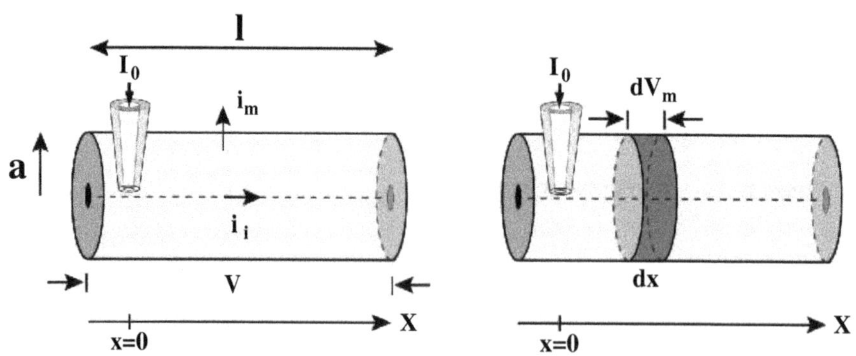

Figure IV.2.2 Left: The nerve axon is modeled as a hollow cylinder of radius a filled by a homogeneous cytosol. At position $x = 0$, a current I_0 is injected. The current i_i is the fraction of the current flowing along the cable at position x, while i_m denotes the current flowing through the membrane from inside to outside. Right: There exists a voltage difference dV between position x and $x + dx$.

The membrane of the axon can also conduct. Some of the current will leak out of the axon toward the outside medium. The cable-specific current leaking through the membrane at position x shall be denoted as i_m (units [A/m]). Kirchoff's first law states that the total current is conserved. Therefore, the current $i_m dx$ (unit [A]) through the membrane must be equal to the reduction in current i_i in the segment dx along the axon:

$$i_m = -\frac{\partial i_i}{\partial x}. \tag{IV.2.4}$$

IV.2.2 Cable Equation

The cable equation describes the time and position dependence of the voltage along the axon, provided that a current I_0 is injected at position $x = 0$.

Differentiating eq. (IV.2.3) and inserting eq. (IV.2.4), we obtain

$$\frac{\partial V_m^2}{\partial x^2} = -r_i \cdot \frac{\partial i_i}{\partial x} = r_i \cdot i_m. \tag{IV.2.5}$$

The current through the membrane depends on the resistance of the membrane. Just as in eqs. (IV.2.1) and (IV.2.2), the total membrane resistance $\mathcal{R}_m$ of a cylinder with radius a and length Δx can be written as

$$\mathcal{R}_m = R_m \frac{1}{\Delta x \cdot 2\pi a} = r_m \frac{1}{\Delta x}. \tag{IV.2.6}$$

This introduces the specific resistance R_m with units [Ωm^2] and the cable-specific resistance r_m (units [Ωm]). This resistance depends on the radius of the axon.

At constant voltage, the transmembrane current i_m depends on the cable-specific resistance of the membrane such that $i_m = V_m/r_m$ (Ohm's law). When voltage changes in time, the total membrane current is the sum of a capacitive current charging the membrane ($i_c = c_m dV_m/dt$) and the Ohmic current through the membrane.

$$i_m(x,t) = c_m \frac{dV_m(x,t)}{dt} + \frac{V_m(x,t)}{r_m}, \tag{IV.2.7}$$

where the cable-specific capacitance c_m displays the units [F/m] and depends on the radius of the cable. The voltage depends on time t and the position x along the cable.

We finally obtain the cable equation

$$\frac{1}{r_i}\frac{\partial V_m^2(x,t)}{\partial x^2} = c_m \frac{\partial V_m(x,t)}{\partial t} + \frac{V_m(x,t)}{r_m}. \tag{IV.2.8}$$

By introducing the length scale $\lambda = \sqrt{r_m/r_i}$ (unit [m]) and timescale $\tau_m = r_m c_m$ (unit [s]), we can rewrite this equation as

$$\lambda^2 \frac{\partial V_m^2(x,t)}{\partial x^2} = \tau \frac{dV_m(x,t)}{dt} + V_m(x,t). \tag{IV.2.9}$$

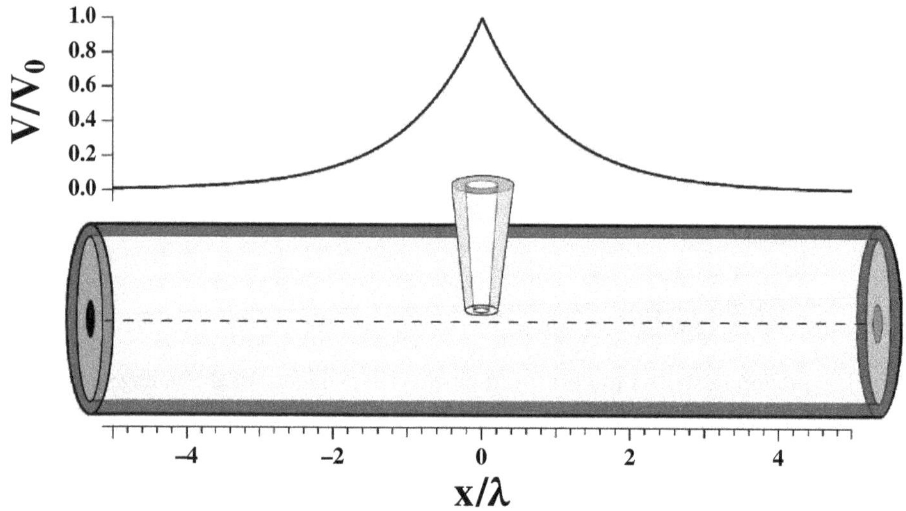

Figure IV.2.3 The voltage decays with distance from the current injection point. The characteristic length $\lambda = \sqrt{r_m/r_i}$ over which the injection of a current is felt depends on the specific resistances R_m and R_i of the membrane and the cytosol.

The solutions are complicated transcendental functions that depend on time and position. In the steady state limit, i.e., long time after the onset of current injection at $x = 0$, the time-dependent term disappears and eq. (IV.2.9) can be written as

$$\lambda^2 \frac{\partial V_m^2(x)}{\partial x^2} = V_m(x) \qquad \text{after long time} \tag{IV.2.10}$$

with the simple single-exponential solution

$$V_m(x) = V_0 \exp\left(-\frac{x}{\lambda}\right), \tag{IV.2.11}$$

see Figure IV.2.3. The constant λ is the decay length of the voltage, and $V_0 \equiv V_m(0)$.

IV.2.3 Entrance Resistance and Conductance of a Cable

If the cable is infinitely extended from $-\infty$ to $+\infty$, we can readily assume that at position $x = 0$, the current is given $i_i(0) = I_0/2$ (half of the total current flows in positive and half in negative direction). In steady state we find according to eqs. (IV.2.3) and (IV.2.11):

$$-r_i i_i(0) = \frac{\partial V_m}{\partial x}(0) = -\frac{1}{\lambda}V_0 \quad \longrightarrow \quad V_0 = \frac{r_i \lambda I_0}{2}. \tag{IV.2.12}$$

Using Ohm's law, we can also define an entrance resistance R_N of the infinitely long neuron (or cable) such that

$$V_0 = R_N i_i(0) \quad \longrightarrow \quad R_N = \frac{V_0}{i_i(0)} = r_i \lambda. \tag{IV.2.13}$$

Let us now introduce specific resistances of the cable, R_i and R_m:

$$R_i = r_i \cdot A = r_i \cdot \pi a^2 \,[\Omega m] \qquad \text{and} \qquad R_m = r_m \cdot 2\pi a \,[\Omega m^2], \tag{IV.2.14}$$

where a is the radius of the cable, and A is the cross-sectional area of the cable. In contrast to r_i and r_m, the specific resistances R_i and R_m are independent of the dimensions of the cable. Further, we introduce a specific capacitance

$$C_m = \frac{c_m}{2\pi a} \qquad \left[\frac{F}{m^2}\right], \tag{IV.2.15}$$

which is also a material property and independent of the dimensions of the cable.

With these definitions, the cable equation assumes the form

$$\frac{2}{aR_i}\frac{\partial V_m^2(x,t)}{\partial x^2} = C_m\frac{\partial V_m(x,t)}{\partial t} + \frac{V_m(x,t)}{R_m}. \tag{IV.2.16}$$

This differential equation is at the core of the Hodgkin–Huxley model. One now has to search for a solution for the space and time dependence of the voltage, $V_m(x,t)$. As obvious from the aforementioned equation, this solution depends on the radius a of the axon. The main modification introduced by Hodgkin and Huxley in 1952 is that the function R_m turns into a complicated function of voltage and time, $R_m(V_m,t)$, that is controlled by voltage-gated ion channel proteins.

The steady-state solution of this equation after infinite time (c.f. eq. (IV.2.11)) is

$$V_m(x) = V_0 \exp\left(-\frac{x}{\lambda}\right) \quad \text{with} \quad V_0 = \sqrt{\frac{R_iR_m}{8\pi a^3}} I_0; \quad \lambda = \sqrt{\frac{R_m a}{2R_i}}. \tag{IV.2.17}$$

Thus, both amplitude, V_0, and the characteristic length of the voltage decay, λ, depend on the radius of the cable.

When injecting a current I_0 at position $x = 0$, the entrance resistance of the cable defined in eq. (IV.2.13) is

$$R_N = r_i\lambda = \sqrt{r_ir_m} = \sqrt{\frac{R_i}{\pi a^2}\frac{R_m}{2\pi a}} = \sqrt{\frac{R_iR_m}{2\pi^2}} \cdot a^{-3/2}. \tag{IV.2.18}$$

The entrance conductance, $G_N = 1/R_N$, is given by

$$G_N = \sqrt{\frac{2\pi^2}{R_iR_m}} \cdot a^{+3/2}. \tag{IV.2.19}$$

Thus, the entrance conductance of a cable depends on its radius. It is proportional to $a^{+3/2}$. The prefactor $\sqrt{\frac{2\pi^2}{R_iR_m}}$ is a constant that is independent of the cable dimensions.

IV.2.4 Rall's Model

Typically, nerves do not consist of single infinite cables. They are rather branched trees of dendrites. Figure IV.2.4 shows the drawing of a pyramid cell drawn by Ramon y Cajal. Can one also apply cable theory to such branched dendritic trees?

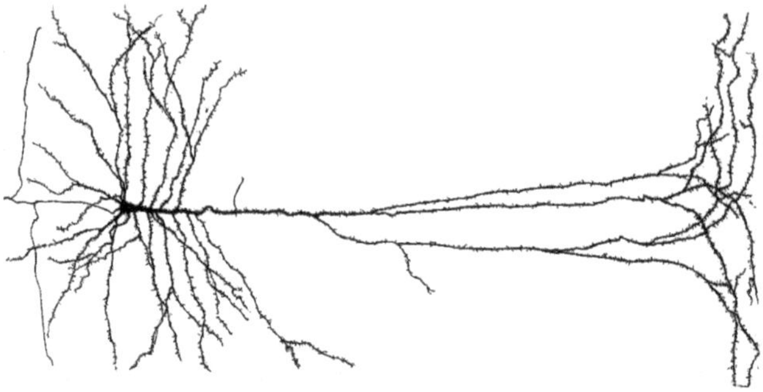

Figure IV.2.4 Giant deep pyramid of the ascending frontal convolution in man (drawing by Ramón y Cajal. 1899/Public Domain).

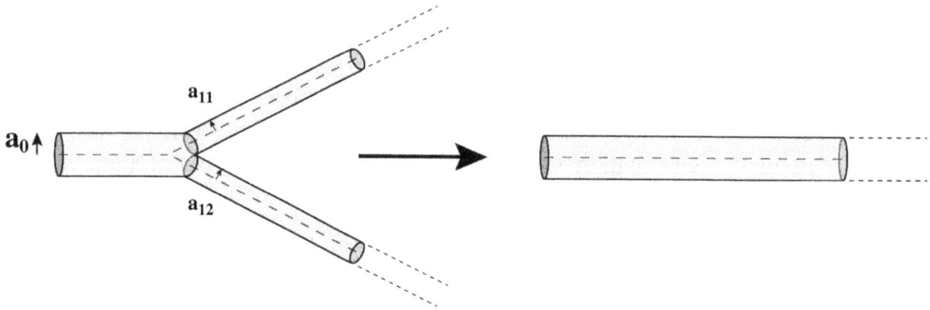

Figure IV.2.5 Left: Branching of an axon with radius a_0 into two dendrites with radius a_{11} and a_{12}. Right: The branched cable can be reduced to a single cable, if the $a^{3/2}$ rule for the entrance conductances is obeyed.

Conductances are additive. Thus, if an axon with radius a_0 branches into two dendrites with radii a_{11} and a_{12} (see Figure IV.2.5), the combined entrance conductances of these dendrites are equal to the entrance conductance of the single axon, if the following relation is obeyed:

$$G_{N,0} = G_{N,11} + G_{N_12} \quad \text{or} \quad a_0^{\frac{3}{2}} = a_{11}^{\frac{3}{2}} + a_{12}^{\frac{3}{2}}. \tag{IV.2.20}$$

More generally, if an axon branches into n infinitely long dendrites, one can reduce the tree to a single cable if

$$G_{N,1} = \sum_i^n G_{N,1i} \quad \text{or} \quad a_0^{\frac{3}{2}} = \sum_i^n a_{1i}^{\frac{3}{2}}. \tag{IV.2.21}$$

This is illustrated in Figure IV.2.5. This requirement for the application of cable theory to dendritic trees is called the "Rall model" [10].

In Figure IV.2.6, we show a dendritic tree with three branching points. In order to be able to reduce it to one "equivalent" cable, one has to demand that:

$$a_{11}^{\frac{3}{2}} = a_{211}^{\frac{3}{2}} + a_{212}^{\frac{3}{2}} + a_{213}^{\frac{3}{2}} \quad \text{and} \quad a_{12}^{\frac{3}{2}} = a_{221}^{\frac{3}{2}} + a_{222}^{\frac{3}{2}}. \tag{IV.2.22}$$

For finite nerves, similar relations can be obtained [10].

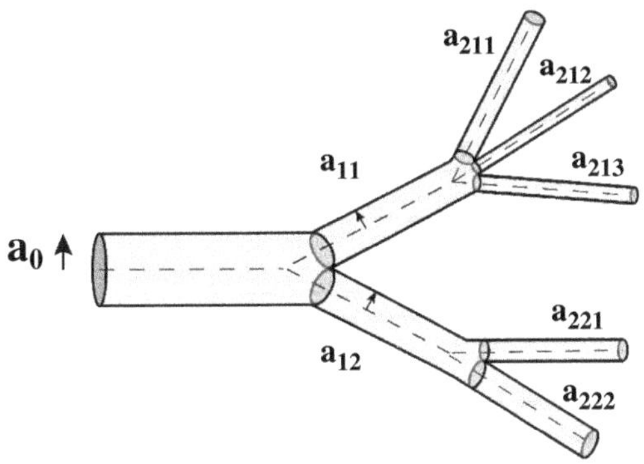

Figure IV.2.6 Dendritic tree with two branching points.

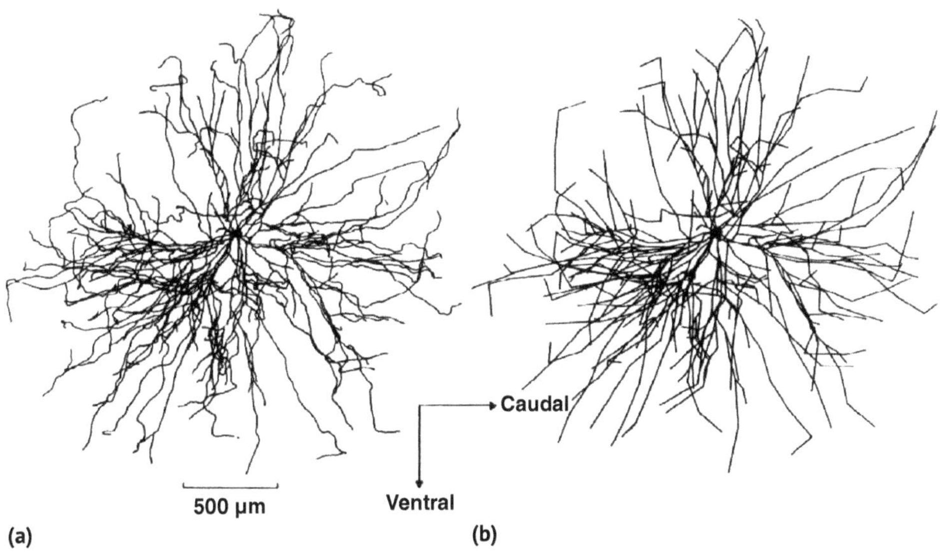

Figure IV.2.7 (a) Projection of the dendritic tree of a motor neuron. (b) Digitized computer representation of the same cell, drawn with the same projection and scale. From [12]/with permission of John Wiley & Sons.

IV.2.4.1 Validity of Rall's Model for Real Nerves

It has been shown experimentally that dendritic trees in real nerves approximately obey the aforementioned relation [11].

In Figure IV.2.7, an image of a motor neuron is shown. The right-hand panel shows a dissection of this tree into cylindrical cable. Figure IV.2.8 shows an analysis of one dendritic tree [12]. It displays the thickness of the individual cables as a function of distance from the soma [13]. The diameter of the cables decreases systematically with increasing distance from the soma after each branching point of the dendritic tree.

Thus, it seems as if one can approximately understand the properties of a dendritic tree by analyzing the properties of a single equivalent cable.

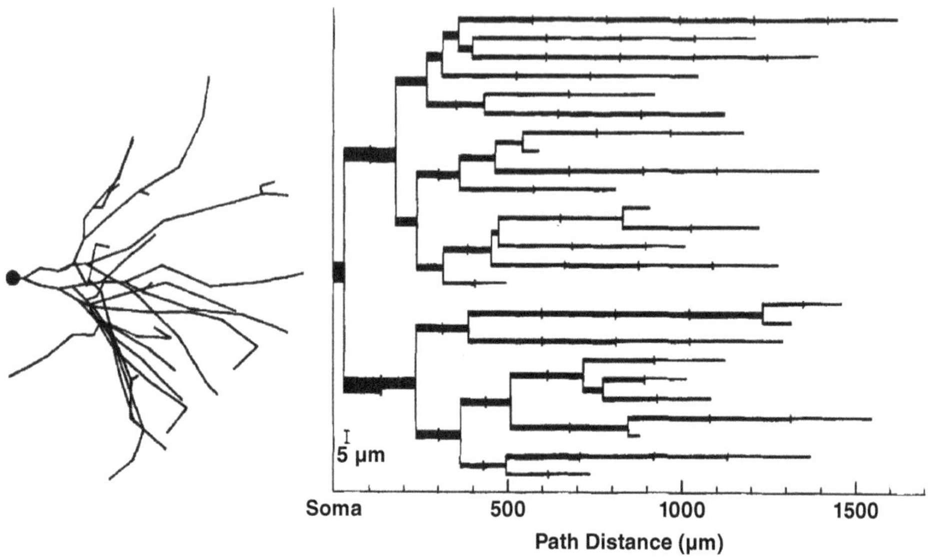

Figure IV.2.8 Left: Two-dimensional projection drawing of a dendritic tree of a motor neuron. Right: "Dendrogram" of the same dendrite showing the cable diameters as a function of the distance from the soma. From [13]/with permission of John Wiley & Sons.

IV.2.5 Summary: Key Ideas of Chapter IV.2

- In cable theory, the static and dynamic properties of a nerve axon are model by considering a hollow membrane cylinder with radius a. The membrane possesses a constant resistance, r_m, and a constant capacitance, c_m. The internal medium is also modeled as a resistor, with a constant resistance r_i.
- A combination of Ohm's law and Kirchhoff's first law leads to the cable equation

$$\frac{1}{r_i}\frac{\partial V_m^2(x,t)}{\partial x^2} = c_m\frac{\partial V_m(x,t)}{\partial t} + \frac{V_m(x,t)}{r_m},$$

  which describes the voltage V_m as a function of space and time.
- When using specific resistances, $R_m = r_m \cdot 2\pi a$ and $R_i = r_i \cdot \pi a^2$, and the specific capacitance, $C_m = \frac{c_m}{2\pi a}$, independent of cable dimensions (a being the radius of the cable), the cable equation assumes the form

$$\frac{2}{aR_i}\frac{\partial V_m^2(x,t)}{\partial x^2} = C_m\frac{\partial V_m(x,t)}{\partial t} + \frac{V_m(x,t)}{R_m}.$$

  The solutions of this equation depend on the radius of the axon. This equation is central to the derivation of the Hodgkin–Huxley model. In this model, the specific membrane resistance, R_m, will be voltage- and time-dependent.

- If an axon with radius a_1 branches into two dendrites with radii a_{11} and a_{12}, the properties of these axons can be condensed into one single equivalent cable with radius a_1 if $a_1^{3/2} = a_{11}^{3/2} + a_{12}^{3/2}$. The assumption that dendritic trees of real nerve cells obey this relation is called "Rall's model."
- Rall's hypothesis was found to be approximately fulfilled in real nerve branches. Thus, the electrical properties of neurons with many branched dendrites can be approximated by a single cable, and the following chapters mainly deal with single cylindrical axons.

IV.3

Voltage Gating

In Chapter IV.2, it was assumed that the neural membrane displays constant resistance and capacitance. However, it was found experimentally for nerves that the resistance or conductance of a nerve membrane does not stay constant when the voltage is changed. Instead, the conductance of a neural membrane is voltage-dependent and changes as a function of time.

IV.3.1 Voltage-clamp Experiments

In the 1930s, Hodgkin and Huxley started using the giant axon of squid for electrophysiological measurements (Figure IV.3.1 (left)). These axons can display diameters of about 1 mm, which makes them perfectly suited to studying the electrical properties of nerve membranes. Due to their large diameter, it is easy to insert electrodes such that the transmembrane currents can be measured (see Figure IV.3.1). For this reason, a significant part of the early literature of neurophysiology deals with squid axons. Figure IV.3.1 (right) shows the action potential of a squid axon [4]. The authors found a voltage pulse of a duration of about 1–2 ms with a voltage amplitude of about 80–100 mV. It displays a large positive signal called a depolarization, followed by a negative phase called a hyperpolarization.

The left panel of Figure IV.3.1 shows a squid axon with inserted electrodes. It allows to perform the so-called voltage-clamp experiments. In these experiments, the voltage difference between the inside and outside of the neuron is kept constant over the whole length of the axon. In such experiments, no action potential can propagate, and only the transmembrane currents are measured. This setup is therefore suitable to investigate the membrane properties in the absence of a propagating pulse. It turns out that the description of the membrane currents reveals the information that is necessary to describe the propagating pulse.

Figure IV.3.2 also shows the insertion of electrodes into a squid axon [14]. The right-hand panel shows voltage-clamp experiments performed on squid axons embedded in seawater. The voltage is changed in a step from the resting potential to the voltage indicated next to the traces. One can recognize that there is a first phase where the current flows against the applied voltage difference and a second phase where the current flows along the

Physics of Nerves and Excitatory Membranes, First Edition. Thomas Heimburg.

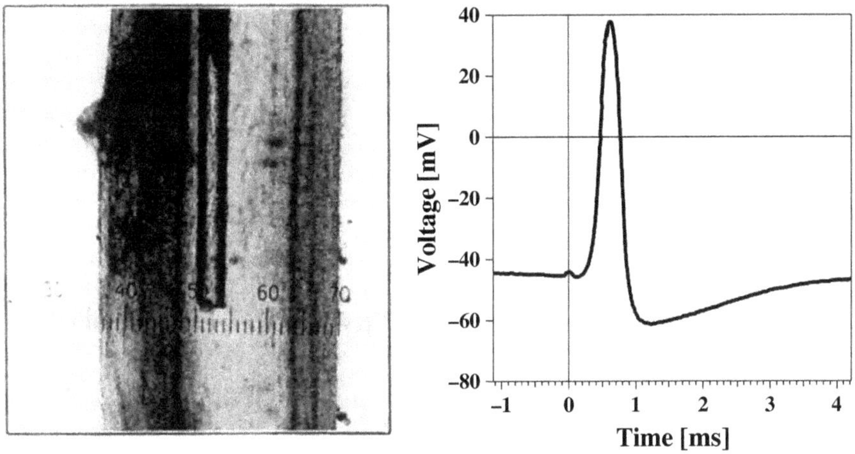

Figure IV.3.1 Left: Giant squid axon with an inserted electrode. Each segment of the scale corresponds to 33 μm [4]/with permission of Springer Nature. Right: Action potential in the squid axon. Adapted from [4].

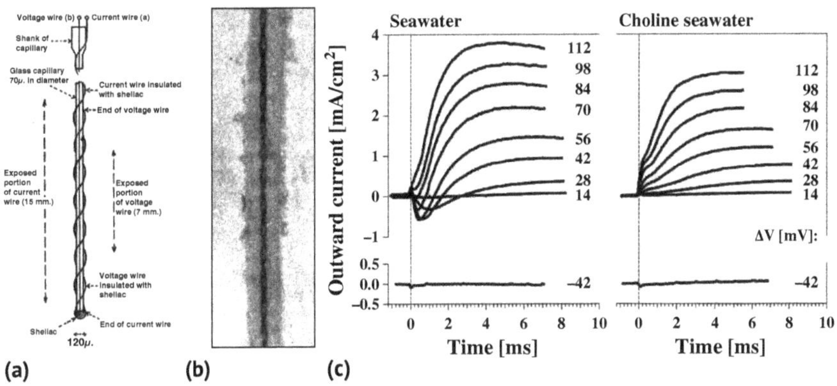

Figure IV.3.2 The voltage-clamp experiment of Hodgkin, Huxley, and Katz. (a) Micro-electrodes wound around a glass capillary. [14]/with permission of John Wiley & Sons. (b) Squid axon (diameter 600 μm) with a micro-electrode inserted [14]/with permission of John Wiley & Sons. (c) Currents through the membrane of a squid axon after depolarizing voltage jumps from the resting potential of the squid axon. In seawater, the current has two phases – a negative phase against the applied field and a positive current along the applied field. In choline water (free of sodium), the negative current contribution does not show up. Data were recorded at 11 °C. Adapted from [15].

voltage gradient. On the right-hand side, the experiment is repeated in choline seawater, i.e., water with the composition of seawater where sodium is replaced by choline. In these experiments, the negative phase cannot be seen. This suggests that the negative current is related to the sodium in the external medium. Figure IV.3.3 shows two traces recorded at $\Delta V = 42\,\text{mV}$. One can well recognize the negative and positive phases of the current (solid trace) and the disappearance of the negative phase of the current in choline seawater.

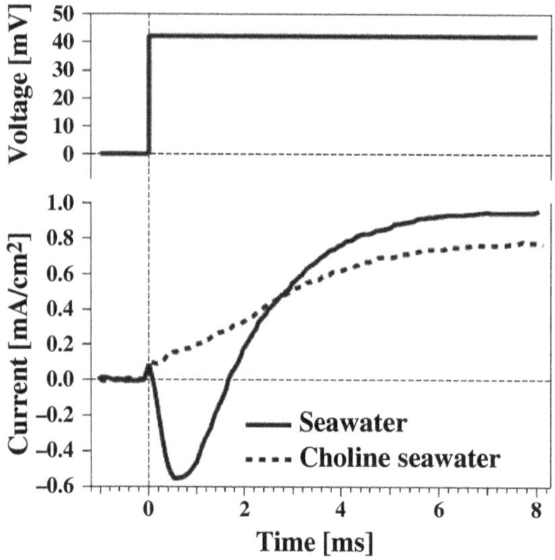

Figure IV.3.3 Current through a squid axon at 11 °C after a voltage jump of 42 mV from the resting potential. Top: Voltage change as a function of time. Bottom: Current in seawater (solid line) and in choline seawater (dashed line). In the latter trace, the inward current disappears. From Figure IV.3.2.

IV.3.2 The Gate Model of Hodgkin and Huxley

In cable theory, the conductance of the membrane is assumed to be constant. However, it is obvious from the previous section that the conductance of the squid axon membrane is not constant but is rather a function of voltage and time. In order to describe the nerve pulse, Hodgkin and Huxley used the following strategy: They adapted the concepts of cable theory but replaced the constant membrane conductance by the parametrized voltage- and time-dependent conductance for sodium and potassium from the voltage-clamp experiment described in the previous section. It turns out that this yields a satisfactory description of the propagating nerve pulse.

To describe the voltage- and time-dependent conductance of the squid axon membrane for potassium and sodium, Hodgkin and Huxley assumed that there exists a first-order reaction between a closed and an open gate in the membrane:

$$\text{closed} \underset{\beta}{\overset{\alpha}{\rightleftharpoons}} \text{open} \tag{IV.3.1}$$

If the gate is open, a current can flow. If the gate is closed, the current is blocked. The rate constant of transition from the open to the closed state shall be given by α and the rate constant from closed to open state by β.

The process of opening a pore will be mediated by a gating particle associated with the gate, which carries a charge ze_0 or zF in molar units, corresponding to an electrostatic free energy difference across the membrane of $zF\Delta V_m$. The variable y shall describe the probability of finding an open channel. The rate constants α and β depend on voltage but not on time: $\alpha = \alpha(V_m)$ and $\beta = \beta(V_m)$. Let us assume that the membrane displays the open gate probability y_0 at a fixed voltage, ΔV_0. At $t = 0$, the voltage difference across a membrane is suddenly changed to a new voltage, ΔV. The time dependence with which the fraction of open gates changes can be determined from the differential equation:

$$\frac{dy}{dt} = \alpha(V_m) \cdot (1 - y) - \beta(V_m) \cdot y. \tag{IV.3.2}$$

The time-dependent solution of this equation is given by:

$$\begin{aligned} y(V_m, t) &= y_\infty - (y_\infty - y_0) \exp\left(-\left[\alpha(V_m) + \beta(V_m)\right] t\right) \\ &\equiv y_\infty - (y_\infty - y_0) \exp\left(-\frac{t}{\tau}\right), \end{aligned} \tag{IV.3.3}$$

where y_∞ is the probability of finding an open state in steady state after infinite time when $dy/dt = 0$, and τ is a time constant. The steady-state open probability is given by

$$y_\infty(V_m) \stackrel{\text{eq.IV.3.2}}{=} \frac{\alpha(V_m)}{\alpha(V_m) + \beta(V_m)}. \tag{IV.3.4}$$

The time constant τ in eq. (IV.3.3) of the single-exponential relaxation process is given by

$$\tau(V_m) = \frac{1}{\alpha(V_m) + \beta(V_m)} \tag{IV.3.5}$$

and also only depends on voltage. The function $y(V, t)$ is shown in Figure IV.3.4 (dashed line).

If the pore possesses k independent identical gates in sequence, the open probability is given by the probability that all gates are open at the same time is given by

$$Y(V_m, t) = \left[y(V_m, t)\right]^k. \tag{IV.3.6}$$

For the example $k = 4$, this is shown in Figure IV.3.4 (solid line). Equation (IV.3.6) is based on the assumption that the probability to finding all gating particles on one side is equal to the product of the individual probabilities for each particle. A time dependence described by four independent gating particles is assumed as follows for the potassium conductance.

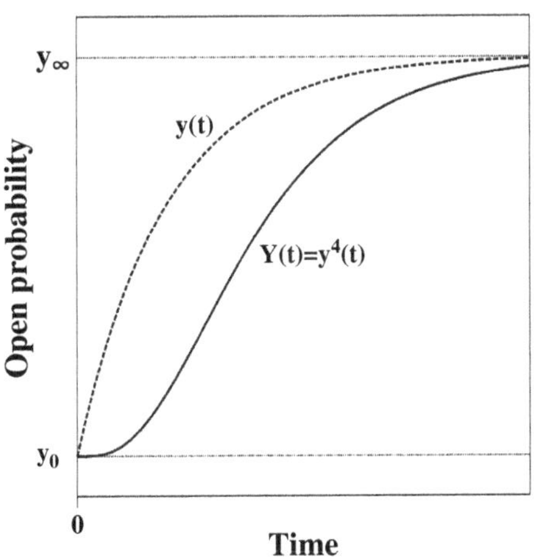

Figure IV.3.4 Time dependence of the probability y and of $Y(t) = y^4(t)$ to find an open state of a pore with four gating particles.

IV.3.3 The Parametrization of Potassium and Sodium Conductance

Hodgkin and Huxley made use of the pore open probabilities described in the previous section. They described a current through a pore

$$I = g(V_m, t)\,(V_m - E) = g_0 Y(V_m, t)(V_m - E), \tag{IV.3.7}$$

where g_0 is the conductance of an open pore, $Y(V_m, t)$ the voltage and time dependence of finding an open pore. E is the Nernst potential, i.e., the voltage created by the difference in ion concentration across the membrane:

$$E = \frac{RT}{zF} \ln \frac{c_{out}}{c_{in}}, \tag{IV.3.8}$$

where c_{in} is the concentration of the ion inside the axon, and c_{out} is the concentration outside the axon. Consequently, the potassium and sodium currents through the membrane are given by

$$\begin{aligned} I_K &= g_{K,0}\ Y_K(V_m, t)\,(V_m - E_K) \qquad \text{and} \\ I_{Na} &= g_{Na,0}\ Y_{Na}(V_m, t)\,(V_m - E_{Na}). \end{aligned} \tag{IV.3.9}$$

In Section IV.3.1, it was shown that the positive current associated with a potassium conductance is not described by a single exponential function but displays a sigmoidal appearance. For this reason, Hodgkin and Huxley described the open probability $Y_K(V_m, t)$t with a gating variable $n(V_m, t)$ (instead of $y(V_m, t)$) and four identical gates, i.e., $k = 4$. The time course of the potassium current is then given by

$$I_K = g_{K,0}\ n^4(V_m, t)\,(V_m - E_K). \tag{IV.3.10}$$

The function $n(V_m, t)$ is determined by rate constants $\alpha_n(V_m)$ and $\beta_n(V_m)$ as described in the previous section.

The negative current in Section IV.3.1, which was associated with the sodium current, shows a sigmoidal rise and then a decay back to zero. Hodgkin and Huxley decided to fit the open probability Y_{Na} with two different gating variables, $m(V_m, t)$ and $h(V_m, t)$, where the m-gate is initially mostly closed and opens upon depolarization ($y_\infty > y_0$), while the h-gate is initially mostly open and closes upon depolarization ($y_\infty < y_0$). They described the sodium by

$$I_{Na} = g_{Na,0}\ m^3(V_m, t)\,h(V_m, t)\,(V_m - E_{Na}). \tag{IV.3.11}$$

The function $m(V, t)$ is determined by rate constants $\alpha_m(V)$ and $\beta_m(V)$, and $h(V, t)$ is determined by rate constants $\alpha_h(V)$ and $\beta_h(V)$, respectively:

$$\begin{aligned} n(V_m, t) &= n_\infty(V_m) - (n_\infty(V_m) - n_\infty(V_{m,0})) \cdot \exp\left(-\frac{t}{\tau_n(V_m)}\right), \\ m(V_m, t) &= m_\infty(V_m) - (m_\infty(V_m) - m_\infty(V_{m,0})) \cdot \exp\left(-\frac{t}{\tau_m(V_m)}\right), \\ h(V_m, t) &= h_\infty(V_m) - (h_\infty(V_m) - h_\infty(V_{m,0})) \cdot \exp\left(-\frac{t}{\tau_h(V_m)}\right), \end{aligned} \tag{IV.3.12}$$

where the relaxation time $\tau_n(V)$ and the steady-state values of n, $n_\infty(V)$, and $n_\infty(V_0)$ are given by

$$
\begin{aligned}
n_\infty(V_m) &= \frac{\alpha_n(V_m)}{\alpha_n(V_m)+\beta_n(V_m)} \quad &\text{and} \quad \tau_n(V_m) &= \frac{1}{\alpha_n(V_m)+\beta_n(V_m)},\\
m_\infty(V_m) &= \frac{\alpha_m(V_m)}{\alpha_m(V_m)+\beta_m(V_m)} \quad &\text{and} \quad \tau_m(V_m) &= \frac{1}{\alpha_m(V_m)+\beta_m(V_m)},\\
h_\infty(V_m) &= \frac{\alpha_h(V_m)}{\alpha_h(V_m)+\beta_h(V_m)} \quad &\text{and} \quad \tau_h(V_m) &= \frac{1}{\alpha_h(V_m)+\beta_h(V_m)}.
\end{aligned} \tag{IV.3.13}
$$

Hodgkin and Huxley now fitted the time- and voltage-dependent profiles in Section IV.3.1 in order to obtain values for the six rate constants, $\alpha_n(V)$, $\beta_n(V)$, $\alpha_m(V)$, $\beta_m(V)$, $\alpha_h(V)$, and $\beta_h(V)$, as a function of voltage. For a temperature of 6.3 °C, they obtain:

$$
\begin{aligned}
\alpha_n(V_m) &= \frac{10^4\cdot(V_m+0.055)}{1-\exp\left(-\frac{V_m+0.055}{0.010}\right)}; \quad &\beta_n(V_m) &= 125\cdot\exp\left(-\frac{V_m+0.065}{0.08}\right),\\
\alpha_m(V_m) &= \frac{10^5\cdot(V_m+0.040)}{1-\exp\left(-\frac{V_m+0.040}{0.010}\right)}; \quad &\beta_m(V_m) &= 4000\cdot\exp\left(-\frac{V_m+0.065}{0.018}\right),\\
\alpha_h(V_m) &= 70\cdot\exp\left(-\frac{V_m+0.065}{0.020}\right); \quad &\beta_h(V_m) &= \frac{1000}{1-\exp\left(-\frac{V_m+0.035}{0.010}\right)},
\end{aligned} \tag{IV.3.14}
$$

where the transmembrane voltage V_m is given in units of [V].[1] Here, the functions $\alpha_i(V_m)$ and $\beta_i(V_m)$ were not derived from first principles but rather parametrized from the experimental voltage-clamp data [14, 15]. The voltage dependence of the rate constants α_i and β_i is shown in Figure IV.3.5.

The open probabilities Y_K and Y_{Na} of the potassium and sodium currents can now be calculated by applying eqs. (IV.3.10), (IV.3.14), and (IV.3.11). This is shown in Figure IV.3.6.

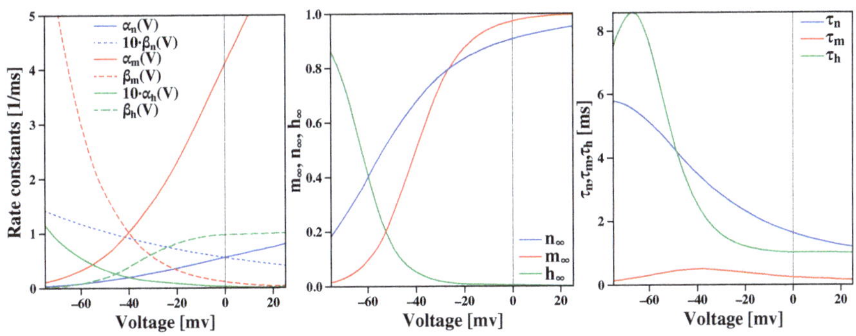

Figure IV.3.5 The six rate constants (left), the steady-state values of the the gating variables n_∞, m_∞, and h_∞ (center), and the relaxation time constants of the gating particles n, m, and h (right) as a function of voltage for a temperature of 6.3 °C.

1 Note that in [16], the voltage was defined relative to the resting potential of −65 mV, which yields different values in eq. (IV.3.14).

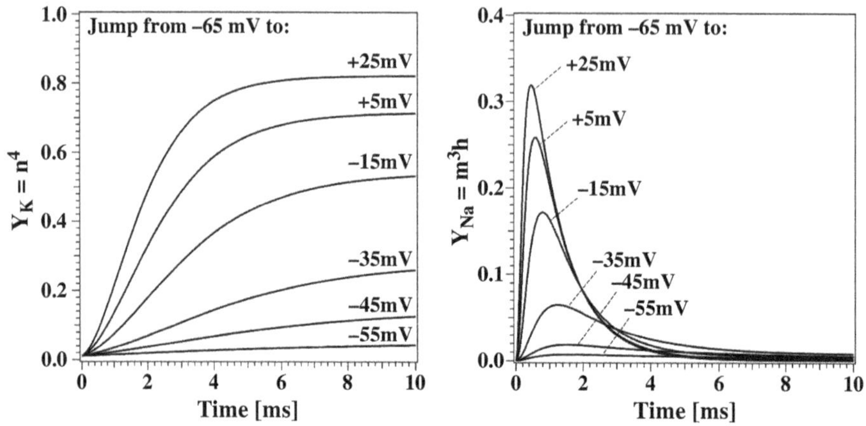

Figure IV.3.6 The open probability as a function of time for the potassium channel (left) and the sodium channel (right) for a voltage jump from −65 mV to the indicated voltage. Calculated for a temperature of 6.3 °C.

Finally, the potassium current I_K, the sodium I_{Na}, and the total current I_{tot} after a voltage jump from the resting potential can be determined by

$$\begin{aligned} I_K &= g_{K,0} Y_K (V_m - E_K), \\ I_{Na} &= g_{Na,0} Y_{Na} (V_m - E_{Na}), \\ I_{tot} &= I_K + I_{Na} + I_L. \end{aligned} \tag{IV.3.15}$$

Here, $E_K = -77$ mV, $E_{Na} = +50$ mV, and the resting potential is $E_R = -65$ mV, $g_{K,0} = 36$ mS/cm^2, and $g_{Na,0} = 120$ mS/cm^2. Hodgkin and Huxley also included a leak current, $I_L = g_L (V_m - E_L)$, with $g_L = 0.3$ mS/cm^2 and a leak conductance that is independent of voltage and time [16].

In Figure IV.3.7, the calculated voltage-clamp data at 6.3 °C are compared with the experimental data recorded at 4 °C. While the temperatures are not exactly the same, the results are quite similar. The right-hand panel shows the ionic membrane current for six different voltage jumps from the resting potential of −65 mV. This calculation may be compared with Figure IV.3.2 (which was recorded at 11 °C).

Summarizing it can be stated that the parametrization introduced in this section describes the ionic currents in a quite satisfactory manner. It is clear that the macroscopic membrane currents correlate with the properties of the action potential. However, this does not mean that the assumptions made to achieve this parametrization are correct. This was pointed out by Hodgkin and Huxley [16]:

> *"The agreement must not be taken as evidence that our equations are anything more than an empirical description of the time-course of the changes in permeability to sodium and potassium. An equally satisfactory description of the voltage clamp data could no doubt have been achieved with equations of very different form, which would probably have been equally successful in predicting the electrical behaviour of the membrane. ... the success of the equations is no evidence in favour of the mechanism of permeability change that we tentatively had in mind when formulating them. "*

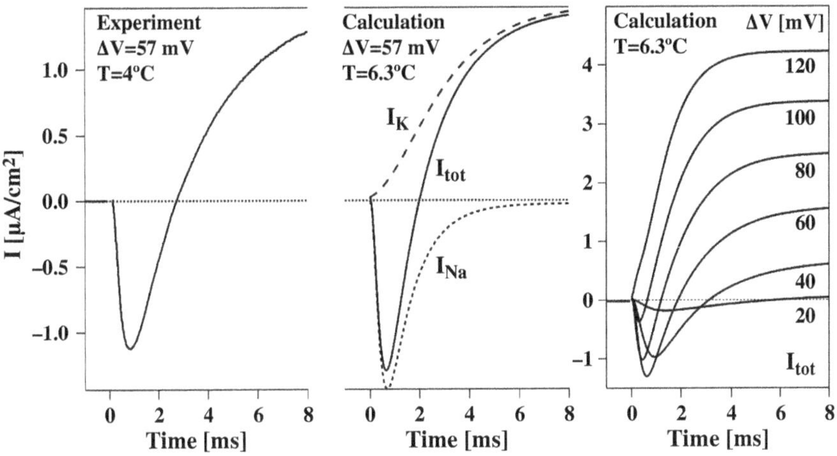

Figure IV.3.7 Left: Experimentally measured current through a squid axon after a voltage jump of 57 mV and a temperature of 4 °C. Adapted from [16]. Right: Calculated using the aforementioned equations for a temperature of 6.3 °C.

IV.3.4 Summary: Key Ideas of Chapter IV.3

This chapter describes the voltage and time dependence of currents through the squid axon membrane, which is called Hodgkin and Huxley's gate model.

- The surprising finding is that one can measure currents after a voltage change across the membrane that first are directed against the voltage gradient, and later along the voltage gradient.
- This was explained by the existence of two different currents carried by potassium and sodium ions that both flow along their gradients. A change in voltage changes the permeability of the membrane for these ions in a voltage- and time-dependent manner. This hypothesis is supported by experiments in choline seawater that does not contain sodium, and by the use of toxins that selectively "block" inward or outward currents.
- Hodgkin and Huxley rationalize this finding by assuming "gates" in the membrane, which exists in an open and closed state. The probability of being in an open state is described by the variable $y(V_m, t)$ that depends on voltage and time.
- The gate can undergo a transition between closed and open described by $\frac{dy}{dt} = \alpha \cdot (1 - y) - \beta \cdot y$. The α and β are voltage-dependent rate constants that are determined by the magnitude and position of an activation barrier located at a position δ in the membrane. They depend on voltage. The open probability y thereby becomes a function and voltage and time, $y(V, t)$.
- After a voltage change, a new equilibrium position is assumed, which is denoted by $y_\infty(V)$.
- The new equilibrium is adopted in a single-exponential equilibration process with a time constant τ.

- In order to analytically describe the experimentally determined potassium and sodium conductances, Hodgkin and Huxley described open probability of the potassium conductance Y_K by a gating variable n such that

$$Y_K = n^4(V_m, t), \tag{IV.3.16}$$

  assuming four identical and independent gates in the same pore (see eq. (IV.3.10)). In order to obtain an open pore, all four gates have to be open. The potassium gates are described by rate constants $\alpha_n(V)$ and $\beta_n(V)$ that depend on voltage only.
- In order to describe the sodium conductance, assume that the open probability of the sodium pore, Y_{Na}, is described by

$$Y_{Na} = m^3(V_m, t) \cdot h(V_m, t), \tag{IV.3.17}$$

  Here, it is assumed that the sodium conductance is described by three identical and independent gates $m(V, t)$ that open upon depolarization, and one gate $h(V, t)$ that closes upon depolarization. The sodium gates are described by rate constants $\alpha_m(V)$, $\beta_m(V)$, $\alpha_h(V)$, and $\beta_h(V)$ that depend on voltage only.
- Hodgkin and Huxley did not provide a theory for the rate constants. Instead, they fitted them by using empirical functions with many different parameters. This is given in eq. (IV.3.14).
- Quite satisfactory descriptions of potassium and sodium conductances as a function of voltage and time are obtained.

Summarizing, the analytical description of the potassium and sodium conductances is a fit to experimental data that involves many parameters. The assumption of gates and gating particles is a working hypothesis that helps to parametrize the conductances.

IV.4

The Hodgkin–Huxley Model

IV.4.1 The Membrane Current of a Nerve

The membrane current is given by the sum of the capacitive current and the ionic current, which is

$$\begin{aligned} I_m &= C_m \cdot \frac{dV_m}{dt} + I_K + I_{Na} + I_L \\ &= C_m \cdot \frac{dV_m}{dt} + g_K(V_m, t)\,(V_m - E_K) + \\ &+ g_{Na}(V_m, t)(V_m - E_{Na}) + g_L(V_m - E_L), \end{aligned} \tag{IV.4.1}$$

with $V_m(x, t)$ being a function of position x and time t, and C_m is the specific capacitance.

In cable theory, we found the following relationship (eq. (IV.2.5)):

$$i_m = \frac{1}{r_i}\frac{\partial^2 V_m}{\partial x^2} \quad \longrightarrow \quad I_m = \frac{a}{2R_i}\frac{\partial^2 V_m}{\partial x^2} \qquad \text{units in } [\mathrm{A/m^2}], \tag{IV.4.2}$$

making use of $r_i = R_i/\pi a^2$ and $i_m = I_m \cdot 2\pi a$, where r_i is the cable-specific internal resistance along the axon, R_i is the corresponding specific resistance, i_m is the cable-specific membrane current, and I_m is the specific current in $\mathrm{A/m^2}$. Hodgkin and Huxley also considered a small leak current with a constant leak conductance g_L that is independent of voltage and time [16].

IV.4.2 The Propagating Pulse

When eq. (IV.4.1) is inserted into eq. (IV.4.2), one obtains a wave equation for the action potential:

$$\begin{aligned} \frac{a}{2R_i}\frac{\partial^2 V_m}{\partial x^2} &= C_m \frac{\partial V_m}{\partial t} + g_K(V_m, t)\,(V_m - E_K) + \\ &+ g_{Na}(V_m, t)\,(V_m - E_K) + g_L(V - E_L). \end{aligned} \tag{IV.4.3}$$

This equation is not easy to solve because I_K and I_{Na} are complicated nonlinear functions of voltage and time (see Section IV.3.3).

Physics of Nerves and Excitatory Membranes, First Edition. Thomas Heimburg.

If one assumes that the action potential may propagate with constant shape and velocity, one can make use of the following relation between position and time $x = \theta \cdot t$ leading to

$$\frac{\partial^2 V_m}{\partial x^2} = \frac{1}{\theta^2}\frac{\partial^2 V_m}{\partial t^2}, \tag{IV.4.4}$$

where θ is the propagation velocity of the action potential. Now eq. (IV.4.3) can be simplified to

$$\begin{aligned}\frac{a}{2R_i\theta^2}\frac{\partial^2 V_m}{\partial t^2} &= C_m\frac{\partial V_m}{\partial t} + g_K(V_m,t)(V_m - E_K) + \\ &+ g_{Na}(V_m,t)(V_m - E_{Na}) + g_L(V_m - E_L),\end{aligned} \tag{IV.4.5}$$

which now only contains voltage and time as variables. The conductances $g_K(V_m, t)$ and $g_{Na}(V_m, t)$ were determined in Section IV.3.3. The leak conductance g_L is independent of voltage and time and leads to a small correction term. In order to obtain the action potential, one has to solve the following coupled set of equations consisting of eq. (IV.4.5) and three equations for the three gating particles n, m, and h originating from eq. (IV.3.2):

$$\begin{aligned}\frac{dn}{dt} &= \alpha_n \cdot (1-n) - \beta_n \cdot n \\ \frac{dm}{dt} &= \alpha_m \cdot (1-m) - \beta_m \cdot m \\ \frac{dh}{dt} &= \alpha_h \cdot (1-h) - \beta_h \cdot h,\end{aligned} \tag{IV.4.6}$$

where the voltage-dependent functions α_i and β_i are given by eq. (IV.3.14). In Figure IV.4.1, a calculated profile of an action potential is compared to an experimental profile. The

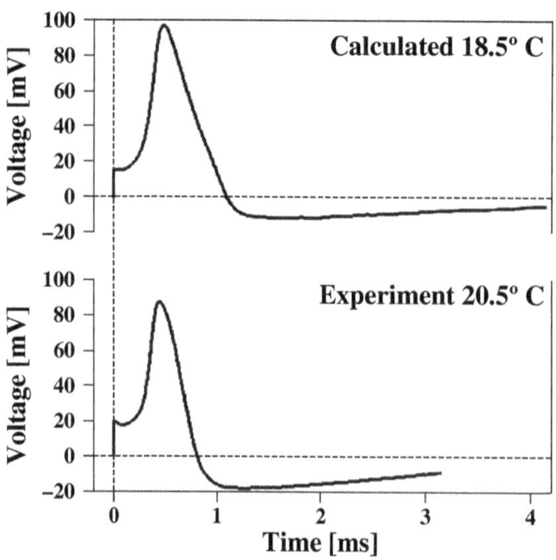

Figure IV.4.1 The action potential relative to the resting potential as determined by Hodgkin and Huxley: Top: Calculated for a temperature of 18.5°C. Bottom: Measured at a temperature of 20.5°C. One can see a voltage depolarization of about 100 mV and a subsequent hyperpolarization under the baseline. Adapted from [16].

profiles are quite similar. With the help of these equations, the pulse shape can be determined by stepwise integration.

Readers trying to reproduce this can make use of software packages such as "NEURON" (https:www.neuron.yale.edu/neuron).

IV.4.2.1 The Pulse Velocity

Hodgkin and Huxley pointed out that only for one value of the velocity θ, the pulse profile returns to the baseline as found experimentally. The propagation velocity of a squid axon was determined to be ~20 m/s. Equation (IV.4.7) indicates that the solution of the differential equation leads to the same result if

$$\frac{a}{2R_i\theta^2} = \text{const.} \quad \longrightarrow \quad \theta \propto \sqrt{a}. \tag{IV.4.7}$$

Thus, the propagation velocity θ, which leads to the solution of the differential equation, depends on the square root of the radius a of the axon. At a fixed temperature, the shape of the action potential in time is the same for all radii. The temperature dependence enters into the picture via the temperature dependence of the rate constants α_i and β_i.

This radius dependence originates from the dimensions of the axon and is unrelated to the voltage and time dependence of the channels. Thus, the conduction velocity of an axon with a given radius is half of that of an axon with four times this radius. This is approximately true for squid axons. However, it is not true for myelinated nerves and earthworm axons (see Section III.3.2 for experimental details).

IV.4.2.2 Hyperpolarization

The action potential in a squid axon consists of a depolarization of the nerve potential from the negative resting potential E_R, i.e., $V_m > E_R$ (Figure IV.4.2 demonstrates this for the data of Hodgkin and Huxley from 1939 where $E_R = -45$ mV [4]). The depolarization is followed by a hyperpolarization phase, where $V_m < E_R$. The Hodgkin–Huxley model describes this behavior for the giant squid axon. The hyperpolarization phase is difficult to reconcile with Bernstein's theory, which assumed an unspecific breakdown of membrane resistance during the action potential [1, 2], but it is consistent with the idea that the breakdown in resistance is carried by two different conductors for sodium and potassium, respectively, with different time and voltage characteristics. It should be noted that a hyperpolarization phase as seen in squids is quite rare. Most nerves do not show the hyperpolarization phase (e.g., lobster connective neurons [17]).

IV.4.2.3 Threshold and All-or-nothing Spiking

In many experiments, there exists a minimum stimulation amplitude before the axon fires. This is called an excitation threshold. Furthermore, the voltage will reach a maximum amplitude of about 100 mV above the resting potential. These phenomena are well described by Noble's model [18], which represents a simplification of the Hodgkin–Huxley model that can help to understand the maximum amplitude of the pulse, the existence of a threshold and the refractory period.

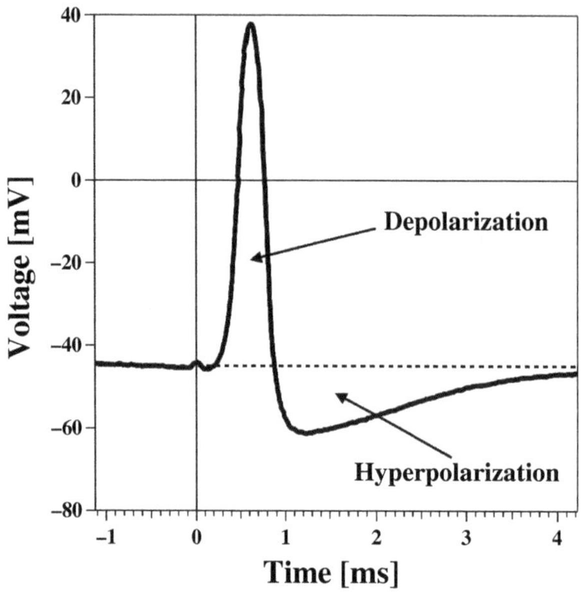

Figure IV.4.2 Depolarization and hyperpolarization. Adapted from [4].

In Noble's model, the emergence of a stimulation threshold is explained by a critical stimulation voltage V_{crit} at which sodium and potassium currents add up to zero. At lower voltages, $V_m < V_{crit}$, the total current is dominated by potassium currents from inside to outside that changes the potential back to the resting potential. No pulse will emerge. At higher voltages, $V_m > V_{crit}$, the current is dominated by sodium currents that are flowing from outside to inside. They lead to a further depolarization of the axon until a maximum pulse amplitude slightly below the Nernst potential of sodium is obtained, at which the net current is zero again and no further depolarization can occur.

The existence of a threshold results in an all-or-none behavior of the action potential after stimulation. Either the signal spikes or there is no stimulation. There are no action potentials with half-amplitude. An example for this is the all-or-none stimulation in earthworm axons discussed in Section III.2.2.

IV.4.2.4 Refractory Period in the Hodgkin–Huxley Model

The refractory period is a period after the passing of an action potential during which one cannot stimulate another action potential. It is described in Section III.2.3.

Hodgkin and Huxley explain this as follows: an action potential starts with an influx of sodium into the axon followed by an outflux of potassium. The sodium current is a function of the fast opening of m-gates and the slow closing of h-gates. After firing of a pulse, the h-gates will be closed and will now relax back to the equilibrium value. In order for sodium to diffuse through the membrane, the h-gates have to be open, which again is a slow process, especially at voltages close to the resting potential. As can be seen in Figure IV.3.5, the relaxation time at voltages close to the resting potential is up to 8 ms, which means that it

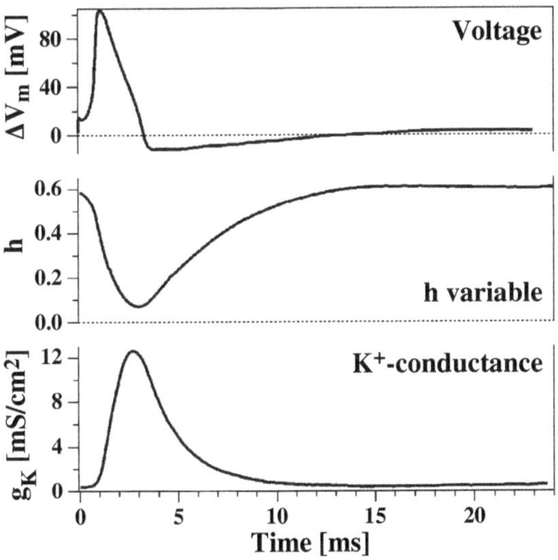

Figure IV.4.3 Voltage, open probability of h particles, and potassium conductance g_K of the squid axon at 6 °C as calculated by Hodgkin and Huxley. The time course of h and g_K extends over a longer time than the depolarization phase of the action potential. Adapted from [16].

takes up to 8 ms for the sodium channels to be excitable again. This corresponds to up to four times the pulse width of 1–2 ms in the squid, during which time the nerve cannot be excited. Figure IV.4.3 shows calculated data for the squid axon at 6 °C [16], which shows that the time course of potassium conductance and h open probability extends over a period of up to 8 ms while the depolarization phase extends over only one-third of that time. During the extended time period, no excitation is possible.

IV.4.3 Free Energy Dissipation During the Action Potential

The ionic fluxes proposed in the Hodgkin–Huxley model represent a very inefficient process for creating an action potential as discussed in depth in [19]. Figure IV.4.4 (left) shows the time course of sodium and potassium currents calculated by Hodgkin and Huxley in their seminal paper [16] (convention: current from inside to outside is positive). One finds that they flow in opposite directions with significant temporal overlap. Only the sum of these currents is the ionic current contributing to the charging of the capacitor (gray-shaded region). The largest part of the ion flux does not contribute to changing the voltage and charging the capacitor because it consists of the flux of positive ions in opposite directions. Figure IV.4.4 (center) shows the integral of the currents, i.e., the total charge transported from inside to outside during the action potential. The free energy change of each ion following its gradient from one side of the membrane to the other is given by

$$\begin{aligned} \Delta\mu_K &= kT\ln[\frac{\mathrm{K}_{in}]}{[\mathrm{K}_{out}]} \quad \text{for potassium,} \\ \Delta\mu_{Na} &= kT\ln(\frac{[\mathrm{Na}_{out}]}{[\mathrm{Na}_{in}]} \quad \text{for sodium,} \end{aligned} \tag{IV.4.8}$$

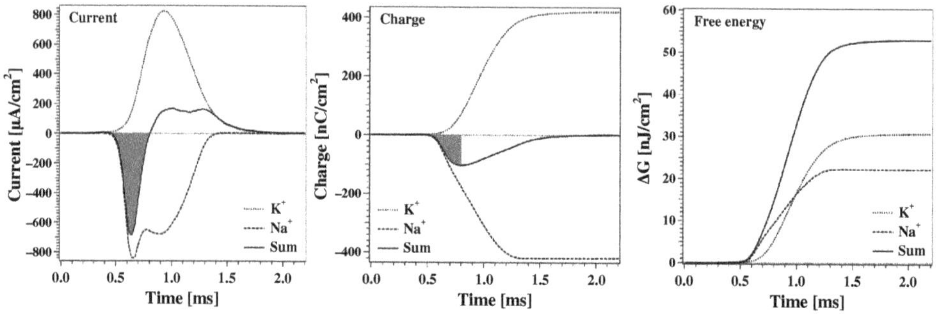

Figure IV.4.4 Left: Predicted potassium and sodium currents during the action potential in squid in the Hodgkin–Huxley model. The data were extracted from the original paper Adapted from [16]. Since potassium and sodium currents have opposite directions, their associated currents mostly compensate each other. Only the current in the gray-shaded region can be used to charge the membrane capacitor. Center: The integral over the currents yields the total number of charges flowing across the membrane. Only the sum of potassium and sodium charges is associated to charging the membrane. It adds to zero after the action potential. Right: Since the concentration differences and the Nernst potentials of potassium and sodium are known, one can calculate the free energy dissipation during the action potential. Since the free energy change displays the same sign for potassium and sodium currents, the dissipation is substantial. [19]/with permission of Elsevier.

which has to be multiplied by the number of ions that flow across the membrane. These numbers can be calculated from the sodium and potassium currents given in the left panel of Figure IV.4.4. The total number of ions flowing across the membrane during an action potential is given in the center panel. We take the textbook values $[K_{in}] = 400$ mM, $[K_{out}] = 20$ mM, $[Na_{in}] = 50$ mM, and $[Na_{out}] = 440$ mM [9]. The dissipated free energy per cm^2 for each ion species and the sum of both is shown in Figure IV.4.4 (right). Even though potassium and sodium currents flow in opposite directions, they both follow the concentration gradient. Therefore, both fluxes dissipate free energy. The maximum of the solid line yields

$$\Delta G_{diss} = 52.9 \frac{\text{nJ}}{\text{cm}^2}. \tag{IV.4.9}$$

This is the free energy dissipated per cm^2 for each segment of the active axon. Therefore, this number has to be multiplied with the complete surface area of the axon in order to obtain the free energy dissipation when an action potential travels from one end to the opposite end of an axon.

Since the total number of positive ions flowing out of the axon (potassium) and flowing into the axon (sodium) adds up to zero, there is no net change of the number of charges on the two sides after completion of the action potential and the membrane voltage returns to the resting potential. Therefore, we do not have to consider electrostatics in the calculation of the dissipation of free energy.

The action potential consists of a charged section of the nerve membrane. In order to calculate how efficient the Hodgkin–Huxley process is, we have to compare this free energy with the free energy necessary to charge the membrane capacitor. The membrane changes its potential by 100 mV relative to the resting potential. The capacitance is assumed being

$C_m = 1\mu\text{F/cm}^2$. This yields a free energy of the charged capacitor at the maximum of the action potential of

$$\Delta G_{cap} = \frac{1}{2} C_m V_{max}^2 = 5 \frac{\text{nJ}}{\text{cm}^2}, \tag{IV.4.10}$$

which is the nerve signal transported from one end to the other end of the axon. This corresponds to just 9.4% of the dissipated free energy over the length of one action potential. The Hodgkin–Huxley model therefore dissipates about 11 times more free energy than necessary to charge the capacitor over the length of one single action potential.

The energy of the capacitor is not stored but has to be recreated by ion flows at any given position of the axon. The velocity of the action potential in the squid axon was calculated to be 18.8 m/s and lasts about $\Delta t = 1.3$ ms (according to Figure IV.4.4 on which this calculation is based), which yields a length of the pulse of $\Delta x = v \cdot \Delta t = 2.5$ cm. For a nerve that is 25 cm long (10 lengths of the action potential),

$$\Delta G_{diss} = 110 \Delta G_{cap} \qquad \text{for a squid axon with a length of 25 cm,} \tag{IV.4.11}$$

i.e., 110 times more free energy is dissipated compared to the energy that is transported in the signal. All of this dissipated free energy has to be provided in the form of ATP in order to pump the potassium and sodium ions back. Thus, the Hodgkin–Huxley action potential is a very inefficient mechanism that dissipates two orders of magnitude more free energy than the energy of the signal.

IV.4.4 The Number of Parameters

It is a general principle in physics that a theory should contain as few parameters as possible (but not less, as Albert Einstein famously remarked). As shown in Section IV.3.3, the Hodgkin–Huxley model requires a large number of parameters that are needed to describe the conductances, for which there exists no theory based on first principles. Any model that led to a similarly successful fit of the voltage- and time dependence of the transmembrane currents would lead to an equally successful description of the action potential. We estimate the number of parameters used in the fits as follows.

The open probability of the potassium channel is given by

$$Y = n^{p_n}, \tag{IV.4.12}$$

where n is the open probability of a gate determined by rate constants α_n and βn, and the parameter p_n is the number of independent gates that Hodgkin and Huxley determined to be 4.

The open probability of the sodium channel is given by

$$Y = m^{p_m} h^{p_h}, \tag{IV.4.13}$$

where m and h are the open probabilities of two gates determined by rate constants $\alpha_m, \beta_m, \alpha_h$, and β_h, which are voltage-dependent functions. The parameters p_m and p_h represent the number of independent m and n gates that Hodgkin and Huxley determined to be 3 and 1, respectively.

The parameters α (α_n, α_m and β_h) introduced in Section IV.3.3 were described by

$$\alpha_i = \frac{A_i \cdot (V + B_i)}{1 - \exp\left(-\frac{V+B_i}{C_i}\right)}. \tag{IV.4.14}$$

One needs nine parameters to describe the three α_is. Similarly, if one describes the three β_i by

$$\beta_i = D_i \cdot \exp\left(-\frac{V + E_i}{F_i}\right) \tag{IV.4.15}$$

requires further nine parameters. Together with the number of gates p_n, p_m, and p_h, and the open channel conductances $g_{K,0}$ and $g_{Na,0}$, one needs 23 parameters to describe the time and voltage dependences of the conductances for potassium and sodium.

The potassium and sodium currents further require the knowledge of the Nernst potentials E_K and E_{Na}, which are functions of the ion concentrations. Two more parameters are needed for the leak conductance g and the leak potential E_L. This yields a total of 27 parameters to fit the experimentally measured currents. Additionally, one needs to know the radius a of the axon, the membrane capacitance C_m, and the resistance R_i of the internal medium. Thus, altogether, we need 30 parameters to describe the action potential of the squid axon. Of those, only E_K, E_{Na}, C_m, a, and R_i and possibly $g_{K,0}$ and $g_{Na,0}$ can directly be measured in independent experiments.

Summarizing, the Hodgkin–Huxley model for a simple axon such as the squid axon requires 30 parameters.

It is important to note that the Hodgkin–Huxley gate model and the above parametrization of the six αs and βs only serve the purpose to reproduce the experimental voltage-clamp measurements on the squid axon. However, as pointed out by Hodgkin and Huxley [16], this does not give any validity to the assumptions made in order to fit the data. Any other model that is able to reproduce the experimental data would be equally valid when describing the action potential. In principle, one could just use the experimental voltage-clamp data in the differential equation described in Chapter IV.4. It is possible that such descriptions could involve fewer parameters.

IV.4.4.1 More Complicated Models for Human Neurons

The Hodgkin–Huxley model relies on the conductance of one class of sodium and one class of potassium channels (or more generally, their conductances) and one membrane capacitance. It is well-suited to describe the currents in squid axons but not in other neurons. However, there exist models with additional channels that can be applied to more complex membranes, such as those of human nerves. Bostock and collaborators developed a model for human motor axons, which is based on four channel proteins: a sodium channel, a fast and a slow potassium channel, and an inward rectifier. They also consider leak currents and a sodium–potassium pump. Further, they allow for inhomogeneous distributions of the channels in different regions and add three different capacitances associated with myelinated and nonmyelinated regions (e.g., [20–23]). These models naturally contain many parameters. While the Hodgkin–Huxley model has about 30 parameters (mostly to describe the conductances, see previous section), the Bostock model requires between 60 and 70 parameters.

IV.4.5 Summary: Key Ideas of Chapter IV.4

- The Hodgkin–Huxley model consists of an application of cable theory. In contrast to standard cable theory, the membrane conductances for potassium and sodium are assumed to be a function of voltage and time.
- The voltage- and time dependences are taken from voltage-clamp experiments. It is not derived but fitted to the experimental data.
- The voltage- and time-dependent membrane current is inserted into the cable equation. The resulting differential equation is

$$\frac{a}{2R_i\theta^2}\frac{\partial^2 V_m}{\partial t^2} = C_m\frac{\partial V_m}{\partial t} + g_K(V_m, t)\,(V_m - E_K)) + g_{Na}(V_m, t)\,(V_m - E_{Na})$$

- The pulse velocity θ is proportional to the square root of the radius of the axon, a.
- Constant capacitance is assumed, i.e., the capacitance does not change during the action potential.
- The model is consistent with a stimulation threshold, i.e., a minimum stimulus to obtain a pulse.
- Due to the relaxation times of the protein conductances, there exist temporal conditions during and directly after an action potential, where the axon cannot be stimulated. This period is called the *refractory period.*
- In a typical squid axon, the amount of free energy dissipated during a single action potential is 100 times larger than the electrical energy of the action potential. The Hodgkin–Huxley mechanism is a very wasteful process.
- What the Hodgkin–Huxley model achieves is to demonstrate that the transmembrane currents measured under voltage-clamp conditions are consistent with the voltage changes during the propagating pulse. However, no theory for the transmembrane currents is provided. These currents are parametrized.
- The Hodgkin model describes the conductances for potassium and sodium by a parametrization of experimental data that require 23 parameters. Together with other properties of the axon, one requires a total of 30 parameters.
- One can extend the Hodgkin–Huxley model to include more conductances for other ions and also different distributions in myelinated and nonmyelinated parts of the axon at the expense of having many more parameters in the model.

IV.5

Protein Ion Channels

The Hodgkin–Huxley (HH) model [16] relies on the voltage- and time dependence of sodium and potassium conductance. The new part of the HH model as, compared to the membrane theory by Bernstein (see Section I.2.3, [2]), was that sodium and potassium currents are controlled by separate mechanisms. Hodgkin and Huxley did not directly attribute the conductance for sodium and potassium to a particular membrane object. While it was speculated that there may exist membrane-soluble carriers for the individual ions, Hodgkin and Huxley's landmark paper did not make use of protein channels. The history of the emergence of the concept of ion channel proteins was briefly reviewed from a personal perspective by Bertil Hille in 1999 [24] Hille's drawing of channels is shown in Figure IV.5.1. It is clear that the concept of a protein channel was not generally accepted until the 1990s.

Hille and Schwarz proposed in 1978 that the potassium conductance could be a single-file carrier molecule [25]. This concept is consistent with the crystal structure of a potassium channel and the mechanism for ion transport by MacKinnon and collaborators [26]. Parallel to the search for a structure, there were important experiments by E. Neher and B. Sakmann who invented the patch-clamp technique to investigate currents through small membrane

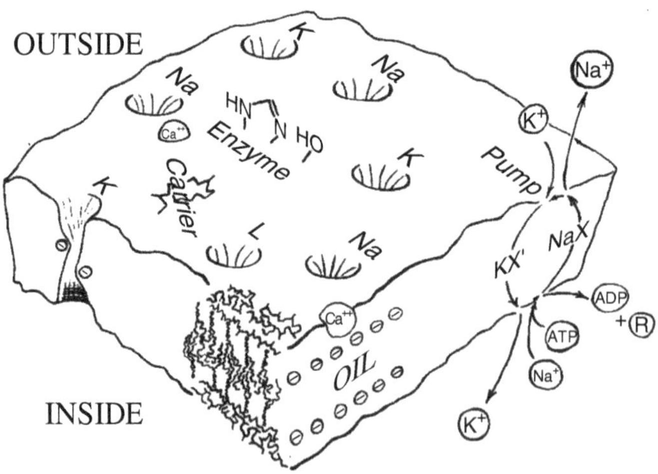

Figure IV.5.1 Drawing of the biological membrane from the PhD thesis of Bertil Hille from 1967. [24]/with permission of Springer Nature.

Physics of Nerves and Excitatory Membranes, First Edition. Thomas Heimburg.

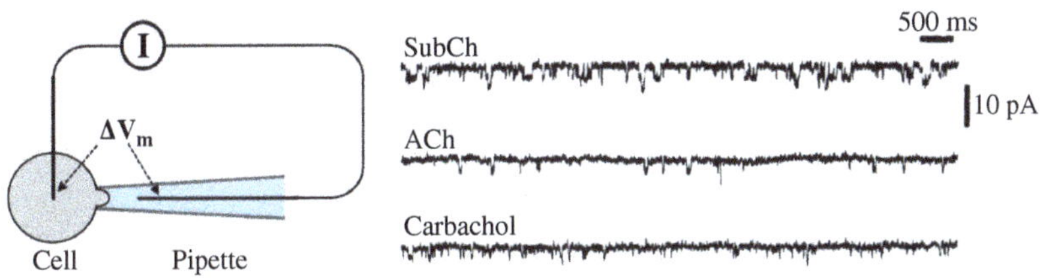

Figure IV.5.2 Left: Simplified schematic drawing of a patch-clamp experiment. A membrane patch on the surface of a cell is sucked into a tiny glass pipette. The current through the patch is measured at constant voltage between electrodes inside a cell and inside the pipette. Adapted from [28]. Right: The first quantized conductance recordings by Neher and Sakmann on an acetylcholine receptor in the presence of suberyldicholine (SubCH), acetylcholine (ACh), and carbachol show that the conductance recordings depend on the chemical nature of choline-containing drugs. Adapted from [27].

patches of cell membranes. In this experiment, a tiny pipette tip is attached to the surface of a cell while applying slight suction. The edges display a very high resistance in the GΩ regime (gigaseal). Currents between the inside and outside of cells can be recorded (see a schematic and simplified drawing in Figure IV.5.2). Neher and Sakmann found that the currents appear quantized and react to the presence of drugs [27]. This supports the concept of single-file channel proteins – or at least a pore that can open and close. Figure IV.5.2 (right) shows the first recordings on the acetylcholine receptor in the presence of three different drugs.

The molecular biology of channel proteins became a huge field. The working hypothesis in electrophysiology is that the thermodynamics of a biological membrane is controlled by these channels. For instance, the change of membrane voltage will activate voltage-gated channels, mechanical forces will activate mechanosensitive channels, and temperature changes will activate temperature-controlled or thermosensitive channels. Nobel Prizes have been given for the description of the structure of voltage-gated channels to R. MacKinnon (2003), for temperature-sensitive channels to D. Julius (2021), and to A. Patapoutian for mechanosensitive channels (2021). This shows how much importance is attributed to understanding the origin of the control of membrane function via the function of single molecules.

In the following, we will present some of the experimental evidence that led to the postulate of protein channels that respond to intensive thermodynamic variables.

IV.5.1 Voltage-gated Protein Channels

The conductance in the HH model does not rely on particular proteins but rather discusses voltage-dependent conductances more generally. However, it was speculated that particular carrier molecules may help transport the ions through the hydrophobic core of the membrane. Today, it is believed that these transport carriers are particular transmembrane proteins called ion channels, with a pore in the center and charged groups that serve as gates for the pore [29–31]. These proteins open and close in a voltage-dependent manner.

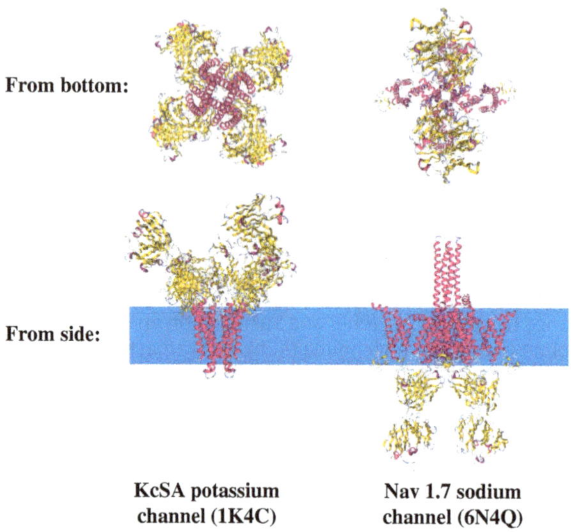

Figure IV.5.3 Molecular structures of voltage-gated ion channels. Left: The KcsA potassium channel from the top and in the membrane plane (pdf-file: 1K4C). Adapted from [32]. Right: The Na_V 1.7 sodium channel from the top and in the membrane plane. The region shaded in blue is a tentative visualization of the membrane. Adapted from [33]. Images generated on http://rcsb.org.

The crystal structure of the voltage-gated KcsA potassium channel and cryo-TEM structure of the voltage-gated Na_V 1.7 sodium channel are shown in Figure IV.5.3.

In the HH picture, the conductance for potassium is controlled by four identical gating particles (see Section IV.3.3). The fact that the crystal structure of potassium channel proteins displays a fourfold symmetry (see Figure IV.5.3 (top, left) for the KcsA channel) is consistent with the notion that the protein might indeed represent the carrier for potassium. The sodium conductance in the HH picture is gated by two different gating particles (three identical and one other particle) – i.e., also four in total (see Section IV.3.3). Therefore, the structure of the sodium channel in Figure IV.5.3 (top, right) is also consistent with such a fourfold symmetry. (Fourfold symmetries of proteins are in fact quite common, even for globular proteins such as hemoglobin.)

The conductance of membranes for potassium and sodium as a function of voltage was first described by [14, 15]. In the presence of K^+-channel proteins, the conductance for potassium in squid axons was first given by [36]. The voltage-gated nature of the sodium channel protein was reviewed by Catterall [37]. Schmidt et al. [34] measured the potassium conductance of the KcsA channel as a function of voltage and found a current–voltage profile consistent with the potassium conductance in the HH model (Figure IV.5.4).

Thus, there is evidence that the conductance in the HH model is consistent with the notion of channel proteins that act as carrier molecules. The assumption is that these molecules change their conformation as a function of voltage. The surprising selectivity of the protein for a particular ion is rationalized by a region in the center of the protein surrounding the pore, which contains amino acids that serve as ligands for potassium ions [26]. This is shown in Figure IV.5.5. The description of the selectivity filter was rewarded with the Nobel Prize to Roderick MacKinnon in 2003.

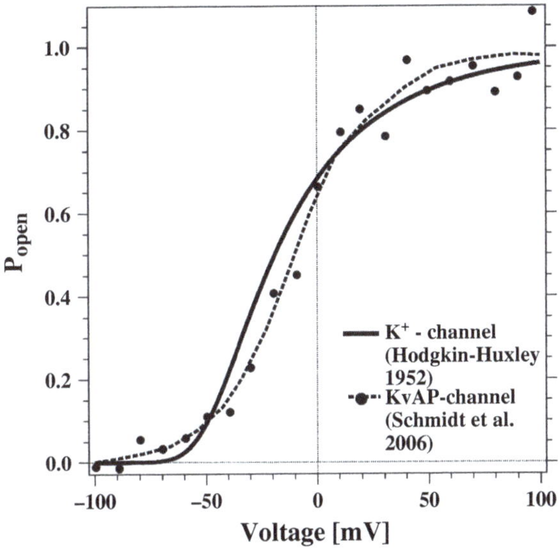

Figure IV.5.4 Potassium conductance as a function of voltage parametrized by Hodgkin and Huxley (solid line) and measured for the KcsA channel by Schmidt et al. (dashed line and symbols). [16], [34], [35].

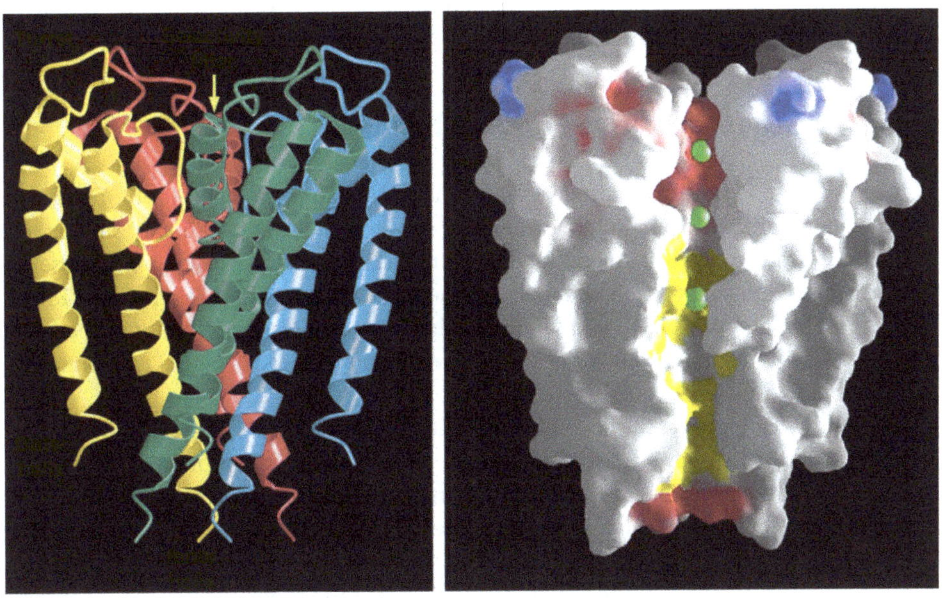

Figure IV.5.5 Left: The KcsA potassium channel protein contains a central region through which potassium ions may pass. It is called the selectivity filter. Right: This representation shows a cross-section through the channel protein including the potassium ions in green. [26]/with permission of American Association for the Advancement of Science.

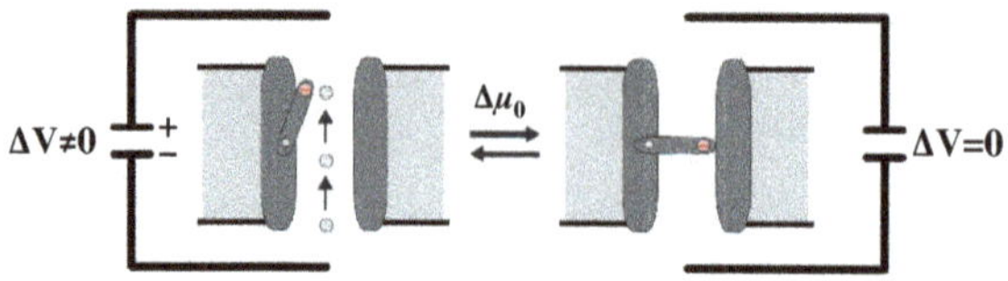

Figure IV.5.6 Schematic drawing of a gate with a negative gating charge in voltage-gated channels.

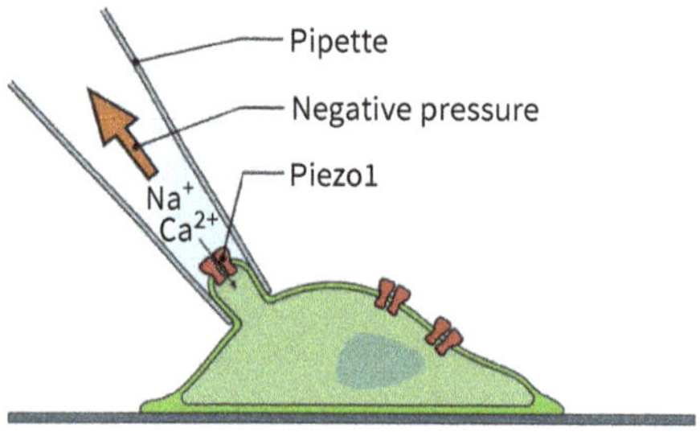

Figure IV.5.7 Schematic drawing of a suction experiment linking conductance of the mechanosensitive channel Piezo1 to suction pressure. [41]/with permission of Elsevier

Voltage gating is thought to be due to some regions in the protein called gates that carry charges. They move in an electrical field and open or close a gate in the pore. This is schematically shown in Figure IV.5.6.

IV.5.2 Mechanosensitive Protein Channels

Voltage is only one of the intensive variables that can change the behavior of membrane conductance. Some channel proteins seem very sensitive to the application of suction to a cell surface, which affects the membrane tension. Such proteins are called mechanosensitive proteins. Examples of such channels are the Piezo channels [38–44] and the TRAAK channel [45]. Figure IV.5.7 shows a schematic drawing of an experiment in which suction is applied to a membrane patch of a cell [41]. The suction will lead to lateral tension within the membrane.

Figure IV.5.8 (left) shows the dependence of the membrane current of the Piezo1 channel from Drosophila as a function of suction pressure. The sigmoidal shape of the current versus pressure profile suggests that the current originates from a single pore that is either open or closed [38] in a pressure-dependent manner. The half-maximum pressure is −31.8 mm Hg.

Figure IV.5.8 (right) shows the cryogenic-EM structure of the Piezo1 channel protein [44]. It has a threefold symmetry. The structure suggests that the protein bends the membrane and that tension in the membrane may lead to forces acting on the protein.

A tentative schematic representation of what might happen during the change in the open-close equilibrium caused by lateral tension is shown in Figure IV.5.9. Here, $\Delta\mu_0$ describes the change in the free energy difference between closed and open conformations, which is a function of lateral tension or curvature.

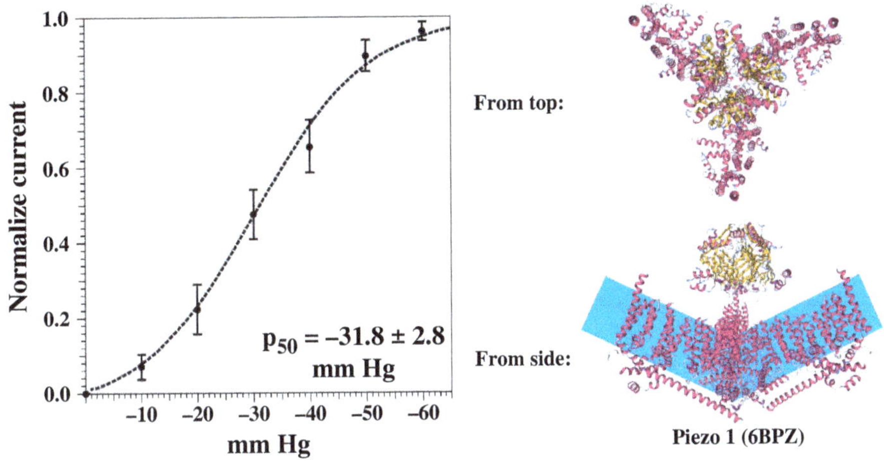

Figure IV.5.8 Left: Current-suction pressure profile of Piezo channels from Drosophila. [38]/with permission of Springer Nature. Right: CryoEM reconstruction of Piezo1. Top: From the top. Bottom: From the side. The region shaded in blue is a tentative visualization of the membrane. Adapted from [44]. Images generated on http://rcsb.org.

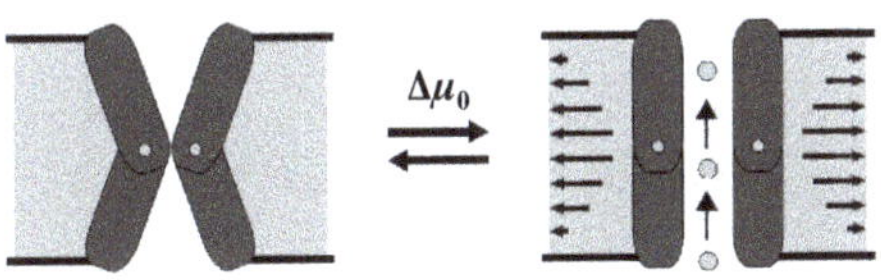

Figure IV.5.9 Tentative drawing of the opening of a pore caused by lateral tension.

IV.5.3 Temperature-sensitive Protein Channels

There are also channels sensitive to temperature changes. This phenomenon is called *thermoreception*. There is a complete family of receptors called the TRP (transient receptor potential) channels that are believed to be temperature-sensitive [46]. They are schematically depicted with their range of temperature sensitivity in Figure IV.5.10.

For instance, the menthol receptor TRPM8 is believed to be a sensor for environmental cold [47]. In contrast, the TRPV1 channel is activated at temperatures above body temperature. The structures of the TRPM8 channel [48] and the TRPV1 channel [49] are given in Figure IV.5.11.

Something that is special about TRP channels is their gigantic activation enthalpy [50]. It can be calculated by fitting the conductance profile to a two-state protein transition characterized by an enthalpy and entropy difference between an open and a closed state:

$$K = \exp\left(-\frac{\Delta\mu_0}{RT}\right) \tag{IV.5.1}$$

where $\Delta\mu_0 = \Delta H_0 - T\Delta S_0$, and ΔH_0 is the van't Hoff enthalpy of a two-state transition (Figure IV.5.12).

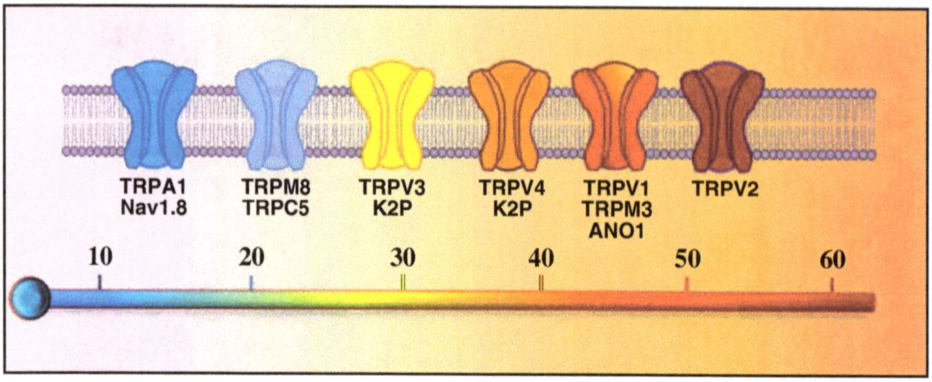

Figure IV.5.10 A schematic diagram depicting the temperature-sensitive ion channels. Ion channels are ordered according to their relative activation threshold to temperatures. [46]/with permission of Taylor & Francis.

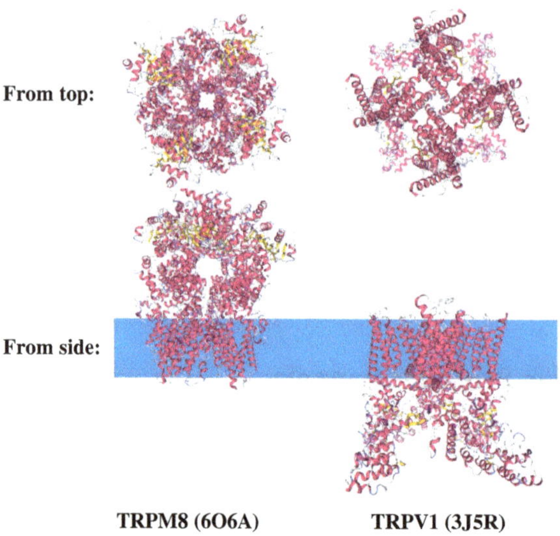

Figure IV.5.11 The structures of the cold receptor TRPM8 and the hot receptor TRPV1 seen from the top and from the side. The region shaded in blue is a tentative visualization of the membrane. [48], [49]. Images generated on http://rcsb.org.

The chemical potential difference between open and closed forms is a temperature-dependent function. The open probability to find an open channel is given by [52]

$$P_{open} = \frac{1}{1+K}. \tag{IV.5.2}$$

The open probability is displayed in Figure IV.5.12b.[1] The open probability of a TRPM8 (adapted from [51]) has been fitted with the analysis in eqs. (IV.5.1) and (IV.5.2). We call this a *van't Hoff analysis*. The temperature dependence of the open probability is dictated by the

1 It should be noted here that ΔH_0 contains a pressure- and voltage dependence as well. Thus, the temperature dependence of the channel is thought to be described at constant lateral tension and voltage.

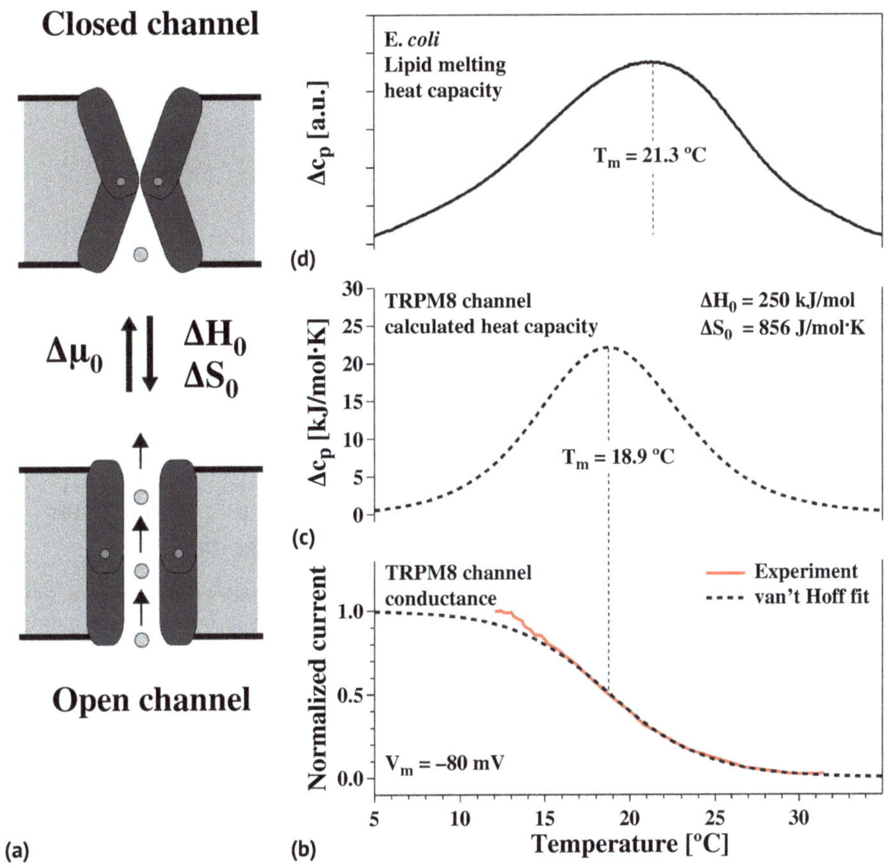

Figure IV.5.12 (a) Schematic drawing of a pore that closes upon temperature increase. (b) The open probability of a TRPM8 channel from [51] and the van't Hoff fit. (c) The heat capacity of the conformational change in the protein calculated from the van't Hoff profile in panel (b). (d) For comparison, the heat capacity of the lipids in *E. coli* membranes. It can be seen that both transition midpoint and the width of the transition is remarkably similar to the open-close behavior of the TRPM8 channels. [52]/with permission of American Chemical Society.

transition enthalpy. The analysis in Figure IV.5.12 yields an enthalpy difference between the open and closed states of 250 kJ/mol (for details see [52]). Clapham and Miller [50] calculated a very similar value of 236 kJ/mol from different data. 250 kJ/mol is a very high enthalpy that is of similar order to the complete unfolding enthalpy of small globular proteins such as cytochrome c and myoglobin. If a protein is to explain the conductance of the membrane, drastic changes in structure upon small temperature changes are required. The fitted profile in Figure IV.5.12b can be used to calculate a heat capacity profile because the total transition enthalpy and the temperature dependence of the change in enthalpy are known. This heat capacity is shown in Figure IV.5.12c. In Figure IV.5.12d, the melting profile of *E. coli* membranes [53] is shown for comparison. We show in Chapter V.2 that many other biomembranes display very similar melting profiles. It can be seen that the activation pattern of the TRPM8 channel is quite similar to the temperature dependence of the

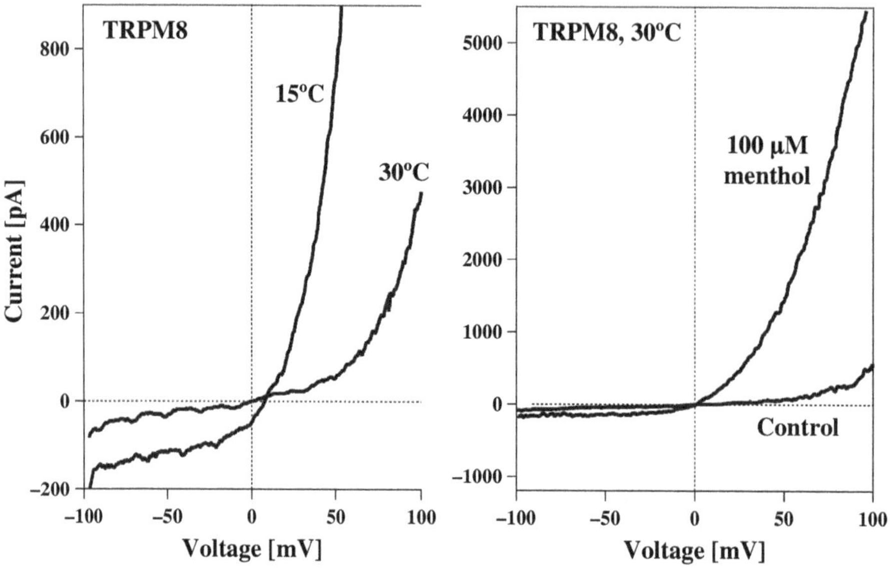

Figure IV.5.13 The TRPM8 channel in HEK293 cells has both a dependence on temperature (left) and on menthol concentration (right). The outward rectification also demonstrates voltage gating of membranes containing TRPM8. Adapted from [51].

biomembrane melting process. Thus, the activation pattern of the TRPM8 channel coincides with the melting process of the lipid membrane. It would be surprising if this were just fortuitous.

The temperature-sensing TRP channels are also dependent on other variables such as pH, voltage, or the chemical potentials of cofactors such as menthol and capsaicin. Figure IV.5.13 shows the TRPM8 channel in HEK293 cell membranes (from [51]). The membrane has a higher conductance at 15 °C than at 30 °C. However, the conductance of the membrane is strongly enhanced by the presence of menthol, a molecule that gives the illusion of cold. Thus, membranes containing TRPM8 channels depend on at least three intensive thermodynamic variables: temperature, voltage, and the chemical potential of menthol. Likewise, membranes containing the TRPV1 channel are activated upon changes of at least four intensive variables, i.e., temperature increase, the chemical potential of capsaicin, pH (related to the chemical potential of protons), and voltage [51, 54, 55].

IV.5.4 The Sensitivity of Protein Channels to the Lipid Environment

Some channels have been reported to display different open-close characteristics when the lipid environment is changed. Schmidt et al. [34] showed that the KvAP potassium channel shows different current–voltage profiles when reconstituted in different lipid mixtures. Figure IV.5.14 shows this for three mixtures: DOTAP:DOPA = 1:1, POPC:POPG = 3:1, and POPE:POPC = 3:1, respectively. Here, DOTAP is a positively charged lipid; DOPA and

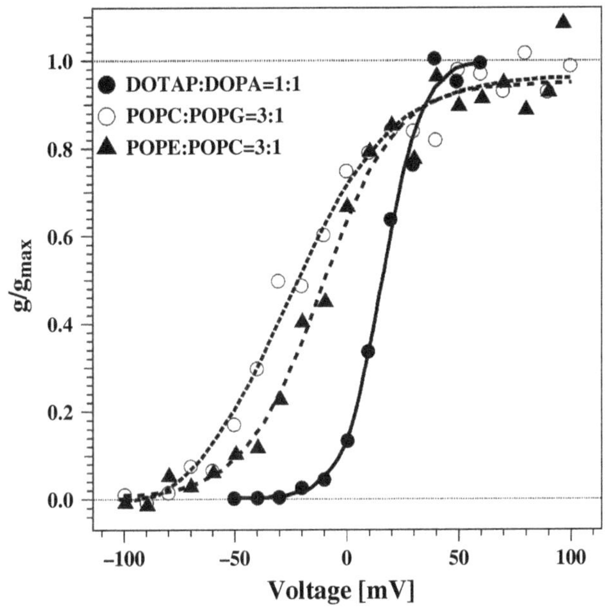

Figure IV.5.14 Current–voltage profiles of the KvAP potassium channel reconstituted into three different artificial membranes: DOTAP:DOPA = 1:1, POPC:POPG = 3:1, and POPE:POPC = 3:1, respectively. The IV curves depend on the host lipid membrane. Adapted from [34].

POPG are negatively charged; and POPE and POPS are zwitterionic (no net charge). The membranes display different charges and also different phase behavior. Lipid-dependent gating of a potassium channel was also reported by [56].

That the conductance of potassium channels depends on the phase state of the host lipid matrix has also been shown. Seeger et al. [57] found that the conductance of the KcsA potassium channel is closely connected to the phase state of the membrane. When the channel proteins are reconstituted into DOPE:DOPG mixtures (3:1 and 1:1), the membrane conductance changes in the melting regime. Figure IV.5.15 (left) shows the relative change in conductance compared to the heat capacity profiles. It can be seen that there is a close correlation between the conductance and the c_p profiles of the lipid membrane. A part of this change in conductance is probably due to the redistribution of proteins in the melting transition range. Figure IV.5.15 (right) shows atomic force microscopy images of KcsA channel proteins in a DOPE:DOPC = 3:1 mixture at two different temperatures from [58]. At 28°C, the membrane is in its liquid-disordered state (fluid), and the proteins are homogeneously distributed in the membrane (Figure IV.5.15 (right, a)). At 26.5°C, the membrane is in the coexistence regime of solid-ordered (gel) and liquid-disordered (fluid) phases. One can recognize the formation of a gel domain that excludes the proteins (Figure IV.5.15 (right, b)). Thus, it is clear that channel proteins redistribute in the melting regime such that the proteins are found close to the gel–fluid interface.

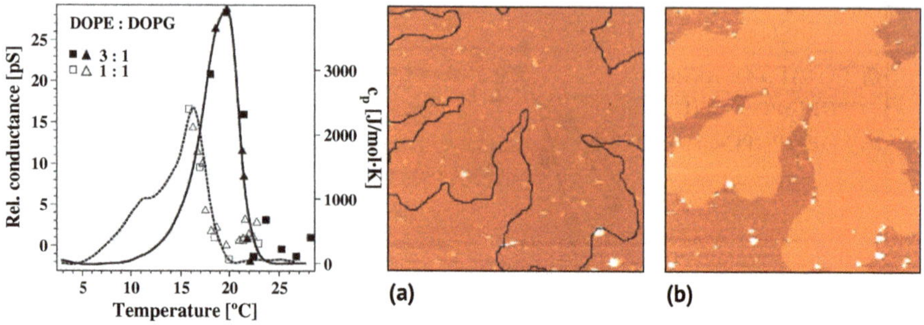

Figure IV.5.15 Left: Change in conductance of KcsA potassium channels caused by changes in temperature in synthetic membranes (DOPE:DOPG = 3:1 and DOPE:DOPG = 1:1) compared to the heat capacity profiles of the host membrane. [57]/with permission of Elsevier. Right: Distribution of the KcsA potassium channel in a DOPE:DOPC = 3:1 lipid mixture as a function of temperature measured by atomic force microscopy. The image size is 5 × 5 μm. (a) 28°C. All lipid is in the liquid-disordered phase. KcsA channel proteins (bright spots) are distributed randomly. For comparison, the black contour line indicates the shape of the solid-ordered domain in panel b. (b) Upon cooling to 26.5°C, a solid-ordered domain forms that contains hardly any proteins. The proteins are only miscible with the liquid-disordered membranes. [58]/with permission of American Chemical Society.

IV.5.5 Summary: Key Ideas of Chapter IV.5

- There exists a class of integral proteins called ion channels. These channels assist the transfer of ions across biological membranes.
- Some of these channels are specific to particular ions. There exist specialized channels for potassium, sodium, calcium, chloride, and other ions.
- The selectivity is surprising but has been attributed to the arrangement of amino acids in a pore through the protein, which is called a selectivity filter.
- These pores must not be thought of as all-or-none filters but rather as pores that display much faster transfer rates for some ions as compared to others.
- Some of these channels display crystal structures that reflect the number of gating particles postulated in the HH model. For example, the KcsA potassium channel described by MacKinnon and collaborators displays a fourfold symmetry, while the HH model uses four independent n-particles to gate the K^+ channel.
- The conductance of membranes in the presence of channel proteins is correlated with the phase behavior of the host lipid membrane. Proteins redistribute at lower temperature, and the conductance shows a strong correlation with heat capacity profiles.
- Ion channels seem to be dependent on intensive thermodynamic variables. There exist voltage-gated channels, temperature-gated channels, and mechanosensitive channels corresponding to the variables' electrostatic potential, temperature, and mechanical force. There also exist channels depending on the chemical potential of protons (pH) or the lipid composition corresponding to the chemical potentials of lipid species.
- Channels have been described that are sensitive to several thermodynamic variables, such as lipid-dependent voltage-gated channels.

Critical note: In electrophysiology, the thermodynamic variables are introduced via specific proteins (channels or receptors) that are controlled by voltage, membrane tension, temperature, and other intensive variables. However, it is questionable whether thermodynamics can be properly introduced on the basis of single molecules. Thermodynamics only works as a theory for the complete system under consideration, e.g., the complete nerve membrane. Only here, the Maxwell relations that describe the couplings between different variables are well defined. The second law of thermodynamics only holds for a system as a whole. It does not hold for parts of a system. It is likely that attributing the sensing of changes in the intensive variables to single channels and receptors leads to an incomplete or even wrong understanding of biological function. This is especially true when it comes to excitatory membranes and nerve pulses that are of macroscopic dimensions. Here, it is clear that the system is much larger than the size of a single molecule.

In Part VIII, it will be shown that channels can also emerge in macroscopic membranes in a manner that is consistent with thermodynamics. Such channels do not require the presence of proteins but can be influenced or modulated by them. These membrane channels show very similar properties to protein channels.

Part V

Properties of Artificial and Biological Membranes

In this part, we describe the macroscopic properties of artificial and biological membranes, with a focus on cooperative transitions in the membrane. We show that the thermodynamic properties depend on voltage, pressure, tension, temperature, pH, calcium concentration, the chemical potentials of the membrane components, and other intensive variables without the need for molecular receptors. These findings will be the prerequisite for a thermodynamic description of the nerve pulse.

V.1

Membrane Structure

Biological membranes are thin layers predominantly composed of lipids and proteins. Both classes of molecules display internal degrees of freedom and can exist in a low-temperature form that is ordered and a high-temperature form that is disordered. Therefore, they do not only have a structure but also display complex thermodynamic behavior. The typical protein-to-lipid ratio of biomembranes by weight is about 1. However, purple membranes display a ratio of about 4, while lung surfactant is mostly composed of lipids with a low protein content of about 1/4 [1]. Significant fractions of the protein are extra-membranous and are thus outside the membrane plane. If one only considers the fraction of the protein in the membrane plane, most of the mass is that of the lipids.

V.1.1 The Membrane Bilayer

V.1.1.1 Early Membrane Models

Early studies on cells by Nägeli, de Vries, and Pfeffer revealed that they are surrounded by a thin layer that allows to maintain differences in osmotic pressure [2–5]. The layer surrounding the cell plasma was called "Plasma Haut (cell skin)" or "plasma membrane" by Pfeffer [5]. Pfeffer believed that the membranes are composed of proteins. Charles Ernest Overton developed the concept of a cell membrane composed of fatty lipid molecules [6], in which other hydrophobic molecules such as general anesthetics can be dissolved. In fact, Overton showed that lipid-soluble substances penetrate membranes at the fastest rate. This early phase of cell membrane studies was nicely reviewed by [7].

In 1925, Gorter and Grendel extracted lipids from erythrocytes and deposited them as monolayers on an aqueous surface [8]. From the overall lipid monolayer area compared with the cell surface area, they concluded that the lipid membrane consists of two monolayers of lipids (see Figure V.1.1 (left)). They stated: "It is clear that all our results fit in well with the supposition that the chromocytes are covered by a layer of fatty substances that is two molecules thick." This, however, was a statement about the lipid membrane rather than about the biological membrane with all its ingredients. Ten years later, Danielli and Davson [9] proposed that the biological membrane contains oil-like substances (lipoids), lipids, and globular proteins. It considered the membrane to be a layer of unknown thickness, thinner than the microscopic resolution of ≈200 nm with proteins adsorbed to its

Physics of Nerves and Excitatory Membranes, First Edition. Thomas Heimburg.

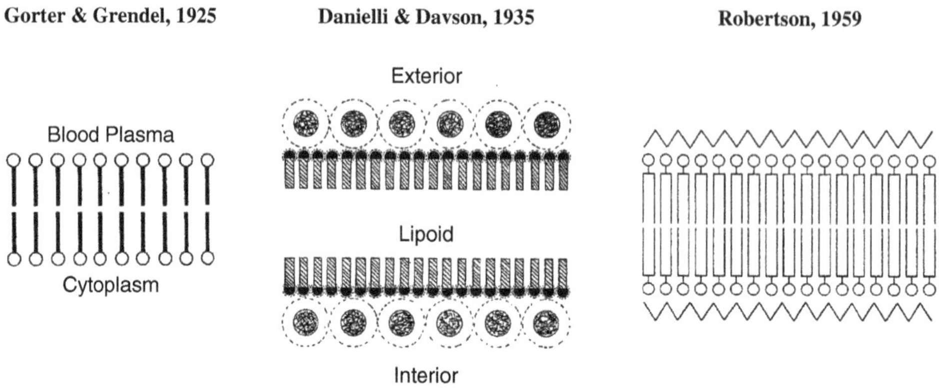

Figure V.1.1 Earlier membrane models. Left: Gorter and Grendel first proposed that lipid membranes consist of two monolayers of lipids. [8, 10]. Center: The Danielli and Davson model of a biological membrane describes the membrane as a oil-like phase with lipids at the interface. Globular proteins are adsorbed to the surface. [9]/with permission of John Wiley & Sons. Right: Robertson used electron microscopy to identify the thickness of the biological membrane and proposed that the biological membranes consist of bilayers with a surface layer of proteins. [10]/Portland Press Limited.

surface (Figure V.1.1 (center)). Using the newly developed electron microscopy, Robertson deduced that cell membranes are around 75 Å thick, including a protein layer on the surface [10]; see Figure V.1.1 (right).

V.1.1.2 The Fluid Mosaic Model

The "fluid mosaic model" by Singer and Nicolson [11] is a landmark publication that has shaped the view of the role and function of the biological membrane until today. It proposes that biological membranes consist of a fluid lipid layer into which integral proteins are embedded (Figure V.1.2). The lipid membrane acts as a two-dimensional liquid in which the proteins can diffuse and interact. Until today, this view seems very reasonable and is the predominant picture in the literature. The biological membrane is considered a fluid lipid layer containing integral proteins. Peripheral (or soluble) proteins are not considered in much detail, because their interactions with membranes are considered to be weak.

The main conclusion of Singer and Nicolson is that (Quote):

> "The fluid mosaic structure is therefore formally analogous to a two-dimensional oriented solution of integral proteins ... in the viscous phospholipid bilayer environment."

In the following, we will significantly modify this view, since lipid membranes are not generally fluid, and there exist lipid–protein interactions that lead to rearrangements of proteins and lipids within the membrane. These rearrangements depend on the thermodynamic variables, e.g., temperature, membrane tension, or pH and ionic strength.

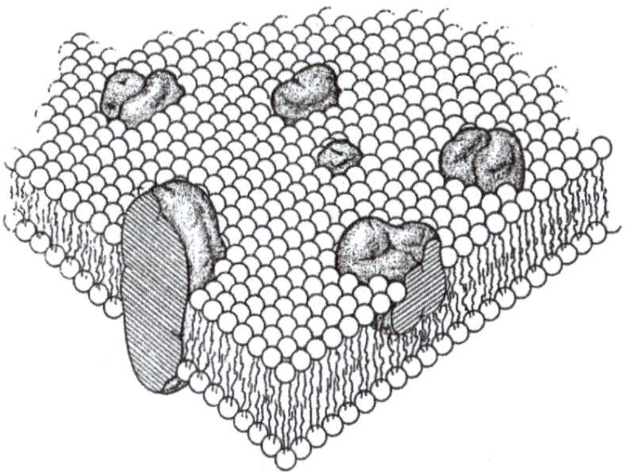

Figure V.1.2 The Singer and Nicolson model [11] or fluid mosaic model suggests that proteins are associated with the lipid membrane. The lipid matrix serves a fluid two-dimensional confinement for proteins that may interact in order to generate biological function. From [12].

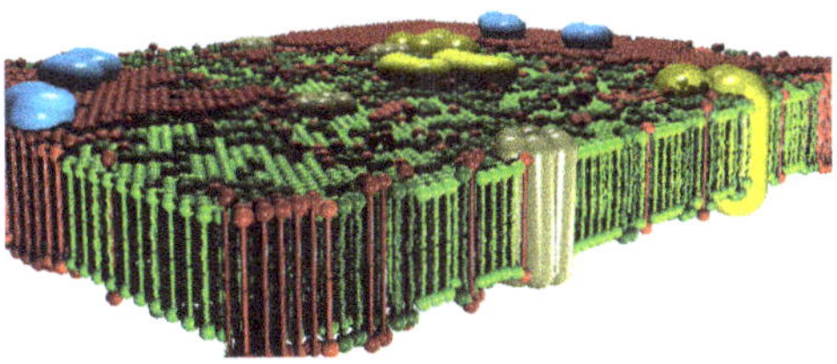

Figure V.1.3 Schematic representation of a biological membrane with lipids in different states and surface and integral proteins.

V.1.1.3 The Membrane More Realistically

Figure V.1.3 shows a schematic representation of the membrane structure. It shows membrane proteins (in white, yellow, and blue) that either penetrate the membrane or are associated with the surface. Penetrating proteins are the *integral proteins*. They possess hydrophobic regions that match the surrounding lipid matrix. Surface proteins are also called “peripheral proteins.” They adhere to the membranes due to electrostatic interactions or hydrophobic hydrocarbon anchors. Lipids can be in the fluid or gel state and are shown in green and red, respectively.

The typical thickness of a biological membrane is of the order of about 5 nm. This is of similar order as the size of many proteins, e.g., cytochrome c with about 3 nm and the potassium channel with about 5 nm. Thus, membranes are mostly insulators that form very thin layers. This makes them good capacitors. This will play an important role in the chapters related to the electrical properties of membranes and also in cable theory and the Hodgkin–Huxley model (see Chapter IV.2). An important property of the membranes is that the lipids may melt, i.e., they may exist in an ordered and an unordered state. Melting of membranes is described in some detail in Chapter V.2. Due to variance in molecular dimensions, membrane components may not mix ideally. Clustering of molecules and domain formation may occur, which is considered in the mattress model by Mouritsen and Bloom ([13], see Section V.3.1).

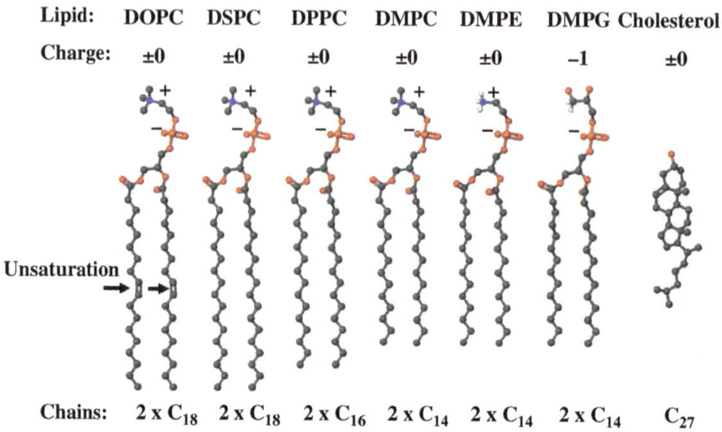

Figure V.1.4 Representation of single lipids: DOPC, DSPC, DPPC, DMPC, DMPE, DMPG, and cholesterol. These lipids display differences in chain length, chain saturation, head group structure, head group charge, and dipole moment. Cholesterol is also counted as a lipid but cannot form membranes on its own.

V.1.2 Chemical Structure of Membrane Lipids

One of the two major classes of membrane molecules is lipids. Most lipids are phospholipids with two hydrocarbon chains. The chains possess internal degrees of freedom, which contribute to the entropy of the lipids and the membrane ensemble.

Figure V.1.4 shows a couple of common lipids:

- Dioleoyl phosphatidylcholine (DOPC) with two C-18 chains and one double bond at the 9–10 position in each chain and a phosphatidylcholine (lecithin) head group. The head group contains a negative charge on the phosphate and a positive charge on the choline group. Thus, the head group is zwitterionic and possesses a dipole moment. Membranes of DOPC display a melting temperature of about −20°C (see next section).
- Distearoyl phosphatidylcholine (DSPC) is nearly identical to DOPC. It possesses two C-18 chains but without double bonds. This small structural difference leads to a difference in melting temperature of about 70°C. DSPC melts at about 55°C.
- Dipalmitoyl phosphatidylcholine (DPPC) is similar to DSPC, but has only 16 carbons in each chain without any double bonds. DPPC is one of the most common lipids. For instance, about 50% of the lipids in lung surfactant are DPPC. DPPC has a melting temperature of about 42°C, which is quite close to the body temperature of mammals.
- Dimyristoyl phosphatidylcholine (DMPC) has two C-14 chains and a melting temperature around 24°C.
- Dimyristoyl phosphatidylethanolamine (DMPE) is similar to DMPC but displays a small modification of the head group. Instead of a $-N(CH_3)_3$ group, it has an $-NH_3$ group. This seemingly small change leads to a change in the melting temperature of nearly 30 degrees to 53°C . The head group is also zwitterionic and possesses a strong dipole moment.

- Dimyristoyl phosphatidylglycerol (DMPG) has a glycerol instead of choline in the head group. Glycerol is uncharged. Therefore, the DMPG head group carries one negative charge from the phosphate but no (significant) dipole moment. Its melting temperature of 23°C is nearly identical to that of DMPC.
- Cholesterol is not a lipid as the phospholipids shown earlier. It is a steroid and a precursor of cortisol and hormones such as testosterone and estrogens, which are structurally very similar. Cholesterol does not form membranes on its own but is very membrane-active. It has a pronounced influence on melting processes, membrane fluctuations, and membrane stability.

The aforementioned list presents only a very small fraction of lipids. However, they have certain key features that are important for the lipid behavior in the cell membrane. What determines the behavior are:

- Chain length: Longer chains lead to higher melting temperatures.
- Chain unsaturation: Lipids with unsaturated chains (i.e., containing double bonds) display much lower melting temperatures. Most unsaturated fatty acids and oils melt far below body temperature and are liquid. Most saturated oils are hard and solid.
- Charge of the head group: Higher charge leads to lower melting points due to electrostatic repulsion with the membrane plane.
- Dipole moment of the head group.
- Number of chains: Lysolipids with only one chain and fatty acids both also belong to the lipid class.

Since there exist enzymes that modify the chemistry of lipids, membranes may adapt their composition to ambient conditions. This is discussed in much more detail in Chapter IX.2.

V.1.3 Membrane Proteins

Singer and Nicolson [11] call their fluid mosaic model a thermodynamic model because it is based on minimizing the free energy of hydrophilic and hydrophobic contacts. Research on the hydrophobic effect was pioneered by Walter Kauzmann in the 1950s [14]. It explores the interactions of hydrophobic groups, e.g., the lipid chains and the hydrophobic amino acids, with water. Kauzmann found that the free energy of the transfer of apolar molecules such as alkanes from an oil-like environment to water is temperature-dependent. This implies that the interface of apolar groups with water possesses both enthalpy and entropy contributions, with an associated heat capacity of the hydrophobic interface. The consequence is that apolar groups prefer apolar environments.

The free energy of electrostatic interactions is described by Debye–Hückel theory [15]. Membranes consist mostly of lipids and proteins. The primary sequence of many proteins contains stretches of amino acids with apolar residues, and other stretches with charged or polar residues. Apolar residues can avoid contact with water when they are embedded into the hydrophobic core of the membrane. At the same time, charged amino acids are located on the outside of the membrane in order to optimize the interaction with the electrolyte (see Figure V.1.5 (left)). When the *fluid mosaic model* was published, crystal structures of membrane proteins did not exist. Therefore, the prediction that integral proteins exist and

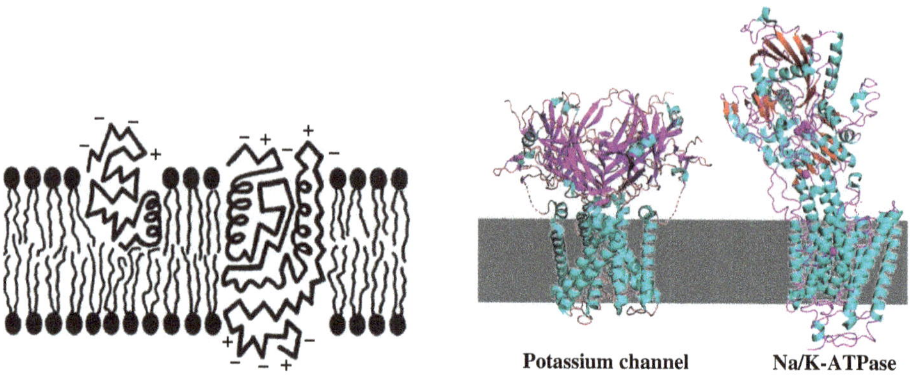

Figure V.1.5 Left: In the fluid mosaic model [11], proteins contain hydrophobic sections that may be embedded in the hydrophobic core of the lipid membrane. From [12]. Right: Crystal structures of the potassium channel and the Na^+, K^+-ATPase. One can recognize that significant parts of the integral proteins are extra-membranous.

fold in a manner that allows them to minimize the free energy of charges and hydrophobic residues at the same time was very important.

Such proteins do in fact exist in biological membranes. Figure V.1.5 (right) shows two prominent integral proteins: the potassium channel, first crystallized by MacKinnon and collaborators [16], and the Na^+, K^+-ATPase, first described by Skou [17].

V.1.4 Summary: Key Ideas of Chapter V.1

This chapter deals with the composition of the biomembrane: its geometry, the structure of lipids, and integral proteins. In summary, we find:

- Biomembranes mostly consist of lipids and proteins.
- They display a thickness of about 5 nm, which is of similar order as the diameter and length of many proteins.
- The most prominent model for a biological membrane is the *fluid mosaic model* by Singer and Nicolson from 1972 [11]. It states that the biological membrane is a fluid matrix made of lipids in which proteins can freely diffuse. In the following chapters, we will show that this is an oversimplification of the thermodynamics of the biological membrane, and that much of the biological function of membranes can be attributed to phenomena that represent deviations from the fluid mosaic model.
- The membrane lipids possess a polar head group that likes to interact with water and hydrophobic chains. The chains differ in length, saturation (i.e., the number of double bonds), dipole moment, and charge.
- There exist two classes of proteins: integral proteins and peripheral proteins. The integral proteins contain hydrophobic segments embedded into the hydrophobic core of the lipid membrane. Peripheral proteins loosely associate with the surface of membranes.

V.2

Membrane Melting

Artificial and biological membranes display melting processes in which the membrane undergoes a change from an ordered to a disordered state of the lipid chains. The transition plays an important role in the physical properties of membranes and is of exceptional importance for the function of biomembranes.

V.2.1 Change in Chain and Lattice Order Upon Heating

The spatial structure of lipids is closely related to the melting of membranes. At low temperature, lipid chains assume an all-trans configuration that is described in more detail in (Figure V.2.1) [1]. This is the configuration with the lowest energy, and the CH_2 groups of the lipid chains arrange in a regular pattern. At high temperature, this regular configuration is lost, and the lipid chains assume a more random or disordered configuration. This configuration (or ensemble of configurations) displays both a larger enthalpy and entropy.

Upon melting, the dimensions of a membrane change. They become thinner (for DPPC, the membrane thickness D is reduced by about 17%, while the area A increases by about 24%). This is shown in Figure V.2.1 (left, top). Further, at low temperature, lipids are arranged on a lattice while they are more randomly distributed above the melting point (Figure V.2.1 (left, bottom)). Thus, the low-temperature phase is often called the solid-ordered (or gel) phase, while the high-temperature phase is called the liquid-disordered (or fluid) phase. In the following, we will, for convenience, often use the terms *gel* and *fluid* phase, or *solid* and *liquid* phase, respectively.

Among others, the following properties of membranes change in the transition:

- membrane enthalpy, H
- membrane area, A
- membrane thickness, D
- membrane capacitance, C_m

and the following susceptibilities assume maxima at the transition temperature:

- heat capacity, c_p
- isothermal area compressibility, κ_T^A
- membrane elasticity, κ_B, i.e., how easy it is to bend the membrane
- membrane permeability, i.e., how easy it is to create pores or defects in the membrane.

All these changes will be discussed in the subsequent chapters.

Physics of Nerves and Excitatory Membranes, First Edition. Thomas Heimburg.

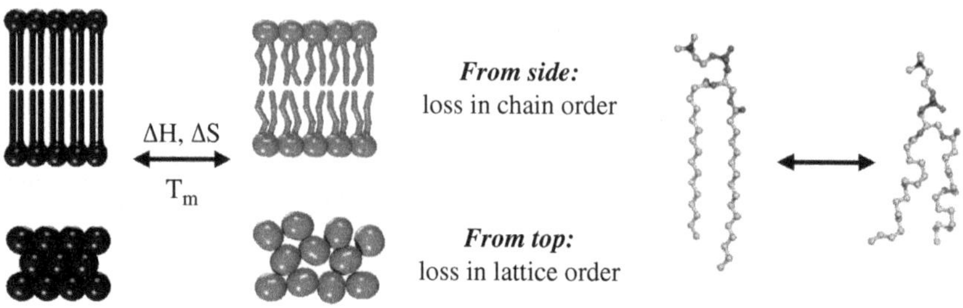

Figure V.2.1 Left: The melting of lipid membranes seen from the side and from the top. Right: The molecular changes in the lipid chains upon melting.

V.2.2 Single Lipids

V.2.2.1 Heat Capacity Profiles

Each material possesses a heat capacity. It corresponds to the amount of heat that is necessary to change its temperature, dQ/dT. For the cases relevant in this book, the heat capacity is recorded at constant pressure,

$$c_p = \left(\frac{dQ}{dT}\right)_p = \left(\frac{dH}{dT}\right)_p . \tag{V.2.1}$$

Since $dH = dQ + Vdp$, it follows that $dQ = dH$ at constant pressure (see Section II.1.9). Traditionally, the unit for the heat is *calories* (cal). A calorie is the amount of heat necessary to change the temperature of one cubic centimeter of water at standard pressure from 14.5 °C to 15.5 °C. This unit is outdated, and since 1948, the unit *Joule* (J) has been used, with 1 cal = 4.184 J. The units of heat capacity are correspondingly J/K (or J/mol·K in molar units). For materials that do not undergo transitions, the heat capacity is mostly due to vibrations in chemical bonds and small isomerization reactions within macromolecules. If there exist transitions in the material, one can find a much larger heat capacity because larger parts of molecules or even ensembles of many molecules can change their state at the same time. In the thermodynamics section, we showed that the heat capacity is proportional to the enthalpy fluctuations, which are especially large in transitions. Since the excess heat capacity of these transitions, i.e., the part of the heat capacity associated with large cooperative events, often displays a sharp peak over a very limited temperature interval, it is easy to measure. An example is given in Figure V.2.2, which shows the melting profiles of multilamellar vesicles of diacyl phosphatidylcholine with three different chain lengths. The area under the excess heat capacity profile is the transition enthalpy, ΔH_0, which is typically given in units Joule per mole (J/mol). The peak maximum is often associated with the transition temperature.

The excess heat capacity shows the change between gel and fluid phases. However, the individual phases of the membranes outside the transition regime also possess a heat capacity $c_{p,0}$ that is due to intramolecular vibrations and isomerizations, which forms the heat capacity baseline under the heat capacity profile. The absolute heat capacity is much more difficult to measure. According to Blume [18], the heat capacity of lipids without the excess

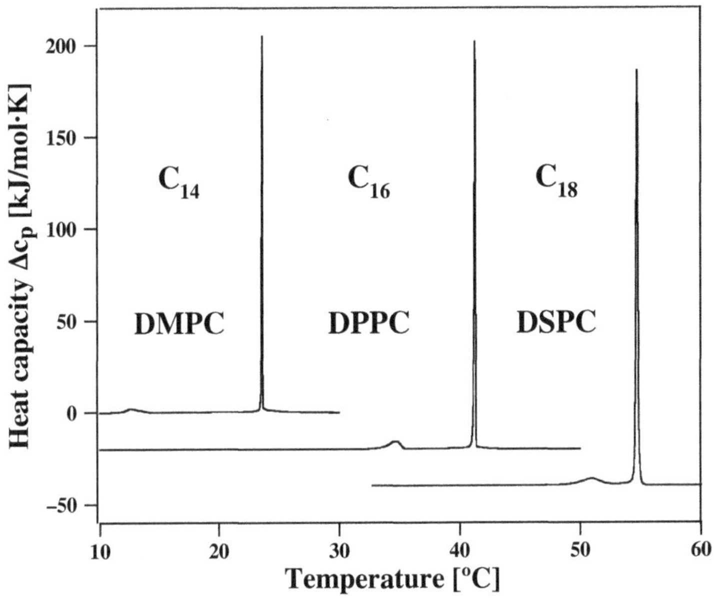

Figure V.2.2 Excess heat capacity profiles of diacyl phosphatidylcholine (multilamellar vesicles, MLV) with three different chain lengths (14, 16, and 18 carbons in each chain, respectively).

Table V.2.1 Specific heat capacities of different lipids outside the transition range. From [18].

Lipid	20 °C	$T_m - 20°$	$T_m + 10°$
Lipid	c_p [J/g]	c_p [J/g]	c_p [J/g]
DPPC	2.113	2.113	2.222
DPPE	2.268	2.598	2.360
DPPA (pH 6)	2.067	2.628	1.820
DPPA (pH 12)	2.356	2.473	2.540

heat capacity $c_{p,0}$ is given in Table V.2.1. The heat capacity is given at 20 °C, and at 20 ° below and 10 ° above the melting transition.

One can estimate the specific heat capacity for lipids with other chain length by using the values per additional methylene group. Blume gives a value of 54.4 J/mol·K for each extra methylene group in phosphatidylcholines at 20 °C.

The total heat capacity is given by

$$c_p = c_{p,0} + \Delta c_p. \tag{V.2.2}$$

For large unilamellar vesicles of DPPC, the heat capacity profiles are shown in Figure V.2.2. The bottom panel shows the total heat capacity, c_p. The values of the lipids outside of the transition given by Blume [18] are denoted as *baseline*. The top panel is the excess heat capacity, Δc_p, which is associated with the melting process.

V.2.2.2 Enthalpy and Entropy Changes

The enthalpy as a function of temperature can be obtained by integrating the heat capacity.

$$H(T) = \int_{T_0}^{T} c_p dT + \text{const.}, \tag{V.2.3}$$

where T_0 is a temperature below the melting transition. The constant in this equation depends on T_0. Figure V.2.4 (left) shows the absolute enthalpy change determined from the heat capacity profile in Figure V.2.3 (bottom). The enthalpy change associated with changes in lipid state in the transition is calculated as the integral of the excess heat capacity, Δc_p:

$$\Delta H(T) = \int_{T_0}^{T} \Delta c_p dT\,. \tag{V.2.4}$$

This is shown in Figure V.2.4 (center).

The heat capacity profile also allows for the calculation of entropy changes, since $dQ = TdS$ and $c_p = TdS/dT$. From this follows that $dS = (c_p/T)dT$. The entropy change is therefore given by

$$S(T) = \int_{T_0}^{T} \frac{c_p}{T} dT + \text{const.}; \quad \Delta S(T) = \int_{T_0}^{T} \frac{\Delta c_p}{T} dT\,. \tag{V.2.5}$$

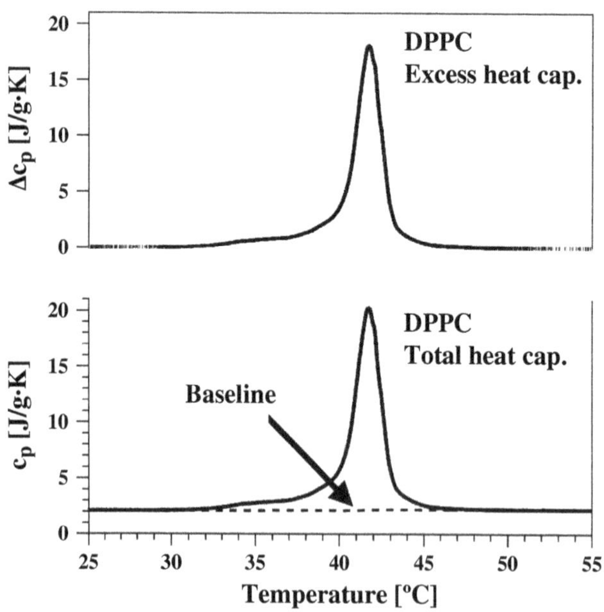

Figure V.2.3 Specific heat capacity profile of large unilamellar vesicles (LUV) of DPPC. Bottom: total heat capacity including the heat capacity of the chains $c_{p,0}$ (indicated by baseline) given by Blume. Adapted from [18].

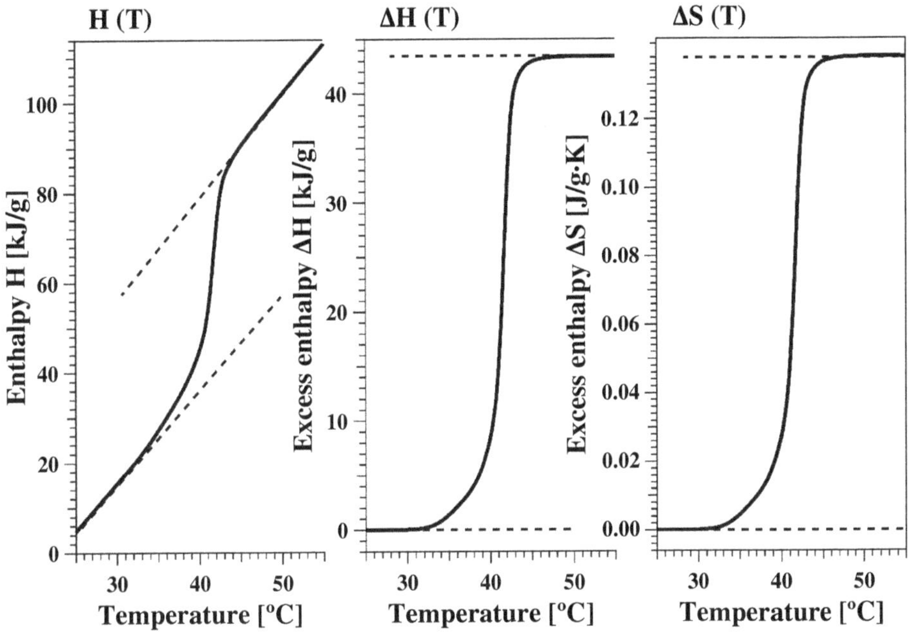

Figure V.2.4 Left: Total enthalpy of DPPC LUV obtained from integrating the heat capacity in Figure V.2.4. Center: Excess enthalpy. Right: Excess entropy. Numbers are per gram.

The excess entropy change, ΔS, associated with the heat capacity profile of the DPPC LUV shown in Figure V.2.2 (top) is shown in Figure V.2.4 (right): It looks comparable to the enthalpy change but there are differences because the entropy is (in contrast to the enthalpy) a logarithmic function of temperature.

The transition enthalpy, ΔH_0, and transition entropy, ΔS_0, can be obtained when integrating from a temperature below to a temperature above the transition events. The respective values for four different diacyl phosphatidylcholines are given in Figure V.2.5 (left). At the transition temperature, the standard chemical potential difference is zero. This allows us to calculate the transition temperature:

$$\Delta\mu_0 = \Delta H_0 - T_m \cdot \Delta S_0 = 0 \quad \longrightarrow \quad T_m = \frac{\Delta H_0}{\Delta S_0}. \tag{V.2.6}$$

The transition temperatures are given in Figure V.2.5 (right). Vice versa, the measurement of the transition temperature allows us to calculate the transition entropy from the transition enthalpy, $\Delta S_0 = \Delta H_0 / T_m$. Both ΔH_0 and T_m are readily available from a calorimetric experiment.

V.2.2.3 Calculating a Heat Capacity Profile

Many excess heat capacity profiles, Δc_p, can be approximated by a simple calculation. We assume that n lipids undergo a transition at the same time and that the melting occurs in one step per unit of n lipids. We call n a cooperative unit size. The rationale for using a

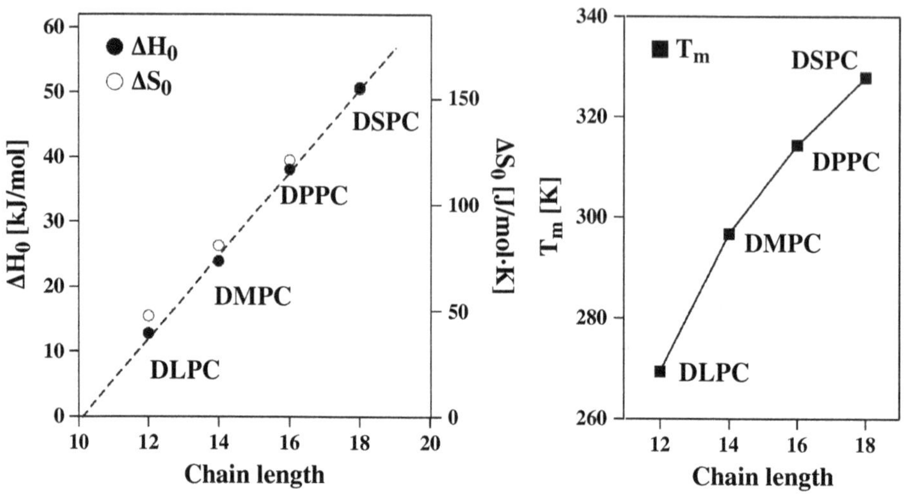

Figure V.2.5 Left: Excess melting enthalpies and entropies of diacyl phosphatidylcholine with different chain lengths. Right: Melting temperatures of these lipids. [1]/with permission of John Wiley & Sons.

cooperative unit size is given in [1]. The melting enthalpy is given by ΔH_0 per mol of lipid ($n \cdot \Delta H_0$ per mol of cooperative unit), and the melting entropy by $\Delta S_0 = \Delta H_0/T_m$ ($n \cdot \Delta S_0$ per mol of cooperative unit). The equilibrium constant between the two states is now given by

$$K = \exp(-n \cdot \frac{\Delta H_0 - T\Delta S_0}{RT}). \tag{V.2.7}$$

The fraction of fluid lipids is given by

$$f = \frac{K}{1+K}, \tag{V.2.8}$$

and the excess enthalpy by

$$\Delta H(T) = f \cdot \Delta H_0. \tag{V.2.9}$$

The excess heat capacity Δc_p is the derivative of the enthalpy:

$$\Delta c_p(T) = \frac{df}{dT} \cdot \Delta H_0 = \frac{K^2}{(1+K)} \cdot \frac{n\Delta H_0^2}{RT^2} \text{ per mole of lipid.} \tag{V.2.10}$$

If we assume that $n = 70$, $\Delta H_0 = 35$ kJ/mol, and $T_m = 314.5$ K (the values for DPPC membranes), we obtain the heat capacity profile in Figure V.2.6, which is very similar to the experimental profile. If n is increased, the profile becomes narrower. Thus, the cooperative parameter n determines the width of the calorimetric profile. For convenience, we will in the following make use of calculated profiles at various places. Since $n \cdot \Delta H_0$ is a van't Hoff enthalpy, we call these profiles *van't Hoff profiles*.

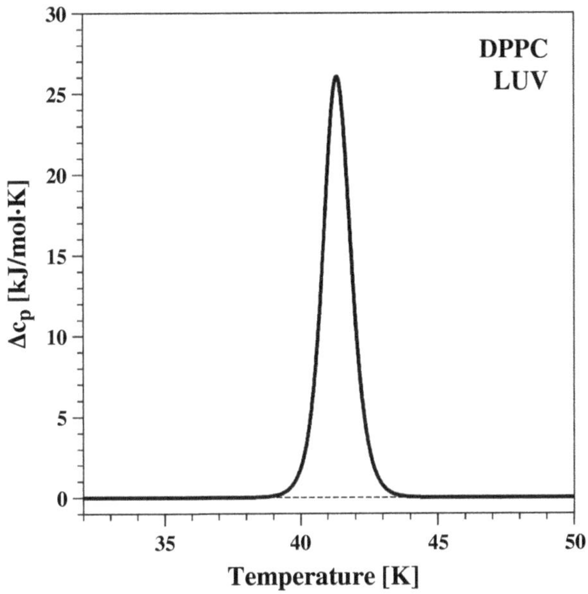

Figure V.2.6 The calculated heat capacity profile for DPPC LUV is very similar to experimental data.

V.2.3 Lipid Mixtures

Let us consider two different lipids, e.g., dimyristoyl phosphatidylcholine (DMPC, C14 chains) and distearoyl phosphatidylcholine (DSPC, C18 chains). They display melting temperatures of 24 °C and 54.7 °C, respectively. When the two lipids are mixed, one might expect that one would find two transition peaks corresponding to the individual transition temperatures of the two lipids. This would be the case when the different lipid species don't influence each other because they do not mix at all. Another possibility is that the properties of the two lipids simply average, and one finds a transition peak in between. For instance, in a 50:50 mol:mol mixture, one might expect a temperature around 39.5 °C. This would happen if the lipids do mix ideally at all temperatures. However, neither of these two cases is found in an experiment. Figure V.2.7 shows the melting profile of the above mixture. It shows two maxima that correspond neither to the melting temperature of DMPC nor to the melting temperature of DSPC. It is therefore clear that the two lipids influence each other in their melting behavior. This behavior can be explained by ideal or regular solution theory [1, 19]. We conclude that the lipids neither separate macroscopically nor do they mix ideally at all temperatures. We find an explanation for this behavior when we discuss the mattress model by Mouritsen and Bloom in Section V.3.1. The exact melting behavior depends on the chain length and head group of the components. For now, we state as an experimental fact that in lipid mixtures, the lipids influence each other in their melting behavior. We will see in the following section that this is also true for integral and peripheral proteins that are dissolved or adsorbed to membranes.

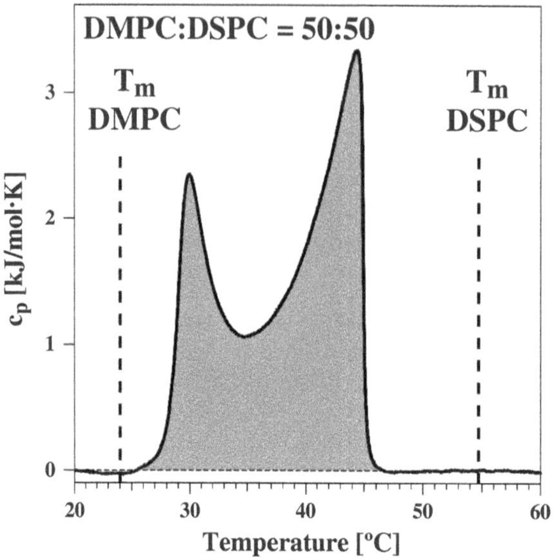

Figure V.2.7 Excess heat capacity profile of a DMPC:DSPC = 50:50 mixture. The melting temperatures of DMPC and DSPC are indicated as dashed lines. The gray-shaded area indicates the melting regime of the membrane.

V.2.4 Melting of Biological Membranes

Biological membranes consist of hundreds of different lipids with different chain lengths, chain saturation, and heat groups. It is to be expected that the melting behavior of such multicomponent lipid mixtures is complex. In the following, we show that there exist lipid melting peaks in biological membranes, which are very important for the line of argument followed in this book.

V.2.4.1 Lung Surfactant

Figure V.2.8 shows an alveolus of the lung. The surface of the alveoli is covered by a lipid film that is called *lung surfactant*. It contains small amounts of some proteins called SP-A, SP-B, and SP-C but mostly consists of lipids. Lung surfactant exists in a dynamic equilibrium between monolayers and bilayers [20].

The major lipid component of lung surfactant is the lipid DPPC with a melting temperature of about 41 °C. For details of the composition of various surfactant films, see [21]. Since the major lipid component melts at 41 °C, one expects a melting temperature of lung surfactant not far away from this temperature. Figure V.2.9 shows the melting of two preparations of lung surfactant (or lipid extracts) from cow and pig. They both display a transition maximum at 26.9 °C, i.e., about 10 ° below the body temperature of the animals. The melting transition is broad. Its upper end is found approximately at body temperature. This implies that a small decrease in temperature, or small shifts in the transition toward higher temperatures caused by changes in surface tension, will partially move the lung surfactant into the melting regime.

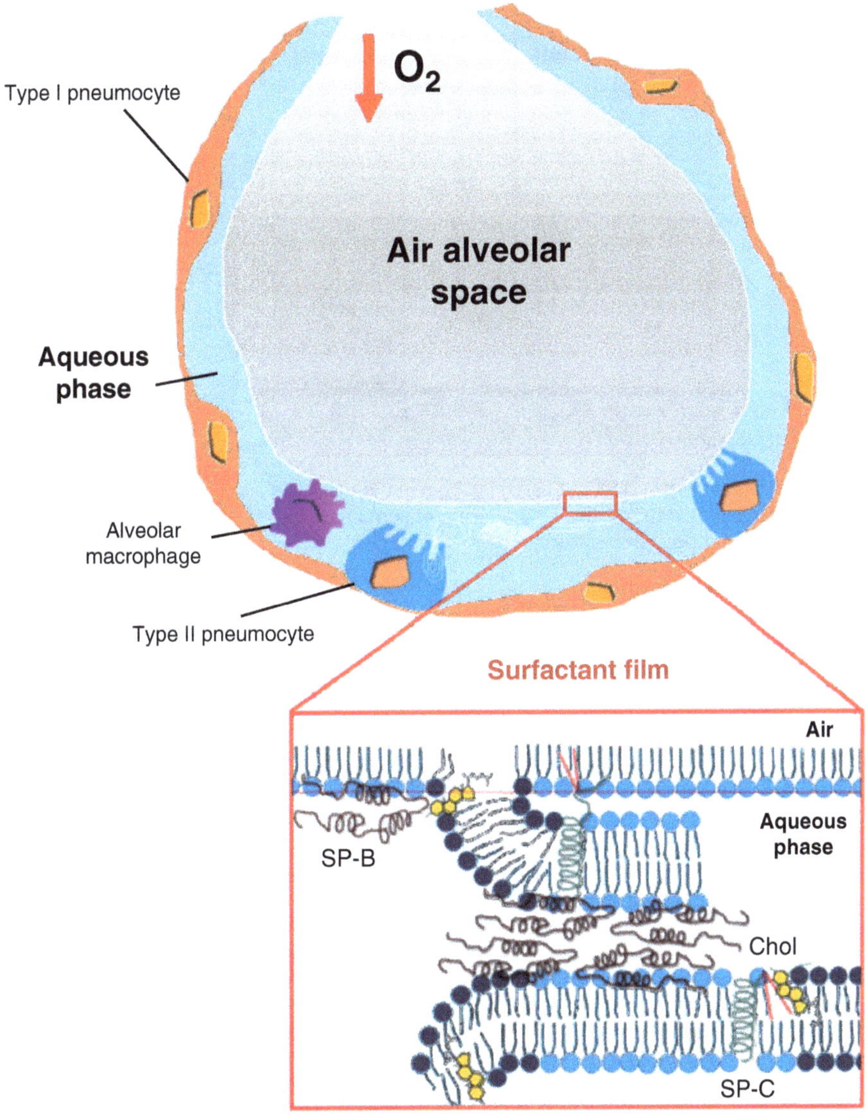

Figure V.2.8 Alveolus of the lung. Its surface is covered by a lipid film containing the proteins SP-B and SP-C. [21]/with permission of Elsevier.

De la Serna et al. [23] reported similar melting transitions in four different lung surfactant preparations from pig lungs with transition maxima between 30.2 °C and 33.2 °C.

V.2.4.2 Bacterial Membranes

Transition peaks can also be found in membrane preparations of bacteria, e.g., from *E. coli* or *Bacillus subtilis* [22]. The membranes in Figure V.2.10 were prepared from living cells that were disrupted in a French press. Thus, they still contain the membrane proteins and can

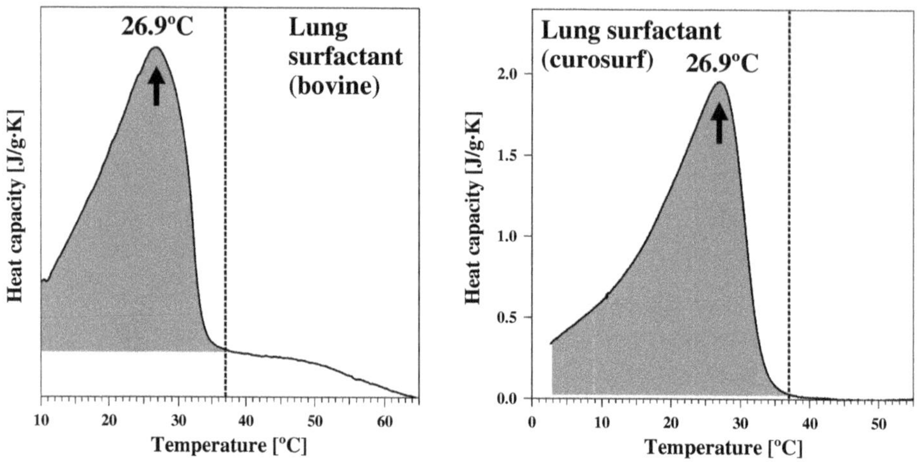

Figure V.2.9 Melting of lung surfactant. Left: From bovine lungs. Right: Lipid extract from pig lungs (Curosurf). Both preparations display a melting maximum at 26.0 °C. The dashed line indicates 37 °C (close to body temperature). Adapted from [22].

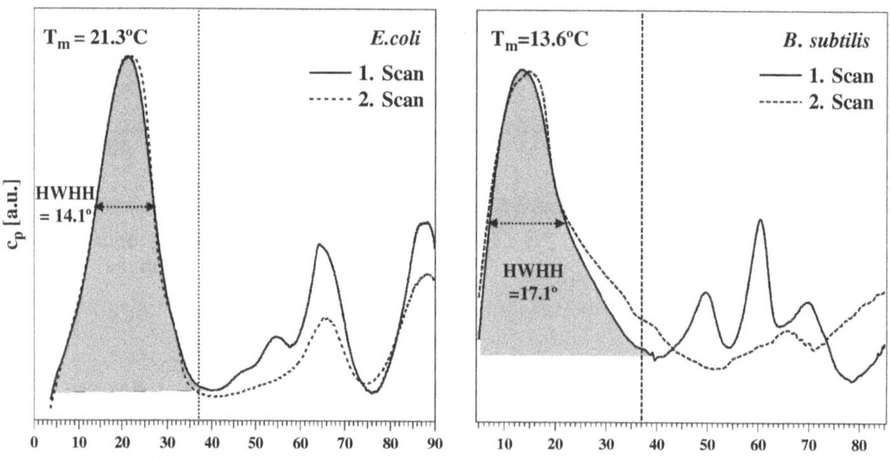

Figure V.2.10 Melting of native *E. coli* and *Bacillus subtilis* membranes. The transition maxima of the lipid melting (gray-shaded) are found at 21.3 °C and 13.6 °C, respectively. The lipid melting peak is shaded in gray. The dashed line indicates 37 °C (close to body temperature). [22]/with permission of Elsevier. The other peaks above body temperature represent protein unfolding peaks. They disappear in the second scan because protein unfolding is mostly irreversible on short timescales.

be considered as more or less native membranes. We see several peaks in the heat capacity scans. The gray-shaded peak below body temperature indicates the lipid melting, and the peaks above body temperature represent the unfolding of membrane proteins. Since protein unfolding is mostly irreversible on short timescales, the protein unfolding peaks disappear in a second scan. This is one of several ways to distinguish lipid and protein transitions (discussed in [22]). Other possibilities are the different pressure dependences of protein and lipid transitions discussed in Section V.4.1 and the investigation of lipid extracts (see next section). The transition maxima of the two bacterial membrane preparations are at 21.3 °C and 13.6 °C, respectively. The growth temperature of the bacteria was at 37 °C. In Section IX.2.1, we show that, in contrast to protein unfolding, the position of the lipid melting peak adapts to growth temperature, which is another way to distinguish lipid melting from protein denaturation.

V.2.4.3 Nerve Membranes

Melting transitions are also found in nerve membranes. Figure V.2.11 shows heat capacity profiles of preparations from pig spine [22]. Just like the melting profiles of bacterial membranes, one can find transition peaks at 23.1 °C in pig spine membranes. Similar transitions were found in membranes from rat spine and sheep spine. In Figure V.8.2, the broad melting of B35 neuroblastoma cell membranes is shown with a transition temperature of about 20 °C that is consistent with the aforemntioned findings [24]. This is of similar order as the transitions in lung surfactant and in *E. coli* membranes. In Figure V.2.11 (bottom), the heat capacity profile of the extracted lipids from porcine spinal nerves is shown. It shows a transition maximum at 27.6 °C. The comparison of transition peaks from intact biomembranes with the transitions of membranes from the extracted lipids is another way to identify lipid transitions.

Fedosejevs and Schneider [25] investigated the membranes of living SH-SY5Y cells with fluorescence spectroscopy methods. They found a surprisingly sharp transition at 17.1 °C,

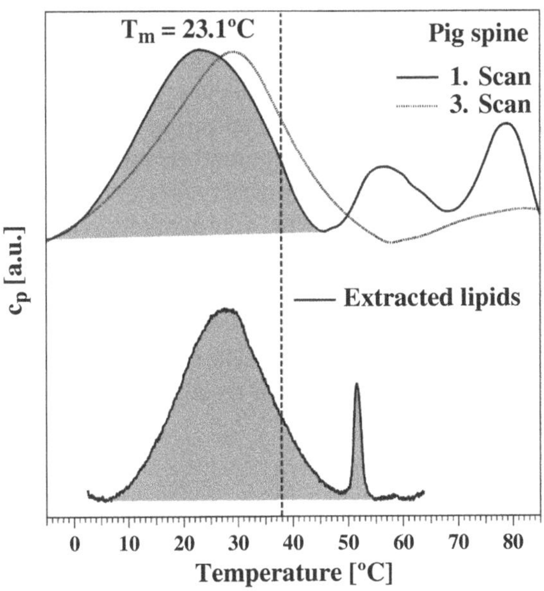

Figure V.2.11 Melting of spinal nerves from pig. The transition maximum of the lipid melting (gray-shaded) is found at 23.1 °C. The lipid melting peak is shaded in gray. The other peaks above body temperature represent protein unfolding peaks. The profile at the bottom is from lipid extracts from porcine spinal nerves. The transition in this lipid system is roughly the same as in the intact membrane. [22]/with permission of Elsevier.

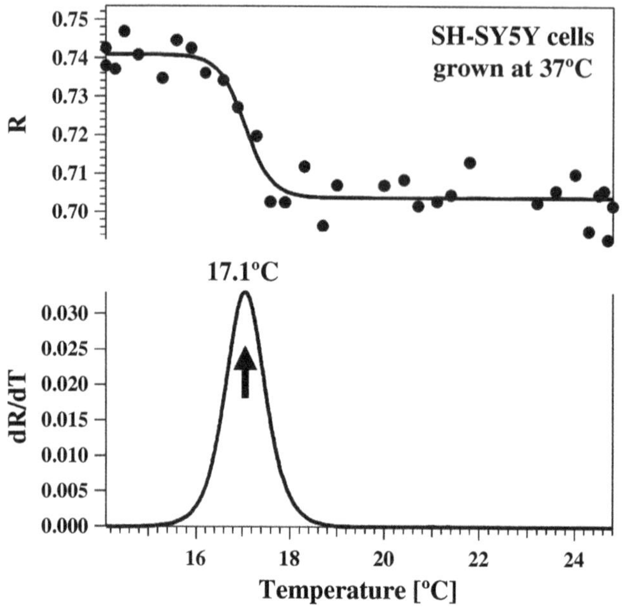

Figure V.2.12 An unusually sharp transition in living neuroblastoma cells (SH-SY5Y human neuroblastoma cell line) indicated by the blue dashed line, which is the derivative of the change of a fluorescence parameter of a lipid label. The transition maximum is found at 17.1 °C. Adapted from [25].

somewhat in between *E. coli* membranes and *Bacillus subtilis* membranes (see the earlier discussion) (Figure V.2.12). The half-width in this preparation is about 1 ° in contrast to a width of 10 ° or more in all other preparations reported in this chapter. There may be various reasons for this. One possible reason is that the fluorescence markers predominantly mix with the fluid membrane phase. Fluorescence markers are local probes and only indicate the state of the membrane at their location. Thus, they often only monitor parts of the membrane. In contrast, calorimetry reports all the membranes at once. On the other hand, the preparation by [25] is to our knowledge the only living system that has been investigated. It is possible that the preparations from lung surfactant and isolated cell membranes are more randomly mixed in their disrupted state, and that living cells have different properties. It is clear that neurons display melting transitions in their membranes both in the case of isolated membranes and in living cells.

V.2.4.4 Cancer Cell Membranes

Lipid transitions have also been found in various cancer cells. Figure V.2.13 (left) shows heat capacity profiles of HT 29 cells from a human colorectal adenocarcinoma [26]. The membranes of these cells display lipid melting transitions at 20.0 °C and 15.3 °C. One can see how in a second or third heating scan, the protein unfolding peaks completely disappear. Similar findings have been reported for MDA-MB231 cells from a human breast adenocarcinom and human dermal fibroblasts (HDF) cells, which are normal neonatal primary dermal fibroblasts [26].

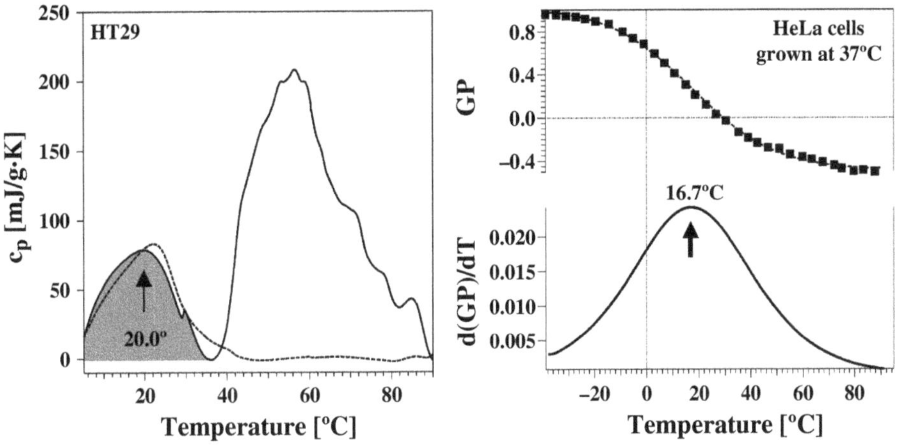

Figure V.2.13 Left: Lipid melting per in native membranes from HT 29 cells from a human colorectal adenocarcinoma. The dashed line is from the second heating scan. The protein unfolding peaks disappear due to irreversible denaturation of proteins. Adapted from [26]. Right: Lipid melting in HeLa cells measured by the generalized polarization (GP) of the fluorescent lipid label Laurdan. The transition maximum is at 16.7 °C. Adapted from [24].

Färber and Westerhausen [24] investigated the melting in the membranes of HeLa cells (cervical cancer cells) with the fluorescence marker Laurdan, which monitors the difference in surface polarization between gel and fluid lipid phases by a parameter called the generalized polarization (GP). Figure V.2.13 (right) shows the GP profile and its derivative that indicates the lipid melting. A transition temperature of 16.7 °C was found, which is in the same range as the cancer cells shown in Figure V.2.13 (left).

V.2.5 Summary: Key Ideas of Chapter V.2

Membrane melting slightly below body temperature is a general finding in all cells:

- The membranes of most lipids display melting transitions in which the chains and the arrangement of the lipid molecules undergo order transitions from a phase of crystalline packing (solid) and ordered hydrocarbons chains (solid-ordered phase) to a random (liquid) arrangement of lipids with disordered chains (solid-disordered phase).
- In a melting transition, the enthalpy and the entropy of the lipids change.
- Upon melting, the membrane becomes thinner (about −17%) and its area becomes larger (about +24%) for the lipid DPPC.
- Due to changes in thickness and area, membrane melting affects the capacitance of the membrane and therefore the electrical properties of membranes.
- In mixtures of lipids, the different lipids influence the melting behavior of each other, which can be recognized in heat capacity profiles.
- Biological membranes generally show melting transitions below physiological temperature (body temperature or growth temperature). All systems shown in this chapter show this behavior.

- The maximum of the melting profiles is found at 10° below physiological temperature for lung surfactant, and 10–20° below physiological temperature for bacterial membranes, nerve membranes, cancer cells, red blood cells, and fibroblasts.
- The transitions are in most cases broad (>10° width) and range up to physiological temperature. Relatively small changes in the intensive variables can bring the membranes into the melting regime.

It is to be expected that the membrane transitions play an important role in biology.

V.3

Phase Behavior, Domains, and Rafts

In complex lipid mixtures, e.g., biological membranes, the individual components display different melting temperatures. As an ensemble, they display complex melting processes called *phase behavior*. It is mapped in phase diagrams. A consequence of phase behavior is phase separation, domain formation, and the emergence of rafts in biomembranes.

V.3.1 The Mattress Model

In 1984, Mouritsen and Bloom proposed the mattress model for integral proteins [13]. It considers the interaction of integral proteins with gel and fluid lipids or lipids of different lengths. It suggests that the thickness of the hydrophobic core of integral proteins and the membrane may be different, and that lipid–protein interactions depend on hydrophobic matching. A gel membrane made of the lipid dipalmitoyl phosphatidylcholine is thicker than a fluid membrane of the same lipid by about 17% in [27]. Thus, an integral protein that fits well into a fluid phase cannot fit perfectly into a gel phase membrane. As a consequence, the lipids at the interface of the proteins may have to adapt their length to the protein in order to minimize hydrophobic interactions, as shown in Figure V.3.1a, b. The deformation of the lipids creates capillary forces that can lead to the aggregation of proteins. This implies that, in contrast to the assumptions of Singer and Nicolson [11] (see Section V.1.1.2), the lipids of the membrane play an active role in controlling the protein distribution, and changes in either the lipid state or the length of the hydrophobic core of the integral protein will lead to rearrangements of the proteins. Since integral proteins must interact differently with gel and fluid membranes, they must influence the melting behavior of membranes, which was also dismissed as a likely possibility in Singer and Nicolson 's paper [11].

A similar line of argument as above can also be applied to mixtures of different lipids or to lipids in different physical states. As long as they display different hydrophobic lengths, they will not mix ideally. This gives rise to the possibility of clusters or domains of lipids in different physical states (Figure V.3.1c). The interface between two regions of different state and thickness possesses a line tension and a free energy. The interfacial energy is a parameter in many lattice models used to describe phase behavior of membranes (e.g., [1, 28–32]).

Physics of Nerves and Excitatory Membranes, First Edition. Thomas Heimburg.

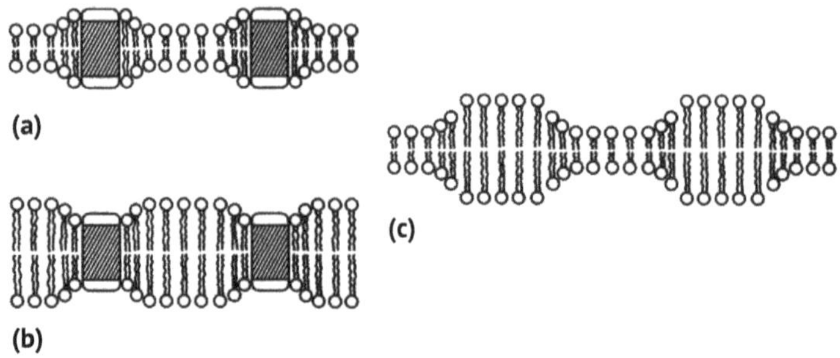

Figure V.3.1 The mattress model by Mouritsen and Bloom suggests lipid–protein interactions that mediate forces between proteins. In cartoon (a), integral proteins dissolve in a membrane with lipids of smaller hydrophobic length. In cartoon (b), integral proteins dissolve in a membrane with lipids of larger hydrophobic length. There are attractive forces between the proteins in both cases. Adapted from [13]. Cartoon (c) represents an adaptation of the mattress model in the absence of proteins with lipids of different hydrophobic lengths or states. The contact line of regions of different hydrophobic lengths creates a line tension and leads to domain formation.

V.3.2 Phase Diagrams

Biological membranes consist of a complex mixture of hundreds of different lipids and integral proteins. The lipids display differences in chain saturation, head group, and the length of their fatty acid chains. Due to differences in their chemical structure, they also display different melting temperatures. Mixtures of such lipids form regions within the membrane plane called phases or domains that display different local compositions and physical states. A simple example of a phase diagram of a binary lipid mixture of the lipids DMPC and DSPC is shown in Figure V.3.2 (from [33]). DSPC has four more CH_2 groups in each chain. Thus, the two lipids display different hydrophobic lengths and different melting temperatures. The left-hand side of Figure V.3.2 shows the heat capacity profiles of different mixtures of the two lipids. Figure V.3.2 (right) shows the phase diagram constructed from the lower and upper temperature limits of the heat capacity profiles. The diagram shows these temperatures versus the composition of the membrane. One can see that there exist regions in the phase diagram that are purely fluid or purely gel, and other regions where one finds a separation of coexisting gel and fluid phases with different compositions. The thermodynamics of phase behavior in membranes was thoroughly derived in [19]. It is shown that one may not only find coexistence between gel and fluid regions, but also the coexistence of two fluid regions of different composition, or two gel regions of different composition.

Biological membranes consist of hundreds of components (both lipids and proteins) and consequentially display a rich phase behavior. To obtain a sense of how many phases can coexist in a mixture, it is useful to consider Gibbs' phase rule, which states for a system at constant pressure that

$$F = K - J + 1, \tag{V.3.1}$$

where K is the number of chemical components, J is the number of coexisting phases, and F is the number of degrees of freedom. F describes how many variables can be changed independently without changing the number of coexisting phases. For instance, in the diagram

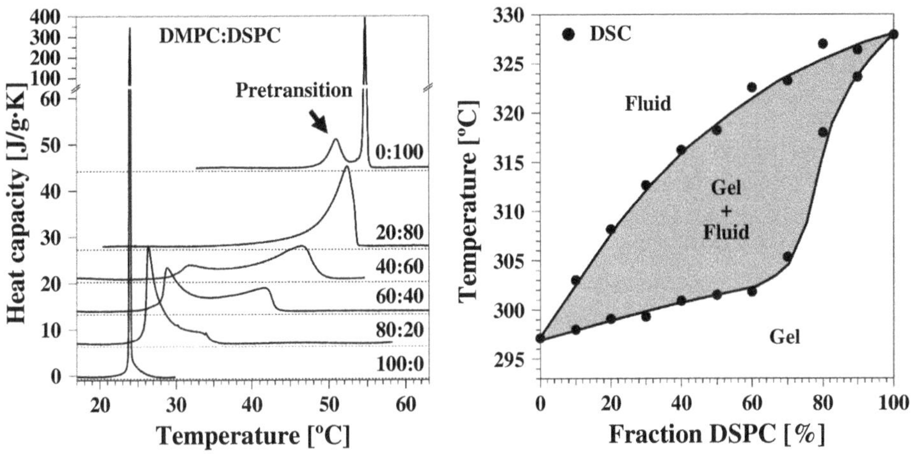

Figure V.3.2 Left: Heat capacity profiles of the melting of DMPC-DSPC mixtures. Right: The phase diagram extracted from the lower and upper boundaries of the heat capacity profiles. The gray-shaded region displays gel-fluid phase coexistence. Outside the gray-shaded area one finds single phase: a fluid phase at high temperature and a gel phase at low temperature.

in Figure V.3.2, for two lipid components ($K = 2$), there exist three regimes: a gel phase in which both composition and temperature can be changed freely ($F = 2$) and it still remains a gel phase, and a fluid phase where the same is true. In the two-phase region, only one variable, e.g., temperature, is free ($F = 1$). For any given temperature, the composition of the phases is fixed and cannot be changed freely. In a diagram with two components, there may also exist a single (critical) point at which $F = 0$ and three phases coexist. This is the maximum number of coexisting phases in a two-component system.

In a biological membrane composed of 200 components (just as an arbitrary example), the maximum number of coexisting phases is 201 ($K = 200$, $J = 201$, and $F = 0$). In a finite system, macroscopic phase separation is unlikely, and one rather expects domains of different states and compositions that exceed the number of possible phases because the lengths of the domain interfaces represent further degrees of freedom. For this reason, it is clear that a biological membrane is unlikely to be homogeneous, especially when one is close to a melting transition. It is more appropriate to consider it as a fluctuating distribution of regions with different states and compositions.

V.3.3 Membrane Domains

In regions where several phases coexist, the lipids are not distributed homogeneously but are rather arranged in domains and lipid clusters of different composition. The biophysical community would call them domains or even phases, depending on the size of the clusters, while the biological community would rather refer to them as rafts. This expresses a schism between the two communities: Biophysicists think of lipid structures in terms of thermodynamics (this is the view taken here), while biologists think of domains as stable platforms with specific functions, i.e., as functional units.

Different domains display different compositions of lipids and contain different proteins. For this reason, the function of the biomembranes can be strongly influenced by changing the domain structure and the mixing of the membrane components. This can be done by changing the intensive variables, e.g., temperature T, pressure p, pH (i.e., the chemical potential of protons), and lipid composition corresponding to changes in the chemical potentials of the membrane components.

The lipids in different domains differ in chain length, chain charge, and melting points, which prevents their uniform mixing. Figure V.3.3 (left and center) shows domains in an artificial lipid mixture. Their size and composition change with temperature and composition. Domains in biological membranes are often much smaller because they contain many more different lipids and the phase behavior is more complex. However, they can also display large domains, as shown for native lung surfactant at physiological temperature in Figure V.3.3 (right). Figure V.3.4 shows fluorescence images of vesicles made of ternary mixtures (sphingomyelin, cholesterol, and DOPC), a mixture that serves as a model for rafts in biological membranes. One can see that not only does one find fluid-fluid domains in such mixtures, but also regions of different curvature [35].

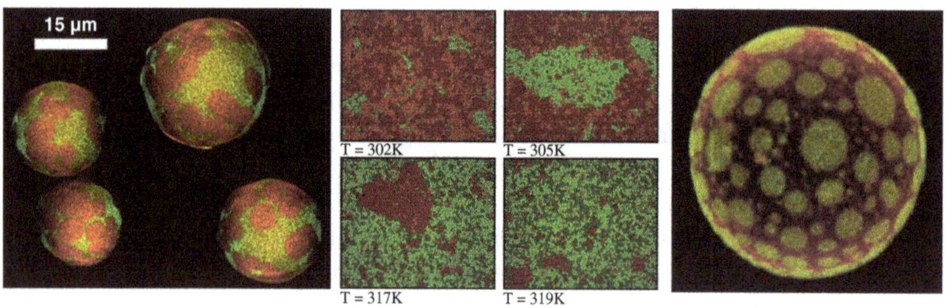

Figure V.3.3 Domains in artificial membrane vesicles made of a DMPC-DSPC mixture. Left: Confocal fluorescence microscopy image of giant vesicles at 33 °C (306 K). Shown in red are solid lipid domains; in green, the fluid regions. Center: Monte Carlo simulations of the same system show the change in domain sizes induced by temperature changes. [33]/with permission of Elsevier. Right: Native pulmonary surfactant at 36 °C showing segregation into gel and liquid domains. [34]/with permission of Elsevier. The vesicle has a diameter of approximately 30 m.

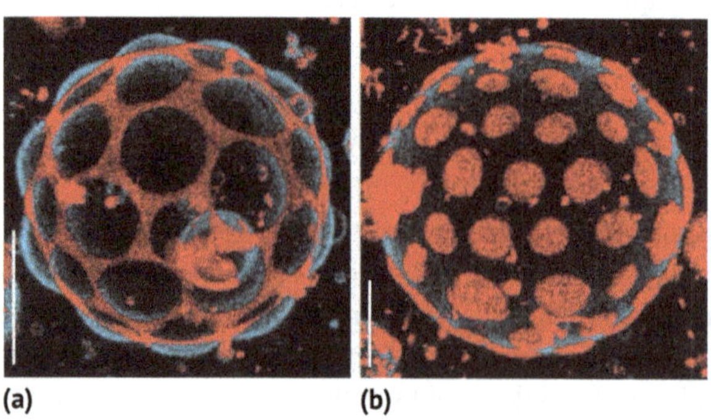

Figure V.3.4 Giant vesicles from mixtures of sphingomyelin, DOPC, and cholesterol with the indicated ratios. Left: 0.56/0.24/0.2. Right: 0.615/0.135/0.25. One can recognize that different domain compositions induce different curvatures. From [35]/with permission of Springer Nature.

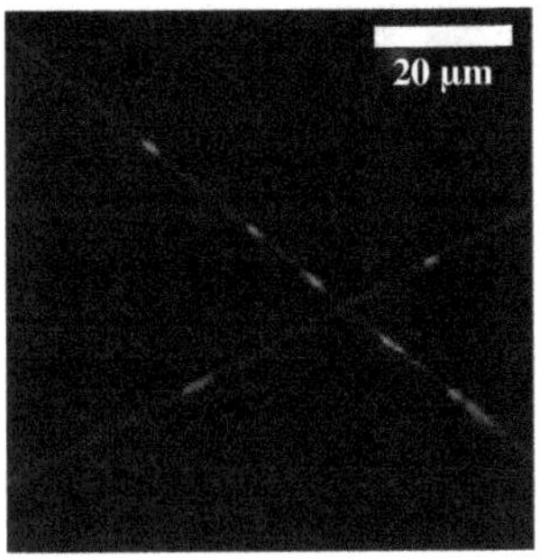

Figure V.3.5 Domains on cylindrical membrane tethers made of a DPPC-DLPC mixture. Courtesy of Marie Lundgaard Daviðsdóttir, NBI.

Figure V.3.5 shows that one can also find membrane domains in cylindrical tethers from lipid mixtures. This is interesting since the whole thermodynamic theory of membrane phase behavior will later be applied to nerve axons.

It should be emphasized that the domain sizes in the above systems (micrometer size) are still much smaller than the size of a nerve pulse (mm to cm). Therefore, on the scale of an action potential, the physical properties of such domains are averaged. This will be assumed in the thermodynamic treatment of the action potential in Part VII.

V.3.4 Domains in Lipid Monolayers

Domains can be nicely imaged on lipid monolayers. Most fluorescence markers prefer the liquid state, and therefore one obtains a contrast between the liquid-condensed and the liquid-expanded phase. A monolayer experiment is performed in the manner indicated in Figure V.3.6 (left). Lipids are deposited on an aqueous surface. The area per molecule is varied by the motion of two Teflon barriers. A needle in the center (called a Wilhelmy balance) is used to monitor the surface pressure of the lipid film.

The fluorescence images of a DPPC lipid film under various states of compression are shown in Figure V.3.6 (right). The lipid area is decreased from left to right and top to bottom. First, small domains of a liquid-condensed phase appear within the liquid-expanded phase. With increasing pressure, they grow and form solid domains with chiral shapes that reflect the chirality of the individual lipids.

Domains in monolayers can also be monitored by atomic force microscopy (AFM). Figure V.3.7 (left) shows a film of the lipid DMPC at its critical point by Nielsen et al. [36, 37]. The size of the image is $25{\times}25\ \mathrm{m}^2$. Different shades correspond to different heights of the film that was deposited on a piece of mica. Nielsen et al. found that the domains in this image display scaling invariance, which is typical for critical phenomena. This implies that the phenomenological shapes of the domains do not depend on the size of the image. On all scales, the domain shapes are self-similar. Figure V.3.7 (right) shows a lattice Monte Carlo simulation of a DPPC layer at the critical point ([32], $200{\times}200$ lipids $\approx 150{\times}150\ \mathrm{nm}^2$). While the simulation snapshot is much smaller than the AFM image, one can see that both images show similar patterns and self-similarity on different scales.

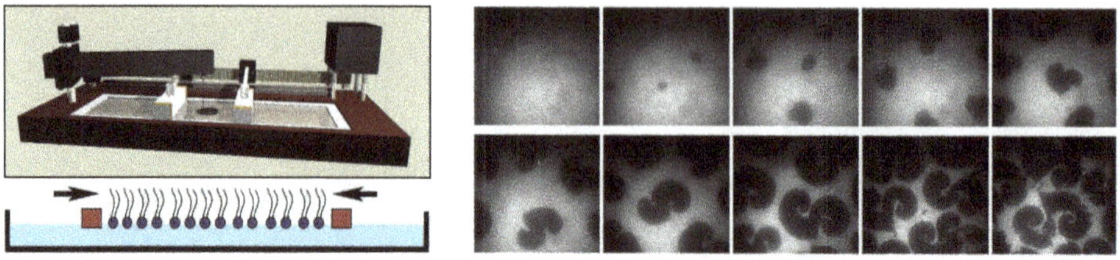

Figure V.3.6 Left, top: Drawing of a monolayer trough. Two teflon barriers can move in order to change the area per molecule. The Wilhelmy balance in the center consists of a needle that measures the surface pressure via capillary forces. Drawing courtesy of Martin Gudmand, Niels Bohr Institute. Left, bottom: Schematic drawing of lipids on the surface of the trough and the two Teflon barriers. Right: Fluorescence images of a DPPC monolayer under compression with 0.85 μm^2/lipid·min. The lateral pressure increases from left to right. Light regions represent lipid in the liquid-expanded state, and dark regions correspond to the liquid-condensed phase. Data from Martin Gudmand, NBI.

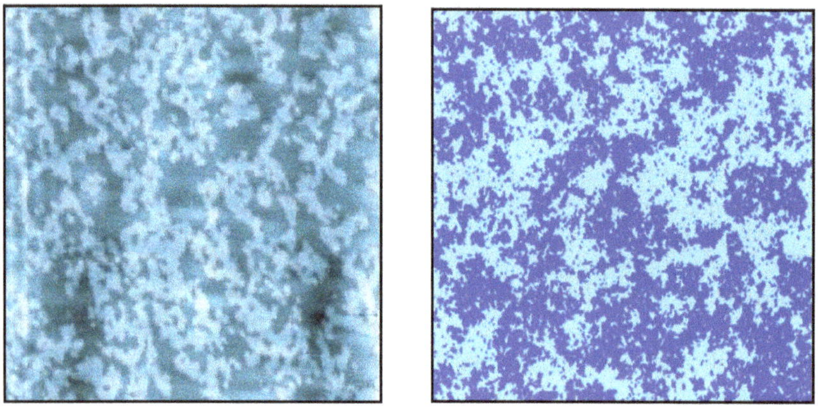

Figure V.3.7 Critical fluctuations in lipid monolayers. Left: AFM image of liquid-condensed/liquid-expanded domains of DMPC at its critical point. Image size is 25×25 m^2. From [36]/with permission of Springer Nature. Right: Critical fluctuations of DPPC from a Monte Carlo simulation. The scale is approximately 150×150 nm^2. See [32] for details on the simulations.

What we learn from these considerations is that one can induce domain formation not only by changing temperature but also by lateral pressure. In general, the state of the membrane and the appearance of domains depend on all intensive variables, including T, p, Π, V, μ_H (i.e., pH), and μ_i (the chemical potentials of the membrane or monolayer components).

V.3.5 Nanodomains or Rafts in Biological Membranes

The appearance of (small) domains in biological membranes is (or was) a hot topic in biophysics. Simons and Ikonen [38] proposed that there exist small domains in biomembranes composed of sphingolipids, cholesterol, and unsaturated lipids to which proteins are associated. Since the authors considered these domains as a kind of functional platform or aggregate (just as a ribosome is a functional aggregate of nucleic acids and proteins), they called these platforms *rafts*. This is schematically shown in Figure V.3.8 taken from [39]. There was a controversial debate on whether these domains represent something like phase

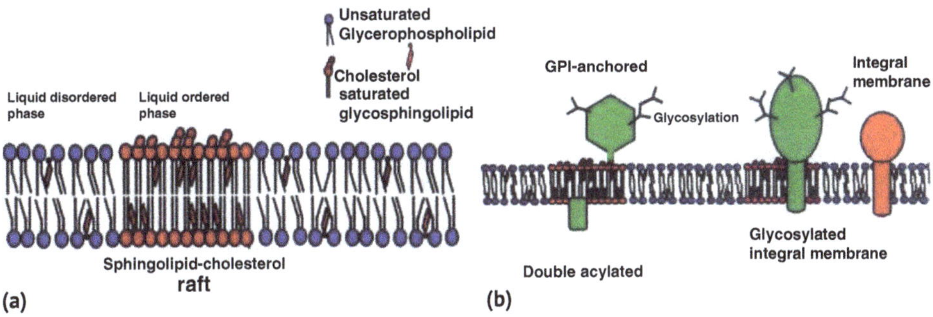

Figure V.3.8 Schematic drawing of lipid rafts in a biological membrane. The raft consists of regions enriched in saturated sphingolipids (with high melting temperature), cholesterol, and GPI-anchored proteins. [39]/with permission of Walter de Gruyter GmbH.

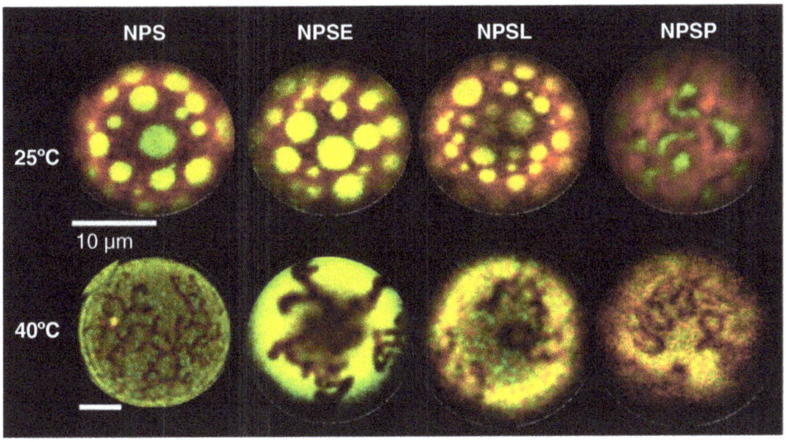

Figure V.3.9 Four pulmonary surfactant preparations at 25 °C and 40 °C showing different segregation into gel and liquid domains. [23]/with permission of Elsevier.

behavior, or whether they are functional and rigid units as the name raft suggests. Today, most physical chemists and biophysicists most likely agree that rafts are in fact domains that occur due to phase behavior. Sphingolipids are mostly lipids with high melting temperature (often above physiological temperature), while unsaturated lipids display low melting temperatures way below physiological temperature. Cholesterol usually increases the transition temperature and makes membranes more rigid (see Section V.8.4). This means that the shape and composition – and thus the function – of rafts should depend on temperature, pressure, and other intensive variables. Thus, *raft* may just be another word for a domain in a biological membrane.

Rafts are believed to be on nanoscale, which are difficult to monitor in normal microscopy. However, some systems display large domains. Figure V.3.9 shows the phase behavior of four different lung surfactant preparations at two temperatures. One can see large circular domains at 25 °C and thin elongated structures at 40 °C.

The temperature dependence of domains in lung surfactant has been shown even more nicely by [40]. In Figure V.3.10 (left), one can see the domain structure of lung surfactant

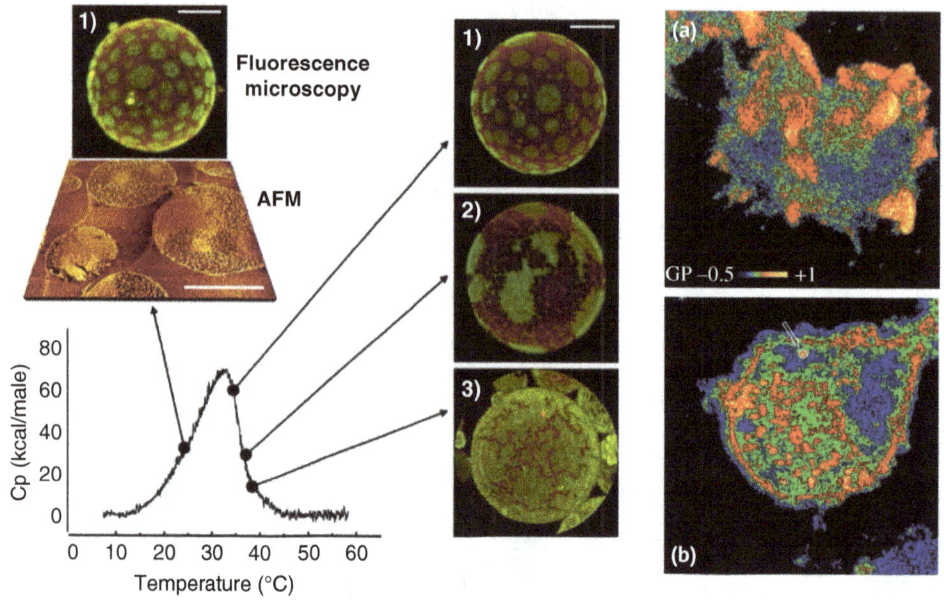

Figure V.3.10 Left: Fluorescence microscopy images of giant vesicles from pulmonary surfactant at different temperatures. [40]/with permission of Elsevier. An AFM image of the vesicular surface also indicates the presence of domains with different membrane thickness. Right: Fluorescence microscopy images of domain formation in living cells. (a) A mouse macrophage (RAW264.7 cell, horizontal width 28.3 μm). (b) A human leukemia monocyte cell (THP-1, horizontal width 39 μm). Different colors indicate general polarization values indicative of different lipid states. Yellow and red shades indicate more solid regions, while blue and green shades indicate more liquid regions. [41]/with permission of National Academy of Sciences.

at four different temperatures in the melting regime. The domains change shape and the liquid area (shown in green) increases with temperature. The domain shapes can also be monitored by AFM measurements. This indicates that both in artificial lipid membranes and biological preparations, the melting of the membrane goes along with rearrangements of molecules and the presence of regions of different composition and physical state at different locations in the membrane. This strongly suggests that the thermodynamic variables will change biomembrane function because the diffusion of membrane components (e.g., proteins) can change as a function of these variables.

Domains can also be seen in some living cells. Gaus et al. [41] investigated mouse macrophages and human leukemia monocyte cells with the fluorescence marker Laurdan that can monitor the different polarizations of gel and fluid membrane phases via a parameter called the generalized polarization (GP) (Figure V.3.10). In both cells, one can recognize different shades of the GP value corresponding to gel and fluid lipid states. Thus, it must be concluded that domain exist in living cells. This implies that cell membranes display phase behavior.

V.3.6 Summary: Key Ideas of Chapter V.3

- Lipids can display different hydrophobic lengths due to a different chemistry of their chains or due to a different physical state (ordered or disordered).
- Lipids of different lengths do not mix perfectly but display the potential of forming domains with different compositions or states that minimize the free energy of the membranes. The minimization of the interfacial contact free energy is called *hydrophobic matching.*
- Proteins will not mix equally well with lipids of different lengths or states. The *mattress model* by Mouritsen and Bloom takes the hydrophobic matching into account. It states that depending on the lipid environment, the interaction between proteins and lipids can lead to attractive forces between molecules with similar hydrophobic length. This was not predicted by Singer and Nicolson's fluid mosaic model.
- The non-ideal mixing of lipids leads to phase behavior, which describes the temperature- and pressure-dependent formation of lipid phases, the composition of the phases, and their coexistence.
- Gibbs' phase rule predicts that in a system with K components, a maximum of K+1 phases can coexist. For a biological membrane that consists of many hundreds of components, the phase behavior will be very complex and practically impossible to resolve.
- In lipid membranes, regions of different composition and state are not macroscopically separated. One rather finds domains of finite size. These domains are found in bilayers of artificial lipids and in monolayers. The composition and size depend on the intensive thermodynamic variables.
- In biological membranes, the existence of nanoscopic domains called *rafts,* consisting of sphingolipids, cholesterol, and unsaturated lipids, has been proposed. Most likely, rafts are not functional multi-molecular platforms (as suggested by the term *raft*). They are rather evidence of phase behavior and domain formation that are dependent on the intensive variables.

V.4

Influence of Hydrostatic Pressure and Lateral Pressure

Pressure is a variable that is important for membrane melting. This can be hydrostatic pressure, lateral pressure, or tension within the membrane plane. Hydrostatic pressure is important for deep-sea animals that can experience pressures of up to 1000 bars. It is also important for the interesting phenomenon of the pressure reversal of anesthesia, discussed in Section IX.1.5. Lateral tension in cell membranes can occur in cells that are subject to osmotic pressure differences inside and outside the cell, in pipette aspiration experiments, in neuronal membranes after the mechanical extension of nerves (see Chapter IX.3), and importantly in the propagation of electromechanical pulses discussed in Part VII.

V.4.1 Hydrostatic Pressure Dependence of Membrane Melting

The melting temperature of a lipid membrane is given by

$$T_m = \frac{\Delta H_0}{\Delta S_0}, \tag{V.4.1}$$

where

$$\Delta H_0 = \Delta E_0 + p\Delta V_0 + \Pi\Delta A_0 - \Psi\Delta q_0 + \dots . - \sum \mu_i \Delta n_i . \tag{V.4.2}$$

Thus, the melting temperature depends on all intensive variables. When only hydrostatic pressure is altered, $\Delta H_0 = \Delta E_0 + p\Delta V_0$. Figure V.4.1 shows a densitometric experiment on large unilamellar vesicles (LUV) of the lipid DMPC adapted. This sample has been chosen because its heat capacity profile shows a lot of detail. One can see the temperature dependence of the specific volume of the lipid membranes, which is close to 1 cm^3/g, i.e., the specific volume of water. Further shown is the volume expansion coefficient, dV/dT at constant pressure. The right-hand panel shows that the temperature dependence of the volume expansion coefficient and the excess heat capacity are completely superimposable. This means that, within error

$$\frac{d\Delta V(T)}{dT} = \gamma_V \frac{d\Delta H(T)}{dT}, \tag{V.4.3}$$

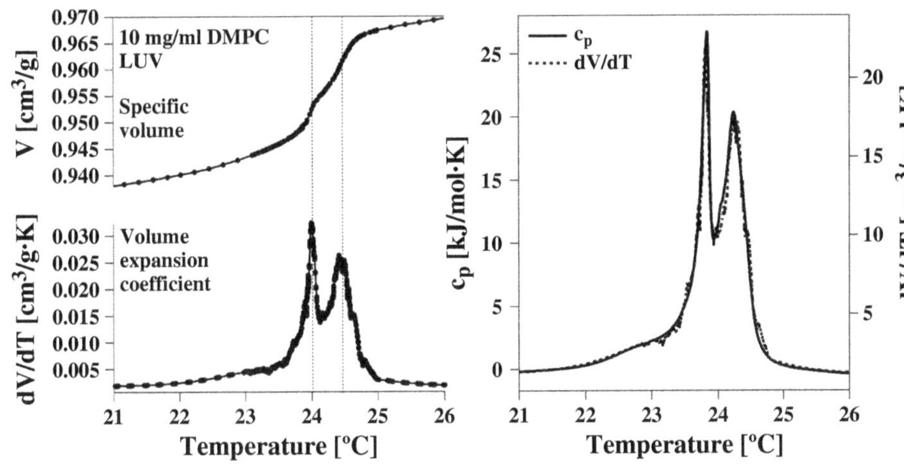

Figure V.4.1 Left: Specific volume and volume expansion coefficient of DMPC LUV. Right: Comparison of the heat capacity and the volume expansion coefficient suggest the enthalpy and volume changes in the transition are exactly proportional functions. [42]/with permission of American Chemical Society.

where $\gamma_V = 9.1 \cdot 10^{10}$ m^2/N in the densitometric experiment. This implies that

$$\int_{T_0}^{T_1} \frac{d\Delta V(T)}{dT} dT = \gamma_V \int_{T_0}^{T_1} \frac{d\Delta H(T)}{dT} dT$$
$$\Delta V_0 = \gamma_V \cdot \Delta H_0. \tag{V.4.4}$$

In Section VI.3.2, we report about pressure calorimetry, which is another, more accurate method to measure γ_V. Here we find $\gamma_V = 7.761 \cdot 10^{10}$ m^2/N for the system in Figure V.4.1. In the following, we will rather use this value.

Thus, the pressure dependence of the melting temperature is given by

$$T_{m,p} = \frac{\Delta E_0}{\Delta S_0} + p\frac{\Delta V_0}{\Delta S_0}, \tag{V.4.5}$$

where ΔV_0 is the excess volume change in the transition. We therefore see that the melting temperature of a lipid membrane is proportional to the hydrostatic pressure.

Let us assume a lipid membrane with a cooperative $n_{coop} = 120$. According to eq. (V.2.9), the heat capacity profile can be calculated as

$$\Delta c_p(T) = \frac{df}{dT} \cdot (E_0 + p\Delta V_0) = \frac{K}{(1+K)^2} \cdot \frac{n(E_0 + p\Delta V_0)^2}{RT^2}. \tag{V.4.6}$$

where $K = \exp(-n \cdot (\Delta E_0 + p\Delta V_0 - T\Delta S)/RT)$. This is shown in Figure V.4.2 for two excess pressures, $p = 0$ bar and $p = 100$ bars. In this calculation, we assumed that $\Delta H_0 = 35$ kJ/mol, $\Delta V_0 = \gamma_V \cdot \Delta H_0 = 2.73 \cdot 10^{-5}$ m^3/mol, $n = 120$, and $\gamma_V = 7.8 \cdot 10^{10}$ m^2/N (see below for justification of γ_V). We see that the heat capacity profile shifts by 2.46 °C upon application of 100 bars.

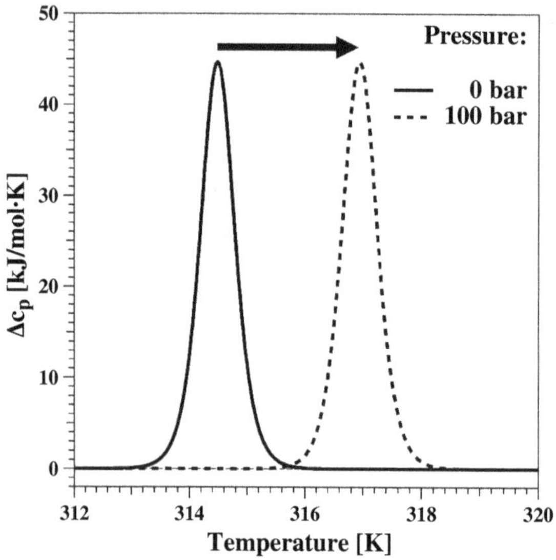

Figure V.4.2 Calculated c_p profiles for DPPC LUV at two different excess pressures: 0 bar and 100 bars.

V.4.1.1 Pressure Dependence of Single Lipid Melting

The pressure dependence of experimental heat capacity profiles closely resembles the calculated profiles shown in Figure V.4.2. For the example of DPPC lipid membranes, the enthalpy at a pressure of 1 bar is $\Delta H_0 = 35\,\text{kJ/mol} \equiv 47.7\,\text{J/kg}$. The melting temperature $T_{m,1bar} = 314.49$ K. Thus, the entropy change in the transition is $\Delta S_0 = 151.8$ J/mol·kg. The change in volume in the transition is $\Delta V_0/V_0 = 0.0406$ and $V_0 \approx 0.001$ m^3/kg for DPPC [27]. This implies that $\Delta V_0/\Delta H_0 = 8.51 \cdot 10^{10}$ m^2/N. If the pressure is changed by $\Delta p = 40$ bars $(= 40 \cdot 10^5\ \text{N/m}^2)$, the shift in the transition temperature is then expected to be

$$\Delta T_{m,p} = \Delta p \frac{\Delta V_0}{\Delta S_0} = T_m \Delta p \frac{\Delta V_0}{\Delta H_0} = 1.07\ \text{K}. \tag{V.4.7}$$

Figure V.4.3 (left) shows the heat capacity profile of DPPC LUV at 1 bar and at 176 bar pressure. The shift induced by this pressure is 4.19 K, which corresponds to 0.95 K/40 bars. The ratio γ_V between ΔV_0 and ΔH_0 is then given by

$$\gamma_V = \frac{\Delta V_0}{\Delta H_0} = \frac{\Delta T_{m,p}}{T_{m,p=1} \cdot \Delta p} = 7.68 \cdot 10^{10} \frac{\text{m}^2}{\text{N}}, \tag{V.4.8}$$

which is slightly lower than the value estimated earlier based on literature values.

V.4.1.2 Pressure Dependence of Lipid Mixtures

The heat capacity profiles of lipid mixtures display a very similar pressure dependence. Figure V.4.3 (right) shows heat capacity profiles of a DMPC:DPPC = 50:50 mixture. The value for γ_V is given by [42]:

$$\gamma_V = 7.91 \cdot 10^{10} \frac{\text{m}^2}{\text{N}}. \tag{V.4.9}$$

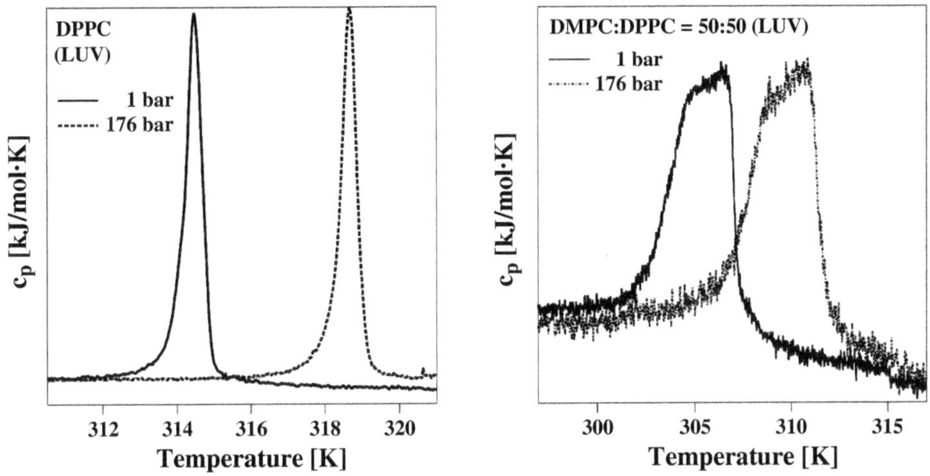

Figure V.4.3 Left: Shift of the heat capacity profile of DPPC LUV induced by a pressure change of 175 bars. Adapted from [42]. Right: Shift of the heat capacity profile of the lipid mixture DMPC: DPPC = 50:50 induced by a pressure change of 175 bars. Adapted from [42].

According to [42], the γ_V value of DMPC LUV is $\gamma_V = 7.761 \cdot 10^{-10}$ m^2/N, for DMPC MLV it is $\gamma_V = 7.841 \cdot 10^{-10}$ m^2/N, for DPPC LUV it is $\gamma_V = 7.543 \cdot 10^{-10}$ m^2/N, and for DPPC MLV it is $\gamma_V = 7.82 \cdot 10^{-10}$ m^2/N. Thus, the shift of the mixture is, within error, the same as for the individual lipids.

V.4.1.3 Pressure Dependence of Phase Diagrams

Just as the heat capacity profiles in mixtures of two lipids shift with pressure, the complete phase diagrams of the lipids also shift with pressure. Figure V.4.4 shows the phase diagrams of DMPC:DPPC mixtures at three different pressures. Generally, it is to be expected that phase diagrams of the melting of arbitrary lipid mixtures will shift with pressure.

Taking the points in Figure V.4.4, one calculates an average value $\gamma_V = 9.97 \cdot 10^{10}$ m^2/N. This value is close to the values mentioned earlier, but slightly higher. However, the data in [43] were determined from electron spin resonance, and absolute temperature values are much less accurate than those determined with calorimetry.

V.4.1.4 Biological Membranes

Biological membranes are complex mixtures of lipids (and proteins). The melting peak of the lipids does not directly monitor the proteins. However, due to hydrophobic matching (Section V.3.1), proteins influence lipid melting. Thus, the heat capacity profile seen in calorimetric experiments reflects the presence of proteins but still displays the melting of the lipids. The influence of pressure on bovine lung surfactant can be seen in Figure V.4.5 (left). The shift of the melting profile is 4.4° for a pressure of 196 bars. According to [42], this corresponds to $\gamma_V = \Delta V_0/\Delta H_0 = 7.809 \cdot 10^{-10}$ m^2/N – very similar to the γ_V of DPPC and DMPC (see eq. (V.4.8)). This is not so surprising because the major lipid in lung surfactant is DPPC.

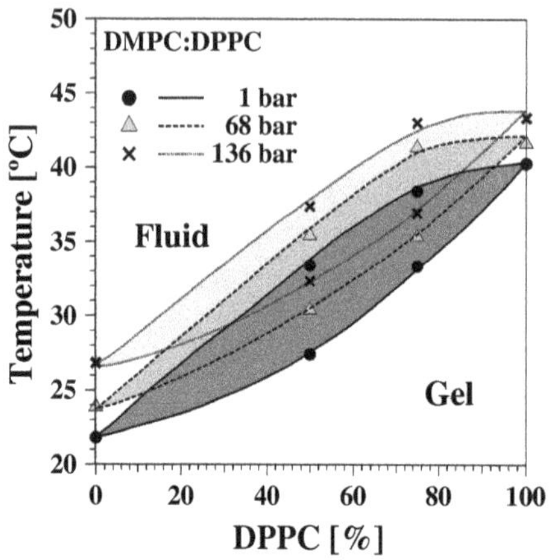

Figure V.4.4 Phase diagram of DMPC:DPPC mixtures as a function of pressure. The complete phase diagram shifts toward higher temperatures. The gray-shaded regions represent the gel-fluid coexistence regime. Adapted from [43].

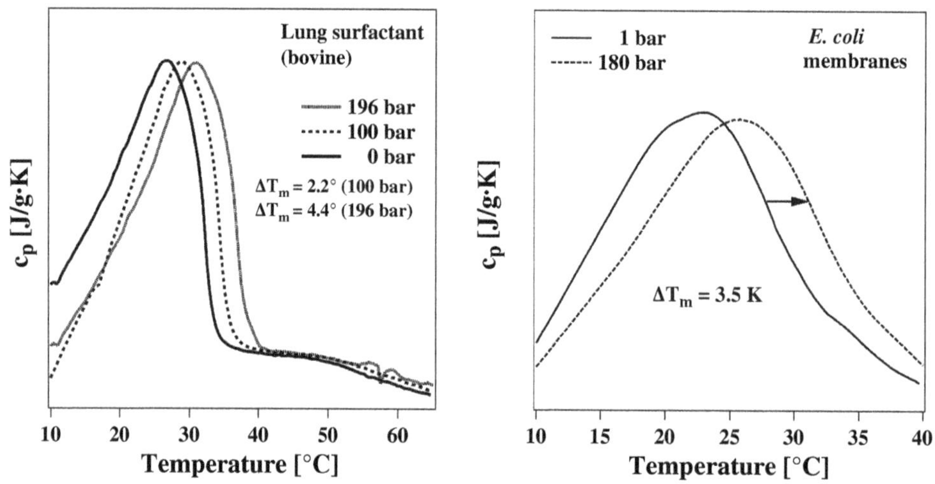

Figure V.4.5 Left: The heat capacity profiles of bovine lung surfactant at three different pressures. Adapted from [22]. The shift of the transition at 196 bars corresponds to a change in the c_p maximum of 4.4 °. Right: Heat capacity profiles of *E. coli* membranes at two pressures. 180 bars of pressure correspond to a shift of the transition maximum by 3.5 °. Adapted from [44].

Another example of the influence of pressure on membrane melting is given in Figure V.4.5 (right). 180 bars of pressure shift the transition profile of *E. coli* membranes by 3.5 °. This corresponds to $\gamma_V = 6.5 \cdot 10^{-10}\ \mathrm{m}^2/\mathrm{N}$, somewhat smaller than the aforementioned preparations but still remarkably close. According to [42], the average γ_V value of different DMPC, DPPC preparations, lipid mixtures, and lung surfactant is $\gamma_V = 7.79 \pm 0.11 \cdot 10^{-10}$, and all values reported earlier are, within error, the same or at least very similar. This allows us to estimate volume changes in biological membranes from their heat capacity profiles. With certain caution, this can also be done for bacterial membranes that contain a lot of uncommon lipids. The volume-to-enthalpy ratio γ_V is about 16% smaller in the bacterium as compared to lung surfactant. However, the difference is relatively small and still allows for meaningful general estimates of the elastic constants (see below). We see that the pressure dependence of biological preparations is remarkably close to that of single lipids.

V.4.1.5 Volume Compressibility

From the above, it is obvious that the transition profile shifts as a function of pressure. This implies that during a measurement at constant temperature, the volume will change as a function of pressure. Figure V.4.6 (left) shows the change in excess volume, ΔV, as a function of pressure at constant temperature. The profile changes for different temperatures. Shown are the excess volume changes at two temperatures – 316.0 K and 317.5 K – both above the melting temperature $T_m = 314.49$ K (41.34 °C).

It is also clear that the change in volume for an incremental change in pressure will be much larger within the transition. The susceptibility of volume changes to a pressure change is called the compressibility. The specific isothermal volume compressibility of a membrane is defined by

$$\kappa_T^V = -\frac{1}{V}\left(\frac{\partial V}{\partial p}\right)_T, \tag{V.4.10}$$

or more explicitly,

$$\Delta\kappa_T^V = -\frac{1}{V}\frac{df}{dp}\Delta V_0 = \frac{1}{V}\frac{K}{(1+K)^2}\cdot\frac{n(\gamma_V \Delta H_0)^2}{RT}, \tag{V.4.11}$$

where $f = K/(1+K)$ and $K = \exp(-n\cdot(\Delta E_0 + p\Delta V_0 - T\Delta S)/RT)$ is the fluid fraction. Figure V.4.6 (right) shows the compressibility as a function of pressure for two temperatures, 316 K and 317.5 K, respectively [27]. The compressibility displays a maximum in the melting transition. The melting peak is found at different pressures for different temperatures.

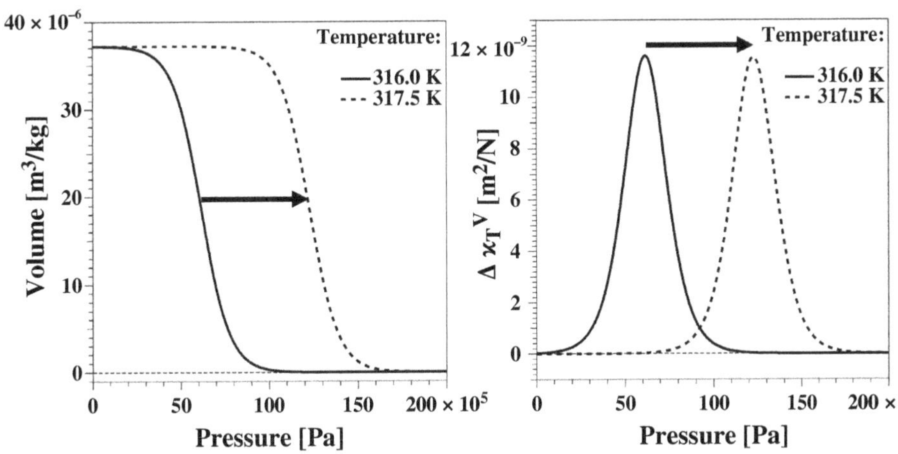

Figure V.4.6 Left: Calculated excess volume change of DPPC LUV at two different temperatures above $T_m = 314.5$ K. Right: Calculated excess isothermal volume compressibility at two different temperatures above $T_m = 314.5$ K.

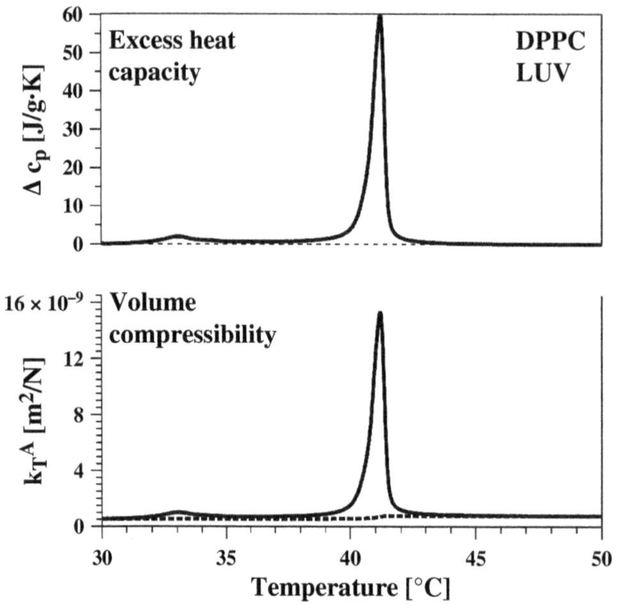

Figure V.4.7 The volume compressibility of DPPC LUV calculated from the heat capacity. Adapted from [27].

Since the compressibility is obviously also a function of temperature, one can also plot κ_T^V as a function of temperature. From eqs. (V.4.6) and (V.4.11), it is clear that

$$\Delta\kappa_T^V = \frac{T \cdot \gamma_V^2}{V}\Delta c_p. \tag{V.4.12}$$

Thus, the excess isothermal compressibility can be calculated from the excess heat capacity. This is shown for DPPC LUV in Figure V.4.7. In this figure, the total compressibility was calculated by taking into account the compressibility of the pure phases. For details, see [27].

V.4.2 Lateral Pressure and Tension

We showed in Sections V.2.2.2 and V.4.1.1 that the enthalpy H is a function of temperature T, and the volume V is a function of hydrostatic pressure p, respectively. In the same manner, the area of a membrane A is a function of lateral pressure Π. Here, we call the pairs of extensive and intensive variables, $(H - T, V - p, A - \Pi)$ pairs of conjugated variables.

The calculations from the previous section can be repeated for lateral compressibility. The change in the transition temperature upon changes in lateral pressure in analogy to eq. (V.4.5) is given by

$$T_m = \frac{\Delta H_0 + \Pi\Delta A_0}{\Delta S_0} \quad \rightarrow \Delta T_m = \frac{\Pi\Delta A_0}{\Delta H_0} \cdot T_{m,0}. \tag{V.4.13}$$

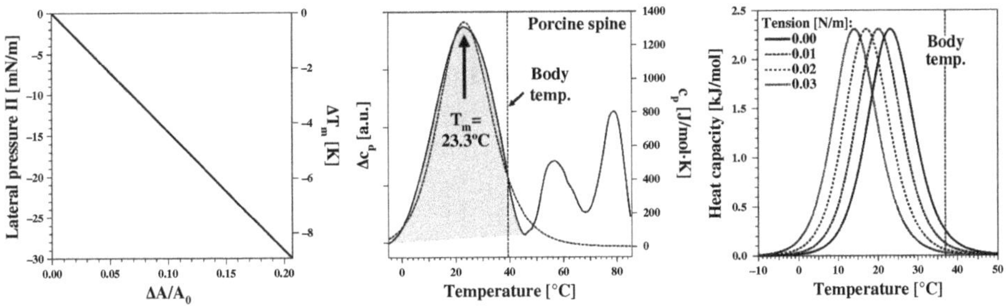

Figure V.4.8 Left: The calculated dependence of membrane tension and the shift in T_m upon a change in membrane area. Center: The heat capacity profile of membranes from pig spine, and a fit according to eq. (V.2.9). Right: The shift of the theoretical profile for different membrane tensions. [45]/with permission of Elsevier.

And the heat capacity profile is given in analogy to eq. (V.4.6) by

$$\Delta c_p(T) = \frac{df}{dT} \cdot (\Delta E_0 + \Pi \Delta A_0) = \frac{K}{(1+K)^2} \cdot \frac{n(\Delta E_0 + \Pi \Delta A_0)^2}{RT^2}. \tag{V.4.14}$$

where $K = \exp(-n \cdot (\Delta E_0 + \Pi \Delta A_0 - T\Delta S)/RT)$. The effect of lateral tension (defined as $-\Delta\Pi$) is shown in Figure V.4.8.

The specific isothermal area compressibility of membranes is given by

$$\kappa_T^A = -\frac{1}{A}\left(\frac{\partial A}{\partial \Pi}\right)_T, \tag{V.4.15}$$

which can readily be calculated from the heat capacity in analogy to eq. (V.4.12) by

$$\kappa_T^A = \frac{T \cdot \gamma_A^2}{A} \Delta c_p. \tag{V.4.16}$$

The factor γ_A between area and enthalpy ($\Delta A(T) = \gamma_A \cdot \Delta H(T)$) was given by $\gamma_A = 0.89$ m/N for DPPC membranes [27]. The area compressibility calculated from the heat capacity is shown for DPPC LUV in Figure V.4.9. The factor γ_A between area and enthalpy ($\Delta A(T) = \gamma_A \cdot \Delta H(T)$) given in [27] is $\gamma_A = 0.893$ m/N. The total compressibility was calculated by taking into account the area compressibility of the pure phases. For details, see [27]. The maximum in area compressibility was also described by [46].

V.4.3 Compressibility of Monolayers

While there do not exist many experiments relating a compressibility to a lateral pressure and the values given in the previous section are inferred indirectly, the lateral isothermal compressibility can be easily measured in monolayer experiments, where the area as a function of pressure is calculated directly. Figure V.4.10 shows compression isotherms of the lipid DPPC at four different temperatures and the corresponding lateral compressibility.

The right-hand panel of this figure corresponds to Figure V.4.6 for volume compressibility as a function of pressure at different temperatures. As in the case of volume compressibility, the transition peak is found at higher pressure when the temperature is higher.

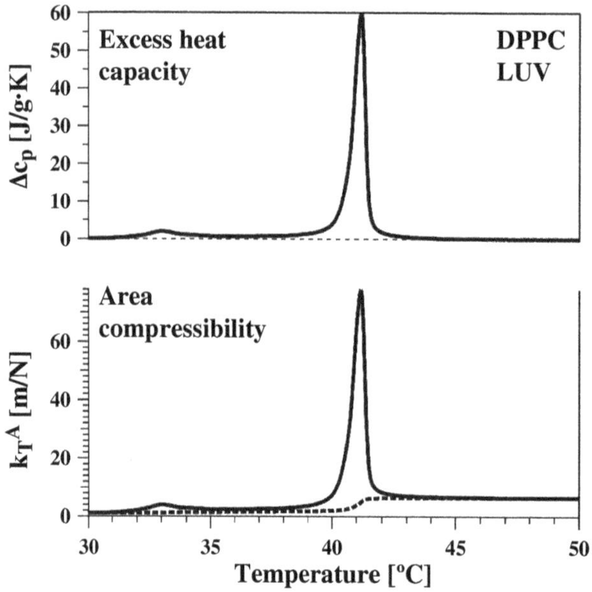

Figure V.4.9 The area compressibility of DPPC LUV calculated from the heat capacity. Adapted from [27].

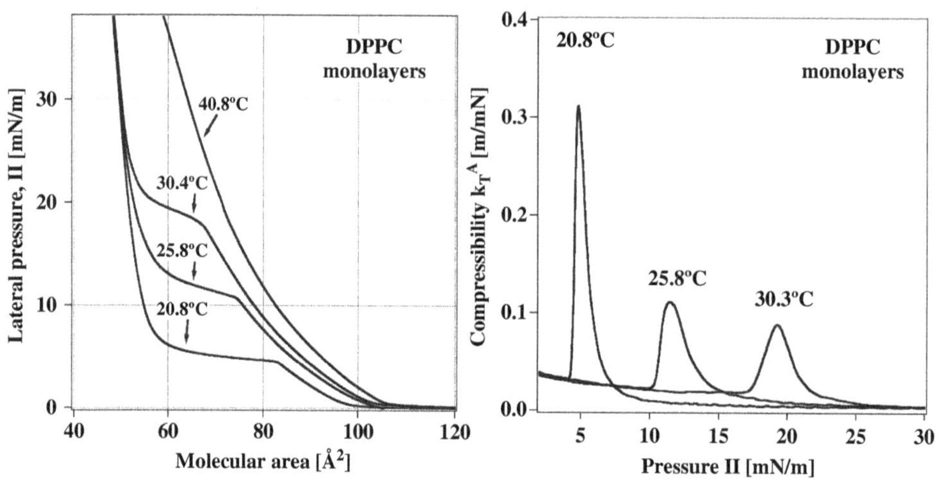

Figure V.4.10 Left: Pressure area isotherms of monolayers of the lipid DPPC for four different temperatures. Right: The monolayer compressibility derived from the isotherms in the left-hand panel. [1]/with permission of John Wiley & Sons.

V.4.4 Summary: Key Ideas of Chapter V.4

- Hydrostatic pressure is an important intensive variable for the behavior of biological membranes.
- It shifts the melting profiles of single lipids, lipid mixtures, and biological membranes in a very similar manner.
- Likewise, lateral pressure shifts melting profiles.

- The excess volume expansion coefficient, $d\Delta V/dT$, is proportional to the heat capacity, i.e., $d\Delta V/dT = \gamma_V \cdot \Delta c_p$. The constant $\gamma_V = 7.8 \cdot 10^{-10}$ m^2/N is surprisingly similar for many preparations, including biological membranes.
- From the heat capacity, changes in isothermal volume compressibility can be obtained:

$$\Delta \kappa_T^V = \frac{T\gamma_V^2}{V}\Delta c_p \tag{V.4.17}$$

- The behavior of the area compressibility is quite similar. Here, $\gamma_A = 0.89$ m/N, and changes in area compressibility can be determined:

$$\Delta \kappa_T^A = \frac{T\gamma_A^2}{A}\Delta c_p \tag{V.4.18}$$

- Area compressibility can be easily measured in monolayer compression experiments.

V.5

Curvature

Many biological membranes are curved. The curvature is especially high for small vesicles, such as synaptic vesicles, or in fusion pores during exocytotic and endocytotic fission and fusion events. The curvature c is the inverse of the radius of curvature. Since a membrane is quasi-two-dimensional, there exist two radii of curvature, $c_1 = 1/R_1$ and $c_2 = 1/R_2$, respectively. If a membrane is not symmetric due to a different state or composition of the outer and inner monolayers, the membrane is even curved in the absence of forces. The equilibrium curvature is called the *spontaneous curvature* c_0.

V.5.1 Helfrich's Description of Curvature Free Energy

In 1973, Helfrich published a famous paper that treats the membrane as a two-dimensional liquid crystal and assigns elastic constants to its deformation free energy [47]. For a two-dimensional fluid membrane in which the elastic constants are uniform and independent of orientation, the elastic free energy density is given by

$$g = \frac{1}{2}K_B\left(c_1 + c_2 - c_0\right)^2 + K_G\,c_1c_2, \tag{V.5.1}$$

where c_1 and c_2 are principal curvatures, and c_0 is the spontaneous curvature, which is the equilibrium curvature of the membrane when no forces are present. K_B is the bending modulus, and K_G is the Gaussian modulus. The total free energy of curvature of a closed surface is $G = \oint g dA$. One can add other terms to eq. (V.5.1), for instance when the membrane is stretched or sheared.

Examples of applications of membrane curvature are vesicular shapes [48–50], adhesion phenomena [51, 52], and the calculation of the energy of fusion pores [53]. Figure V.5.1 shows several cell membrane nanovesicles (panels a–e) and lipid vesicles (panels f–i), along with the corresponding minimum free energy shapes (from [50]). Interestingly, the different vesicular shapes in Figure V.5.1(f–i) were generated by slightly adjusting temperature (and thus membrane area); see also [54].

Elastic behavior was not considered in the fluid mosaic model. However, the elastic behavior of membranes is important for everything that is related to changes in the shape of cells, vesicles, and membrane surfaces. If there were thermodynamic ways to control the magnitude of the elastic constants, one would provide a possibility to influence the shapes

Physics of Nerves and Excitatory Membranes, First Edition. Thomas Heimburg.

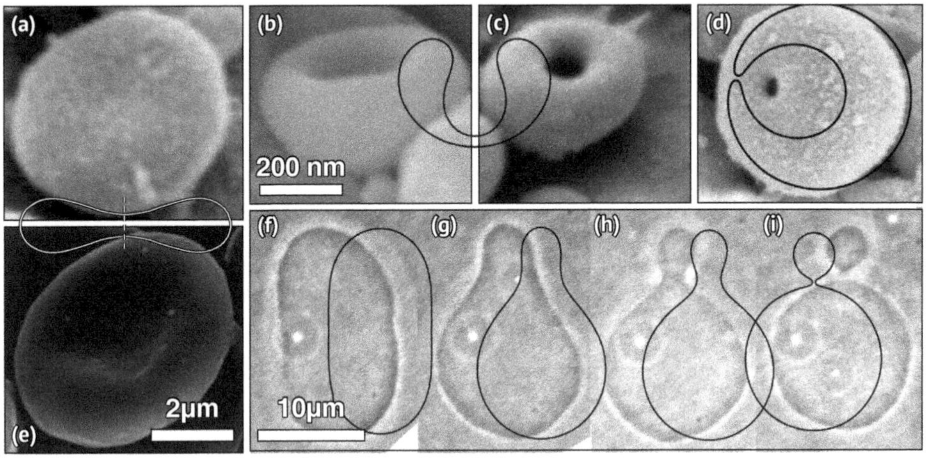

Figure V.5.1 Cell membrane nanovesicles imaged by electron microscopy (a–e) and lipid vesicles imaged by light microscopy (f–i) and shapes minimizing the free energy (solid lines). From [50]/PUBLIC LIBRARY OF SCIENCE (PLOS)/CC BY 4.0.

of biological membranes. If these changes occur locally, one would provide the means to create local mechanical perturbations.

V.5.2 The Bending Modulus and the Lateral Compressibility

Upon membrane bending, the outer monolayer is expanded while the inner monolayer is compressed. This is indicated schematically in Figure V.5.2. Therefore, the curvature bending modulus K_B is related to the area compression modulus, K_A.

The bending modulus K_B is the inverse of the bending elasticity κ_B, and the area compression modulus K_A is the inverse of the area compressibility κ_T^A. In [27], the relation between area compressibility and bending elasticity was calculated to be

$$\kappa_B = \frac{16}{D^2}\kappa_T^A, \tag{V.5.2}$$

where κ_T^A is proportional to the heat capacity given in eq. (V.4.16). For the bending modulus, this implies that $K_B = D^2/16\kappa_T^A$. The bending modulus displays a minimum when the bending elasticity displays a maximum (Figure V.4.9). Figure V.5.3 shows the experimentally measured bending modulus of dipalmitoyl phosphatidylcholine (DPPC) membranes. It is compared to the values predicted from heat capacity or area compressibility (solid line). As indicated earlier, the modulus displays a minimum at the melting temperature. This means that the membrane is softest at the melting point. The agreement between experiment and calculation is a proof of concept of the coupling between lateral compressibility and bending modulus.

Since compressibility is a function of temperature, pressure, and other intensive variables, the same is also true for bending elasticity. This implies that the likelihood of highly curved membrane regions, e.g., during membrane fusion, pore formation, and endo- and exocytotic

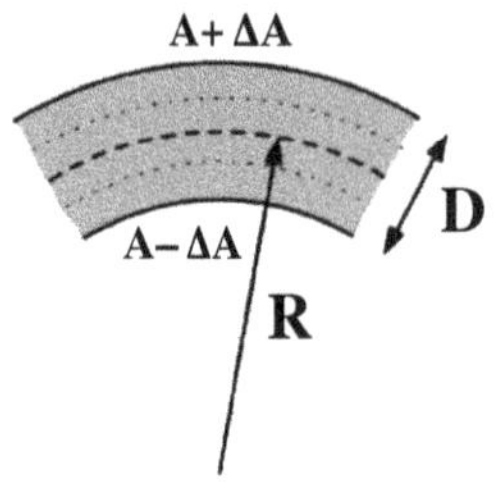

Figure V.5.2 Curvature of a membrane. The outer monolayer is stretched while the inner monolayer is compressed. From this, it follows a relation of the curvature elasticity $\varkappa_B$ and the area compressibility $\varkappa_T^A$. [1]/with permission of John Wiley & Sons.

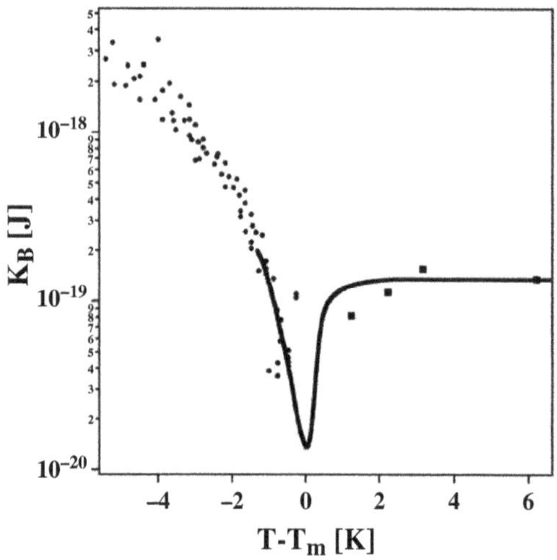

Figure V.5.3 The bending modulus K_B of DPPC membranes as a function of the temperature distance from the melting point. Symbols reflect experimental measurement by [55] and [56].The solid line is the prediction from the heat capacity of DPPC large unilamellar vesicles (LUV).

events, is also a function of the intensive variables and depends on temperature, pressure, and other variables.

V.5.3 Summary: Key Ideas of Chapter V.5

- The elastic free energy of a membrane is described by the so-called Helfrich Hamiltonian, which describes the dependence of the free energy on the principle curvatures.
- The elastic free energy depends on two elastic constants: the bending modulus K_B and the Gaussian modulus K_G, as well as the value of the spontaneous curvature that represents asymmetry between the inner and outer monolayers.
- The value of the bending modulus is inversely proportional to the area compressibility and therefore also inversely proportional to the heat capacity.
- At a heat capacity maximum, the bending modulus will display a minimum, i.e., the membrane is easiest to bend.

- Since the value of the heat capacity depends on temperature, hydrostatic pressure, and lateral pressure due to shifts in the melting temperature, the bending modulus also depends on these intensive variables.
- Knowledge of the bending modulus will help to determine the likelihood of membrane pore formation (see Part VIII) and fusion and fission events of membrane vesicles, for instance, synaptic fusion.

V.6

Influence of pH and Ionic Strength

Protons and ions interact with charged lipids because they neutralize or shield electric charges. If a membrane is charged, there act repulsive forces within the membrane plane that favor the lipid phase with the larger area, i.e., the fluid phase. As a consequence, the melting temperature increases when the surface charge density is reduced. For details, see [1]. Here, we are mostly interested in phenomenology to get trends.

V.6.1 pH and Salt Dependence of Single Lipid Melting

Träuble et al. [57] performed a careful theoretical and experimental study on the melting temperatures of charged lipid membranes at different pH and different concentrations of NaCl. This is shown for the lipid methylphosphatidic acid (MPA) in Figure V.6.1. They find that there is a change of the melting temperature by about 14–15° upon lowering the pH by three units. This is due to a neutralization of lipid charges by the binding of protons. The midpoint of the pH-dependent shift in transition temperature is around pH 7 for an ionic strength of 0.2 mM NaCl, while it is around pH 4 at 300 mM NaCl. We see that Figure V.6.1

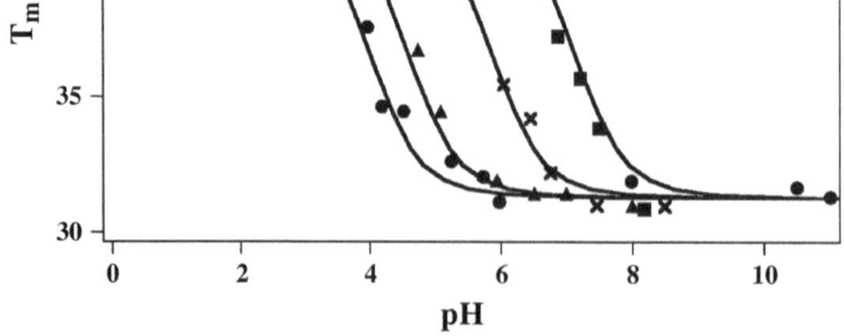

Figure V.6.1 Melting temperature T_m of membranes made of the charged lipid MPA as a function of pH at different NaCl concentrations. Adapted from [57].

suggests the counterintuitive result that an increase in ionic strength leads to a lowering of the melting temperature, although one would expect an increase due to the shielding of electrical charges. There are two phenomena that play a role.

- Neutralization of charges by binding of protons
- Electrostatic screening by Na^+ and Cl^- ions that reduce the association of protons.

Thus, salts can reduce protonation and therefore decrease the melting temperatures. See [1, 57] for details. A consequence is that lowering of pH can make membranes solid and lead to a release of the latent heat of the membrane.

V.6.2 pH Dependence of Biomembrane Melting

Biological membranes typically contain negatively charged lipids (10–40%, [1]) and no positively charged lipids. The pK_A of zwitterionic lipids is far away from physiological pH (e.g., the protonation of phosphatidylcholine occurs around pH 1). The protonation of the negative lipids of biological membranes leads to an increase in melting temperature in a wide pH range around neutral pH. Figure V.6.1 shows the pH dependence of the melting temperature for a synthetic lipid that is charged. Figure V.6.2 (left) shows the pH dependence of native *E. coli* membranes. It can be seen that a change in pH from 9 to 5 changes the transition temperature by about 7.3°, i.e., as expected, the transition temperature increases upon lowering of pH.

Färber and Westerhausen [24] investigated the melting in the membranes of red blood cell ghosts with the fluorescence marker Laurdan, which monitors the difference in surface polarization of gel and fluid lipid phases by a parameter called the generalized polarization (GP). Figure V.6.2 (right) shows the results obtained at a pH of 7 and a pH of 5. The bottom panel shows the derivative of the GP values and resembles a transition profile. The authors

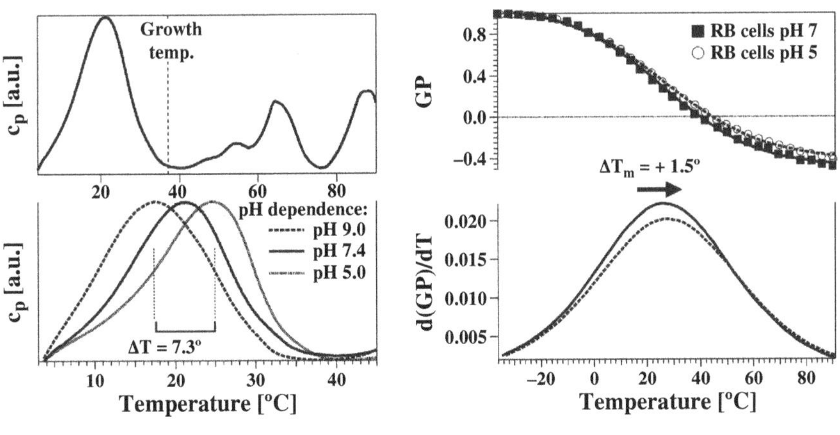

Figure V.6.2 Left: The pH dependence of membrane melting of *E. coli* membranes. A change of the pH from 9 to 5 leads to a shift of 7.3° toward higher temperatures. Adapted from [22]. Right: Transitions in red blood cell ghosts measure by generalized polarization of Laurdan (a fluorescence marker) at two different pH values, pH 7 and pH 5, respectively. The bottom panel shows the derivative of the fits to the top curves. Adapted from [24].

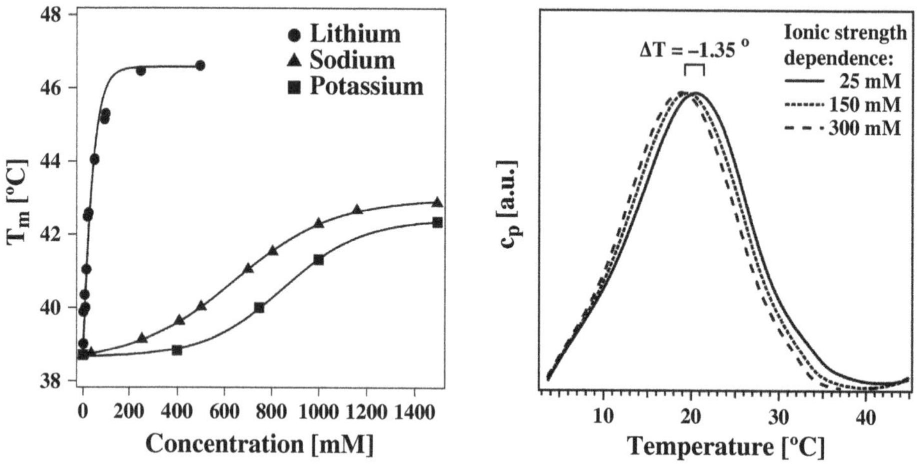

Figure V.6.3 Left: The influence of lithium, sodium, and potassium on the transition of the lipid DMPS in the presence of 25 mM ammonium phosphate at pH 6.8. It can be seen that lithium affects phase transition much more than sodium and potassium. [58, 59]. Right: The effect of NaCl on the melting temperature of *E. coli* membranes. Adapted from [22].

found that the transition temperature increases by about 1.5° upon a decrease of pH from 7 to 5. This is consistent with the single lipid results shown in Figure V.6.1.

V.6.3 Ionic Strength Dependence of Lipid Membrane and Biomembrane Melting

Figure V.6.3 (left, also shown in Part IX) displays the influence of potassium, sodium, and lithium (which are all monovalent positive ions) on the melting of membranes made of the charged lipid dimyristoyl phosphatidylserine (DMPS) (adapted from [58]). The pH is constant at 6.8. It can be seen that potassium and sodium shift the transition temperature by up to 5°. However, such a shift requires very high concentrations above 1 molar of the two ions. While the effects of potassium and sodium are similar, sodium interacts slightly stronger. Lithium, however, interacts much stronger with the membranes and can cause a shift in the transition by up to 10° at concentrations of 100–200 mM. This is closer to the behavior of protons described in the previous section. Figure V.6.3 (right) shows that the ionic strength displays a very mild influence on the transition temperature of *E. coli* membranes [22].

V.6.4 Summary: Key Ideas of Chapter V.6

In this chapter, we discussed the effect of pH (protons) and ions on the melting of membranes. We found that

- Lipids are either zwitterionic (neutral) or negatively charged. A biomembrane contains 10–40% negatively charged lipids.

- Negatively charged lipids increase their transition temperature upon protonation (neutralization) of the charges in the head groups. Electrostatics in general reduces transition temperatures in membranes and provides a negative enthalpy contribution to the melting of membranes. Protonation reduces the electrostatic free energy difference between gel and liquid phase, which leads to an increase of the transition temperature.
- This is also observed in biological membranes.
- Ions influence the melting temperature of charged membranes and also affect the pH dependence.

V.7

Influence of Voltage

One of the most important variables in electrophysiology is voltage, i.e., the electrostatic potential difference across a nerve membrane. In the previous chapter, we showed that the electrostatic of membranes is influenced by pH and ions. Voltage applied across membranes can also directly influence the state of membranes, as will be shown in the following sections.

V.7.1 Voltage Dependence of Monolayers

In Sections V.3.4 and V.4.3, it was shown that a change in lateral pressure generates domains because pressure can shift monolayers through their liquid-expanded (LE) to liquid-condensed (LC) transition. Since electrostatics also influence the melting transition, one can induce melting transitions in monolayers by voltage. Figure V.7.1 (left) shows a setup where a voltage difference is applied to a monolayer. Using an electrode diameter of 15 μm, a glass insulation of the electrode of 50 μm, and assuming that the electrolyte in

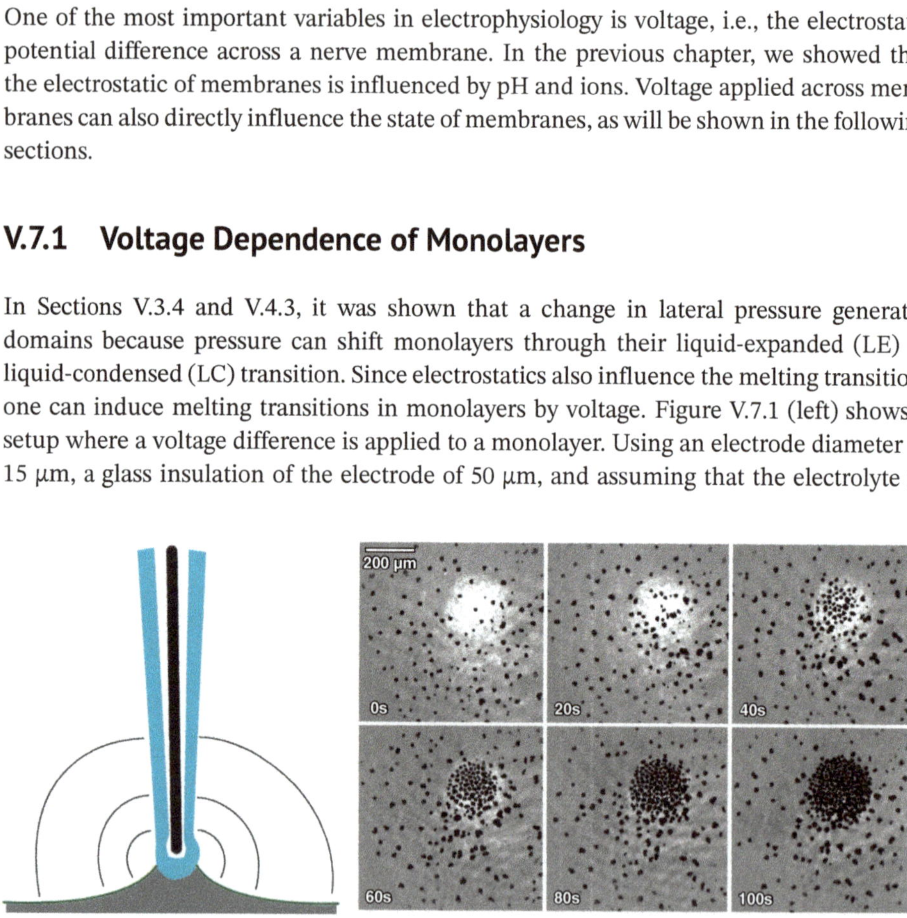

Figure V.7.1 Left: Schematic drawing of glass-insulated electrode (gray and blue) touching a water surface (green) allowing for investigation of the effect of high electric fields (visualized as curved lines) in the area around it. Right: Image series of LC domains being attracted to the area beneath an electrode. Gray areas represent the LE phase. The bright shade in the center is the position of the electrode. [60]/with permission of Niels Bohr Institute.

Physics of Nerves and Excitatory Membranes, First Edition. Thomas Heimburg.

the trough is electrically conducting, one calculates a field strength of about 10 MV/m at a voltage of 500 V, about half the typical physiological field across a biological membrane (100 mV across a membrane of thickness of 5 nm yields 20 MV/m).

Figure V.7.1 (right) shows the accumulation of LC domains at the position of the electrode as a function of time after the field was switched on. It is clear the LC domains prefer regions of high fields across the monolayer. Other experiments show that domains dissolve when the voltage is switched off. This indicates that the voltage gradient influences the physical state of the monolayer. High gradients favor solid domains. This most likely originates from different dipole moments of zwitterionic lipids in the LC and LE phases. Most important is here that the physical state of monolayers depends on translayer voltage differences.

V.7.2 Voltage Dependence of Bilayers

Voltage across cell membranes is frequently applied in electrophysiology. However, there are hardly any studies that directly visualize the influence of voltage on membrane transitions or the physical state of cell membranes. However, there exist theoretical studies predicting an effect of voltage. Thomas and Agrawal [61] showed with molecular dynamics (MD) simulations that one can induce a transition in lipid membranes upon application of in-plane electrical fields. Figure V.7.2 shows the structure of DPPC membranes in the presence and absence of a lateral field of 0.075 V/nm at a temperature of 330 K. It can be seen that the application of the voltage can render the membrane solid, resulting in a reduction of membrane area and an increase of the membrane thickness.

In MD, the simulation time plays an important role. The relaxation time in DPPC unilamellar membranes in the transition is of the order of 1–2 seconds [62, 63]. The simulation time is rather in the range of a few hundred nanoseconds. Thus, simulations of phase transitions display a large hysteresis during the simulation. Figure V.7.3 shows heating and cooling scans of DPPC membranes with different in-plane fields. The experimental transition

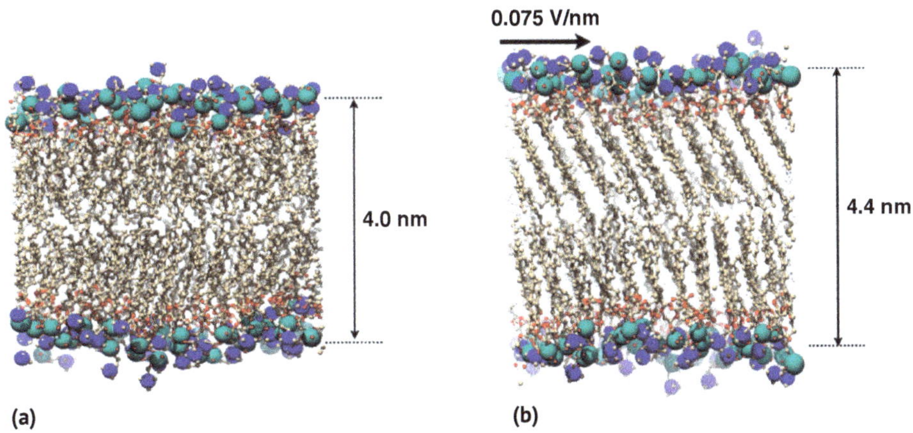

Figure V.7.2 MD simulation of DPPC membranes at 330 K in the absence and presence of an electrostatic field in the membrane plane of 0.075 V/nm. [61]/with permission of Royal Society of Chemistry.

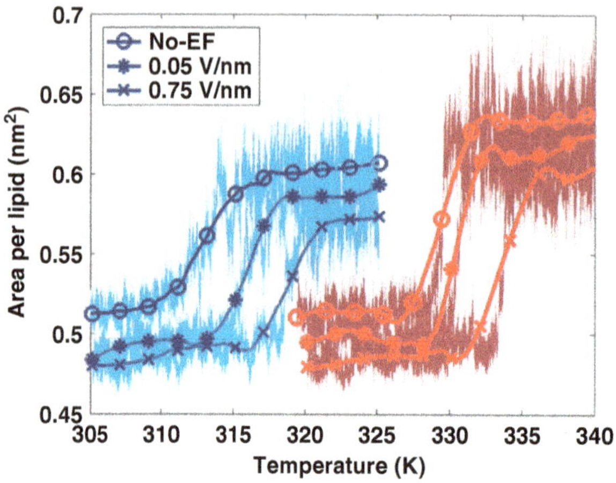

Figure V.7.3 Area per lipid of DPPC membranes as a function of temperature at three different fields in the membrane plane: 0 V/nm, 0.05 V/nm, and 0.075 V/nm, respectively. The MD simulations display significant hysteresis due to the short simulation times. Red profiles represent heating curves and the blue profiles represent cooling curves. [61]/with permission of Royal Society of Chemistry.

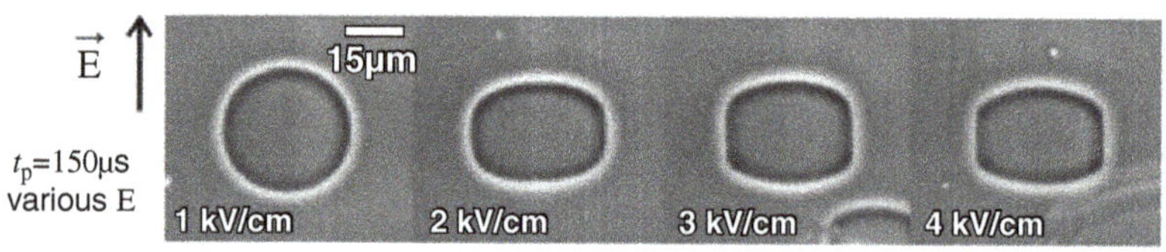

Figure V.7.4 Deformation of giant vesicles of egg-yolk PC by different electrical fields. [64]/with permission of Elsevier.

temperature of DPPC is around 314.5 K. The heating profiles in the absence of a field rather display a transition temperature of 329 K and around 312 K in the cooling scan. Thus, the cooling scan is closer to the experimental transition temperature. However, in both heating and cooling scans, a field of 0.075 V/nm shifts the transition by 5–6 K toward higher temperature. Thus, in-plane voltage, as often applied in electrophysiological extracellular recordings of nerve pulses (where the nerve is stimulated by two electrodes touching the surface), has the potential to shift the membrane through its transition and render the membrane solid. These findings are in agreement with the monolayer experiments shown in the previous section.

The presence of physical forces on membrane bilayers caused by voltage gradients is also evident in microscopy. Figure V.7.4 shows the deformation of giant lipid vesicles from zwitterionic phosphatidylcholines in an electric field. The deformation indicates that the interaction of the vesicles with an electric field must depend on the relative orientation of the membrane surface and the applied field.

V.7.3 Membrane Capacitance

The capacitance C_m of a membrane is a constant relating the charge q to a transmembrane voltage, V_m.

$$q = C_m \cdot V_m. \tag{V.7.1}$$

The capacitance depends on the area and thickness of the membrane, and on the dielectric constant of its insulating core:

$$C_m = \epsilon_0 \epsilon \frac{A}{D}, \tag{V.7.2}$$

where A is the membrane area and D is the thickness of the membrane, $\epsilon_0 = 8.8541878$ F/m is the vacuum permittivity, and $\epsilon = 2$–4 is the dielectric constant of the hydrocarbon chains. For a membrane with a hydrophobic core thickness of 4 nm, we obtain a capacitance of $C_m = 0.844$ μF/cm^2 for $\epsilon = 4$. In fact, a value of $C_m = 1$ μF/cm^2 has often been used for theoretical considerations on nerve membranes (e.g., [65]).

In transitions, both membrane area and membrane thickness change as a function of temperature and other intensive variables (see Figure V.7.5). Generally, the membrane area increases and the membrane thickness decreases. For DPPC membranes, the membrane area increases by 24.6%, and the thickness decreases by 16.3%. Therefore, the capacitance change can be determined by

$$C_m^f = \epsilon_0 \epsilon \frac{A^g \cdot (1 + 0.246)}{D^g \cdot (1 - 0.163)} = 1.49\, C_m^g, \tag{V.7.3}$$

where A^g, D^g, and C_m^g are the area, thickness, and capacitance of the gel phase membrane, and C_m^f is the capacitance of the fluid phase membrane. We find that the capacitance in the fluid phase of DPPC membranes is calculated to be 1.46 times higher than in the gel phase. This prediction can be tested in experiments. Figure V.7.6 shows capacitance measurements on palmitoyl-oleoyl phosphatidylethanolamine (POPE):palmitoyl-oleoyl phosphatidylglycerol (POPG) = 80:20 membranes measured on a patch pipette [66]. Shown are heat capacity profiles of vesicles of this mixture in heating and cooling scans (left panel), the enthalpy changes obtained by integration of the heat capacity profile (center panel), and the temperature-dependent change of the capacitance (right panel). It changes by roughly a factor of two, even more than predicted for DPPC. Thus, changes in membrane state by changes in the intensive variables can significantly influence the capacitance. In the presence of transmembrane voltage, changes in temperature can induce capacitive currents.

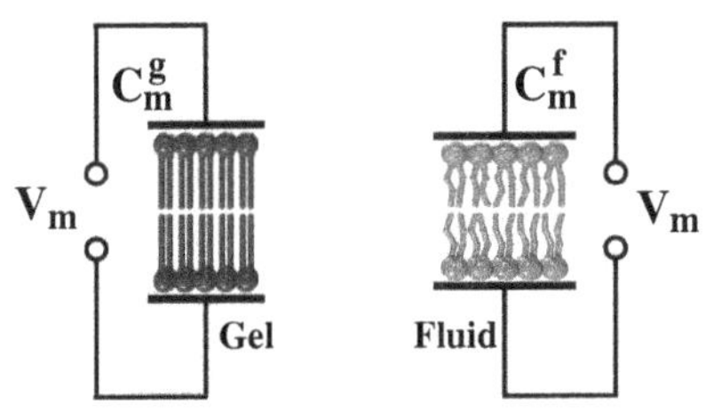

Figure V.7.5 Schematic drawing of the change in geometry and the associated changes in area and thickness of the membrane during a transition.

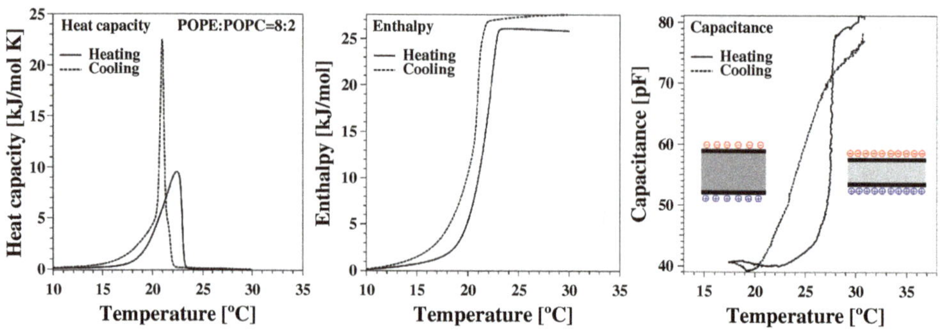

Figure V.7.6 Experiments on DOPE:DOPC = 80:20 membranes. Left: Heat capacity in an up and a down scan. Center: The corresponding enthalpy profiles as a function of temperature. Right: Capacitance changes measured on a patch pipette. [66]/IOP Publishing/CC BY 3.0.

In Section VI.6.1, it will be shown that voltage can alter the capacitance as well, and that one can therefore define a capacitive susceptibility $\widehat{C}_m$ that displays a maximum during the chain melting transition. This is not considered here.

V.7.4 Electrostriction

The capacitance is a static property of a layer. It has a numerical value even when the membrane is not charged. When the membrane capacitor is charged, there are opposite charges associated with the surface that attract each other. Thus, the charges give rise to a force, on the membrane and the associated changes in membrane geometry are called *electrostriction*. This is shown schematically in Figure V.7.7.

The force $\mathcal{F}$ is given by

$$\mathcal{F} = \frac{1}{2}E \cdot q = \frac{1}{2}\frac{V_m}{D} \cdot q = \frac{1}{2}\frac{C_m}{D}V_m^2, \tag{V.7.4}$$

where V_m is the voltage, $E = V_m/D$ is the electrical field inside the capacitor, and the charge is given by $q = C_m \cdot V_m$.

When the membrane undergoes a transition, work is performed on the membrane:

$$\Delta W = \int_{D^g}^{D^f} \mathcal{F} dD = \frac{1}{2}\epsilon_0 V_m^2 \int_{D^g}^{D^f} \epsilon \frac{A}{D^2} dD. \tag{V.7.5}$$

From this, the voltage dependence of the excess enthalpy H_0 can be calculated (see [67] for details). After evaluating the integral in eq. (V.7.5), it turns out that the voltage-dependent transition enthalpy is given by

$$\Delta H(V_m) = \Delta H_0 + \alpha_0 V_m^2, \tag{V.7.6}$$

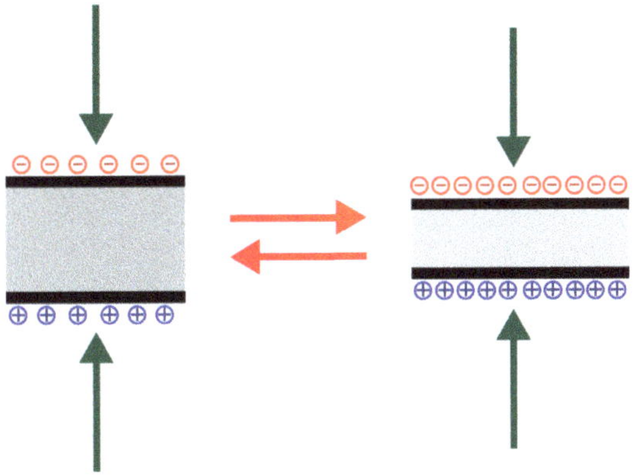

Figure V.7.7 Forces on a charged membrane capacitor. The force $\mathcal{F}$ changes when going from gel to fluid phase.

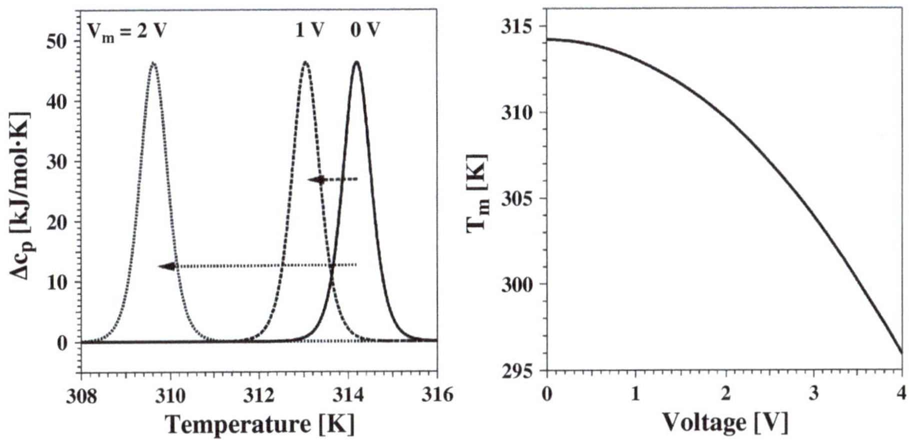

Figure V.7.8 Top: Theoretical transition profiles of DPPC membranes as a function of voltage. Bottom: Transition temperature as a function of voltage. [67]/with permission of Elsevier.

where α is a constant that depends on the geometry of the gel and fluid phase of the membrane. The transition temperature of a membrane is given by $T_m = \Delta H_0/\Delta S_0$. Thus, the voltage dependence of the transition temperature is given by

$$\begin{aligned} T_m(V_m) &= \frac{\Delta H(V_m)}{\Delta S_0} = \frac{\Delta H_0 + \alpha_0 V_m^2}{\Delta S_0} \\ &= T_{m,0} + \frac{\alpha_0 V_m^2}{\Delta S_0} = \left(1 + \frac{\alpha_0}{\Delta H_0} V_m^2\right) \\ &\equiv \left(1 + \alpha V_m^2\right) T_{m,0}. \end{aligned} \tag{V.7.7}$$

This is plotted in Figure V.7.8. Thus, the transition temperature depends in a quadratic manner on the transmembrane voltage. For DPPC membrane, the constant $\alpha = \alpha_0/\Delta H_0 = -0.003634$ [1/V^2] (see [67] for details) and $T_{m.0} = 314.2$ K.

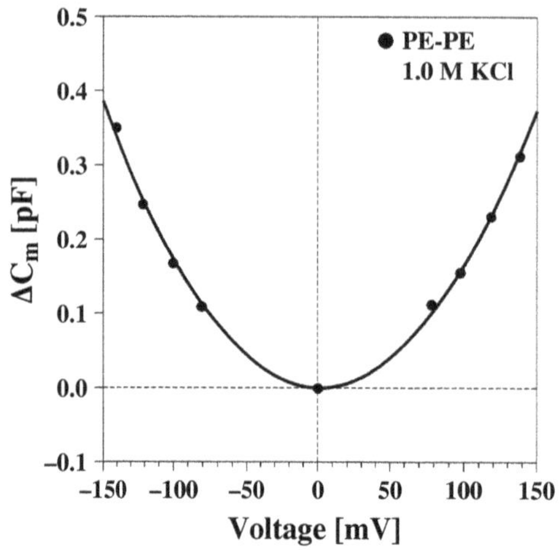

Figure V.7.9 The capacitance as a function of voltage for a symmetric membrane made of bacterial phosphatidylethanolamine in 1 M KCl. The relative capacitance change ΔC_m is at a minimum for a voltage of V = 0 mV. [69] and [70].

Figure V.7.4 shows calculated transition profiles for DPPC unilamellar vesicles as a function of voltage. For instance, a transmembrane voltage or 2 V would shift the transition by 4.57 K toward lower temperatures. At physiological voltages of about 100 mV, the shift would just be 11 mK, i.e., it would be quite small. However, large voltages of 2 V are often used to stimulate nerves, e.g., in giant worm axons [68].

What is clear from what was discussed in this section is that the capacitance itself must be a function of voltage because voltage can potentially change the state of the membrane, make it thinner, or induce gel-to-fluid transitions. Figure V.7.9 shows capacitance measurements as a function of voltage for a symmetric membrane made of bacterial phosphatidylethanolamine (data adapted from [69]). One can see that the capacitance displays a minimum at zero voltage, and the capacitance increases in value at higher absolute voltages, independent of the sign of the voltage. The capacitance change is also a quadratic function. This is a consequence of electrostriction.

V.7.5 Polarization of Membranes

Membranes can display a permanent electric polarization. For a liquid membrane, dipoles in the plane average, but they may still possess a total value normal to the membrane plane. Polarization implies that the charge distribution normal to the membrane surface is not symmetric. This can have several origins. For instance, the lipids on the two sides of the membrane may carry different charges, as shown in Figure V.7.10a. The membrane could also contain polar molecules such as integral proteins with an asymmetric charge distribution (Figure V.7.10b). Zwitterionic lipids possess a head group with a dipole. A different orientation of the dipoles on the two sides will also result in a net polarization of the membrane (Figure V.7.10c). An origin of this might be that lipids on one side are fluid while they are solid on the other side. This yields different head group orientations in the two layers.

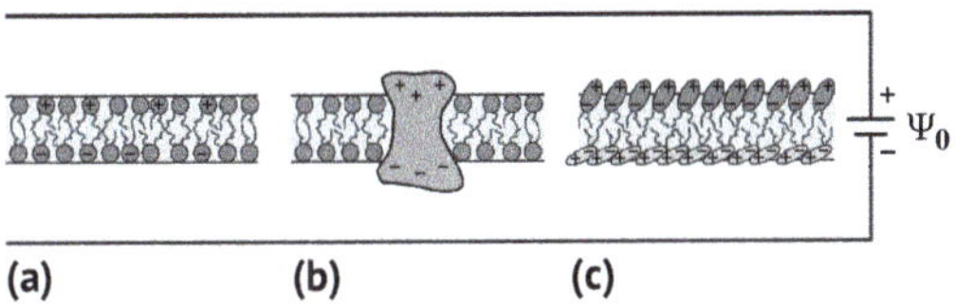

Figure V.7.10 Three different scenarios leading to membrane polarization. (a) The upper monolayer contains a difference charge density than the bottom monolayer. (b) The membrane contains polar molecules such as integral proteins with asymmetric charge distribution. (c) The membrane only contains zwitterionic molecules, but due to different chemistry or physical states of the lipids, the dipoles of the head group display a different orientation on the two sides of the membrane.

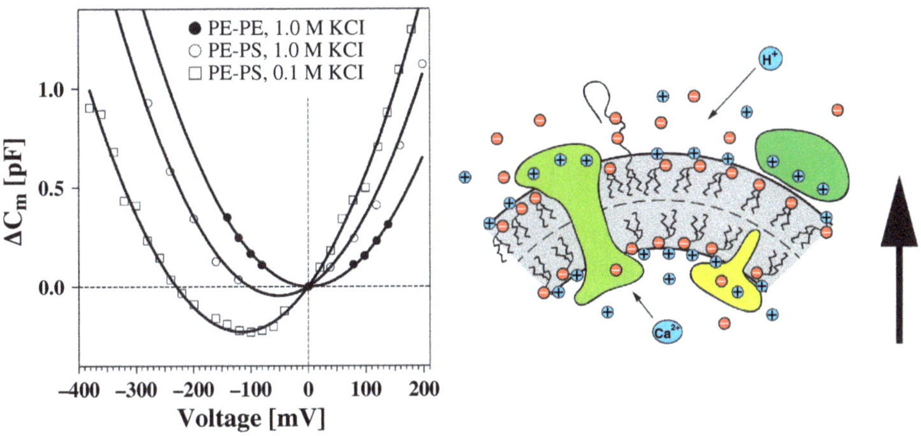

Figure V.7.11 Left: The capacitance as a function of voltage for symmetric and asymmetric membranes made of bacterial PE and bovine brain PS in 1 M KCl or 0.1 M KCl. The relative capacitance change ΔC_m is at a minimum for a voltage of V = 0 mV. [69] and [70]. Right: The polarization of biomembranes can have various reasons. For instance, it can originate from curvature, from an asymmetry of zwitterionic and negatively charged lipids, from polarized or charged integral and peripheral proteins (green objects), from concentration differences of ions such as Na^+, K^+, Cl^-, and Ca^{2+}, or from pH differences. [70]/with permission of Royal Society of Chemistry.

The Gibbs free energy of a membrane with area A and thickness D is given by [70]

$$G = -\frac{\epsilon_0 \epsilon}{2} \frac{A}{D} \left((\Psi + \Psi_0)^2 - \Psi_0^2 \right). \tag{V.7.8}$$

Here, Ψ_0 is the electrostatic potential caused by the polarization of the membrane, and Ψ is the electrostatic potential applied from the outside via electrodes (see [70] for details). Here, as a convention, $\Delta\Psi \equiv -\Delta V$. The charge on the capacitor is given by

$$q = C_m(\Psi + \Psi_0), \tag{V.7.9}$$

i.e., the number of charges on a capacitor depends on the polarization of the membrane. It is at a minimum if $\Psi = -\Psi_0$.

Let us consider the example in Figure V.7.10a. If the membrane displays a net polarization, one finds electrostrictive forces acting on the membranes. These forces affect the capacitance. When a net voltage of opposite sign is applied, the polarization and the externally applied voltage may cancel, and electrostrictive forces vanish. This is shown in Figure V.7.11 (left). The voltage dependence of three different membrane is displayed.

The first one is a symmetric membrane made of bacterial phosphatidylethanolamine (PE). It shows a minimum in the capacitance at V = 0. The second membrane is made of phosphatidylserine (PS, a negatively charged lipid) on one side and PE on the other side at an ionic strength of 1 M KCl. Here, one expects a net polarization. Consequently, the minimum of the capacitance is not at V = 0 mV but rather at −52 mV. If the ionic strength is reduced to 150 mM, the screening of the charged head groups of PS is reduced, and the polarization of the membrane increases. The voltage at minimum capacitance shifts to −116 mV. The shift of the capacitance vs. voltage profiles is direct evidence for membrane polarization.

There are many possible origins for polarization in biomembranes. They are summarized in Figure V.7.11 (right). One possibility to induce polarization is inequality in the ion distribution on the two sides of a membrane. The Nernst potentials discussed in Part IV can create a polarization. This is the resting potential of the membrane. Another one is curvature, introducing an effect called *flexoelectricity*.

V.7.6 Piezoelectricity and Flexoelectricity

If a membrane displays a net polarization, for instance, due to charging in an external field, any change in area and thickness will change the capacitance and the charge on the membrane. Thus, mechanical forces on membranes will influence the charge on the capacitor and one obtains a coupling between mechanical and electrical phenomena. This effect is called piezoelectricity. If the membrane consists of two monolayers with an equal magnitude of the polarization but opposite sign, then the membrane does not display a net polarization. However, if the membrane is curved, the dipole density on the outer layer is smaller than on the inner layer. Therefore, bending of a membrane may induce a net polarization. This effect is called flexoelectricity and is schematically described in Figure V.7.12. It was pioneered by A.G. Petrov (e.g., [71]).

Let us assume we have a system with four variables: the electrostatic potential Ψ, the membrane area A, the curvature c, and the temperature T. The change of the charge upon a change of these variables is given by

$$\begin{aligned} dq = & \left(\frac{\partial q}{\partial \Psi}\right)_{A,c,T} d\Psi + \left(\frac{\partial q}{\partial A}\right)_{\Psi,c,T} dA \\ & + \left(\frac{\partial q}{\partial c}\right)_{\Psi,A,T} dc + \left(\frac{\partial q}{\partial T}\right)_{\Psi,A,c} dT \,. \end{aligned} \qquad \text{(V.7.10)}$$

When using eq. (V.7.9), we find

$$\begin{aligned} & \left[(\Psi+\Psi_0)\left(\frac{\partial C_m}{\partial \Psi}\right)_{A,c,T} + C_m + C_m\left(\frac{\partial \Psi_0}{\partial \Psi}\right)_{A,c,T}\right] d\Psi \\ dq = & + \left[(\Psi+\Psi_0)\left(\frac{\partial C_m}{\partial A}\right)_{\Psi,c,T} + C_m\left(\frac{\partial \Psi_0}{\partial A}\right)_{\Psi,c,T}\right] dA, \\ & + \left[(\Psi+\Psi_0)\left(\frac{\partial C_m}{\partial c}\right)_{\Psi,A,T} + C_m\left(\frac{\partial \Psi_0}{\partial c}\right)_{\Psi,A,T}\right] dc \\ & + \left[(\Psi+\Psi_0)\left(\frac{\partial C_m}{\partial T}\right)_{\Psi,A,c} + C_m\left(\frac{\partial \Psi_0}{\partial T}\right)_{\Psi,A,c}\right] dT \end{aligned} \qquad \text{(V.7.11)}$$

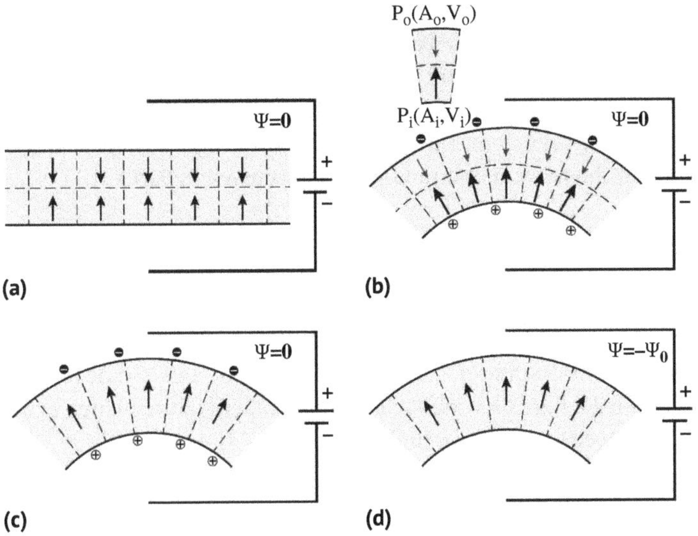

Figure V.7.12 Illustration of polarization by curvature. (a) The two monolayers of the symmetric membrane display opposite polarization. (b) Upon bending (flexing) the membrane, the polarization in the two layers changes. (c) Effective polarization of the membrane. (d) In order to discharge the capacitor, a potential of $\Psi = -\Psi_0$ has to be applied.

or in abbreviated form as

$$
\begin{aligned}
& \left[(\Psi + \Psi_0)\alpha_{A,c,T} + C_m + C_m\beta_{A,c,T}\right] d\Psi \\
dq \equiv & + \left[(\Psi + \Psi_0)\alpha_{\Psi,c,T} + C_m\beta_{\Psi,c,T}\right] dA \\
& + \left[(\Psi + \Psi_0)\alpha_{\Psi,A,T} + C_m\beta_{\Psi,A,T}\right] dc \\
& + \left[(\Psi + \Psi_0)\alpha_{\Psi,A,c} + C_m\beta_{\Psi,A,c}\right] dT
\end{aligned}
\qquad \text{(V.7.12)}
$$

V.7.6.1 Piezoelectricity

Piezoelectricity is the area or thickness dependence of the charge on a capacitor, or more generally, the influence of geometry on the charge of a capacitor. The following is taken from [70]. Let us assume that in eq. (V.7.12) Ψ, c, and T are constant. We then obtain

$$dq = \left[(\Psi + \Psi_0)\alpha_{\Psi,c,T} + C_m\beta_{\Psi,c,T}\right] dA. \qquad \text{(V.7.13)}$$

At $\Psi = 0$, we obtain for a small change in area ΔA,

$$\Delta q \approx \left(\Psi_0\alpha_{\Psi,c,T} + C_m\beta_{\Psi,c,T}\right) \Delta A. \qquad \text{(V.7.14)}$$

If $\Psi_0(\Delta A = 0)$ is zero, the capacitor is uncharged for $\Psi = 0$. Then the charge on the capacitor after a change in area of ΔA is given by

$$q(\Delta A) = C_m\beta_{\Psi,c,T}\Delta A \qquad \text{or} \qquad \Psi_0(\Delta A) = \beta_{\Psi,c,T}\Delta A. \qquad \text{(V.7.15)}$$

The magnitude of the coefficient $\beta_{\Psi,c,T}$ depends on the composition and structure of the membrane. A similar relation was given by Petrov and Usherwood [72]. As evident in eq. (V.7.13), the piezoelectric effect is a function of the externally applied potential if the capacitance is influenced by area changes (which usually is the case).

The piezoelectrical effect leads to electromechanical coupling during the propagation of a mechanical pulse (soliton) in the nerve membrane. This is discussed in Part VII.

V.7.6.2 Flexoelectricity

Flexoelectricity describes the dependence of the charge on the membrane capacitor induced by membrane curvature, c. Let us assume that Ψ and T are constant. It is questionable whether one can define experimental conditions where the membrane area is independent of curvature. Therefore, we consider area a variable depending on curvature, which is not independently controlled. Using eq. (V.7.12), we find

$$dq = \left[(\Psi + \Psi_0)\alpha_{\Psi,T} + C_m\beta_{\Psi,T}\right] dc. \qquad \text{(V.7.16)}$$

This is the (direct) *flexoelectric effect*. If we make the simplifying assumption that the capacitance C_m does not depend on curvature and that the coefficient $\beta_{\Psi,T}$ is constant, we obtain

$$q(c) = C_m\left(\Psi + \Psi_0(0)\right) + C_m\beta_{\Psi,T} \cdot c, \qquad \text{(V.7.17)}$$

where $C_m(\Psi + \Psi_0(0))$ is the membrane charge at $c = 0$. If further the applied potential, Ψ, is zero and the polarization in the absence of curvature is also assumed to be zero, we obtain

$$q(c) = C_m\beta_{\Psi,T} \cdot c \qquad \text{or} \qquad \Psi_0(c) = \beta_{\Psi,T} \cdot c. \qquad \text{(V.7.18)}$$

Thus, the offset potential Ψ_0 is proportional to the curvature. This relation is a special case of the flexoelectric effect described in eq. (V.7.16). It was previously discussed by Petrov [71]. He introduced a flexoelectric coefficient, f, which is given by $f \equiv \varepsilon \cdot \beta_{\Psi,T}$. Petrov found experimentally that $f = 10^{-18}$[C], or $\beta_{\Psi,T} = 2.82 \cdot 10^{-8}$ [m] for $\varepsilon = 4\varepsilon_0$, respectively.

V.7.7 Summary: Key Ideas of Chapter V.7

Voltage is a very important intensive variable electrophysiology. One uses voltage pulses to stimulate nerves, and there exist electrostatic potential differences across the nerve membrane. The nerve pulse itself consists of a variation in transmembrane potential. In this chapter, we found

- Voltage gradients in the plane of a monolayer can attract and stabilize LC domains in an LE film.
- Electric fields applied in the plane of a bilayer increases the transition temperature of zwitterionic membranes.
- The capacitance of a membrane is different in the gel and fluid states of a membrane. The difference can be up to a factor of two.

- Charging the membrane capacitor by applying a transmembrane voltage leads to forces on the membrane. Since membranes are soft materials with transitions, they can be deformed and change their state. This effect is called *electrostriction*. Charging of a membrane lowers their melting temperature.
- When membranes are not symmetric, i.e., the two monolayers have different composition, state, or environment, they can possess a permanent dipole. This phenomenon is called polarization of the membrane. The Nernst potentials in electrophysiology cause such polarization. The polarization caused by ions is called the resting potential in electrophysiology.
- When membranes display polarization, the charging profile of the membrane capacitor displays a minimum at a voltage different from zero.
- Membranes can be charged by changing the membrane area. This phenomenon is called piezoelectricity.
- Membranes can be charged by curvature. This phenomenon is called flexoelectricity.

V.8

Influence of Drugs and Proteins

The thermodynamics of membranes depends on all intensive variables. These include the chemical potentials of molecules associated with membranes or dissolved in the membrane. There are many molecules in a membrane. Besides all the lipids, these include the integral and peripheral proteins, membrane peptides, and small molecules such as anesthetics, neurotransmitters, and cholesterol, all of which contribute to the Gibbs free energy of a membrane.

V.8.1 Dependence of Melting on Anesthetics

Anesthetics are small, membrane-soluble drugs that are partially hydrophobic. Their solubility in the membrane (i.e., their oil-water partition coefficient) determines their anesthetic efficiency. There exists a surprising linear correlation between the partition coefficient in the membrane and the effective dose ED_{50}, at which 50% of the subjects are anesthetized. It is known as the Meyer–Overton correlation [73]. It has long been known that anesthetic drugs lower the melting temperatures of lipid membranes (e.g., [44, 74, 75, 76]. The physiological role of this phenomenon is described in Chapter IX.1. There are two major classes of anesthetics: general anesthetics, which distribute throughout the body, and local anesthetics, which are injected and don't rapidly redistribute. Figure V.8.1 (left) shows the influence of the general anesthetic octanol on the melting of DPPC membranes [76]. One can see that the melting profile shifts toward lower temperatures and is broadened. This was explained in [44] as a consequence of the melting-point depression law, which requires that the anesthetics dissolve in the fluid membrane phase but not in the gel phase. Figure V.8.1 (right) shows that the local anesthetic lidocaine has a similar effect. Lidocaine also shifts transitions toward lower temperatures. Differences in the shape of the transition profiles can be explained by the magnitude of the partition coefficient, which is higher for octanol than for lidocaine. The effects of different general anesthetics are additive [77], and this is also true for the effect of general and local anesthetics administered at the same time (see Chapter IX.1 for details). Thus, it seems that from a thermodynamic perspective, general and local anesthetics are very similar.

Anesthetics have a similar effect on lipid mixtures as well. Figure V.8.3 shows the phase diagram of DMPC-DPPC mixtures (compare with Figure V.4.4). The addition of 60 mM methoxyflurane (a general anesthetic) lowers the transition events of the complete phase

Physics of Nerves and Excitatory Membranes, First Edition. Thomas Heimburg.

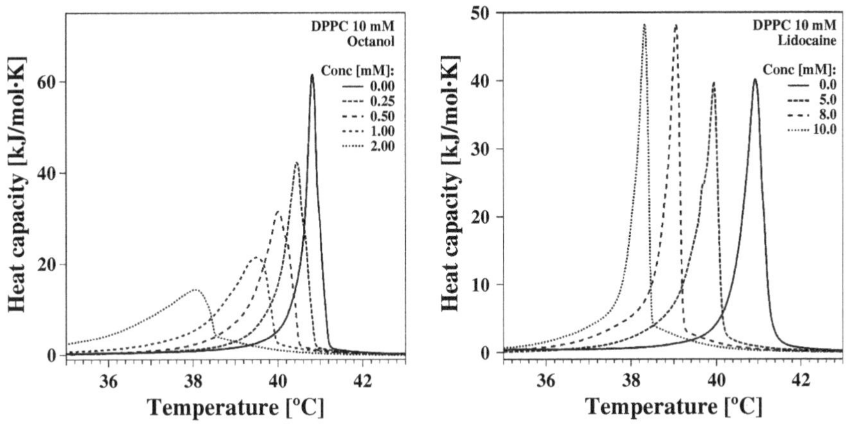

Figure V.8.1 The influence of the general anesthetic octane on the melting of DPPC membranes. Adapted from [76].

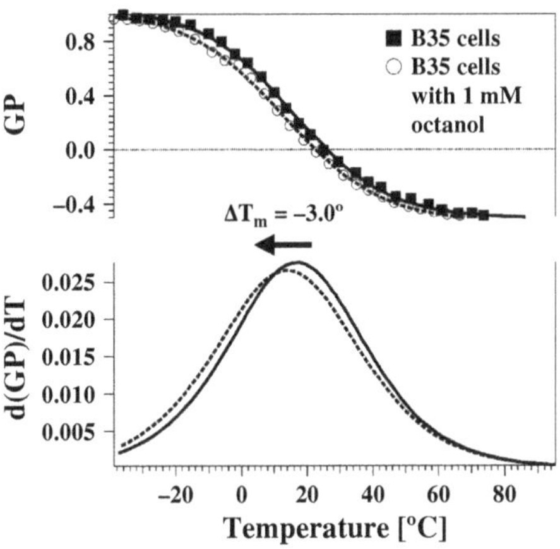

Figure V.8.2 Influence of the general anesthetic octanol on the melting of native B35 cell membranes. Adapted from [24].

diagram. Anesthetics also act on the melting temperature of biological membranes, which are complex mixtures of lipids and proteins. The transition events in B35 neuroblastoma cells were shown to shift toward lower temperatures upon application of the general anesthetic octanol (Figure V.8.2, [24]).

In Figure V.4.4, it was shown that hydrostatic pressure increases the temperature range of transition events. In Figure V.8.3, it is shown that the simultaneous application of 60 mM methoxyflurane and 137 bars of pressure more than reverses the effect of the anesthetics. It is known that the phenomenon of anesthesia can be reversed by hydrostatic pressure [78–83]. This suggests that the effect of both anesthetic drugs and hydrostatic pressure on the melting transitions in biomembranes is related to the working mechanism of anesthesia. This effect is explained in Chapter IX.1.

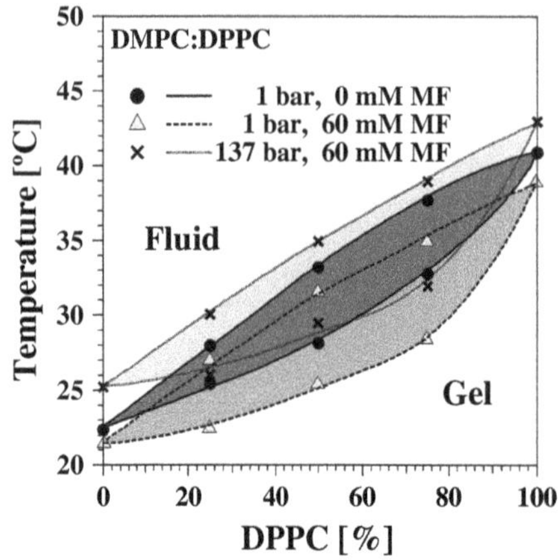

Figure V.8.3 Phase diagram of DMPC:DPPC mixtures as a function of methoxyflurane concentration and pressure. The complete phase diagram shifts downward in the presence of 60 mM methoxyflurane and shifts toward higher temperatures when a pressure of 137 bars is applied. Pressure can counteract the effect of anesthetics. The gray-shaded regions represent the gel-fluid coexistence regime. Adapted from [43].

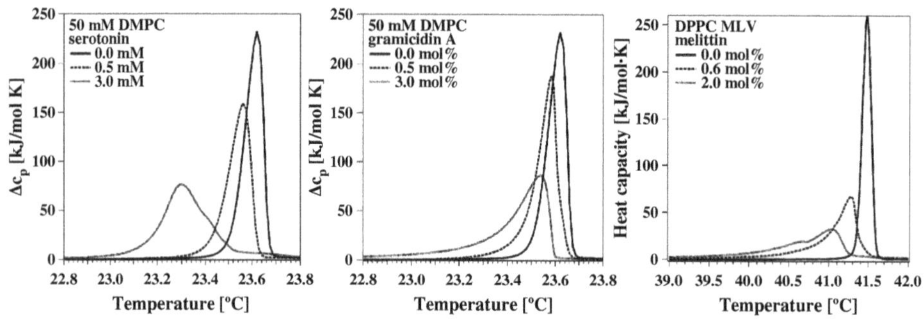

Figure V.8.4 Left: Influence of the neurotransmitter serotonin on the melting of an DPPC membranes. Adapted from [63]. Center: The influence of the antibiotic peptide gramicidin A on the melting of DMPC membranes. Right: The influence of the bee-venom peptide melittin on the melting of DPPC membranes. Adapted from [85].

V.8.2 Dependence of Melting on Neurotransmitters

The lowering of the transition temperature caused by anesthetics can be attributed to the melting-point depression law [44]. Generally, transition events are moved toward lower temperatures by any substance that inserts into or associates with the fluid phase better than with the gel phase. This includes some membrane-soluble neurotransmitters such as serotonin (Figure V.8.4 (left)). For this reason, some authors have speculated whether neurotransmitters may have anesthetic effects. R.S. Cantor called them *endogenous anesthetics* [84].

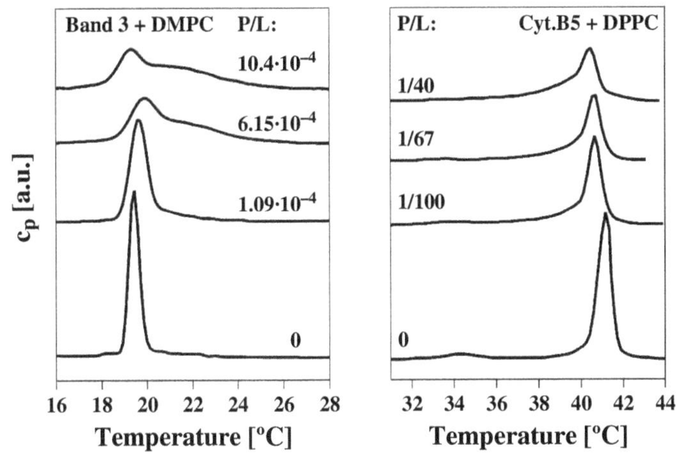

Figure V.8.5 Left: The influence of the integral proteins band 3 from erythrocytes on the melting of DPPC membranes. Adapted from [86]. Right: Influence of cytochrome b_5 in DPPC membranes. Adapted from [87].

V.8.3 Dependence of Melting on the Presence of Peptides and Proteins

Biological membranes consist of 50% protein by mass on average. While a lot of these proteins have extra-membranous domains or are associated with the surface via α-helical or lipid anchors, they still represent significant fractions of the membrane core. Other amino acid polymers that dissolve in membranes are peptides, which are small proteins. Figure V.8.4 (center, right) shows that the pore-forming peptide gramicidin A (a 15 amino acid peptide antibiotic) and the bee-venom peptide melittin (a 26 amino acid cationic peptide that comprises about 50% of the bee venom) both lower the transition temperature in surrounding DMPC and DPPC lipid membranes, which indicates that both peptides dissolved better in fluid membranes than in gel membranes. Similarly, integral proteins also affect melting events. The band 3 protein of erythrocytes increases melting events (Figure V.8.5) indicating that it interacts favorably with the gel membrane. In contrast, cytochrome b_5 from rabbit liver (which possesses a C-terminal membrane-soluble domain) lowers transition events just like the peptides in Figure V.8.4 (right). In summary, it is clear that membrane proteins influence the thermodynamics of the overall membranes and thereby the physical state of the lipid membrane.

V.8.4 Dependence of Melting on Cholesterol

Most small drugs such as anesthetics and membrane-active neurotransmitters dissolve better in the fluid phase membrane. This is different for cholesterol and related steroids. Cholesterol is very abundant in biological membranes (around 40% in erythrocyte membranes [88]). It mixes with both the fluid and gel membrane. At low concentrations, it broadens the transition profile, and at higher concentrations above 10%, the transition events are

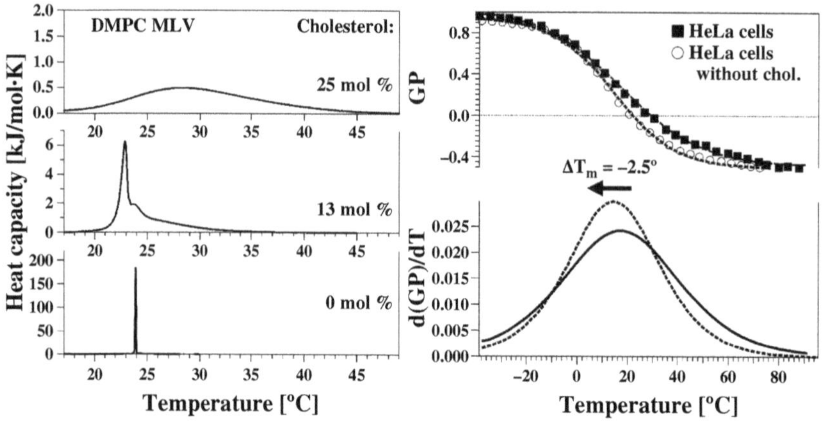

Figure V.8.6 Left: Influence of cholesterol on the melting of DMPC membranes. Note the broadening and shift toward higher temperature at higher cholesterol concentrations. [89]/Springer Nature/CC BY 4.0. Right: The influence of the removal of cholesterol from native HeLa cells by exposure to cyclodextrin. Adapted from [24].

moved to higher temperatures (Figure V.8.6 (left)). There have been considerable efforts in the literature to understand the role of cholesterol in membranes and to describe their influence on phase behavior, e.g., [30]. Cholesterol is also thought to be a major component of domains in biological membranes (rafts; see Section V.3.5).

Since cholesterol is abundant in biomembranes, one can test its role in the membrane phase behavior by removing it. Färber and Westerhausen [24] depleted the cholesterol in HeLa cells by adding cyclodextrin. Since cholesterol shifts melting events toward higher temperatures, cholesterol depletion is expected to lower the temperature range of transition events. This is in fact what is found in experiments (see Figure V.8.6 (right)) where the influence of temperature on the generalized polarization of the fluorescence marker Laurdan was investigated.

V.8.5 Summary: Key Ideas of Chapter V.8

The melting transition in both synthetic lipid membranes and native biomembranes is influenced by numerous membrane components, including the following:

- Both general and local anesthetics
- Membrane-soluble neurotransmitters such as serotonin
- Antibiotic and lytic peptides such as gramicidin A, melittin, and alamethicin
- Integral proteins such as the band 3 protein of erythrocytes and cytochrome b_5 from rabbit liver
- Cholesterol at high concentrations. It shifts transitions upward and makes membranes more ordered. At low concentration (up to 10%), cholesterol does not have this effect.

In general, any molecule that is added to membranes affects its melting profile and thus the physical state and the thermodynamics of the membrane. This includes the lipids themselves and the membrane proteins.

Part VI

Fluctuations and Susceptibilities

Susceptibilities are the partial derivatives of an extensive variable with respect to an intensive variable. Examples include the following:

- Heat capacity: $c = \partial Q/\partial T$, where Q (extensive) is heat and T is the temperature (intensive)
- Volume compressibility: $\partial V/\partial p$, where V (extensive) is volume and p is the pressure (intensive)
- Area compressibility: $\partial A/\partial \Pi$, where A (extensive) is area and Π is the lateral pressure (intensive)
- Volume expansion coefficient: $\partial V/\partial T$
- Area expansion coefficient: $\partial A/\partial T$
- Capacitive susceptibility: $\partial q/\partial V$, where q (extensive) is the charge on a capacitor and V is the voltage (intensive)
- Chemical susceptibility: $\partial n_i/\partial \mu_j$, where n_i is the amount of molecular species i, and μ_j is the chemical potential of species j

There exists a deep connection between the susceptibilities and the thermal fluctuations of the extensive variables. All susceptibilities are proportional to the magnitude of the fluctuations of the respective extensive variables. These couplings are called the "fluctuation relations." In lipid membranes and biomembranes, many of the susceptibilities are related to the heat capacity, which allows one to predict them in a simple manner. Knowing the susceptibilities allows one to describe the dynamics of nerve treated in Part VII and also the emergence of channels described in Part VIII. This part is somewhat technical. It serves as a reference guide on how to deduce the susceptibilities and characteristic time constants from simple thermodynamic experiments. This is important in Part VII because the membrane compressibility enters into the wave equation. However, you can skip this part if you simply trust the results.

VI.1

Entropy and Fluctuations

In a closed system, the total energy is constant. At equilibrium, it displays well-defined values for temperature, pressure, and other intensive variables. The only thing that can change in a closed system is the order and arrangement of the molecules and their state. Any given subsystem of the ensemble can therefore fluctuate in its extensive variables while the overall average is maintained. The starting point in nonequilibrium thermodynamics is that there exists a single thermodynamic potential, the entropy S, that describes the magnitude of the fluctuations [1, 2, 3, 4].

The second law of thermodynamics states that the entropy possesses a maximum. Due to Boltzmann's principle, the entropy is related to a probability. One can phrase the second law as:

> In a closed system there exists a most likely state.

However, there exist other states close to the most likely state that are also possible with a finite likelihood, i.e., one finds fluctuations around the most likely state. As Albert Einstein phrased it:

> "... the irreversibility of physical processes is only apparent." [1]

If one prepares an unlikely initial state at the beginning of an experiment, i.e., if the system is out of equilibrium, it will approach a more likely state with high probability and a less likely state with a much smaller probability. Thus, the equilibrium consists of the most likely distribution of states rather than one single most likely state. Therefore, thermal fluctuations around the most likely state are part of the equilibrium distribution. Nonequilibrium thermodynamics considers the functional form of this distribution and how it depends on the extensive variables.

At the point of the entropy maximum, the value for the extensive variables ξ_i shall be given by $\xi_{I,0}$, which could be $E_0, H_0, V_0, A_0, x_0, q_0, n_{i,0}$, and so on. Fluctuations occur around this point. A fluctuation in enthalpy is given by $\alpha_H = (H - H_0)$, a fluctuation in volume is given by $\alpha_V = (V - V_0)$ or generally, $\alpha_i = (\xi_i - \xi_{i,0})$.

Physics of Nerves and Excitatory Membranes, First Edition. Thomas Heimburg.

Einstein first introduced the concept of entropy as a potential that depends on the thermodynamic variables. The entropy can be expanded into a Taylor series around the most likely state. The linear terms vanish because the derivatives of the entropy at the maximum are all zero. Therefore, we can approximate:

$$\begin{aligned} S &= S_0 + \frac{1}{2}\sum_i\sum_j\left(\frac{\partial^2 S}{\partial\xi_i\partial\xi_j}\right)_0 \alpha_i\alpha_j + \text{higher order terms} \\ &\equiv S_0 - \frac{1}{2}\sum_i\sum_j g_{ij}\alpha_i\alpha_j + \text{higher order terms} \end{aligned} \tag{VI.1.1}$$

where $\alpha_i = \xi_i - \xi_{i,0}$ is the fluctuation of the extensive variable ξ_i around the equilibrium point. The g_{ij} are constants. The leading term of the Taylor expansion is quadratic. This is the reason why so many quadratic laws (such as Hooke's law) are found in physics. If one only considers one variable, one finds

$$S = S_0 - \frac{1}{2}g\alpha^2 + \ldots \tag{VI.1.2}$$

For small fluctuations, we will in the following neglect the higher order terms. For constant overall energy, the relative probability to find a particular microstate of a system is

$$P = \exp\left(\frac{S}{k}\right) = P_0\exp\left(-\frac{g\alpha^2}{2k}\right), \tag{VI.1.3}$$

where P_0 is the probability of the maximum entropy state with $\alpha = 0$. This can be compared to a Gaussian distribution:

$$P = \frac{1}{\sqrt{2\pi\sigma^2}}\exp\left(-\frac{\alpha^2}{2\sigma^2}\right) \tag{VI.1.4}$$

We find that the standard deviation, σ, is given by

$$\sigma^2 = \frac{k}{g}. \tag{VI.1.5}$$

We recognize that the frequent occurrence of Gaussian distributions in physics is the consequence of the quadratic nature of the entropy potential close to equilibrium. Thus, it is a consequence of the second law. If the magnitude of the fluctuation is equal to the standard deviation, i.e., $\alpha = \sigma$, the probability is given by $P(\sigma) = (1/\sqrt{2\pi\sigma^2})\exp(-0.5)$. Thus, the value of the entropy at the standard deviation is $S - S_0 = -k/2$. The shape of the entropy potential, the Gaussian distribution, and the standard deviation are shown in Figure VI.1.1.

The fluctuations can be measured and they show up in experiments. But they are not noise in the sense of an unwanted signal perturbation from the outside. They are rather noise in the sense of Brownian motion. Fluctuations are an intrinsic element of thermodynamics, and it is important to appreciate this aspect. In order to illustrate this, Figure VI.1.2 (right) shows fluctuations of the enthalpy of a membrane at three different temperatures – below, within, and above the melting transition of a membrane – as determined during a Monte Carlo simulation. The details of this simulation do not matter

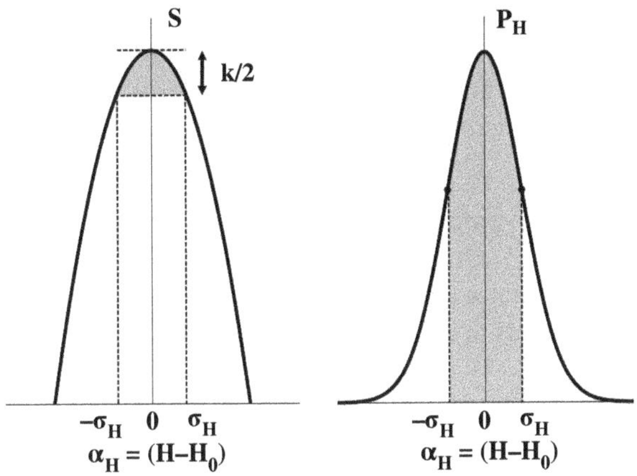

Figure VI.1.1 Enthalpy fluctuations: The entropy S and the corresponding Gaussian distribution P_H. Left: Entropy as a quadratic potential depending on the enthalpy. Right: The corresponding Gaussian distribution. Indicated is the standard deviation of the enthalpy fluctuations, σ_H, which corresponds to a value of the entropy $S - S_0 = k/2$.

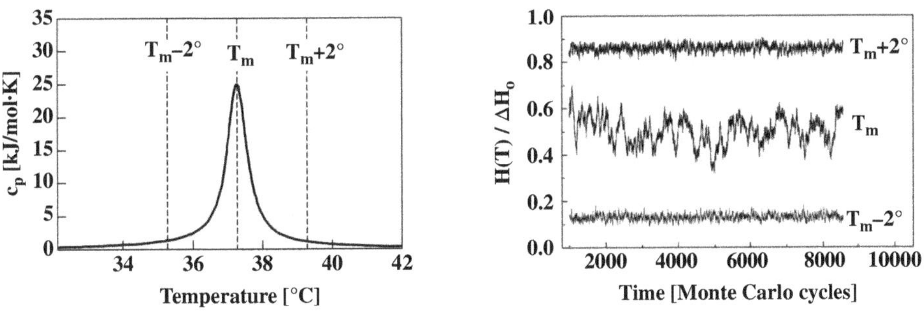

Figure VI.1.2 Left: Monte Carlo simulation of the heat capacity profile for a lipid with melting temperature $T_m = 37.3$°C. Right: Enthalpy fluctuations as a function of computer time at three temperature: $T_m - 2°$, T_m, and $T_m + 2°$. Both the heat capacity and the enthalpy fluctuations are at a maximum and T_m. Adapted from [8].

here (they can be found in [5–7]). One can see that at the three temperatures, the magnitude of the enthalpy fluctuations around the most likely enthalpy value is different. The width of the distribution around the average turns out to be closely related to the heat capacity, c_p, as we show in Chapter VI.2.

In a similar manner, the fluctuations in volume are proportional to the isothermal volume compressibility, and the fluctuations in the charge on a capacitor are related to the capacitance. These functions are the so-called susceptibilities, which represent derivatives of an extensive variable with respect to an intensive variable. For each susceptibility, there exists a set of fluctuations. One can learn a lot from looking at these fluctuations because a wealth of information is hidden in them that usually is not exploited.

In the next section of this part, we will illustrate this for the heat capacity.

VI.1.1 Summary: Key Ideas of Chapter VI.1

In an ensemble, there exists a most likely state that corresponds to the maximum entropy state. However, surrounding the most likely microstate, there exist other states with finite probability and lower entropy. One finds:

- Close to equilibrium, the entropy can be approximated by a harmonic function of the extensive variables.
- This leads to a Gaussian distribution of states around the most likely state.
- The width of this distribution corresponds to the mean square magnitude of fluctuations of an extensive variable.
- These fluctuations can be measured and one also finds them in Monte Carlo and molecular dynamics simulations.
- The fluctuations are an elementary part of the thermodynamics of a system.
- Each susceptibility is related to the magnitude of the fluctuations of one or more intensive variables.

VI.2

Heat Capacity

The heat capacity is a susceptibility that tells us how much heat a system absorbs when the temperature is changed. The heat capacity is high during phase transitions because a small change in temperature leads to a large absorption of heat.

VI.2.1 Heat Capacity at Constant Volume

The heat capacity at constant volume, c_V, describes the temperature dependence of the internal energy, E, because at constant volume, $dE = TdS - pdV = TdS = dQ$. In an ensemble, it describes the temperature dependence of the average internal energy, $\langle E \rangle$:

$$c_V = \left(\frac{d\langle E \rangle}{dT}\right)_V . \tag{VI.2.1}$$

According to eq. (II.4.6), the mean square fluctuations of the internal energy are given by

$$\langle (E - E_0)^2 \rangle = -k\left(\frac{\partial X_i}{\partial \alpha_i}\right)^{-1} = kT^2 \cdot c_V, \tag{VI.2.2}$$

where $\alpha_E = E - E_0$, and $X_E = 1/T$. This yields

$$c_V = \frac{\langle (E - E_0)^2 \rangle}{kT^2}. \tag{VI.2.3}$$

One can also determine the heat capacity from the Boltzmann distribution. The mean internal energy $\langle E \rangle$ of a membrane is given by the thermodynamic average of the internal energy E over all microstates. It is given by $\langle E \rangle = \sum P_i E_i$, where E_i is the energy of the specific microstate i, and $P_i = \exp(-E_i/kT)$ is the corresponding Boltzmann probability (see Section II.2.4):

$$\langle E \rangle = \frac{\sum_i E_i \exp(-\frac{E_i}{kT})}{\sum_i \exp(-\frac{E_i}{kT})} \equiv \frac{\sum_i E_i \exp(-\frac{E_i}{kT})}{Q}, \tag{VI.2.4}$$

where Q is the partition sum. The heat capacity can be obtained from the temperature derivative of this function. After short calculations, one finds that

$$c_V = \frac{d\langle E \rangle}{dT} = \frac{\langle E^2 \rangle - \langle E \rangle^2}{kT^2}, \tag{VI.2.5}$$

Physics of Nerves and Excitatory Membranes, First Edition. Thomas Heimburg.

which is identical to the result in eq. (VI.2.3) because $\langle E \rangle \equiv E_0$ and $\langle (E - E_0)^2 \rangle \equiv \langle E^2 \rangle - \langle E \rangle^2$. This implies the heat capacity at constant volume is proportional to the fluctuations of the internal energy.

VI.2.2 Heat Capacity at Constant Pressure

For most experiments in a laboratory, the pressure is constant while the volume may change. Under such conditions, $dH = TdS + Vdp = TdS = dQ$. Thus, the heat absorption is equal to the change in enthalpy. The heat capacity at constant pressure is given by

$$c_p = \left(\frac{d\langle H \rangle}{dT} \right)_p , \tag{VI.2.6}$$

where $\langle H \rangle$ is the mean enthalpy given by

$$\langle H \rangle = \frac{\sum_i H_i \exp(-\frac{H_i}{kT})}{\sum_i \exp(-\frac{H_i}{kT})} . \tag{VI.2.7}$$

This leads to

$$c_p = \frac{d\langle H \rangle}{dT} = \frac{\langle H^2 \rangle - \langle H \rangle^2}{kT^2} . \tag{VI.2.8}$$

Thus, the heat capacity at constant pressure is proportional to the fluctuations in enthalpy. Since most experiments in biophysics and physiology are measured at constant pressure, we will use the heat capacity c_p in all considerations that will follow.

VI.2.3 Summary: Key Ideas of Chapter VI.2

The heat capacity is a susceptibility that tells us how much heat is absorbed upon a small change in temperature.

- At constant volume, the heat capacity c_V is proportional to the fluctuations in internal energy.
- At constant pressure, the heat capacity c_p is proportional to the fluctuations in enthalpy.
- During transitions, the heat capacity is high, and so are the fluctuations.

The heat capacity has the advantage that it is very easy to measure in a calorimeter.

VI.3

Relation Between Enthalpy, Volume, and Area Changes

Enthalpy, volume, and area are not independent functions. Both are smaller in the gel phase of a membrane as in the fluid phase. Thus, all three of them are increasing functions of temperature. We show in the following sections that there exist simple correlations between the changes of these three extensive variables.

VI.3.1 Relation Between Heat Capacity and Volume Expansion Coefficient as Measured by Densitometry

The mass-density of membranes can be measured with a vibrating capillary (the so-called Kratky balance [10, 11]) filled with a lipid dispersion [9]. Let us assume a glass capillary with a magnetic bead at the end that is subject to a periodic force. The eigenfrequency of this capillary will depend on its mass. If a glass capillary is filled with a substance that changes its mean density, the vibrational eigenfrequency changes. If this is calibrated with a substance of known density (e.g., water density as a function of temperature), one can deduce the changes in lipid density. This is shown in Figure VI.3.1 for a 10 mM dispersion of large unilamellar vesicles (LUV) of dimyristoyl phosphatidylcholine (DMPC). Shown are the volume expansion (bottom), the density (top), and the specific volume of the lipid.

From the specific volume of the lipids, one can derive the volume expansion coefficient at constant pressure $(d\langle V\rangle/dT)_p$. For four different single lipid membranes and lipid mixtures, volume expansion coefficients are compared with the heat capacity in Figure VI.3.2. One finds that the heat capacity is proportional to the volume expansion coefficient within experimental accuracy. This implies that volume changes and enthalpy changes are proportional functions of temperature:

$$\frac{d\langle\Delta V(T)\rangle}{dT} = \gamma_V \cdot \frac{\langle\Delta H(T)\rangle}{dT}, \tag{VI.3.1}$$

where $\gamma_V = 7.8 \cdot 10^{-10}\,\mathrm{m^2/N}$ is a coefficient, which is, within error, the same for all systems shown in Figure VI.3.2 [9]. ΔV and ΔH are the excess volume and enthalpy for a melting transition between the gel and fluid states of a membrane. This implies that volume changes and enthalpy changes are also proportional functions of temperature:

$$\Delta V(T) = \gamma_V \cdot \Delta H(T). \tag{VI.3.2}$$

Physics of Nerves and Excitatory Membranes, First Edition. Thomas Heimburg.

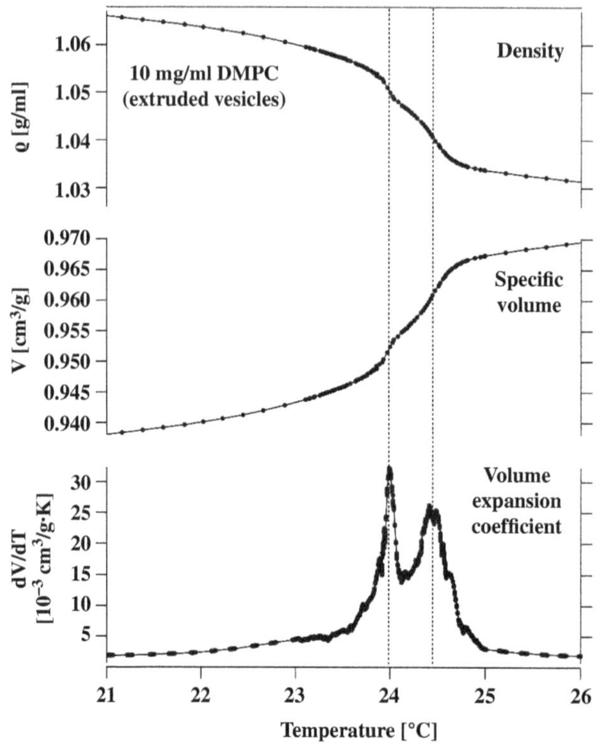

Figure VI.3.1 Changes in specific density, specific volume, and volume expansion coefficient for extruded vesicles of the lipid DMPC. [9]/with permission of American Chemical Society.

VI.3.2 Relation Between Heat Capacity and Volume Expansion Coefficient as Measured by Pressure Calorimetry

One can also obtain the relation between temperature-dependent volume changes and enthalpy changes from pressure calorimetry. The enthalpy $\langle \Delta H \rangle = \langle \Delta E \rangle + p \cdot \langle \Delta V \rangle \equiv \langle \Delta H_0 \rangle + \Delta p \cdot \langle \Delta V \rangle$ is pressure-dependent. Here, H_0 is the enthalpy at a pressure of 1 bar, and Δp is the difference between the applied pressure and atmospheric pressure. As a consequence, the transition temperature $T_m = \Delta H/\Delta S$ is also temperature-dependent. Let us assume that not only $\langle \Delta V \rangle = \gamma_V \cdot \langle \Delta H \rangle$ but also $\Delta V_i = \gamma_V \cdot \Delta H_i$ for each microstate of the lipid membrane where the parameter γ_V is the same for each microstate – an assumption that we will show to be correct below. Then it follows that [9]

$$\langle \Delta V(T) \rangle = \gamma_V \cdot \langle \Delta H(T) \rangle, \tag{VI.3.3}$$

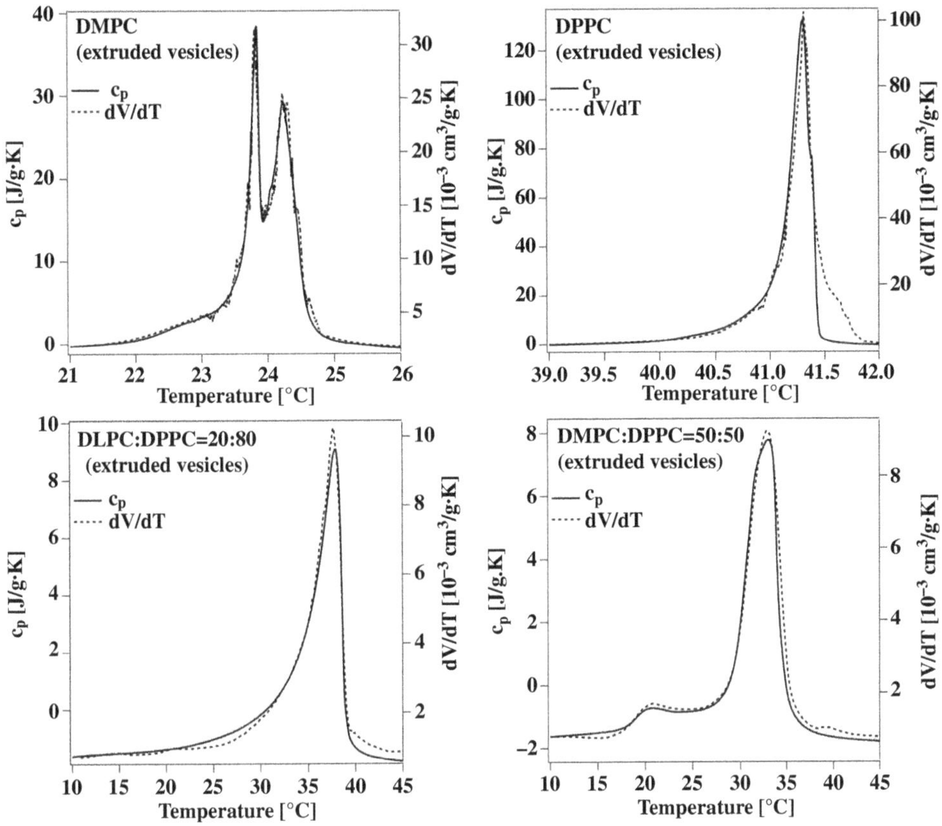

Figure VI.3.2 The volume expansion coefficient compared to the specific heat capacity for four different lipids and lipid mixtures. Both expansion coefficient and heat capacity are proportional functions of the temperature. [9]/with permission of American Chemical Society.

and

$$\begin{aligned}\langle \Delta H(T)\rangle_{\Delta p} &= (1+\gamma_V\cdot\Delta p)\frac{\sum_i \Delta H_{0,i}\exp\left(-(1+\gamma_V\cdot\Delta p)\Delta H_{0,i}/kT\right)}{\sum_i \exp\left(-(1+\gamma_V\cdot\Delta p)\Delta H_{0,i}/kT\right)}\\ &= (1+\gamma_V\cdot\Delta p)\frac{\sum_i \Delta H_{0,i}\exp\left(-\Delta H_{0,i}/kT^*\right)}{\sum_i \exp\left(-\Delta H_{0,i}/kT^*\right)}.\end{aligned} \tag{VI.3.4}$$

Here, we define T^* as

$$T^* \equiv \frac{T}{1+\gamma_V\cdot\Delta p}. \tag{VI.3.5}$$

It follows that

$$\langle \Delta H(T)\rangle_{\Delta p} = (1+\gamma_V\cdot\Delta p)\,\langle \Delta H(T^*)\rangle_0. \tag{VI.3.6}$$

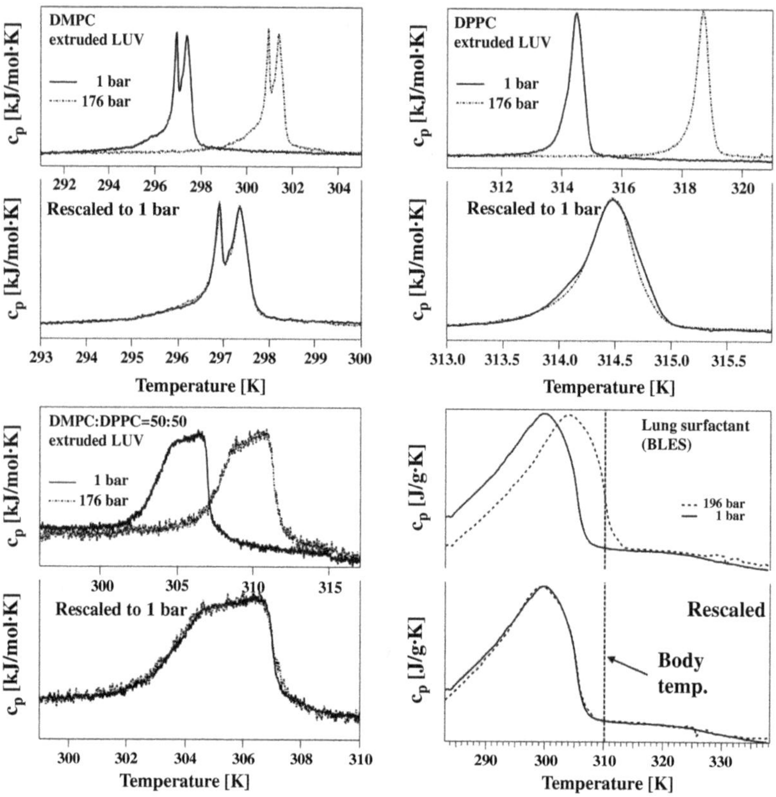

Figure VI.3.3 Calorimetric profiles of lipids with and without excess pressure. Rescaling the temperature axis for profiles with pressure allows to superimpose the profiles on those at a pressure of 1 bar. Top, left: Unilamellar vesicles of DMPC. Top, right: Unilamellar vesicles of DPPC. Bottom, left: A lipid mixture of DMPC:DPPC = 50:50. Bottom, right: Lung surfactant (BLES). See text for details on the rescaling. [9]/with permission of American Chemical Society.

These equations imply that the heat capacity profiles at pressure Δp have the same shape as those measured at atmospheric pressure when the absolute temperature axis is rescaled by a constant factor $(1 + \gamma_V \cdot \Delta p)$. Figure VI.3.3 shows the heat capacity profiles of four different systems with and without pressure. The profiles shift upon application of pressure. 40 bars of pressure cause a shift of about 1°C. If one rescales the c_p profiles measured at pressure Δp by dividing the heat capacity by $(1 + \gamma_V \Delta p)$ and replacing the temperature T by T^*, one finds a profile that is nearly identical to the profile recorded at atmospheric pressure, as expected from the theoretical treatment shown earlier. This identity proves the assumption that $\Delta V_i = \gamma_V \cdot \Delta H_i$ is true for each microstate of the system (see [12]). This can be used to determine the constant γ_V from pressure calorimetry even if the absolute amount of lipid is not known. This is especially useful for biological membranes where it is often difficult to determine absolute lipid concentrations. The constant γ_V determined from Figure VI.3.3 results in $\gamma_V = 7.8 \cdot 10^{-10}$ m^2/N for all four systems.

Remarkably, this also works for the melting profiles of biological membranes. The bottom right panel of Figure VI.3.3 shows it for bovine lipid extract surfactant (lung

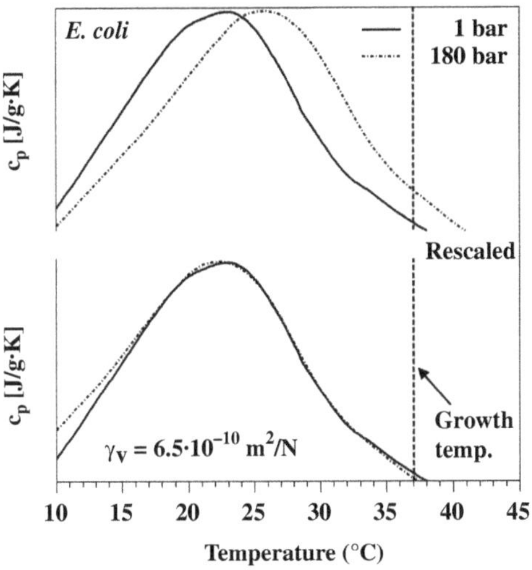

Figure VI.3.4 Pressure effect for the melting transition in native *E. coli* membranes. [13]/with permission of Elsevier.

surfactant, $\gamma_V = 7.8 \cdot 10^{-10}$ m^2/N), and Figure VI.3.4 shows it for native membranes of *E. coli* bacteria. For *E. coli*, one finds $\gamma_V = 6.5 \cdot 10^{-10}$ m^2/N. This is slightly smaller than for the other preparations but of similar order. One can summarize that the factor γ_V is of similar order for widely different lipid systems, including biological membranes that contain their membrane proteins.

VI.3.3 Relation Between Heat Capacity and Volume Expansion Coefficient as Measured by Pressure Perturbation Calorimetry

Another method for obtaining relations between volume expansion coefficient and heat capacity is the pressure perturbation technique. Here, a pressure jump of about 3 bars is applied to a lipid dispersion. 3 bars of pressure cause a shift of the transition by about 0.06°. As a consequence, the heat capacity profile shifts to higher temperatures. Close to the melting transition, some heat is released. When lowering the pressure, the same heat is readsorbed. This is shown in Figure VI.3.5.

The magnitude of the heat changes depends on the exact temperature and displays a maximum close to the transition. There exists a Maxwell reaction that states that $(\partial S/\partial p)_T = -(\partial V/\partial T)_p$. Thus, the change in heat upon change in pressure is

$$\left(\frac{\partial Q}{\partial p}\right)_T = \left(T\frac{\partial S}{\partial p}\right)_T = -T \cdot \left(\frac{\partial V}{\partial T}\right)_p . \tag{VI.3.7}$$

Thus, the volume expansion coefficient α is given by

$$\alpha \equiv \frac{1}{V}\left(\frac{\partial V}{\partial T}\right)_p = -\frac{\Delta Q}{TV\Delta p}, \tag{VI.3.8}$$

where ΔQ is the heat release after a small pressure jump Δp [14].

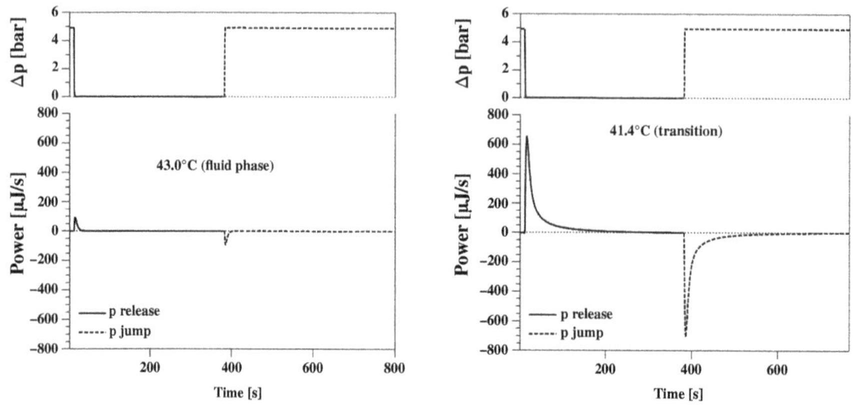

Figure VI.3.5 Pressure jumps in a calorimetric experiment. Left: Jumps from 5 bars to 0 and back and the corresponding power (heat per time) recorded 1.6°C above the melting transition of DPPC. Right: The same experiment made at the transition maximum. One recognizes that the response displays a much higher amplitude at the transition than above. The area under the peaks corresponds the heat ΔQ released or absorbed after a change in pressure, Δp. From this experiment, one can also determine the relaxation time, τ (see Chapter VI.7).

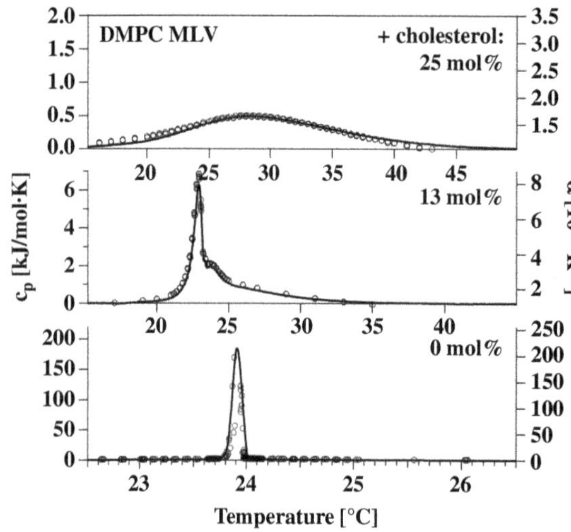

Figure VI.3.6 Pressure perturbation calorimetry of DMPC mixtures with cholesterol also yields the volume expansion coefficient, α. [15]/Springer Nature/CC BY 4.0.

Figure VI.3.6 shows the comparison between the heat capacity profiles of DMPC multilayers at three different cholesterol concentrations. The constant γ_V was found to be between $7.86 - 8.3 \cdot 10^{-10}$ m^2/N, similar to the numbers obtained from the other methods above. It should be noted that the experiment shown in Figure VI.3.5 also allows one to determine the relaxation timescale. This is discussed in Chapter VI.7.

VI.3.4 The Area Expansion Coefficient

It seems plausible to assume that the aforementioned proportional relation between volume and enthalpy also holds for the area, such that

$$\frac{d\langle \Delta A(T)\rangle}{dT} = \gamma_A \cdot \frac{\langle \Delta H(T)\rangle}{dT}, \tag{VI.3.9}$$

where ΔA and ΔH are the excess area and enthalpy in a transition between gel and fluid states of a membrane. We also assume that this relation is true for each microstate:

$$\Delta A_i = \gamma_A \Delta H_i. \tag{VI.3.10}$$

If true, the relationship would allow us to determine the area compressibility and the bending elasticity from the heat capacity (see Chapter VI.4). This leads to agreement with experiments and molecular dynamics simulations [16], and therefore we will assume in the following that this relation holds for lipid systems in general. Heimburg [12] determined

$$\gamma_A = 0.89\,\frac{\mathrm{m}}{\mathrm{N}}. \tag{VI.3.11}$$

We have shown above that the constant γ_V is very similar for a large range of lipid systems, including biological membranes. We will therefore assume in the following that $\gamma_A = 0.89$ m/N is a good approximation for artificial and biological systems with experimental values being of similar order.

VI.3.5 The Consequences of the Relations Between Enthalpy, Volume, and Area

Previously, we have shown that there exist three methods to obtain the relation between volume and enthalpy changes. The exact proportionality is surprising and is not rooted in first principles. It has to be taken as an experimental fact for lipid membranes. However, the finding is a very fortunate circumstance because it allows us to determine various thermodynamic functions from calorimetry.

In the rest of the book, we will assume that the relation between volume and enthalpy is given by:

$$\begin{aligned} \langle \Delta V(T)\rangle &= \gamma_V \langle \Delta H(T)\rangle \qquad \text{with} \qquad \gamma_V \approx 7.8\cdot 10^{-10}\mathrm{m}^2/\mathrm{N} \\ \Delta V_i &= \gamma_V \Delta H_i, \end{aligned} \tag{VI.3.12}$$

where V_i is a microstate of the membrane system. One can show that if the first relation is true for all temperatures, the second relation must also be true [12]. Further, the relation was found to be true in molecular dynamics simulations by [16]; see Figure VI.3.7.

We will extend this concept to the area of a membrane and postulate that

$$\begin{aligned} \langle \Delta A(T)\rangle &= \gamma_A \langle \Delta H(T)\rangle \qquad \text{with} \qquad \gamma_A \approx 0.89\,\mathrm{m}/\mathrm{N} \\ \Delta A_i &= \gamma_A \Delta H_i. \end{aligned} \tag{VI.3.13}$$

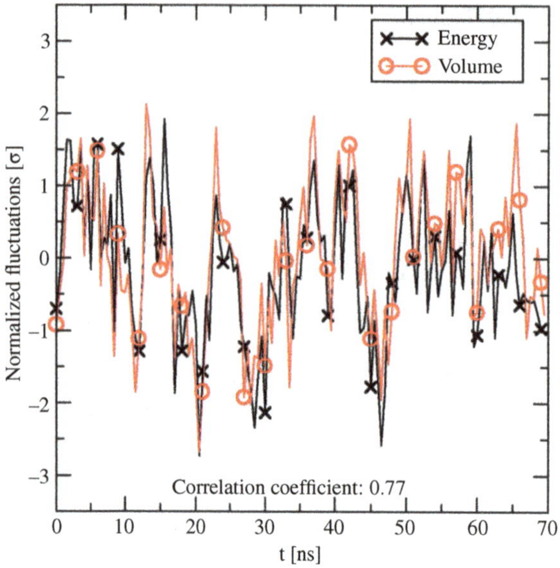

Figure VI.3.7 The slow macroscopic fluctuations in volume and enthalpy of a DMPC membrane are found to be proportional in a molecular dynamics simulation. From [16]/with permission of American Chemical Society.

The value of $\gamma_A = 0.89$ m/N was deduced for DPPC membranes. The justification for this assumption are findings by Evans and Kwok about area and lateral compressibility changes [17]. It is shown in Section VI.4.2 that one can successfully predict the bending elasticity of lipid vesicles by making the assumption in eq. (VI.3.13) [12, 18]. The relation was also found to be approximately true in molecular dynamics simulations by [16].

These findings imply that not only are the changes in volume, area, and enthalpy proportional functions of temperature, but also this hold true for the fluctuations:

$$\left(\langle\Delta V^2\rangle - \langle\Delta V\rangle^2\right) = \gamma_V^2\left(\langle\Delta H^2\rangle - \langle\Delta H\rangle^2\right) \tag{VI.3.14}$$

$$\left(\langle\Delta A^2\rangle - \langle\Delta A\rangle^2\right) = \gamma_A^2\left(\langle\Delta H^2\rangle - \langle\Delta H\rangle^2\right) \tag{VI.3.15}$$

This is one of the most fundamental results of this part of the book, with profound implications for the theory of the elastic constants of membranes.

VI.3.6 Calculating Volume, Area, and Thickness from Heat Capacity

Outside the transition, lipid membranes display a nearly linear temperature dependence of enthalpy, volume, and area both in gel and fluid states. The parameters are given in [12]. The biggest change, however, is in the transition itself. Since the excess volume, area, and enthalpy are proportional, one can calculate the temperature dependence of these functions from the excess heat capacity. We assume that

$$\begin{aligned}\langle V(T)\rangle &= V_0(T) + \gamma_V\langle\Delta H(T)\rangle \\ V_0(T) &= (1-f)\cdot V_g(T) + f\cdot V_f(T)\end{aligned} \tag{VI.3.16}$$

where $V_0(T)$ is the volume change of the transition outside the transition, $V_g(T)$ and $V_f(T)$ are the temperature dependencies of the pure phases (with linear temperature dependence), and f(T) is the fluid fraction of the lipids. $\Delta H(T) = \int_{T_0}^{T} c_p dT$ is the excess enthalpy. We now assume that we can do the same calculation for the membrane area:

$$\begin{aligned} \langle A(T) \rangle &= A_0(T) + \gamma_A \langle \Delta H(T) \rangle \\ A_0(T) &= (1-f) \cdot A_g(T) + f \cdot A_f(T) \end{aligned} \tag{VI.3.17}$$

For details of the calculation, see [12]. Since $V(T) = A(T) \cdot D(T)$, the temperature-dependent membrane thickness can be calculated when volume and area are known. The temperature dependence of enthalpy, entropy, volume, area, and thickness of a DPPC membrane is shown in Figure VI.3.8.

The heat capacity profile of a biological membrane is much broader. However, the same calculation as above can be performed when assuming that γ_V and γ_A are the same as in the DPPC membrane. While this might not be exactly true, it will be very close to the values of membranes with more complex composition. The predicted changes in the structural variables for lung surfactants are shown in Figure VI.3.9.

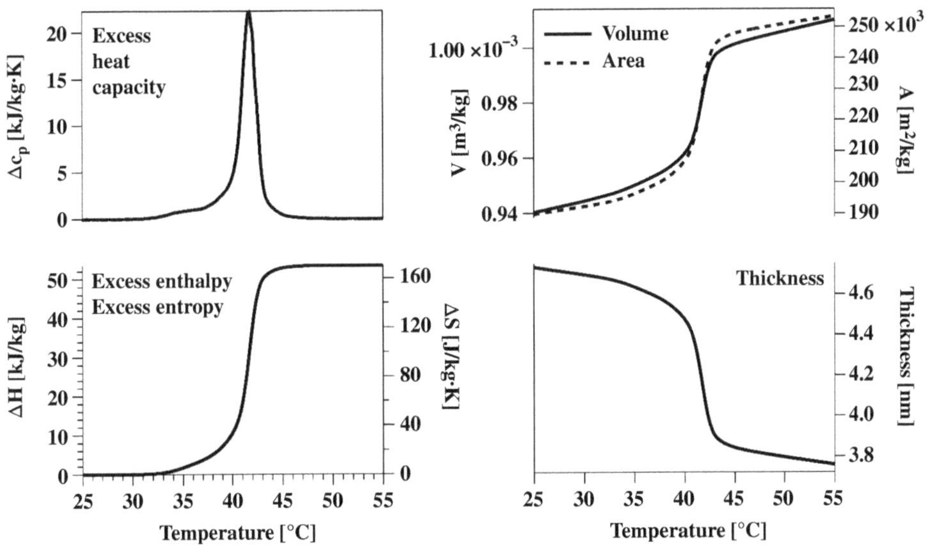

Figure VI.3.8 Left, top: Excess heat capacity of LUV of DPPC with a transition at 41.6°C. Left, bottom: Excess enthalpy and entropy determined from the heat capacity. Right, top: Volume and area of the lipid membrane calculated from the enthalpy and the values of $\gamma_V = 7.8 \cdot 10^{-10}$m²/N and $\gamma_A = 0.89$ m/N determined in the previous section. Right, bottom: Changes in membrane thickness determined from the volume and area changes.

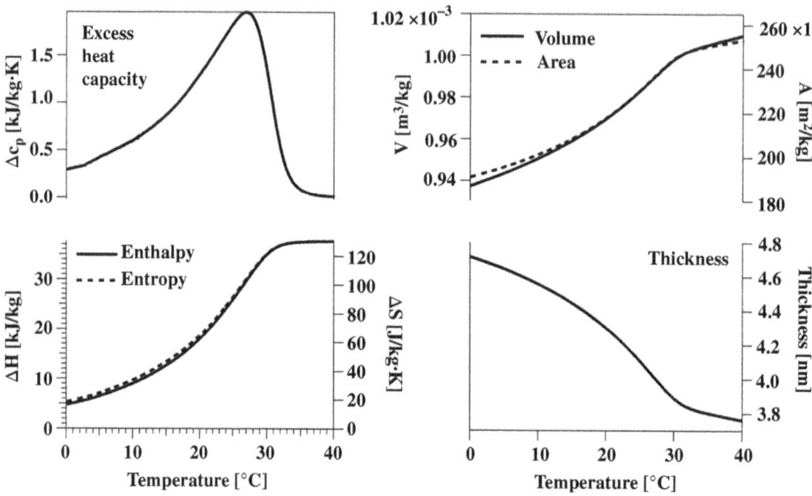

Figure VI.3.9 Left, top: Excess heat capacity of Curosurf lung surfactant with a transition at 41.6 °C. Left, bottom: excess enthalpy and entropy determined from the heat capacity. Right, top: Volume and area of the lipid membrane calculated from the enthalpy and the values of $\gamma_V = 7.8 \cdot 10^{-10}$ m²/N and $\gamma_A = 0.89$ m/N determined in the previous section. Right, bottom: Changes in membrane thickness determined from the volume and area changes.

VI.3.7 Summary: Key Ideas of Chapter VI.3

It is an experimental finding that

- Excess enthalpy, volume, and area changes are proportional functions of temperature. $\Delta V(T) = \gamma_V \Delta H(T)$, where $\gamma_V = 7.8 \cdot 10^{-10}$ m^2/N is a value representative for many membrane systems, i.e., DMPC, DPPC, lipid mixtures, and some biological membranes.
- $\Delta A(T) = \gamma_A \Delta H(T)$, where $\gamma_A = 0.89$ m/N is assumed to be equally representative for different lipids.
- The volume expansion coefficient and the area expansion coefficient are proportional to the excess heat capacity.

$$\frac{1}{V}\left(\frac{\partial \Delta V}{\partial T}\right)_p = \frac{\gamma_V}{V} \cdot \Delta c_p$$
$$\frac{1}{A}\left(\frac{\partial \Delta A}{\partial T}\right)_p = \frac{\gamma_A}{A} \cdot \Delta c_p.$$

- Additions of drugs and small molecules such as cholesterol change the heat capacity and the volume and area expansion coefficients in a similar manner.
- Knowledge of the excess heat capacity allows us to calculate the temperature dependence of membrane enthalpy, volume, area, and thickness. All of these changes will be useful to approximate the temperature dependence of the elastic constants in the next chapter.

VI.4

Transitions and Elastic Constants

In the elastic theory of membranes, the free energy density of a membrane that is stretched and curved is given by [19–21]

$$g = \frac{1}{2A_0} K_A (A - A_0)^2 + \frac{1}{2} K_B (c_1 + c_2 - c_0)^2 + K_G c_1 c_2, \qquad \text{(VI.4.1)}$$

where $(A - A_0)/A_0$ is the change in specific area relative to the equilibrium area A_0, K_A is the area compression modulus, $c_1 = 1/R_1$ and $c_2 = 1/R_2$ are the two principal curvatures of the membranes, c_0 is the spontaneous curvature caused by an asymmetry between the two leaflets of the membrane, R_1 and R_2 are the corresponding radii of curvature, K_B is the bending modulus, and K_G is the Gaussian curvature modulus. A large body of literature uses this equation to predict the shapes of membrane vesicles under various constraints, e.g., for constant surface area to volume ratios (e.g., [20–25]).

While such calculations are helpful and describe shapes well that have been found experimentally, they miss an important point when applied to biological membranes close to physiological temperature: near the melting transition, the elastic "constants" are in fact not constant. The art of dealing with geometrical problems dynamically in biological systems is to understand the thermodynamics of the elastic constant. Unfortunately, this is rarely done, although it is likely to be the most important aspect of elastic theory in living matter.

VI.4.1 Volume and Area Compressibility

Volume and area are fluctuating variables like internal energy and enthalpy. There exists a most likely volume, V_0, and a most likely area, A_0, with fluctuations around these values. This is shown in Figures VI.4.1 and VI.4.2. In this chapter, we show that volume and area fluctuations are directly related to the isothermal volume and area compressibilities – and that they can be determined from the heat capacity. Membranes turn out to be very compressible close to their phase transitions.

The mean volume $\langle V \rangle \equiv V_0$ of a membrane is given by the thermodynamic average of the volume V_i

$$\langle V \rangle = \frac{\sum_i V_i \exp(-\frac{H_i}{kT})}{\sum_i \exp(-\frac{H_i}{kT})} \qquad \text{(VI.4.2)}$$

Physics of Nerves and Excitatory Membranes, First Edition. Thomas Heimburg.

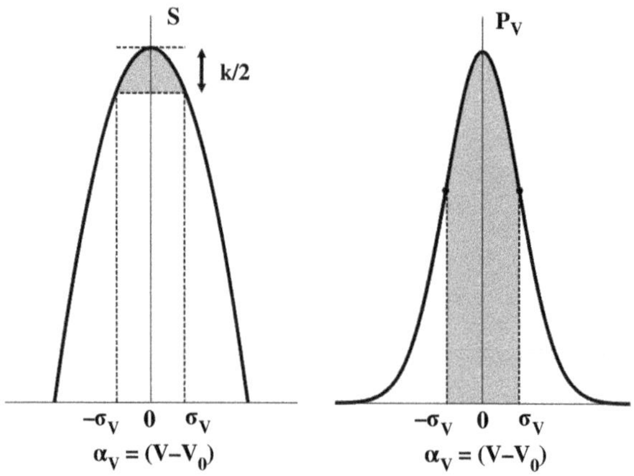

Figure VI.4.1 Volume fluctuations: The entropy S and the corresponding Gaussian distribution P_V plotted as a function of volume. Left: Entropy as a quadratic potential. Right: The corresponding Gaussian distribution. The gray-shaded area marks the standard deviation of the volume fluctuations, σ_V, which corresponds to a value of the entropy $S - S_0 = k/2$.

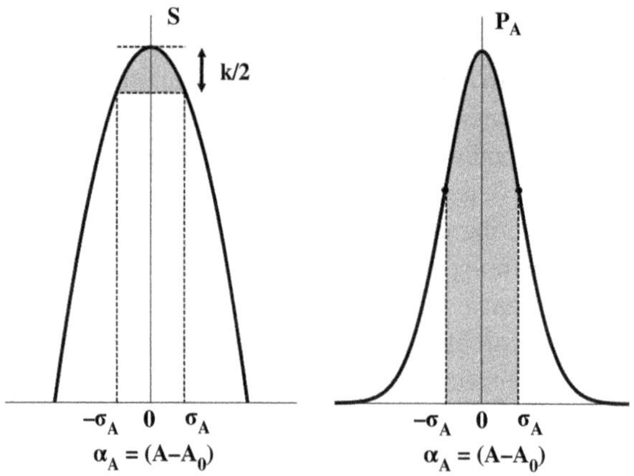

Figure VI.4.2 Area fluctuations: The entropy S and the corresponding Gaussian distribution P_A. Left: Entropy as a quadratic potential depending on the area. Right: The corresponding Gaussian distribution. The gray-shaded area marks the standard deviation of the area fluctuations, σ_A, which corresponds to a value of the entropy $S - S_0 = k/2$.

and the specific isothermal volume compressibility κ_T^V is defined as

$$\kappa_T^V = -\frac{1}{\langle V \rangle}\left(\frac{d\langle V \rangle}{dp}\right)_T \tag{VI.4.3}$$

where p is the hydrostatic pressure. Evaluation of this equation yields

$$\kappa_T^V = \frac{\langle V^2 \rangle - \langle V \rangle^2}{\langle V \rangle kT} \tag{VI.4.4}$$

This fluctuation relation yields the volume compressibility.

Let us use the experimental finding that close to transitions $\langle \Delta V(T) \rangle = \gamma_V \langle \Delta H(T) \rangle$ for all temperatures around the melting transitions, which leads to $\Delta V_i = \gamma_V \Delta H_i$ (see eq. (VI.3.12)). We obtain due to eq. (VI.3.14)

$$\Delta \kappa_T^V = \gamma_V^2 \frac{\langle \Delta H^2 \rangle - \langle \Delta H \rangle^2}{\langle V \rangle kT} = \frac{\gamma_V^2 T}{\langle V \rangle} \Delta c_p, \tag{VI.4.5}$$

where $\Delta \kappa_T^V$ is the excess isothermal volume compressibility, and Δc_p is the excess heat capacity of the melting transition. Thus, with the knowledge of γ_V, the isothermal compressibility can be readily determined. It is at maximum in the transition. This is shown in Figure VI.4.3.

We use the same line of argument for the area compressibility. We assume that $\langle \Delta A(T) \rangle = \gamma_A \langle \Delta H(T) \rangle$ for all temperatures around the transition, which yields $\Delta A_i = \gamma_A \Delta H_i$. In analogy with the calculation above, we obtain for the isothermal excess area compressibility:

$$\Delta \kappa_T^A = \frac{\langle \Delta A^2 \rangle - \langle \Delta A \rangle^2}{\langle A \rangle kT}, \tag{VI.4.6}$$

and together with eq. (VI.3.14)

$$\Delta \kappa_T^A = \gamma_A^2 \frac{\langle \Delta H^2 \rangle - \langle \Delta H \rangle^2}{\langle A \rangle kT} = \frac{\gamma_A^2 T}{\langle A \rangle} \Delta c_p. \tag{VI.4.7}$$

Thus, we conclude that changes in both volume and area compressibility can be determined from changes in the heat capacity close to melting transitions. If a biological membrane exists in the fluid state slightly above a melting transition, a slight lowering of

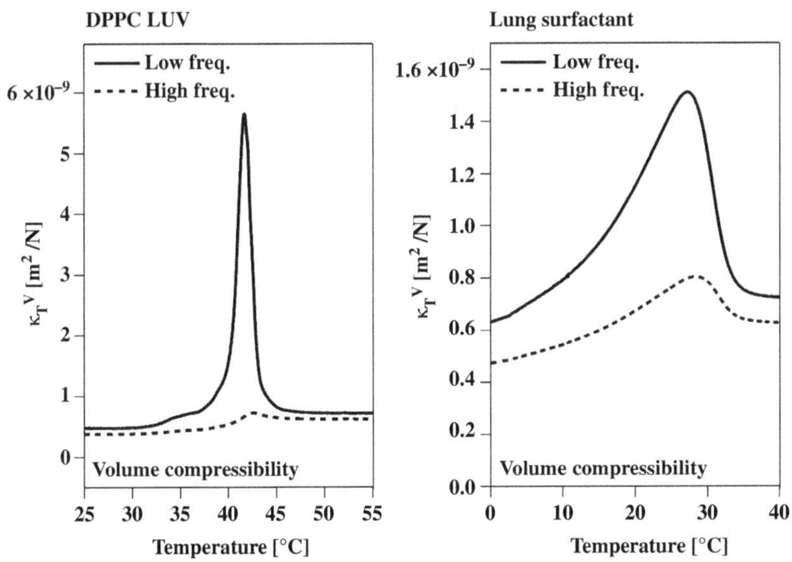

Figure VI.4.3 Calculated isothermal volume compressibility of DPPC LUV (left) and lung surfactant (right). They display a maximum at the melting temperature. The dashed lines indicate the adiabatic compressibilities calculated for 5 MHz frequency (see Section VI.4.1.1).

ambient temperature will render the membrane softer when the membrane is moved into the transition. The nonlinear dependence of the compressibility on temperature and pressure will be of high importance for the emergence of solitons in nerves, as discussed in Part VII. This is shown in Figure VI.4.4.

VI.4.1.1 Adiabatic Volume and Area Compressibility

Lipid membranes and biomembranes are special because they are nearly two-dimensional systems embedded into a three-dimensional liquid medium. This medium can exchange heat with the membrane and thereby influence the elastic properties of the membrane. The isothermal compressibility is recorded under conditions when a membrane has enough time to exchange heat with the environment. For fast adiabatic changes, the system may not have enough time to exchange heat with the environment during compression. Under such conditions, the temperature of the membrane does not stay constant, and the compressibility depends on the rate of compression. Generally, the isothermal compressibility is larger than the adiabatic compressibility. Here, we deal with the adiabatic compressibility because it is essential for determining the sound velocity in Part VII. While the isothermal compressibility is written as κ_T, the adiabatic compressibility is called κ_S (because $dQ = TdS = 0$ implies $dS = 0$).

The adiabatic compressibility is defined as:

$$\kappa_S^V = -\frac{1}{V}\left(\frac{\partial V}{\partial p}\right)_S \tag{VI.4.8}$$

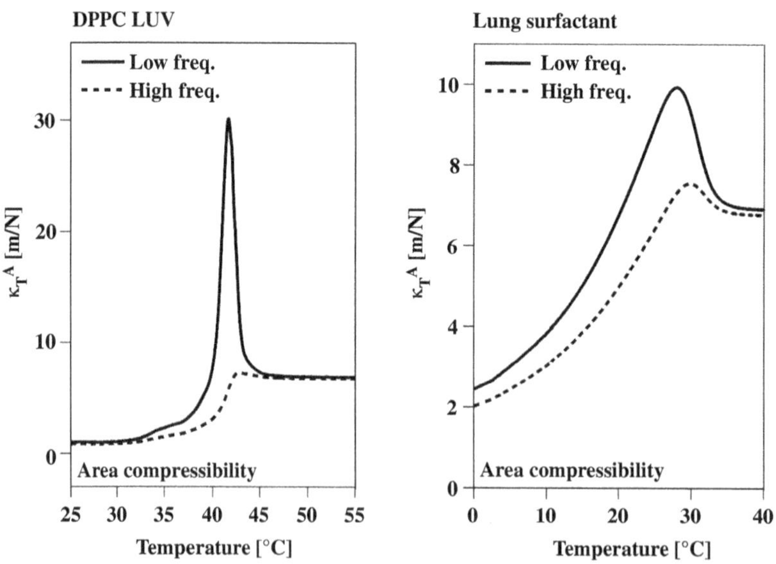

Figure VI.4.4 Calculated isothermal area compressibility of DPPC LUV (left) and lung surfactant (right). Like the volume compressibility, they display a maximum at the melting temperature. The dashed lines indicate the adiabatic compressibilities calculated for 5 MHz frequency (see Section VI.4.1.1).

where the entropy S is constant. It can be derived from the isothermal compressibility (see [26] for a derivation)

$$\kappa_S^V = \kappa_T^V - \frac{T}{V \cdot c_p}\left(\frac{\partial V}{\partial T}\right)_p^2. \quad \text{(VI.4.9)}$$

Lipid membranes are surrounded by water, which displays an isothermal volume compressibility of $4.6 \cdot 10^{-10}$ m^2/N at 25 °C [27], very similar to the volume compressibility of lipid membranes in gel and fluid phases. In the transition range, however, the membrane is much more compressible than water. Due to a conversion between fluid and gel lipids caused by changes in pressure, heat is transferred from the membrane into its aqueous environment and into vibrations in the bonds of the lipids themselves. Thus, the heat capacity in eq. (VI.4.9) denotes the heat capacity of the lipid chains plus that of the surrounding water that forms the reservoir for the heat. The larger the water volume which can absorb the heat released from the membrane, the larger the c_p. However, it takes time for the heat to diffuse into the aqueous medium. For a periodic perturbation of a membrane with frequency ω, the volume of water that is accessible as a heat reservoir will depend on frequency, i.e., $c_p \rightarrow c_p(\omega)$, where ω is a frequency:

$$c_p(\omega) = c_p(\text{lipids}) + c_p(\text{water}, \omega). \quad \text{(VI.4.10)}$$

In principle, c_p(lipids) is also frequency-dependent – but relaxation phenomena on the molecular scale are much faster and we do not consider them here. If the aqueous reservoir communicating with the membrane is very large, the heat capacity in eq. (VI.4.9) is also very large. The interaction with water is schematically illustrated in Figure VI.4.5. Here, a membrane in an aqueous environment of different size is shown, which acts as a heat reservoir for heat released and absorbed by the membrane. In the low-frequency limit of $\omega \rightarrow 0$, the heat can diffuse into the complete aqueous volume. If this is very large, the heat capacity is very large too. In the high-frequency limit (but still slower than the nano- to picosecond timescale of chain isomerizations), the heat capacity of the system is only that of the lipid chains.

Detailed calculations of how the heat capacity depends on the number of associated water molecules are shown in Figure VI.4.6; see [28]. It can be seen that the magnitude of the heat

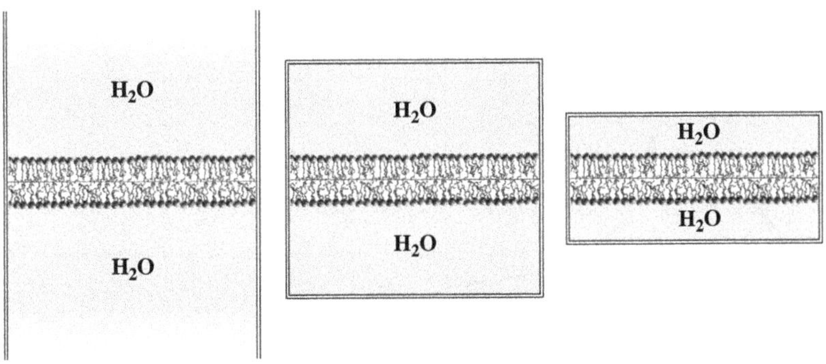

Figure VI.4.5 Diffusion of heat from the membrane into the aqueous volume. Left: A membrane surrounded by an infinite aqueous volume. Center and right: Two finite volumes of water close to the membrane. From [28]/with permission of AIP Publishing.

capacity profile decreases when fewer water molecules contribute to the total heat capacity. This is also found in frequency-dependent calorimetric measurements [29]. The inset in Figure VI.4.6 shows experimental results for DPPC LUV. One can recognize significant changes in the heat capacity when the frequency is changed from 0.01 Hz to 10 Hz. This frequency range corresponds to the relaxation times of such vesicles; see Chapter VI.7. The frequency dependence of the accessible heat reservoir leads to a frequency dependence of the heat capacity and of the adiabatic compressibility of the membrane. This phenomenon is called "dispersion." Dispersion will be discussed in Chapter VII.3.

The result of such considerations is that for very low frequencies, $\omega \to 0$, the heat capacity of the associated water reservoir is very high and thus the adiabatic compressibility in eq. (VI.4.11) approaches the isothermal compressibility:

$$\kappa_S^V(\omega \to 0) = \kappa_T^V, \tag{VI.4.11}$$

as shown in Figure VI.4.3.

By high-frequency limit, we mean the frequencies in the MHz range used in ultrasonic determination of membrane properties. In this limit, $c_p(\omega) \to c_p(\text{lipids})$ (meaning no heat is transferred from the membrane to the aqueous surrounding and only the lipid chains can absorb heat), one obtains the temperature dependence of the adiabatic volume compressibility of the lipids shown in Figure VI.4.7 (left). The profile was calculated from the excess heat capacity and the heat capacity of the lipid chains. It displays a maximum at the transition temperature with a value of about $7.2 \cdot 10^{10}$ m^2/N at 42.5 °C. For comparison, the compressibility of water is $4.478 \cdot 10^{10}$ m^2/N at 30 °C and 25 °C and $4.418 \cdot 10^{10}$ m^2/N at 50 °C. Thus, the ratio of the compressibility of the lipids and of water is slightly below one at low temperature, but above one at the transition temperature. This is the case that was obtained in the ultrasonic experiments at frequencies in the MHz regime described below.

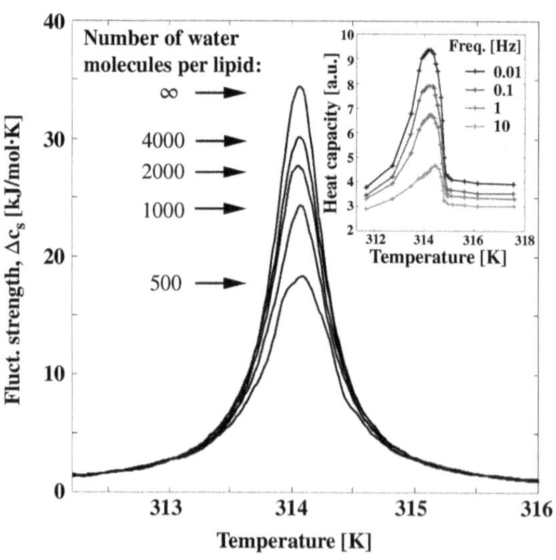

Figure VI.4.6 Calculated heat capacity for different number of water molecules per lipid, mimicking the effect of different frequencies. A large number of membrane-associated water molecules corresponds to a low perturbation frequency. Inset: Frequency-dependent heat capacity. [28, 29].

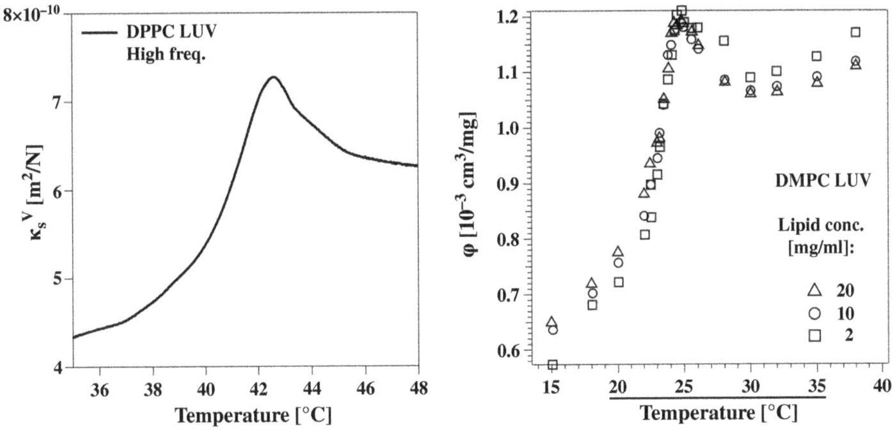

Figure VI.4.7 Left: Adiabatic volume compressibility of DPPC LUV for the high-frequency limit calculated from the heat capacity. Right: The function ϕ from eq. (VI.4.12) is proportional to the adiabatic lipid volume compressibility. The experimental profile for DMPV LUV (determined for three different lipid concentrations) has a very similar shape as the theoretical profile of DPPC LUV shown in the left-hand panel. Adapted from [30].

The adiabatic volume compressibility can be measured by ultrasound methods in a resonator. In such experiments, the wavelength of sound in a lipid dispersion is measured as a function of temperature. Halstenberg et al. [30] used a resonator frequency of 7.2 MHz. The wavelength of sound at this frequency in water is about 0.2 mm, i.e., much larger than the dimensions of lipid vesicles. In an ultrasonic experiment, one therefore measures a compressibility of the lipid dispersion, which consists of an average of the compressibilities of lipids and water weighted by their relative volumes. In order to measure the membrane compressibility, it is therefore convenient to determine the difference between the compressibility of the dispersion and pure water, and normalize it by the compressibility of water and the lipid concentration. This yields the function ϕ (which represents a normalized compressibility) defined as:

$$\phi = \frac{1}{\kappa_{S,0}^{V}} \frac{\kappa_S^V \cdot V - \kappa_{S,0}^V \cdot V_0}{V \cdot [L]}, \qquad \text{(VI.4.12)}$$

where $\kappa_{S,0}^V$ is the compressibility of water, $\kappa_{S,0}^V$ is the compressibility of the dispersion, V_0 is the water volume, V is the total volume, and $[L]$ is the lipid concentration. This number is normalized for the lipid concentration. It has units of an inverse concentration and is independent of lipid concentration. The function ϕ determined for DMPC vesicles is shown in Figure VI.4.7 (right). One recognizes that it has a maximum at the transition temperature of DMPC. The function ϕ is proportional to the adiabatic lipid volume compressibility. Its shape is very similar to the predicted shape in Figure VI.4.7 (left). We see that the relative compressibility is below one at low temperatures and above one at the transition temperature of 24.5 °C, just like in the calculated case of DPPC. This similarity gives us the confidence that it is possible to calculate both low- and high-frequency limits of the adiabatic compressibility from the heat capacity.

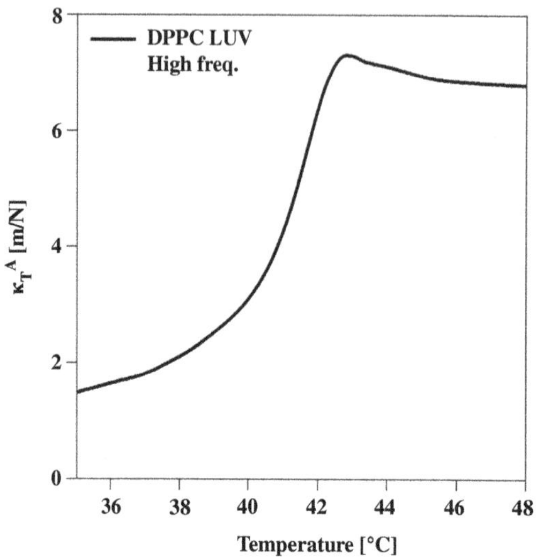

Figure VI.4.8 Adiabatic volume compressibility of DPPC LUV for the high-frequency limit calculated from the heat capacity.

The calculation for the adiabatic area compressibility in the high-frequency limit is analogous:

$$\kappa_S^A = \kappa_T^A - \frac{T}{A \cdot c_p}\left(\frac{\partial A}{\partial T}\right)_p^2 \tag{VI.4.13}$$

It leads to the adiabatic compressibility shown in Figure VI.4.8. In the low-frequency limit, the adiabatic area compressibility is equal to the isothermal area compressibility, κ_T^A (see Figure VI.4.4).

VI.4.2 Bending Elasticity

The bending elasticity is also a susceptibility. It is proportional to the fluctuations in curvature.

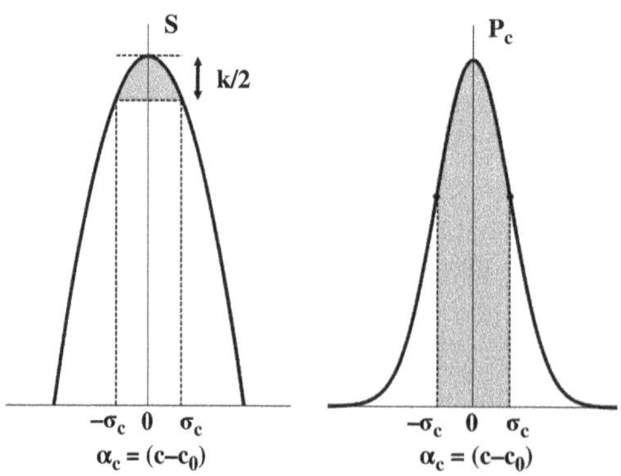

We demonstrated in Part II (Section II.4.2.3) that the bending elasticity κ_B is given by

$$\kappa_B = \frac{\langle c^2 \rangle - \langle c \rangle^2}{kT}. \tag{VI.4.14}$$

This implies that the bending elasticity can be determined from the curvature fluctuations, which is in fact a common method to determine the bending modulus [31–37]. The bending modulus K_B used in eq. (VI.4.1) is the inverse of the bending elasticity.

Another way to derive a good value for the bending elasticity is to deduce it from the lateral compressibility. If a flat membrane is bent, the outer monolayer is stretched and the inner layer is compressed. This relates the curvature fluctuations to the area fluctuations of the monolayers. It was shown in [12] that the change in bending elasticity is related to the change in area compressibility in the following manner:

$$\Delta\kappa_B = \frac{16}{\langle D \rangle^2} \Delta\kappa_T^A = \frac{16}{\langle D \rangle^2} \frac{\gamma_A^2 T}{\langle A \rangle} \Delta c_p, \tag{VI.4.15}$$

and the total bending elasticity by

$$\begin{aligned} \kappa_B &= \kappa_{B,0}(T) + \frac{16}{\langle D \rangle^2} \frac{\gamma_A^2 T}{\langle A \rangle} \Delta c_p \\ \kappa_{B,0}(T) &= (1-f)\kappa_{B_0}^{gel} + f\kappa_{B_0}^{fluid} \end{aligned} \tag{VI.4.16}$$

where $\langle D \rangle$ is the membrane thickness, and $\langle A \rangle$ is the membrane area. The bending elasticities of the pure phase $\kappa_{B_0}^{gel}$ and $\kappa_{B_0}^{fluid}$ have to be taken from experiment; see [12]. They represent the elastic constants in the absence of a transition, where the solid lipid membrane is somewhat stiffer than the liquid phase.

Equation (VI.4.16) shows that the excess bending elasticity is proportional to the excess heat capacity. This calculation leads to the temperature dependence of the bending elasticity shown in Figure VI.4.9 given for DPPC LUV and lung surfactant. It can be seen that the bending elasticity displays a pronounced maximum at the melting transition. This is the condition where it is the most easiest to deform a membrane.

Such predictions can be tested in experiments. Figure VI.4.10 shows the measured bending modulus (which is the inverse of the bending elasticity) of DMPC vesicles in comparison to the bending modulus determined theoretically from the heat capacity changes in such vesicles. The minimum in the bending modulus corresponds to the maximum in the bending elasticity. One can recognize that the prediction and the experimental outcome are very similar. This confirms the coupling of the bending elasticity with area compressibility and heat capacity. Thus, it is likely that any process involving changes in curvature – such as cell division, endocytosis, and exocytosis – will be largely facilitated by everything that moves a biomembrane into the melting transition range. Temperature is only one variable that allows us to do this. Lowering pH would be another change that brings a biological membrane at physiological temperature closer to the melting transition.

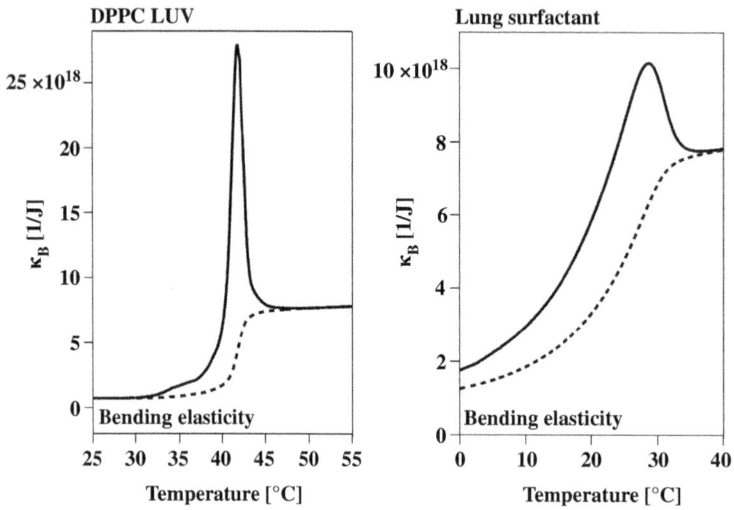

Figure VI.4.9 Bending elasticity predicted from eqs. (VI.4.15) and (VI.4.16) for DPPC LUV and lung surfactant.

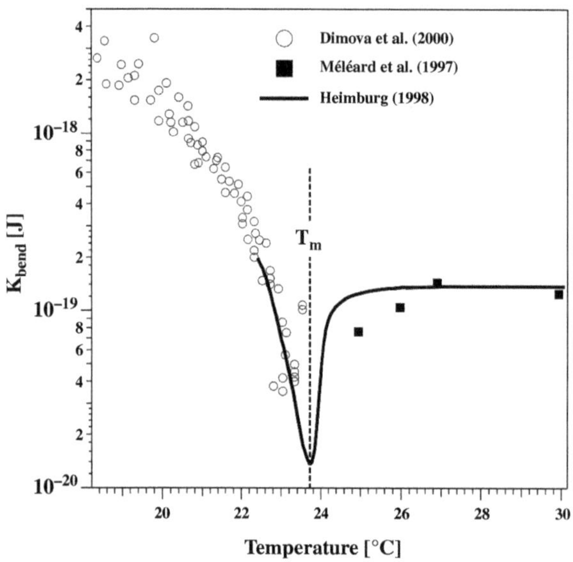

Figure VI.4.10 Bending modulus K_B (the inverse of the bending elasticity κ_B) of DMPC LUV. The solid line is the prediction from the heat capacity. Experiment and prediction are consistent with the predictions from eqs. (VI.4.15) and (VI.4.16). [12, 18, 38, 39].

VI.4.3 Permeability

The permeability of membranes for ions or molecules is not a susceptibility. However, it is related to the area compressibility (which is a susceptibility) because in order to create a pore in a membrane, work has to be done by compressing the membrane surrounding a created pore. Nagle and Scott [40] showed that the membrane permeability is proportional to the area fluctuations. They found that

$$\mathcal{P} = \mathcal{P}_0 + C_2 \cdot \kappa_T^A + \dots , \tag{VI.4.17}$$

where $\mathcal{P}$ is the permeability of the membrane, and C_2 is a constant. Since it is proportional to the area compressibility (which is proportional to the excess heat capacity), the membrane permeability itself is proportional to the heat capacity. At the melting transition, where the heat capacity is at maximum, the membrane permeability is also at maximum. This has been confirmed in numerous experiments (e.g., [41–43]). This phenomenon is also related to the emergence of lipid ion channels. These are membrane pores that appear in an electrophysiological experiment as open-close events. Electrophysiological current recordings look very similar or even identical to protein channel traces. Since this is very important in the context of a thermodynamic theory of nerves, a complete part of this book is dedicated to the emergence of lipid ion channels (see Part VIII). They can be gated by everything that influences the melting transition (see Part V). This results in voltage-gated, temperature-sensitive, and mechanosensitive channel events even in the complete absence of channel proteins.

VI.4.4 Summary: Key Ideas of Chapter VI.4

The elastic constants are proportional to the fluctuations in volume, area, or curvature.

We have shown in eq. (VI.3.14) that volume and area fluctuations are proportional to the enthalpy fluctuation. This results in a simple dependence of the elastic constants on the heat capacity. In particular, we find the following:

- The isothermal volume compressibility is proportional to the volume fluctuations.
- The isothermal area compressibility is proportional to the area fluctuations.
- The bending elasticity is proportional to the curvature fluctuations.
- All of these functions are proportional to the excess heat capacity and are at a maximum in the melting transitions of a lipid or a biological membrane.
- The aqueous environment of a membrane serves as a heat sink.
- Both adiabatic volume and area compressibility can be calculated from the isothermal heat capacity when considering the size of the aqueous volume surrounding the membrane. There exist two limits important for this book: the low-frequency limit, where all water serves as a heat sink and the adiabatic compressibility is identical to the isothermal compressibility, and the high-frequency limit, where the aqueous environment of the membrane does not contribute.
- The calculations for the latter case led to very similar results for the adiabatic compressibility as found in 7 MHz ultrasound experiments.
- Thus, one can confidently assume that the adiabatic compressibility can be determined from the heat capacity.
- The membrane permeability for ions or small molecules is proportional to the area fluctuations and thus to the heat capacity. It is also at maximum in the melting transition.

The melting transition is influenced not only by temperature, pressure, and lateral tension changes but also by the presence of drugs or changes in pH. Therefore, the elastic constants can be fine-tuned by changes of the intensive variables. We show in Part VII that this leads to the emergence of nerve pulse, and in Part VIII that this leads to the emergence of ion channel events in nerve membranes that are voltage-gated, mechanosensitive, and thermosensitive. In Part IX, we show how drugs can influence the elastic constants, and why this explains the phenomenon of anesthesia.

VI.5

Sound Propagation

The sound velocity in membranes will play a central role in Part VII, where it is at the basis of a formalism that leads to the propagation of electromechanical solitons in nerve membranes. As shown in Chapter VII.1, the sound velocity c is given by the inverse of $\sqrt{(\partial\rho/\partial p)}$, which is the square root of a susceptibility.

VI.5.1 Sound Propagation in Three Dimensions

The sound velocity c is given by

$$c = \sqrt{\frac{1}{\kappa_S^V \cdot \rho^V}}, \tag{VI.5.1}$$

where κ_S^V is the adiabatic compressibility determined in the previous chapter and ρ^V is the volume density determined in Section VI.3.6. This sound velocity can be determined experimentally in ultrasonic experiments. In the previous chapter, we have indicated that the wavelength of ultrasound at 7.2 MHz (as in [30]) is 0.2 mm. Since membranes and vesicles are much smaller than this scale, the compressibility of a lipid dispersion is an averaged quantity of the lipid dispersion consisting of lipid membranes and water. In analogy to the calculation in the previous chapter, it is convenient to define a reduced sound velocity number u that takes the water volume and the lipid concentration into account:

$$u = \frac{c - c_{H_2O}}{c_{H_2O} \cdot [L]}, \tag{VI.5.2}$$

where c is the sound velocity of the aqueous lipid dispersion, c_{H_2O} is the sound velocity in water, and $[L]$ is the lipid concentration in mg/ml.

Figure VI.5.1 shows the heat capacity of large unilamellar vesicles (LUV) of dipalmitoyl phoshatidylcholine (DPPC), the measured sound velocity number u, and the theoretical profile predicted from the heat capacity of such vesicles. This graph shows that the agreement between experiment and prediction is excellent. The underlying assumption when calculating the theoretical profile was that water does not contribute to the total heat capacity and that only the lipid chains contribute as a heat reservoir (see Section VI.4.1.1). This was called the high-frequency limit. The success of this calculation justifies this attempt and also proves that in the MHz-regime no heat is transferred between the membrane and

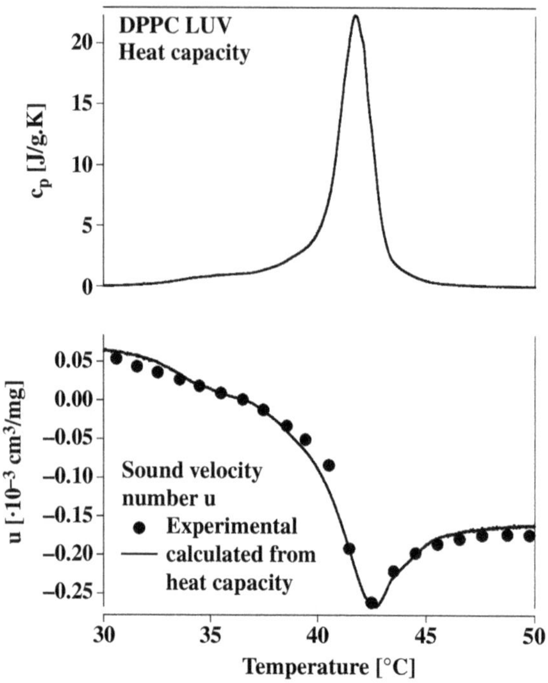

Figure VI.5.1 Heat capacity and sound velocity number for dispersions of large unilamellar vesicles of DPPC. Filled circles represent measurements at 2.5 MHz and solid lines the predictions from the heat capacity. From [26], data adapted from [44].

surrounding water. However, it also suggests that the low-frequency limit case, where the adiabatic compressibility approaches the isothermal compressibility (if there is a large aqueous reservoir), can also be reliably calculated.

The high-frequency limit calculation also yields very good predictions for lipid mixtures. Figure VI.5.2 shows the results for different lipid mixtures (from [44]). The left panel shows five different dimyristoyl phosphatidylcholine (DMPC):dipalmitoyl phosphatidylcholine (DPPC) mixtures (100:0, 83:17, 50:50, 17:83, and 0:100 mol:mol). The right-hand panels display five dimyristoyl phosphatidylcholine (DMPC):dioleoly phosphatidylcholine (DOPC) mixtures (75:25, 90:10, 95:5, 97.5:2.5, and 100:0 mol:mol, respectively). The top panels show the heat capacity, the middle panels the measured sound velocity number, and the bottom panel the prediction from the heat capacity. Schrader et al. [44] also provide results for dilauroyl phosphatidylcholine (DLPC):dipalmitoyl phosphatidylcholine (DPPC) mixtures, dimyristoyl phosphatidylcholine (DMPC):dimyristoyl glycerol (DMG) mixtures, and Halstenberg et al. [30] presented similar results for DMPC:cholesterol mixtures. In all cases, the predictions from the heat capacity are very similar to the experiments, both qualitatively and quantitatively.

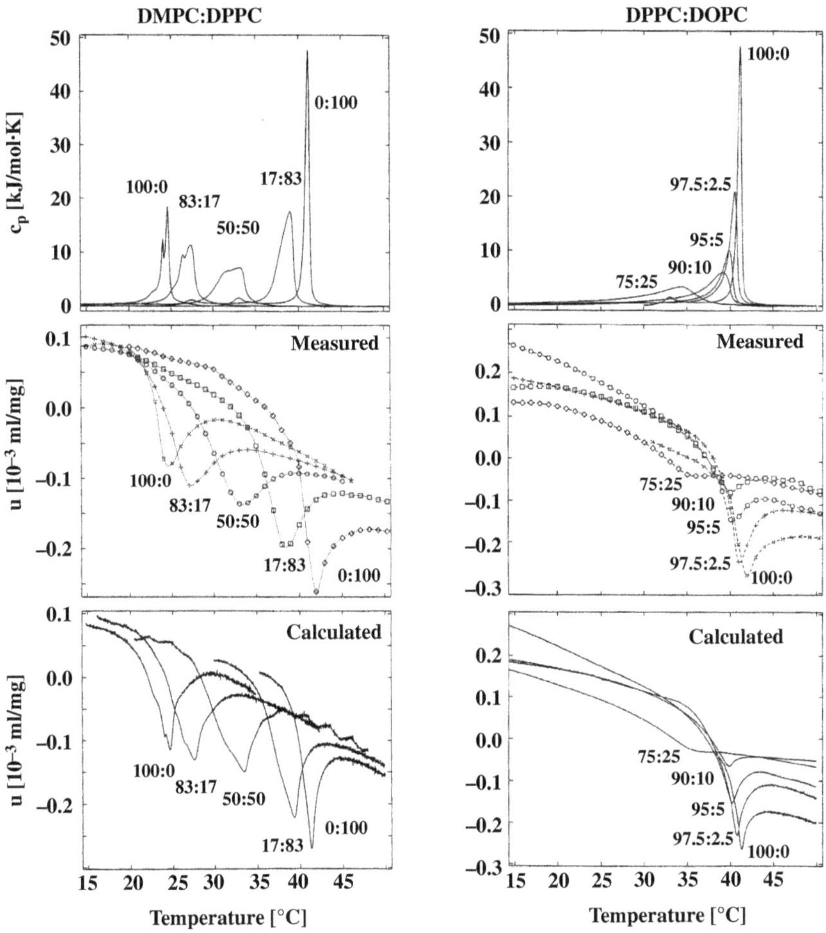

Figure VI.5.2 Sound velocity experiments on lipid mixtures. Top panels: Heat capacity profiles of dispersions of two lipid mixtures: DMPC:DPPC (left) and DPPC:DOPC (right) in different mixing ratios. Middle panels: The corresponding measured sound velocity numbers. Bottom panels: The sound velocity numbers calculated from the heat capacity profiles in the top panels. [44]/with permission of American Physical Society.

VI.5.2 Sound Propagation in Two Dimensions

In analogy to the calculation of the three-dimensional sound velocity, we can also determine the high- and low-frequency limit of the two-dimensional sound velocity theoretically. This is the velocity with which a perturbation in the membrane density will propagate within the membrane plane. As in the three-dimensional case, the two-dimensional sound velocity c in the membrane plane can be written as

$$c = \sqrt{\frac{1}{\kappa_S^A \cdot \rho^A}}, \tag{VI.5.3}$$

where κ_S^A is the two-dimensional adiabatic compressibility and ρ^A is the area mass density, which can be determined along the lines discussed in Section VI.3.6 (see Figure VI.4.8).

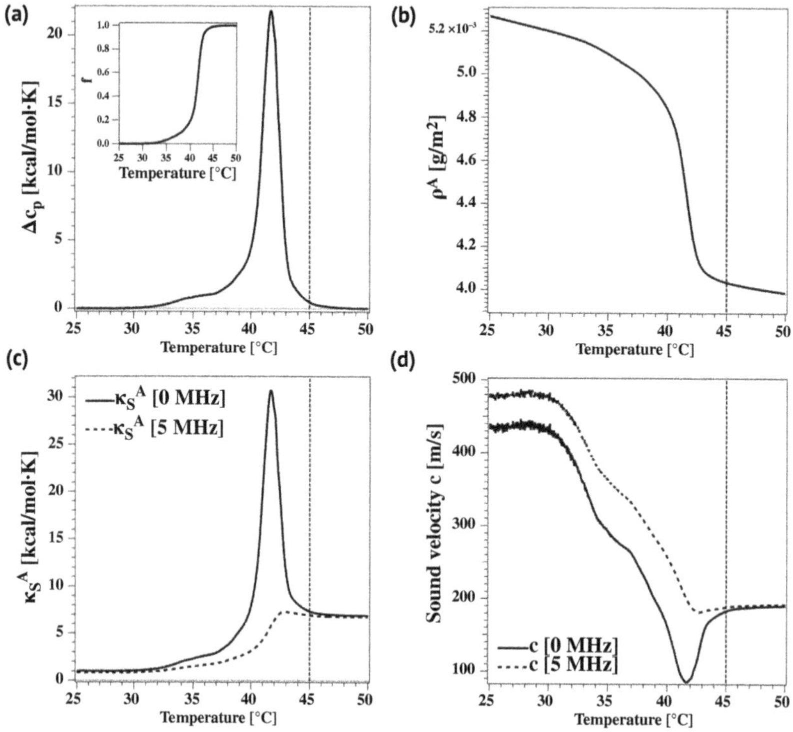

Figure VI.5.3 The determination of the lateral sound velocity from the heat capacity of DPPC LUV. (a) Excess heat capacity Δc_p as a function of temperature. The inset shows the fluid fraction f determined from integrating and normalizing the heat capacity. (b) The area density of the membrane. (c) The adiabatic area compressibility of the membrane at a frequency of 0 MHz (solid line) and at 5 MHz (dashed line). (d) The lateral sound velocity at a frequency of 0 MHz (solid line, low-frequency limit) and at 5 MHz (dashed line, high-frequency limit).

Figure VI.5.3 shows the steps of the calculation of dipalmitoyl phosphatidylcholine membranes using the heat capacity c_p of DPPC LUV, the fluid fraction f, the area density ρ^A, the high- and low-frequency limits of the adiabatic compressibility κ_S^A, and the high- and low-frequency limits of the two-dimensional sound velocity. As in the three-dimensional case, the sound velocity displays a minimum at the transition temperature, which is much more pronounced in the low-frequency limit. The low-frequency limit denotes timescales longer than the relaxation timescale of lipid perturbations. For the calculation of the adiabatic compressibility, it was considered that the membrane is embedded in a three-dimensionally extended aqueous solvent. The frequency dependence of the heat reservoir again leads to a frequency dependence of the sound velocity, i.e., to dispersion. We used the same frequency dependence of the heat capacity as in the previous section. The minimum of the low-frequency velocity is about 85 m/s, quite similar to the pulse velocity in myelinated motor neurons. The sound velocity in the fluid state is about 176 m/s for DPPC. The latter was deduced from the experimentally measured compressibility of fluid lipid membranes [12, 17].

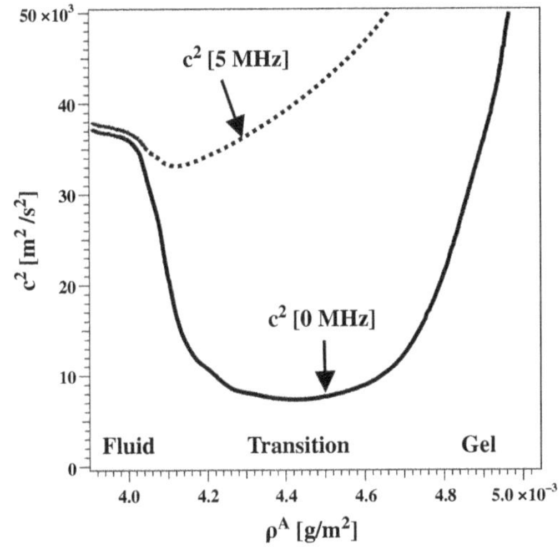

Figure VI.5.4 High- and low-frequency limits of the square of the two-dimensional sound velocity as a function of area density. The minimum of the curves indicates the melting regime of the membrane. Low densities correspond to the fluid phase and high densities to the gel phase of the lipid membrane. Adapted from [45].

In Part VII, we will consider the excitation of sound pulses within the membrane caused by a local density change in the membrane. For the hydrodynamic calculation, it is convenient to represent the square of the sound velocity as a function of density rather than as a function of temperature. This can easily be done with the data given in Figure VI.5.3. Figure VI.5.4 shows the high- and low-frequency limits of c^2 as a function of area density. The minimum of this curve represents the transition regime; lower densities represent the fluid membrane phase and higher densities the gel phase of the membrane. This plot shows that the sound velocity is a nonlinear function of density.

VI.5.3 Viscoelastic Coupling

In the calculation in the previous section, we estimated a value for the lateral density of the membrane in the gel and fluid states by ignoring the coupling of the membrane to its aqueous surroundings. We took the mass density of the lipid membrane and tacitly assumed that no other material is associated with the membrane that contributes to its moving mass. In three dimensions (sound velocity in dispersions of vesicles), this problem does not arise. However, since membranes are two-dimensional objects embedded in a three-dimensional fluid, the coupling of the membrane to the environment needs to be considered.

The density plays an important role in the propagation of sound because, due to Newton's second law, the force in the membrane is equal to the acceleration of the inertial mass in the derivation of the wave equation (see eq. (VII.1.6)). However, there are materials that can couple to the membrane and contribute to its density. These can be surface proteins and, in particular, water. A coupling of the membrane to water was discussed in [46, 47]. The principle is shown in Figure VI.5.5. If there is a time-dependent longitudinal perturbation

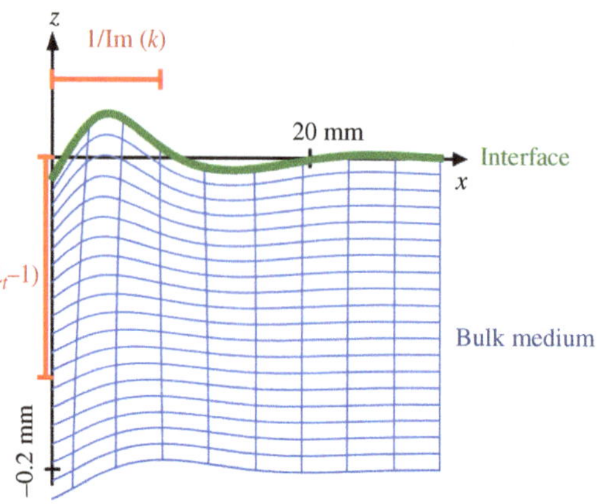

Figure VI.5.5 Schematic representation of the coupling of a longitudinal surface wave to the bulk aqueous medium. The perturbation couples to a layer of the bulk medium. [46]/with permission of American Physical Society.

in the membrane, water moves with it. The amount of water that is coupled to the membrane and thus moved around during a membrane perturbation depends on the frequency of the perturbation.

The overall process of water coupling is quite complex [46]. However, Schneider [47] approximated the sound velocity of a membrane that couples to its aqueous environment by

$$c_0' = c_0 \cdot \sqrt{\frac{d}{\delta}}, \tag{VI.5.4}$$

where d is the thickness of the monolayer (≈ 2 nm, and δ is the thickness of the two-dimensional layer including the surface-associated water. This approximation assumes that the volume densities of the membrane and the aqueous medium are about the same. The penetration depth is a function of the frequency of the surface motion, i.e., it contributes to the dispersion. Schneider calculated the penetration depth δ as:

$$\delta = 2\sqrt{\frac{\eta_w}{\rho_w \cdot \omega}} + d = 2\sqrt{\frac{\mu_w}{\omega}} + d. \tag{VI.5.5}$$

Here, η_w and $\mu_w = \eta_w / rho_w$ are the dynamic and kinematic bulk viscosities (of water), d is the thickness of the layer, and ω is the frequency of the perturbation. Schneider gave the following values for a lipid monolayer: $\mu_w = 10^-6\ \mathrm{m^2/s}$, $d \approx 2$ nm using one single lipid–water interface because a monolayer was discussed. A fluid membrane has twice the thickness and two interfaces. Thus, the change in sound velocity should be the same in monolayers and bilayers. When coupling to the aqueous layer is present, the density and sound velocity convert in the following manner:

$$\begin{aligned} \rho^A &\rightarrow \rho^A \cdot \frac{\delta}{d} \\ c_0 &\rightarrow c_0 \cdot \sqrt{\frac{d}{d \cdot \delta}}. \end{aligned} \tag{VI.5.6}$$

When a typical nerve pulse displays a duration of 1 ms, then the frequency of the perturbation is of the order of 1000 Hz. Thus, the total depth of the aqueous layer is about 63 μm. The two-dimensional sound velocity in the fluid phase of the lipid membrane at 1 kHz is then lowered from about 176 m/s (the order of magnitude of a myelinated motor neuron) to 1 m/s (the order of magnitude for a nonmyelinated nerve, e.g., in worm axons). The above depth of the aqueous layer is of a scale similar to or larger than the distances between cells in organisms. It can therefore be expected that the sound velocity depends on the distance and packing of membranes, and, in the case of motor neurons, on their myelination. In myelinated nerves, one expects a higher velocity.

In a similar manner, surface proteins can influence the membrane density and thereby reduce the lateral sound velocity.

VI.5.4 Summary: Key Ideas of Chapter VI.5

The sound velocity is a function of the adiabatic compressibility and the mass density.

- In three dimensions, the sound velocity of lipid solutions can be measured with ultrasonic resonators typically working in a frequency range around 5 MHz. Data presented here were recorded with frequencies of 7.2 MHz and 2.5 MHz, corresponding to wavelengths of 0.2–0.6 mm. indexsound velocity!3D indexsound velocity!2D
- The experimental profiles at these frequencies can be well predicted from the heat capacity, assuming that there is no heat transfer between the membrane and the environment. Vice versa, the success of this calculation can be taken as strong evidence that there is no heat transfer between membranes and their aqueous surroundings on a timescale from 1.4 to 4 μs (the timescale of the experiment).
- In analogy to the three-dimensional case, the sound velocity can be predicted for the two-dimensional case. During the transition, the sound velocity in the membrane plane is about 85 m/s for DPPC membranes. This is close to the nerve pulse velocity in myelinated nerves.
- If viscoelastic coupling of the membrane to its aqueous surroundings is taken into account, the total mass density of the membrane increases. As a consequence, the sound velocity is rather 1–10 m/s. This is the range of the nerve pulse velocity for nonmyelinated nerves.

VI.6

Capacitance and Capacitive Susceptibility

The capacitance is important in the electrophysiological treatment of the nerve pulse. In this chapter, we show that it is not a constant but a susceptibility that can change as a function of the intensive variables.

VI.6.1 Capacitive Susceptibility

The lipid membrane is a good capacitor. It has a hydrophobic core with a dielectric constant of 2–4. The voltage V across a membrane is minus the potential difference, $V = -\Delta\Psi$. The charge on the surface of a membrane is given by

$$q = -C_m \Delta\Psi = C_m \cdot V, \tag{VI.6.1}$$

where the capacitance C_m is the constant that relates the charge to the transmembrane voltage (in the absence of a permanent membrane polarization). The capacitance is given by

$$C_m = \epsilon_0 \epsilon \frac{A}{D}, \tag{VI.6.2}$$

where $\epsilon_0 = 8.854 \cdot 10^{-12}$ F/m is the vacuum permittivity, ϵ is the dielectric constant of the membrane interior (approximately 2–4), A is the area, and D is the membrane thickness of the capacitor. In this definition, the capacitance depends on the membrane geometry but not on voltage. In most applications (e.g., the Hodgkin–Huxley model, see Part IV.4) it is therefore assumed that the capacitance is a constant:

$$C_m = \frac{\partial q}{\partial V} = \text{const.} \tag{VI.6.3}$$

However, the capacitance of a lipid membrane is not constant. It has been shown in Section V.7.3 that the capacitance of a lipid membrane changes nearly twofold when going from a gel to a fluid state of the membrane.

VI.6.2 The Capacitive Susceptibility

A charged capacitor is covered by charges of opposite sign on the two surfaces. These changes attract each other and thus create a force on the membrane. This force is called the *electrostriction* [48]. Since a membrane is made of soft matter, it can be deformed by this

Physics of Nerves and Excitatory Membranes, First Edition. Thomas Heimburg.

force. It will become thinner and its area increases when more charges are found on the surface of the membrane. A thinner capacitor with a larger area displays a larger capacitance due to eq. (VI.6.2). Thus, the membrane capacitance becomes voltage-dependent. We call the voltage-dependent capacitance the *capacitive susceptibility*, $\widehat{C}_m$:

$$\hat{C}_m(V) \equiv \frac{\partial q}{\partial V_m} = \frac{\partial (C_m V)}{\partial V} = C_m + V \cdot \frac{\partial C_m}{\partial V}. \tag{VI.6.4}$$

We see that there is now an extra term that describes the voltage dependence of the capacitance. The capacitive susceptibility is (as all the other susceptibilities) the derivative of an extensive variable (q) with respect to an intensive variable ($V = -\Delta\Psi$). The theory of capacitive susceptibility has been derived in detail in [48]. We describe the most important findings as follows.

In [49], the mechanical force $\mathcal{F}$ acting on the membrane at voltage V for a given thickness of the membrane is given by

$$\mathcal{F} = \frac{1}{2}\frac{C_m V^2}{D}, \tag{VI.6.5}$$

which means that it is quadratic in voltage.

When we consider a melting transition, this force favors a more fluid state because the liquid membrane displays a smaller thickness and a larger area. As a consequence, the transition temperature becomes voltage-dependent. Furthermore, the fluid state of the membrane has a higher capacitance than the gel state. These two dependencies result in the possibility that an increase in voltage renders a gel membrane more fluid. Thus, an increase in voltage can induce a phase transition from gel to fluid. In this transition, some extra charge is absorbed due to the change in capacitance. This is the excess charge, Δq, of the transition. The changes in capacitance and capacitive susceptibility as a function of voltage are shown in Figure VI.6.1.

Figure VI.6.2 shows the voltage dependence of the charge on a DPPC membrane close to the melting transition for different temperatures. In the transition, there is a sudden increase in the membrane change that corresponds to the excess charge. The voltage derivative of the charge is the capacitive susceptibility, $\widehat{C}_m$, which displays a maximum in the transition. It is found at different voltages for different temperatures because the distance from the melting transition changes. Figure VI.6.2 also shows how the melting transition changes with different voltages. For more details, see [48].

It is clear that if the transmembrane voltage is different from zero, a sudden change in membrane state will lead to a change in the membrane charge, i.e., one will measure a significant capacitive current that is due to the time-dependent change of the charge on the capacitor. This current depends on the magnitude of the transmembrane voltage because the excess charge is voltage-dependent. This capacitive current can be induced by a sudden change in voltage, but also by lateral pressure changes in the membrane and a sudden drop in temperature.

The coupling between the voltage-dependent membrane geometry and the excess charge represents an electromechanical coupling. Heimburg [48] discusses several biologically

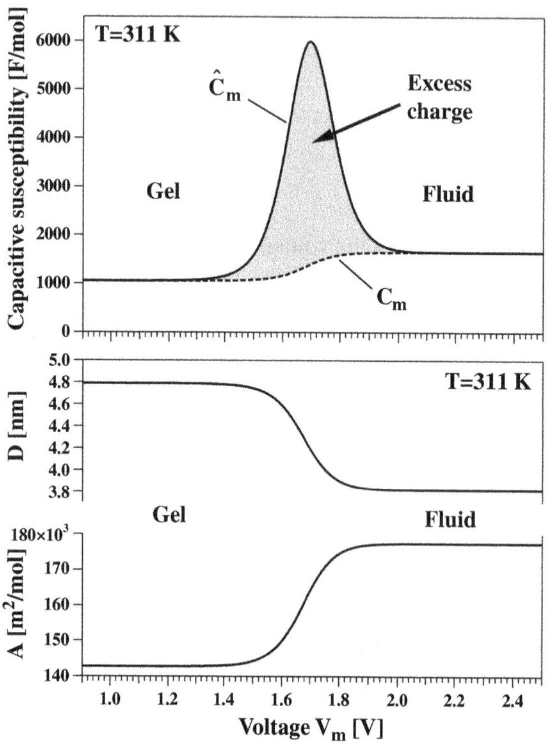

Figure VI.6.1 Change in membrane dimensions as a function of voltage calculated for a DPPC membrane at $T = 311$ K (3° below the melting transition). Bottom: The increase in voltage leads to melting of the membrane, during which the membrane area increases and the membrane thickness decreases in a voltage-dependent manner. Top: Since thickness and area are voltage-dependent functions, the state of the membrane becomes voltage-dependent and $\partial C_m/\partial V$ in eq. (VI.6.4) is no longer zero. This leads to a maximum of the capacitive susceptibility in the transition regime. The area under the peak (gray-shaded) is the excess charge of the transition. The dotted line indicates the change in the capacitance C_m if the geometry changes were not voltage-dependent. [48]/with permission of Elsevier.

relevant applications of this concept. In particular, electromechanical coupling has been intensely discussed in the outer hair cells of the ear., e.g., by Rabbitt et al. [50] and Sachs et al. [51]. Brownell et al. [52] showed that tethers pulled from hair cells contract upon application of a voltage difference across the cell body, which might be explained by a voltage-induced phase transition. The electromechanical coupling described in this section also has pronounced importance for the nerve pulse. In Part VII, we describe the nerve pulse as an electromechanical soliton, during which the melting transition plays an important role. Thus, the couplings described here will lead to capacitive currents, and voltage changes will cause changes in membrane state. Thus, voltage changes can excite mechanical waves and pulses in membranes.

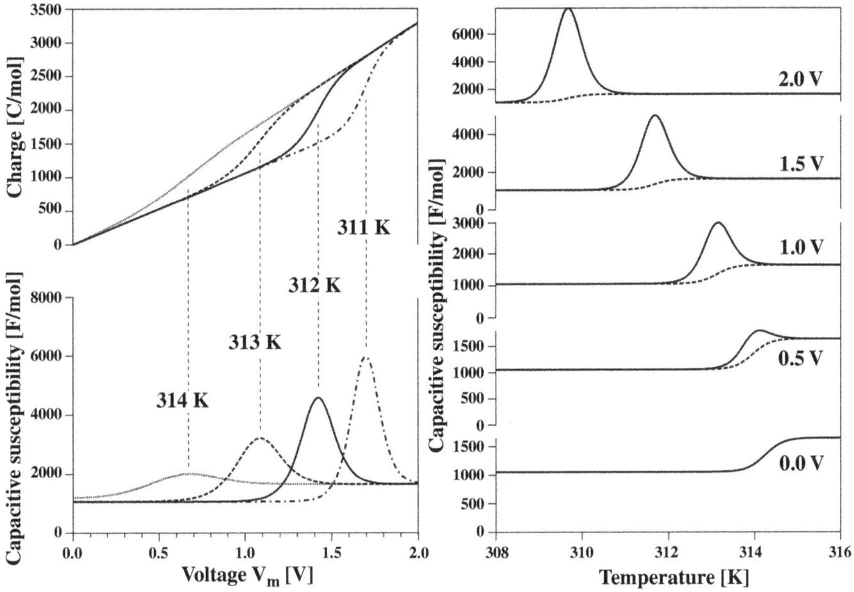

Figure VI.6.2 Left: The charge on the membrane is linear in voltage below and above the transition. In the transition, an excess charge is absorbed. This excess charge is voltage-dependent. The voltage-dependent charge is given for different temperatures. The derivative of the membrane charge yields the capacitive susceptibility with a maximum in the transition (left, bottom). Right: Since electrostriction (the force on the membrane created by charging the membrane) is voltage-dependent, the melting transition temperature between gel and fluid is voltage-dependent. [48]/with permission of Elsevier.

VI.6.3 The Fluctuation Relation of the Capacitance

As with other susceptibilities, the capacitive susceptibility is related to a fluctuation relation. The mean charge on a capacitor is given by

$$\langle q \rangle = \frac{\sum_i q_i \exp\left(-\frac{H_i}{kT}\right)}{\sum_i \exp\left(-\frac{H_i}{kT}\right)}, \tag{VI.6.6}$$

where $H_i = H_{i,0} - \Psi q_i$, where Ψ is the electrostatic potential and q_i is the charge on the capacitor in microstate i. $H_{i,0}$ is the enthalpy of the uncharged membrane. The capacitive susceptibility $\hat{C}_m$ of the membrane under consideration is given by

$$\hat{C}_m = -\left(\frac{\partial \langle q \rangle}{\partial \Psi}\right)_{T,p,\ldots}. \tag{VI.6.7}$$

Evaluation of this equation yields

$$\hat{C}_m = \frac{\langle q^2 \rangle - \langle q \rangle^2}{kT}, \tag{VI.6.8}$$

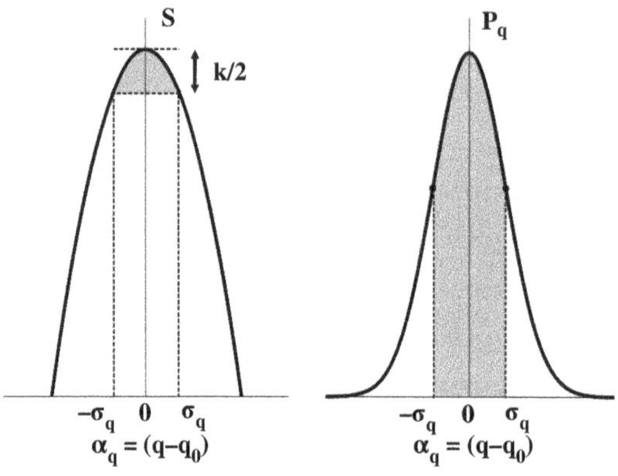

Figure VI.6.3 Charge fluctuations: The entropy S and the corresponding Gaussian distribution P_q. Left: Entropy as a quadratic potential depending on the charge. Right: The corresponding Gaussian distribution. Indicated is the standard deviation of the enthalpy fluctuations, σ_q, which corresponds to a value of the entropy $S-S_0 = k/2$.

where $(q - q_0) \equiv \alpha_q$ is a fluctuation of the charge on the capacitor. Figure VI.6.3 shows the quadratic entropy potential as a function of $\alpha \equiv (q - q_0)$, where $q_0 \equiv \langle q \rangle$. Also shown is the corresponding charge distribution on the capacitor.

The aforementioned fluctuation relations suggest that the capacitive susceptibility at a given voltage can be deduced from the fluctuations in the charge on the capacitor or from the fluctuations of the capacitive current.

VI.6.4 Summary: Key Ideas of Chapter VI.6

The change in the charge on a membrane as a function of voltage is a susceptibility because it is the derivative of the extensive quantity q with respect to the intensive variable V_m. Capacitance changes are important during nerve pulses where both the transmembrane voltage and the membrane dimensions change.

- The capacitance C_m is usually considered to be a constant defined by the geometry of the capacitor. It is given by $C_m = \epsilon_0 \epsilon A/D$, where A is the area and D is the thickness of the capacitor. Its voltage dependence is usually not considered.
- In the case of the membranes, however, the capacitor is soft and changes its dimensions upon charging of the membrane due to electrostatic forces. This renders the capacitance voltage-dependent. In order to distinguish it from a static capacitance C_m we call it the capacitive susceptibility $\hat{C}_m = C_m + V \cdot (\partial C_m)/\partial V)$. The voltage dependence of the capacitance leads to an additional term that is usually (and incorrectly) not considered in the electrophysiological literature.
- The capacitive susceptibility $\hat{C}_m$ displays a maximum in the melting transition.
- The melting transition can be caused by changes in voltage. Vice versa, changes in membrane state will cause changes in membrane charge and voltage.
- During this transition, an excess charge (or latent charge) is absorbed due to the voltage-induced changes in capacitance.

- At constant voltage, a change in the dimensions of a membrane as caused by mechanical waves will cause significant changes in the charge absorbed by the membrane.
- Sudden changes in geometry lead to capacitive currents.
- The capacitive susceptibility is related to a fluctuation relation:

$$\hat{C}_m = \frac{\langle q^2 \rangle - \langle q \rangle^2}{kT}$$

Changes in capacitance and capacitive susceptibility during a transition have major implications for electrical phenomena that occur during the passage of solitary mechanical waves or pulses in membranes. They are responsible for the electromechanical behavior of membranes and nerves. This is discussed in Part VII.

VI.7

Relaxation Timescales

The relaxation timescale determines how fast a membrane relaxes to a new equilibrium after a perturbation. It also determines the fluctuation lifetimes that show up in the characteristic timescale of the nerve pulse (Part VII) and the characteristic open time of channels (Part VIII).

VI.7.1 Theory of Relaxation Timescales Close to Equilibrium

As described in the previous chapters, the experimental finding for membranes is that the extensive variables H, V, and A are proportional functions of temperature and other intensive variables. Therefore, there is only one independent extensive variable (e.g., the enthalpy). This implies that after a perturbation of the membrane, not only the enthalpy but also all other extensive variables relax with the same time dependence toward a new equilibrium state.

Close to equilibrium, the entropy can be considered a quadratic function of the enthalpy (Section II.3.2) such that

$$S = S_0 - \frac{1}{2} g \alpha_H^2, \tag{VI.7.1}$$

where $\alpha_H = H - H_0$ is the deviation of the enthalpy from its equilibrium value (see Section II.4.3). The entropy is analogous to a potential energy in mechanics, and such is the definition of the force as a derivative of the potential. The thermodynamic force X_H driving the system back to equilibrium is given by the gradient of the entropy potential (see Section II.3.4 and Figure VI.7.1).

$$X_H = \frac{\partial S}{\partial \alpha_H} = -g\alpha_H, \tag{VI.7.2}$$

where $\alpha_H = (H - H_0)$. Linear nonequilibrium thermodynamics (see Sections II.3.5 and II.4.3) tells us that the heat flux back to equilibrium, $J_H = \partial \alpha_H / \partial t$, is proportional to the force, $J_H = L \cdot X_H$ (with a phenomenological coefficient L), and

$$\frac{d(H - H_0)}{dt} = -Lg(H - H_0). \tag{VI.7.3}$$

Physics of Nerves and Excitatory Membranes, First Edition. Thomas Heimburg.

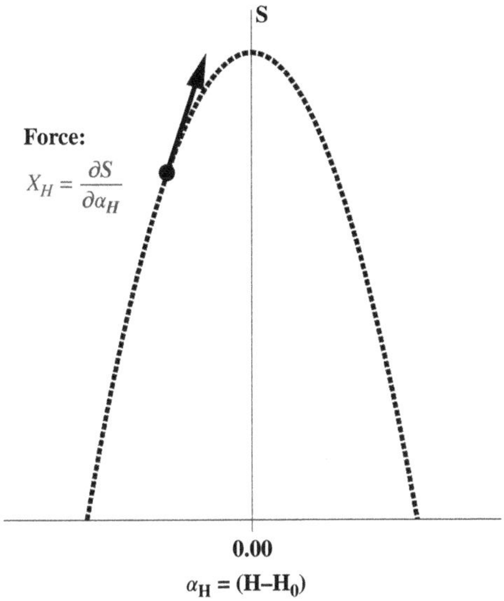

Figure VI.7.1 The entropy of the membrane is a quadratic function of the enthalpy, H. Its derivative X_H is the thermodynamic force driving the system back to equilibrium. The equilibrium entropy is found at $H = H_0$.

This is a simple linear differential equation with a single-exponential solution:

$$(H - H_0)(t) = (H - H_0)_0 \exp(-Lgt) \equiv (H - H_0)_0 \exp\left(-\frac{t}{\tau}\right) \quad \text{with} \quad \tau \equiv \frac{1}{L \cdot g}, \tag{VI.7.4}$$

where $\tau = 1/Lg$ is the relaxation time.

We will now show that the constant g (the curvature of the entropy potential) has a simple relation to the heat capacity. According to eq. (VI.1.5), $\sigma^2 = \langle H^2 \rangle - \langle H \rangle^2 = k/g$. Further, according to eq. (VI.2.8), the heat capacity is given by $c_p = (\langle H^2 \rangle - \langle H \rangle^2)/kT^2$. Therefore,

$$c_p = \frac{1}{kT^2} \cdot \frac{k}{g} = \frac{1}{g \cdot T^2} \to g = \frac{1}{c_p \cdot T^2}, \tag{VI.7.5}$$

and after inserting eqs. (VI.7.5) and (VI.7.4), we finally obtain

$$\boxed{\tau = \frac{1}{L \cdot g} = \frac{T^2}{L} \cdot c_p \qquad \text{relaxation time}} \tag{VI.7.6}$$

meaning that the relaxation time is proportional to the excess heat capacity. Generally, relaxation times are longer in transitions than outside of transitions because the heat capacity reaches a maximum. This phenomenon is known as *critical slowing down*. These relaxation times will reappear in the timescale of the nerve pulse (Part VII) and in the open lifetime of membrane channels (Part VIII).

VI.7.2 Simulation of Relaxation Timescales

Fluctuation lifetimes were theoretical calculated using Monte Carlo simulations [8]. Monte Carlo simulations generated snapshots of a lipid matrix with gel and fluid lipids. For each of these snapshots, an enthalpy can be determined. Thus, the time evolution of the enthalpy can be plotted and analyzed. Figure VI.1.2 in Chapter VI.1 shows that the magnitude of the fluctuations in a Monte Carlo simulation of a lipid membrane is proportional to the heat capacity. In Figure VI.7.2 (top), it can be seen that at different temperatures close to the melting transition, it is not only the magnitude of the fluctuations but also the

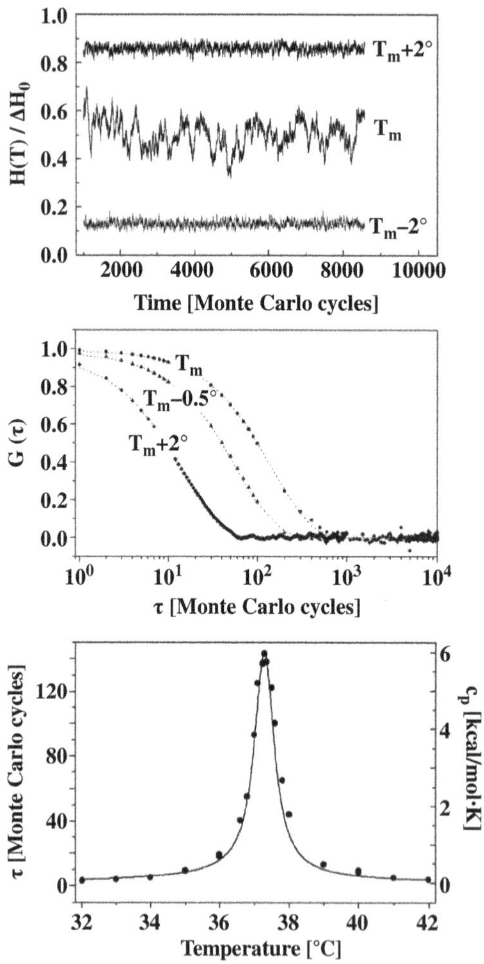

Figure VI.7.2 Fluctuations in a Monte Carlo simulation of a lipid membrane (c.f., Figure VI.1.2). Top: Both fluctuation amplitude and fluctuation timescale are at maximum at T_m and smaller at temperatures below and above T_m. Middle: The autocorrelation function $G(\tau)$ as a function of the number of Monte Carlo cycles. The correlation time is larger at T_m as compared to temperatures below and above T_m. Bottom: When both the heat capacity and the correlation timescale are plotted as a function of temperature, one finds that they are proportional functions as predicted by eq. (VI.7.6). From [8]/with permission of Elsevier.

characteristic timescale of the fluctuations that changes as a function of the heat capacity (and of temperature). It is a central assumption in linear nonequilibrium thermodynamics that the relaxation timescale after a perturbation is identical to the fluctuation lifetime [53]. The fluctuation lifetime in the simulation can be found by autocorrelating the enthalpy fluctuations. In Figure VI.7.2 (middle), the autocorrelation function $G(\tau)$ is shown, where τ is the correlation time given in numbers of Monte Carlo cycles (see [8] for details). If the characteristic correlation time is plotted as a function of temperature, it is seen that it is proportional to the heat capacity. This confirms the theoretical analysis in the previous section (eq. (VI.7.6)), and confirms the close coupling between relaxation timescale and the heat capacity (Figure VI.7.2) (middle).

VI.7.3 Measurement of Relaxation Timescales

It is shown in Figure VI.3.3 that the application of pressure shifts the transition temperature of a membrane. Thus, if a membrane system is above its transition temperature, it can be moved into the transition regime by a pressure jump. One obtains a state with a different equilibrium enthalpy. In a calorimeter, this process will lead to heat release or absorption and one can follow this process as a function of time, as shown in Figure VI.3.5. These measurements allow one to determine the timescale of the relaxation process experimentally (Figure VI.7.3). Here, we show the calorimetric response of a lipid dispersion to pressure jumps close to the melting transition. It can be seen that even small changes ($0.067° \leq \Delta T \leq 0.157°$) in temperature can lead to large changes in the total heat release (the integral over the power traces) and also to significant changes in the timescale of the heat release up to several tens of seconds in very sharp transitions.

Figure VI.7.4 shows the relaxation times determined for multilamellar vesicles (MLV) of DMPC and DPPC (from [8]). They are within the accuracy of the experiment proportional to the excess heat capacity. This confirms the simple thermodynamic theory derived in Section VI.7.1, which predicts that $\tau = (T^2/L)c_p$. From these experiments, the phenomenological coefficient L can be determined. Figure VI.7.4 shows that the heat capacity is proportional to the relaxation times of DMPC and DPPC MLV with approximately the same proportional constant T^2/L between heat capacity and relaxation time for both lipids. The maximum relaxation time for the DPPC MLV is more than 40 seconds, i.e., relaxation can be very slow for preparations with a sharp heat capacity profile. The coefficients L are given in Table VI.7.1.

The melting enthalpy of lipid vesicles is mostly independent of the size of the vesicles. However, small vesicles display broader melting profiles. Therefore, the maximum relaxation timescale is smaller for membranes with broader transitions because the heat capacity at maximum is lower. This can be seen in Figure VI.7.5 (left). The heat capacity profile of DPPC LUV is several times broader than for the MLV above, and the heat capacity maximum is about 4 times smaller. This also leads to a considerable reduction in the relaxation time. The maximum relaxation time for the DPPC LUV is about 3 seconds, more than 10 times faster than for the DPPC MLV. In Figure VI.7.5 (right), two different LUV preparations are shown with a relaxation time of 1.45 and 2.1 seconds, respectively. It can be seen, though, that for all the preparations, the heat capacity and the relaxation timescale are proportional functions, i.e., $\tau \propto c_p$.

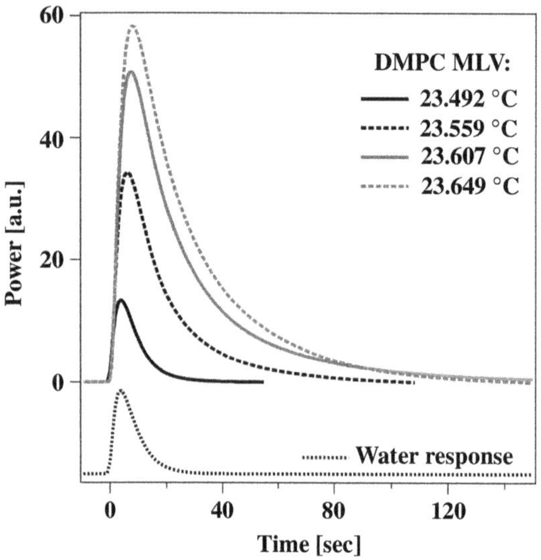

Figure VI.7.3 Calorimetric response of a lipid dispersion to a pressure jump at four different temperatures close to or within the transition of DMPC MLV. It can be seen that even small changes in the temperature have a large effect on the magnitude of the heat release and the timescale. For comparison, the response of pure water to a pressure jump is shown. Adapted from [8].

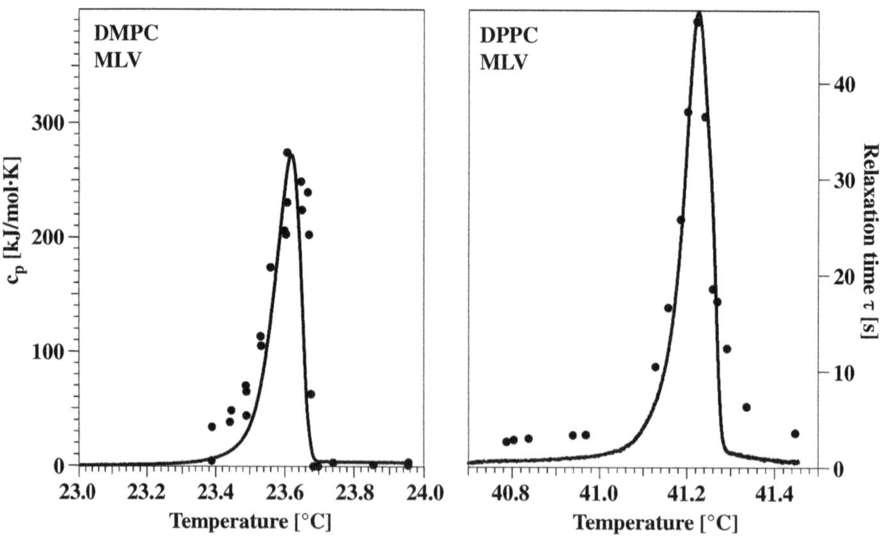

Figure VI.7.4 Relaxation times in comparison with heat capacity profiles. Left: For MLV of DMPC. Right: For MLV of DPPC. Note that the scaling of both the heat capacity axis and the relaxation time axis are the same. Adapted from [8].

Table VI.7.1 The relaxation time is given by $\tau = (T^2/L)c_p$. The table indicates the phenomenological coefficient L determined for different lipid preparations from calorimetric pressure jump experiments. For all MLV preparations, one obtains very similar values, independent of the drug added. The value for LUV is 2–3 times higher, which corresponds to a somewhat faster relaxation. For technical details, see [8, 54].

Lipid	Phenomenological Coefficient L (10^8 J·K/s·mol)	Source
MLV:		
DMPC MLV	7.35	[8]
DMPC MLV	6.6	[54]
DPPC MLV	8.45	[8]
DPPC MLV + 1 mol% cholesterol	8.37	[8]
DMPC MLV + 0.1 wt% octanol	6.3	[54]
DMPC MLV + 0.5 wt% octanol	6.3	[54]
DMPC MLV + 0.5 mM serotonin	6.5	[54]
DMPC MLV + 3.0 mM serotonin	7.5	[54]
DMPC MLV + 0.5 mol% gramicidin A	7.9	[54]
DMPC MLV + 1.5 mol% gramicidin A	5.8	[54]
LUV:		
DMPC LUV	15.7	[54]
DPPC LUV	13.9	[8]
DPPC LUV	18.4	[54]
DMPC LUV + 20 wt% halothane	18.4	[54]

VI.7.4 Influence of Drugs on Relaxation Timescales

As shown earlier, the relaxation time is given by $\tau = (T^2/L) \cdot c_p$. The larger L, the faster the relaxation. The previous sections showed that the theoretically predicted proportional relation between the (excess) heat capacity and the relaxation time is confirmed experimentally. In particular, anything that changes the heat capacity profile will change the relaxation time. A temperature shift of the c_p profile will lead to a shift in the relaxation timescale profile, and a broadening of the c_p profile will lead to a broadening of the τ-profile. Figures VI.7.6 and VI.7.7 show that the addition of cholesterol, the general anesthetics halothane and octanol, the neurotransmitter serotonin, and the antibiotic peptide gramicidin A all lead to a broadening and a shift of the heat capacity profile. The relaxation timescale follows these changes such that heat capacity and relaxation time remain proportional functions with a phenomenological constant that is basically unchanged (see Table VI.7.1). Thus, drugs can change the relaxation times and can also change the mean open time of membrane channels as discussed in Part VIII.

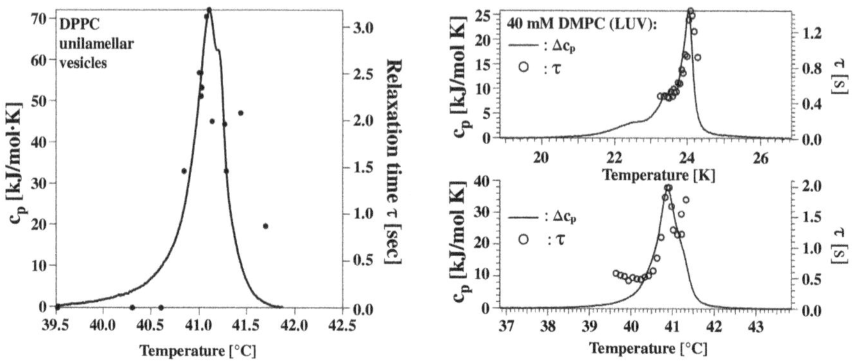

Figure VI.7.5 Relaxation timescales in unilamellar vesicles (LUV). Left: Relaxation times in comparison with heat capacity profiles for LUV of DPPC. From [8]/with permission of Elsevier. Right: The relaxation times for DMPC and DPPC LUV measured by a modified (and faster) calorimetric method. The relaxation time for DMPC LUV at the maximum is about 1.45 seconds, and it is about 2.1 seconds for DPPC LUV. [54]/with permission of American Chemical Society.

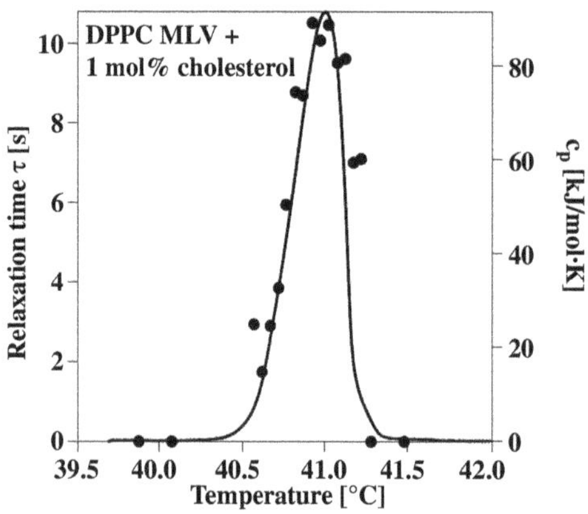

Figure VI.7.6 The maximum relaxation timescale in DPPC MLV in the presence of 1 mol% cholesterol is about 12 seconds, while it is about 44 seconds without cholesterol (see Figure VI.7.4 (right)). The solid profile represents the heat capacity. From [8]/with permission of Elsevier.

VI.7.5 Phenomenological Coefficient *L*

Table VI.7.1 shows the phenomenological coefficient L for the preparations shown in the previous sections. The values for L are surprisingly similar for all the experiments. For the MLV preparations, it is $6.85 \pm 0.80 \cdot 10^8$ J·K/s·mol. For the LUV preparations, it is $16.6 \pm 1.9 \cdot 10^8$ J·K/s·mol. This means that the coefficient L is about 2–3 times larger in LUV. But within one vesicular preparation, the constant L is nearly constant, independent of the drug added, or the nature of the lipid. This shows that the relaxation time is determined by the magnitude of the macroscopic fluctuations and the curvature of the entropy potential and not by molecular properties.

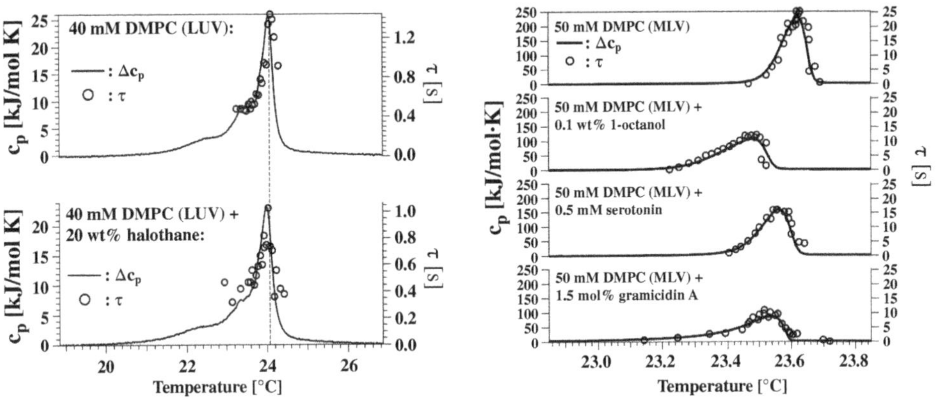

Figure VI.7.7 Effect of drugs on the relaxation timescale. Left: A DMPC LUV dispersion with a halothane/lipid ratio of 0.2 (weight%) displays a slightly lowered c_p profile and a slight reduction of the maximum relaxation time. Adapted from [54]. Right: Relaxation times in comparison with heat capacity profiles for MLV of DPPC in the presence of the indicated amounts of the anesthetic octanol, the neurotransmitter serotonin, and the pore-forming peptide gramicidin A. Adapted from [54].

VI.7.6 Prediction of Relaxation Timescales in a Biological System

Table VI.7.1 shows that the phenomenological coefficient L is very similar for different lipid preparations with and without the addition of drugs and proteins. Let us assume that this is also true for biological preparations such as lung surfactant. Since lung surfactant is rich in DPPC (more than 50 mol%), we assume that the constant L is similar to that of DPPC MLV and LUV. Figure VI.7.8 shows the prediction of relaxation times for lung surfactant. The figure shows a broad heat capacity profile with a maximum at 27 °C (i.e., about 10 °C below body temperature). A phenomenological constant of $L = 7 \cdot 10^8$ J·K/s·mol (the value for DPPC MLV) was assumed. This leads to the prediction of a maximum relaxation time around 95 ms. If the value for LUV is taken, the maximum relaxation time would be around 2.4 times shorter (40 ms). This implies that at the heat capacity maximum, one expects a fluctuation lifetime of several ten milliseconds, and at physiological temperature a few milliseconds. That is also the approximate timescale of ion channel open/closed fluctuations and the temporal timescale of the nerve pulse. At a temperature above T_m, the heat capacity is smaller and the fluctuation lifetime is shorter. Under physiological conditions (at the upper edge of the transition), we expect fluctuation lifetimes of the order of milliseconds. It is unlikely that the similarity of channel lifetimes and relaxation timescales was accidental. When we discuss the soliton theory in Part VII and channels in Part VIII, we will find out that both channel lifetimes and the temporal length of the nerve pulse are related to the relaxation behavior of membranes.

VI.7.7 Summary: Key Ideas of Chapter VI.7

Not only the elastic constant but also the relaxation and fluctuation timescales display a simple relation to the heat capacity. The extensive variables display a proportional behavior close to the melting transition such that it is sufficient to consider only one single

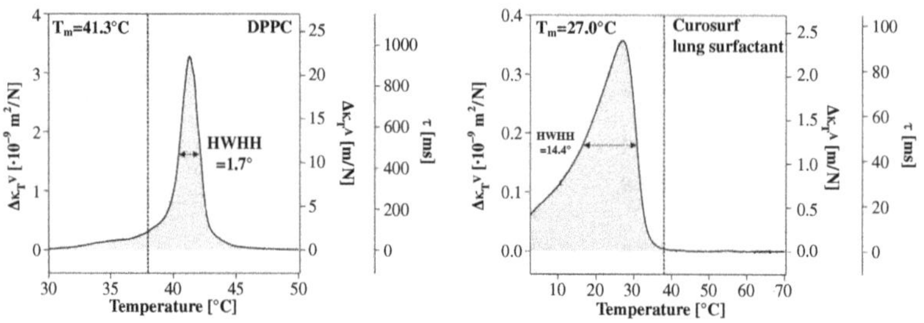

Figure VI.7.8 The relaxation timescale of DPPC LUV is about 1 second. Assuming a phenomenological constant $L = 7 \cdot 10^8$ J·K/s· mol for lung surfactant, the relaxation timescale is predicted to be slightly below 100 ms, while it is in the millisecond regime at physiological temperature (37 °C). From [13]/with permission of Elsevier.

independent variable with a single relaxation time. All extensive variables will then change with the same time dependence. We find that:

- Relaxation times are identical to fluctuation lifetimes in linear nonequilibrium thermodynamics.
- For lipid membranes close to transitions, relaxation times range from milliseconds to nearly a minute. It is maximum at the heat capacity maximum. This effect is known as *critical slowing down*.
- If the entropy is a harmonic function around the equilibrium, the thermodynamic force driving the membrane back to equilibrium after a perturbation is proportional to the magnitude of the perturbation. This leads to single-exponential relaxation behavior.
- The relaxation timescale is proportional to the excess heat capacity in the transition. Sharp transitions display slower relaxation, and broad profiles display faster relaxation.
- For different multilamellar lipid systems, the proportional constant between heat capacity and relaxation time is quantitatively very similar and independent of temperature. This is due to the fact that the timescale is not a molecular property but a property of the magnitude of cooperative fluctuations of the membrane.
- Everything that changes the heat capacity profile, for instance, the adsorption of drugs to membranes, will lead to predictable changes in the relaxation time profile. If a drug lowers the transition temperature, the relaxation timescale profile is similarly moved toward a lower temperature.
- If the heat capacity profile gets broader and the heat capacity at maximum is lower, this leads to the analogous change in the relaxation time profile.
- The effect of cholesterol, anesthetics, neurotransmitters, and small peptides can successfully be predicted by the correlation between excess heat capacity and relaxation timescale.
- Biological membranes are predicted to display a relaxation time of several tens of milliseconds at their transition maximum, and a timescale of milliseconds at the physiological temperature above the transition.
- This is the timescale of the nerve pulse and typical open times of membrane channels.
- The relaxation timescale determines the dynamical response of the membrane to any perturbation.

Part VII

The Soliton Theory

In this part, we explore sound propagation in membranes under physiological conditions. We have learned in Part III that the nerve pulse is adiabatic, i.e., within experimental error, it does not dissipate heat over the length of a single nerve pulse. Further, mechanical dislocations can be observed in the pulse region. These are features reminiscent of sound propagation, which is adiabatic and displays thermal and mechanical variations comparable to those of the nerve pulse (see Figure VII.1 for illustration). In fact, entropy conservation enforces that all thermodynamic variables will display synchronized changes. Besides mechanical and temperature changes, this includes chemical and electrical properties of a membrane.[1]

Here, we will derive the theory for the propagation of sound in membrane cylinders under conditions found in living nerves at physiological temperature. The derivation will lead to solitary waves (which we will loosely call solitons) that share remarkable similarities with action potentials in nerve axons. The goal of this part is to demonstrate that the properties of nerves arise from their thermodynamic state. We also demonstrate how these pulses are excited, what happens when they collide and how they behave under constraint of nerve length. Furthermore, we demonstrate how to prove the existence of such solitary pulses in lipid monolayer experiments.

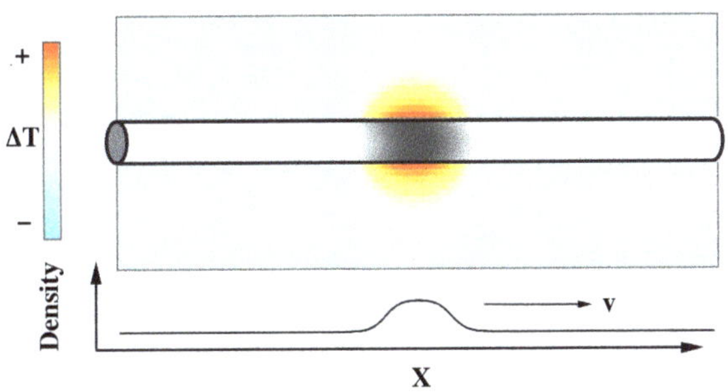

Figure VII.1 Illustration of the thermal and mechanical changes during a nerve pulse as observed in experiments.

---

1 This part partially relies on several joint publications with Andrew D. Jackson (in particular [1]) and his critical proofreading.

VII.1

Hydrodynamics and Sound Propagation

We want to describe the propagation phenomena in a nerve axon using hydrodynamics. We will assume that the axon is a cylinder surrounded by a membrane. Under physiological conditions, the biomembrane consists of a compressible fluid. In order to describe the propagation of sound in such a medium, we must therefore introduce the hydrodynamics of compressible fluids. This chapter cannot replace a textbook on hydrodynamics. However, we will try to introduce the most important concepts necessary for the treatment of propagation phenomena in nerves. These include the Euler equations, the Navier–Stokes equation, the wave equation, the sound velocity, and viscous friction.

VII.1.1 Hydrodynamics in the Absence of Friction

VII.1.1.1 Continuity Equation

The main feature of a moving incompressible fluid is that its density ρ at any given position is constant. This can be expressed as

$$\frac{\partial \rho}{\partial t} = 0 \qquad \text{local change in mass density} \tag{VII.1.1}$$

This automatically implies that the total amount of fluid flowing into a volume equals the total amount flowing out of the volume. This can be expressed as

$$div(\rho \underline{v}) = 0, \qquad \text{zero net flow of mass into a volume} \tag{VII.1.2}$$

where $\underline{v}$ is the velocity of the fluid. For a compressible fluid, the density is not constant, but the overall mass is still conserved. Thus, any change in local density must be compensated by a flow of mass in and out of the volume. The leads to the continuity equation [2]:

$$\frac{\partial \rho}{\partial t} + div(\rho \underline{v}) = 0. \tag{VII.1.3}$$

For incompressible fluids, this is equal to $\rho\, div(\underline{v}) = 0$. The resulting condition

$$div(\underline{v}) = 0 \tag{VII.1.4}$$

is called the incompressibility condition.

Physics of Nerves and Excitatory Membranes, First Edition. Thomas Heimburg.

VII.1.1.2 Hydrostatics, Dynamics, and Euler Equations

Let us assume an incompressible liquid in a steady state, meaning that no changes in time occur. We consider external forces, F, acting on different volume elements. Such forces are, for instance, gravitational forces or centrifugational forces in rotating containers. In hydrostatics, the force acting on a volume element is compensated by a gradient of the pressure of the fluid, p:

$$grad\,p = \underline{F}. \tag{VII.1.5}$$

In the presence of motion, one has to take the inertial forces (mass times acceleration) acting on each volume element into account, and eq. (VII.1.5) becomes

$$\rho\frac{d\underline{v}}{dt} + grad\,p = \rho\left(\frac{\partial}{\partial t} + (\underline{v}\cdot grad)\right)\underline{v} + grad\,p = \underline{F}, \tag{VII.1.6}$$

which together with the incompressibility condition eq. (VII.1.4) is called the *Euler equations for incompressible fluids.*

VII.1.1.3 Euler Equations for Compressible Fluids and Simple Acoustics

The aforementioned equations change somewhat when the fluid is considered being compressible, meaning that the density is not constant in time and space. This implies that it is more convenient to consider forces on mass elements instead of volume elements. We describe this by denoting the external force by $\underline{F}_m = \underline{F}/\rho$. We start again with eq. (VII.1.6):

$$\frac{d\underline{v}}{dt} + \frac{1}{\rho}grad(p) = \underline{F}_m$$

and the continuity equation for compressible fluids given by eq. (VII.1.3)

$$\frac{\partial\rho}{\partial t} + div(\rho\underline{v}) = 0.$$

These two equations combined are called the *Euler equations of compressible fluids.*

This book deals with the propagation of pulses in nerves that we will consider as acoustic phenomena. In the absence of external forces or perturbations, $\underline{F}_m = 0$, and

$$\rho\frac{d\underline{v}}{dt} + grad\,p = 0. \tag{VII.1.7}$$

This equation implies that inertial forces are exactly compensated by changes in internal pressure, which is the hydrodynamic version of Newton's second law. We will limit ourselves to small density variations, $\Delta\rho \ll \rho_0$, with ρ_0 being the equilibrium density. We will further assume that variations of velocity and density in time are much larger than variations of velocity in space and that

$$\frac{d\underline{v}}{dt} \approx \frac{\partial\underline{v}}{\partial t}. \tag{VII.1.8}$$

The Euler equation and the continuity equation can now be simplified to

$$\rho_0 \frac{\partial \underline{v}}{\partial t} + grad\, p = 0 \qquad \text{(VII.1.9a)}$$

$$\frac{\partial \rho}{\partial t} + \rho_0\, div\, \underline{v} = 0 \qquad \text{(VII.1.9b)}$$

Pressure and density gradients of the fluid are related, and one can approximate for small variations in pressure:

$$grad\, p = \left(\frac{dp}{d\rho}\right)_0 grad\, \rho \qquad \text{(VII.1.10)}$$

with

$$\left(\frac{dp}{d\rho}\right)_0 \equiv c^2, \qquad \text{(VII.1.11)}$$

where c is a constant defined for a given ρ_0 and p_0. With this definition, we obtain from eq. (VII.1.9a)

$$\frac{\partial \underline{v}}{\partial t} = -\frac{c^2}{\rho_0} grad\, \rho. \qquad \text{(VII.1.12)}$$

Partial differentiation of eq. (VII.1.9b) with respect to time yields

$$\frac{\partial^2 \rho}{\partial t^2} = -\rho_0\, div \frac{\partial \underline{v}}{\partial t}. \qquad \text{(VII.1.13)}$$

Upon insertion of eq. (VII.1.12) into eq. (VII.1.13), we finally arrive at

$$\frac{\partial^2 \rho}{\partial t^2} = div\left(c^2\, grad\, \rho\right) \qquad \text{wave equation} \qquad \text{(VII.1.14)}$$

This equation governs small-amplitude waves in compressible fluids and is called the "wave equation." If $c = c_0$ is independent of the density, this equation simplifies to

$$\frac{\partial^2 \rho}{\partial t^2} = c_0^2 \triangle \rho. \qquad \text{(VII.1.15)}$$

This is the common equation for sound propagation in linear media and is used for the propagation of sound in air (see Figure VII.1.1 for a periodic solution). However, we will see later that the sound velocity in lipid membranes is a nonlinear function of membrane density, $c = c(\rho)$. Under such conditions, one cannot move the sound velocity out of the divergence term in eq. (VII.1.14).

VII.1.1.4 Sound Propagation in One Dimension and Sound Velocity

Eq. (VII.1.14) describes density changes of matter in three dimensions. The wave equation simplifies if we have a one-dimensional problem, for instance, the propagation of density

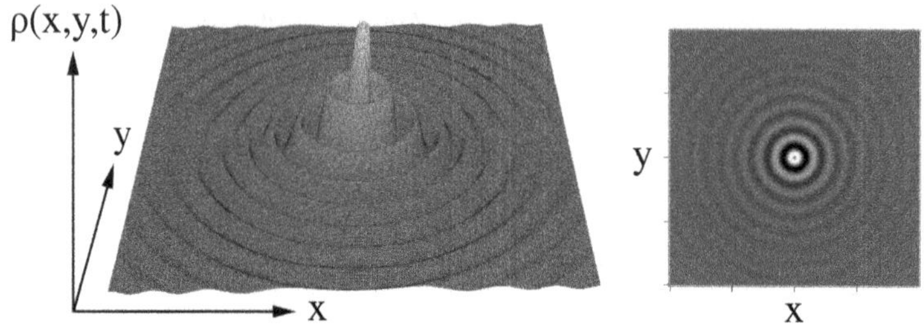

Figure VII.1.1 Sound propagation from a perturbation at a center in 3D (shown is a slice in the x,y plane).

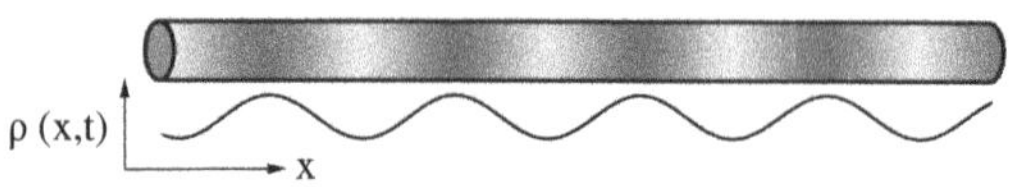

Figure VII.1.2 Sound wave with periodic density variations in a one-dimensional cable.

waves in long cables along the x-axis, such as a lipid cylinder, in which the wavelength is larger than the diameter. Then we obtain (see Figure VII.1.2):

$$\frac{\partial^2 \rho}{\partial t^2} = \frac{\partial}{\partial x}\left(c^2 \frac{\partial \rho}{\partial x}\right) \quad \text{wave equation in one dimension} \qquad \text{(VII.1.16)}$$

If c is independent of density,

$$\frac{\partial^2 \rho}{\partial t^2} = c_0^2 \frac{\partial^2 \rho}{\partial x^2}, \qquad \text{(VII.1.17)}$$

which represents the equation for one-dimensional sound propagation in linear media. The density can be written as $\rho = \rho_0 + \Delta\rho(x,t)$, where ρ_0 is the equilibrium density.

Since both sides of the equation contain derivatives, we can ignore the constant term ρ_0 and write instead

$$\frac{\partial^2 \Delta\rho}{\partial t^2} = c_0^2 \frac{\partial^2 \Delta\rho}{\partial x^2}. \qquad \text{(VII.1.18)}$$

This partial differential equation possesses solutions of nature:

$$\Delta\rho(x,t) = \Delta\rho_0 \sin(kx - \omega t), \qquad \text{(VII.1.19)}$$

where k is the wave vector given by $k = 2\pi/\lambda$ (introducing the wavelength λ), and $\omega = 2\pi/T$ is the frequency. Inserting this expression into eq. (VII.1.18) yields

$$c_0 = \frac{\omega}{k}. \qquad \text{(VII.1.20)}$$

Using this expression in eq. (VII.1.19) finally leads to

$$\Delta\rho(x,t) = \Delta\rho_0 \sin\left(k(x - c_0 t)\right), \qquad \text{(VII.1.21)}$$

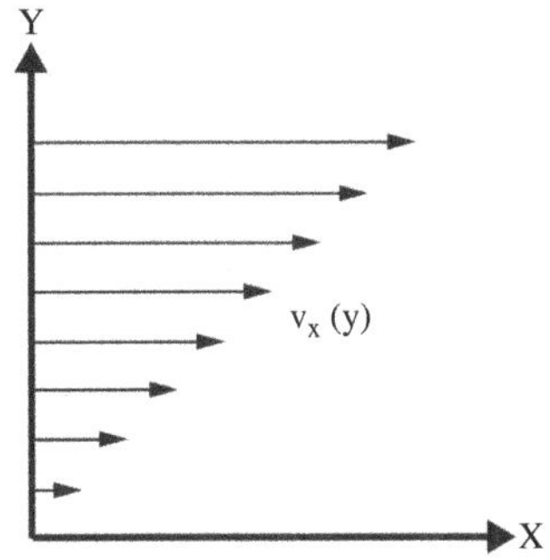

Figure VII.1.3 Viscous friction in a fluid. The velocity of the fluid in x-direction increases as a function of y. Relative velocity differences lead to a force in x-direction.

and it is now obvious that the quantity $c_0 = \sqrt{(\partial p/\partial \rho)_0}$ defined in eq. (VII.1.11) is the propagation velocity of the wave or, in brief, the *sound velocity*.

VII.1.2 Hydrodynamics in the Presence of Friction

So far, we have considered fluids without internal friction. In the following, we extend the aforementioned hydrodynamical concepts by taking friction into account.

VII.1.2.1 Viscosity

If a fluid has internal friction, a force can be measured when two parallel planes in the fluid move against each other. The forces display the opposite sign as the direction of the relative motion. An example is shown in Figure VII.1.3. The velocity of the fluid in x-direction changes as a function of the y-direction. The relative velocity difference leads to a force in x-direction. If one normalizes the force with respect to an area segment, we obtain the analog of a pressure.

For the incompressible fluid, we describe this by following Ansatz

$$p_{xy} = -\eta \frac{\partial v_x}{\partial y}. \tag{VII.1.22}$$

In a similar manner, one can define $p_{xx} = -\eta \partial v_x/\partial x$ and $p_{xz} = -\eta \partial v_x/\partial z$, etc.

VII.1.3 Navier–Stokes Equations

In analogy to eq. (VII.1.5) ($grad\, p = \underline{F}$), one can write

$$\begin{aligned}
\frac{\partial p}{\partial x} + \frac{\partial p_{xx}}{\partial x} + \frac{\partial p_{yx}}{\partial y} + \frac{\partial p_{zx}}{\partial z} &= F_x, \\
\frac{\partial p}{\partial y} + \frac{\partial p_{xy}}{\partial x} + \frac{\partial p_{yy}}{\partial y} + \frac{\partial p_{zy}}{\partial z} &= F_y, \\
\frac{\partial p}{\partial z} + \frac{\partial p_{xz}}{\partial x} + \frac{\partial p_{yz}}{\partial y} + \frac{\partial p_{zz}}{\partial z} &= F_z,
\end{aligned} \tag{VII.1.23}$$

which means that in a steady flow, the external forces F are compensated by internal pressure gradients and gradients in viscous forces. Inserting eq. (VII.1.22) into eq. (VII.1.23) yields

$$grad\, p - \eta \triangle \underline{v} = \underline{F} \tag{VII.1.24}$$

where inertial forces are not yet considered.

The Navier–Stokes equations are the equivalent of the Euler equation for compressible fluids given by eqs. (VII.1.7) and (VII.1.9) in the presence of viscous friction. If eq. (VII.1.24) is taken into account, i.e., replacing $grad\, p$ by $grad\, p - \eta \triangle \underline{v}$, and we combine this with the continuity eq. (VII.1.3) of compressible fluids, we obtain:

$$\rho\left(\frac{\partial}{\partial t} + (\underline{v}\cdot grad)\right)\underline{v} + grad\, p - \eta \triangle \underline{v} = \underline{F} \qquad \text{Navier-Stokes equations}$$
$$\frac{\partial \rho}{\partial t} = -div\,(\rho\, v) \tag{VII.1.25}$$

which are called the *Navier–Stokes equations of compressible fluids*. In one dimension and in the absence of external forces ($F = 0$), these equations assume the form

$$\begin{aligned} \rho_0 \frac{\partial v}{\partial t} + \frac{\partial p}{\partial x} - \eta \frac{\partial^2 v}{\partial x^2} &= 0 \\ \frac{\partial \rho}{\partial t} + \rho_0 \frac{\partial v}{\partial x} &= 0 \end{aligned} \tag{VII.1.26}$$

where we have again neglected the quadratic terms in velocity (i.e., $v(\partial v/\partial x)$ is small).

VII.1.3.1 Sound Propagation in the Presence of Friction

In the following, we modify the wave equation in order to include friction. Assuming small-amplitude waves and going through the argument leading to eq. (VII.1.9), we obtain

$$\begin{aligned} \rho_0 \frac{\partial \underline{v}}{\partial t} + grad\, p - \eta \triangle \underline{v} &= 0 \\ \frac{\partial \rho}{\partial t} + \rho_0\, div\, \underline{v} &= 0 \end{aligned} \tag{VII.1.27}$$

Following the procedure in Section VII.1.1.3, we arrive at the wave equation

$$\frac{\partial^2 \rho}{\partial t^2} = div\left(c^2\, grad\, \rho - \eta \triangle \underline{v}\right) \qquad \text{wave equation with friction} \tag{VII.1.28}$$

By making use of the continuity equation, we can replace the friction term by

$$\eta\, div \triangle \underline{v} = \frac{\eta}{\rho_0} \triangle \left(\frac{\partial \rho}{\partial t}\right), \tag{VII.1.29}$$

and we finally obtain:

$$\frac{\partial^2 \rho}{\partial t^2} = div\left(c^2\, grad\, \rho\right) - \frac{\eta}{\rho_0} \triangle \frac{\partial \rho}{\partial t} \qquad \text{wave equation with friction} \tag{VII.1.30}$$

which in one dimension assumes the form

$$\frac{\partial^2 \rho}{\partial t^2} = \frac{\partial}{\partial x}\left(c^2 \frac{\partial \rho}{\partial x}\right) - \frac{\eta}{\rho_0}\frac{\partial^2}{\partial x^2}\frac{\partial \rho}{\partial t} \quad \text{1D wave equation with friction} \qquad \text{(VII.1.31)}$$

Eqs. (VII.1.16) and (VII.1.31) are important for the propagation of electromechanical nerve pulses.

VII.1.4 Summary: Key Ideas of Chapter VII.1

In this chapter, we applied hydrodynamics in order to describe the propagation of density waves in fluids. In the context of this book, we are mostly interested in small-amplitude sound propagation in one-dimensional cables that represent the nerve axons. Therefore, this summary will focus on the respective results. The limitation to small-wave amplitudes considerably reduces the complexity of the equations.

- Density propagation phenomena in the absence of friction are described by the Euler equations. The Euler equations describe the balance of inertial forces, pressure gradients, and external forces on fluids. For compressible fluids in the absence of external forces, they are given by eqs. (VII.1.9). In one dimension, they assume the form

$$\rho_0 \frac{\partial v}{\partial t} + \frac{\partial p}{\partial x} = 0$$

$$\frac{\partial \rho}{\partial t} + \rho_0 \frac{\partial v}{\partial x} = 0 \qquad \text{Euler equations for compressible fluids in 1D}$$

  The first equation states that pressure gradients must be equal to inertial forces in the fluid. The second of these equations is the continuity equation that describes the conservation of mass.
- Propagation phenomena in viscous fluids (with internal friction) are described by the Navier–Stokes equations. These equations are similar to the Euler equations, but they contain additional forces that result from friction between adjacent layers of the fluid that display different velocities. For small-amplitude density variations, they are given by

$$\rho_0 \frac{\partial v}{\partial t} + \frac{\partial p}{\partial x} - \eta \frac{\partial^2 v}{\partial x^2} = 0$$

$$\frac{\partial \rho}{\partial t} + \rho_0 \frac{\partial v}{\partial x} = 0 \qquad \text{Navier-Stokes equations in 1D}$$

- The resulting partial differential equations governing the propagation of waves in 1D are

$\frac{\partial^2 \rho}{\partial t^2} = \frac{\partial}{\partial x}\left(c^2 \frac{\partial \rho}{\partial x}\right)$	wave equation in 1D without friction
$\frac{\partial^2 \rho}{\partial t^2} = \frac{\partial}{\partial x}\left(c^2 \frac{\partial \rho}{\partial x}\right) - \frac{\eta}{\rho_0} \frac{\partial^2}{\partial x^2} \frac{\partial \rho}{\partial t}$	wave equation in 1D with friction

These equations will be the starting point for the following chapters. In Chapter VII.4, these wave equations will be used to derive a theory for nerve pulse propagation.

VII.2

Sound Velocity in Nerve Membranes

In the previous chapter, we tacitly assumed that the sound velocity is a constant. In this chapter, we describe the two-dimensional (2D) sound velocity in membranes close to melting transitions where it becomes a function of the mass density of the membranes. The purpose is to explore the density dependence of sound velocity in nerve membranes.

VII.2.1 The Density Dependence of Sound Velocity

We have shown in Chapter VI.5 how one can determine the 2D sound velocity c from the heat capacity and represent it as a function of both temperature and lateral density. The sound velocity is given by

$$c = \sqrt{\frac{\partial \Pi}{\partial \rho^A}} = \sqrt{\frac{1}{\kappa_S^A \cdot \rho^A}}, \tag{VII.2.1}$$

where ρ^A is the area density of the membrane and $\kappa_S^A = -(1/A)(\partial A \partial \Pi)_S\, \kappa_S^A$ is the adiabatic compressibility. The latter is a frequency-dependent function due to the time dependence of heat transfer from the membrane into the aqueous environment. Therefore, the sound velocity is also frequency-dependent. This phenomenon is called *dispersion*. It will be discussed in Chapter VII.3.

In the following, we demonstrate that close to transitions, the square of the sound velocity is a nonlinear function of the area density. We show an analysis from Heimburg and Jackson [1]. We illustrate the principle for the thermodynamic parameters of large unilamellar vesicles of DPPC. In Section VI.5.2, we showed that both density and sound velocity are functions of temperature. Therefore, one can also plot the square of the sound velocity as a function of density. This is given in Figure VII.2.1 for the low-frequency limit (0 MHz) and for 5 MHz.

Our goal is to describe sound propagation in nerve membranes at physiological conditions (see Sections III.1.3 and V.2.4), i.e., for the fluid membrane slightly above the melting transition. For the DPPC membrane, we chose a temperature of 45 °C, which is about 3° above the transition maximum. At this temperature, one can determine the density ρ_0^A and the sound velocity c_0 (Section VI.5.2). We see in Figure VII.2.1 that $c^2 = f(\rho^A)$ displays a minimum at a density above the equilibrium density representing a fluid membrane state. This minimum in c^2 corresponds to the transition maximum because the compressibility

Physics of Nerves and Excitatory Membranes, First Edition. Thomas Heimburg.

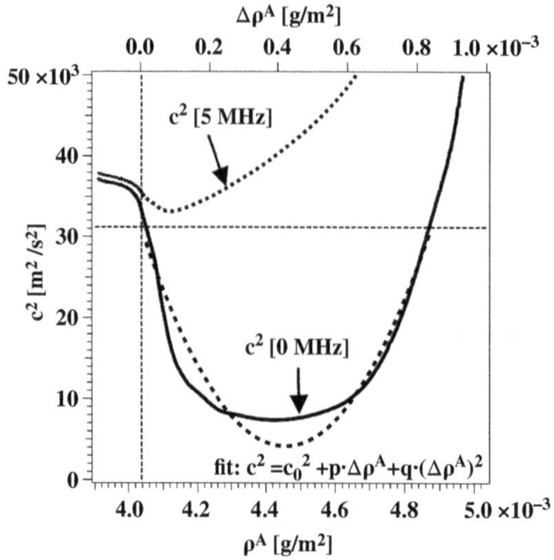

Figure VII.2.1 The square of the lateral sound velocity as a function of lateral density for a frequency of 0 MHz and 5 MHz. When the experimental temperature is at 45°C (slightly above the transition temperature T_m), the area density outside of the pulse is $\rho_0^A = 0.004035$ g/m^2. The density change relative to this density is denoted as $\Delta\rho^A$. The most simple fit to $c^2(\rho^A)$ is a quadratic function indicated by the dashed line. Adapted from [1].

is maximum at the heat capacity maximum. At even higher densities, c^2 increases again when the membrane is moved into its gel state with a very low compressibility. In order to explore the nonlinear features of this curve, we can perform a Taylor expansion of the function around the equilibrium density ρ_0^A up to the second-order term to obtain

$$c^2 = c_0^2 + p\Delta\rho^A + q(\Delta\rho^A)^2 + \dots\,. \tag{VII.2.2}$$

Here, $\Delta\rho^A$ is the density difference from the equilibrium density at 45 °C. The fit provides the following parameters:

$$c_0 = 176.6\ \text{m/s}, \qquad \rho_0^A = 0.004035\text{g/m}^2,$$
$$p = -16.6\, c_0^2/\rho_0^A, \qquad q = +79.5\, c_0^2/(\rho_0^A)^2.$$

Heimburg and Jackson [1] performed a similar analysis for a bovine lung surfactant, which is a biological membrane preparation. Lung surfactant is in its fluid state at body temperature but displays a melting transition around 30 °C [3, 4]. This is shown in Figure VII.2.2.

Here, we chose 37 °C as the reference temperature (body temperature of cows) to determine the equilibrium density. The fit to the square of the sound velocity leads to

$$c_0 = 171.4\ \text{m/s}, \qquad \rho_0^A = 0.004107\ \text{g/m}^2$$
$$p = -6.86\, c_0^2/\rho_0^A, \qquad q = +32.32\, c_0^2/(\rho_0^A)^2.$$

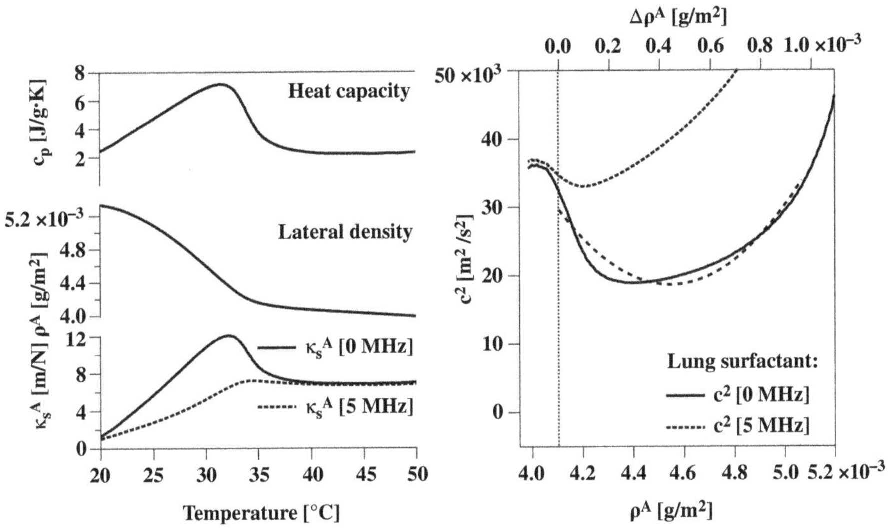

Figure VII.2.2 Left: Heat capacity, lateral density, and the adiabatic compressibility of bovine lung surfactant. Right: The calculated sound velocity squared at 0 MHz and at 5 MHz. Adapted from [1].

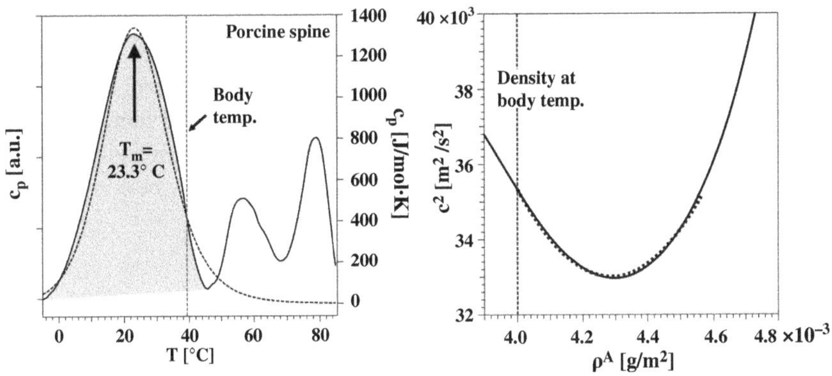

Figure VII.2.3 A similar analysis for nerve membranes. Left: The heat capacity profile of nerve spine membranes, and a fit to the lipid peak using a van't Hoff profile as described in Ref. [5]. Right: The square of the sound velocity calculated from the theoretical profile. [5]/Elsevier/CC BY 4.0.

A similar analysis for the nerve membranes of porcine spine was given by Heimburg [5]; see Figure VII.2.3. The reference temperature of 39 °C was chosen, which is the body temperature of pigs. The analysis leads to the following parameters.

$$c_0 = 188 \text{ m/s}, \qquad \rho_0^A = 0.0040\text{g/m}^2$$

$$p = -1.8291c_0^2/\rho_0^A, \qquad q = +12.7193c_0^2/(\rho_0^A)^2.$$

We can see that the analysis leads to comparable results for synthetic membranes and for biological preparations. In the biological preparations, the minimum of the c^2 profile is less pronounced because the heat capacity profile is broader and, as a consequence, the maximum heat capacity is lower. However, independent of the nature of the preparation, one finds a minimum of c^2 above the equilibrium density.

VII.2.2 Sound Velocity in the Presence of a Coupling to the Aqueous Interface

In the previous section, we calculated the area density of the lipid membrane without considering the possibility that water couples to the membrane surface. However, the 2D membrane is embedded into a three-dimensional (3D) medium that may interact with the membrane. Surface-associated water will increase the membrane area density and lower the sound velocity due to the increase in inertial forces. This is described in Section VI.5.3. There we found that coupling of the medium to the membrane leads to the following changes [6, 7]:

$$\begin{aligned} \rho^A &\rightarrow \rho^A \cdot \frac{\delta}{d} \\ c_0 &\rightarrow c_0 \cdot \sqrt{\frac{d}{\delta}}, \end{aligned}$$

where δ is the frequency-dependent thickness of the layer when water is coupled to it, and d is the thickness of a lipid monolayer ($\approx$ 2nm). The approximate timescale of a nerve pulse is 1 ms, corresponding to a frequency of about 1000 Hz. This leads to δ = 63 μm (for the calculation of δ as a function of frequency, see Section VI.5.3). The 2D sound velocity i is then lowered from about 176 m/s (the order of magnitude of a myelinated motor neuron) to 1 m/s (the order of magnitude for a nonmyelinated nerve, e.g., of worm axons). When taking the data in Figure VII.2.1, the parameters of the fits to c^2 can now be converted in the following manner:

$$\begin{aligned} c_0 &= 176.6\ \text{m/s} \cdot \frac{d}{\delta}, \quad \rho_0^A = 0.004035\ \text{g/m}^2 \cdot \frac{\delta}{d}, \\ p &= -16.6 c_0^2/\rho_0^A, \qquad q = +79.5 c_0^2/(\rho_0^A)^2. \end{aligned}$$

This means that there are no major changes in the analysis except for a drastic reduction in the sound velocity. In a similar manner as above, the c^2 profiles of Figures VII.2.2 and VII.2.3 can be altered.

VII.2.3 The Wave Equation for a Nonlinear Behavior of the Sound Velocity

In the presence of a nonlinear density dependence of the square of the sound velocity, the one-dimensional wave (1D) equation derived in Section VII.1.1 (see eq. (VII.1.16)) can now be written as

$$\frac{\partial^2 \Delta\rho^A}{\partial t^2} = \frac{\partial}{\partial x}\left[\left(c_0^2 + p\Delta\rho^A + q(\Delta\rho^A)^2 + \ldots\right)\frac{\partial \Delta\rho^A}{\partial x}\right]. \qquad \text{(VII.2.3)}$$

The parameters c_0, p, and q alter the shape of the solutions of the wave equation.

VII.2.4 Summary: Key Ideas of Chapter VII.2

In this section, we showed that for the biological membrane:

- The sound velocity c in the membrane is a nonlinear function of the area mass density.
- At rest, the biological membrane is in the fluid state with an equilibrium density ρ_0^A and a sound velocity c_0.
- Upon compression of the membrane, the density increases and the sound velocity decreases, because the membrane is shifted into its transition regime where the membrane is more compressible, and the sound velocity is smaller. Upon further compression, the membrane is moved into its gel state.
- The square of the sound velocity, c^2, can be approximated by a quadratic function of density.
- The coupling of water to the membrane does not change this analysis other than by lowering the sound velocity due to an increase in the area density of the membrane. At 1000 Hz, the sound velocity is in the range of $c_0 = 1$ m/s. In the absence of the coupling to water, the sound velocity is rather $c_0 = 176$ m/s.
- The coupling to water shifts the sound velocity from the typical range of the pulse propagation velocity of a myelinated nerve to that of a nonmyelinated nerve.

In Chapter VII.4, we will show that the density dependence of the sound velocity in biological membranes will result in the possibility of solitary pulses in biomembranes. This pulse describes the action potential in nerves.

VII.3

The Frequency Dependence of Sound Velocity

In Chapter VII.2, we demonstrated that the sound velocity c is a nonlinear function of density if one is close to the melting transition of membranes. This chapter deals with the possibility that the sound velocity is also a function of frequency. This effect is called *dispersion*.

Let us assume a container with a fluid in which we excite a standing wave, for instance, with a piezoelectric transducer that periodically moves on the walls with a frequency ω. One only obtains standing waves when the distance between the outer walls is a multiple of the wavelength (for illustration see Figure VII.3.1). The frequency is a fixed parameter in our experimental setup. The presence of standing waves allows us to easily measure the wavelength λ. This can be done by moving one wall by one wavelength, which again leads to a resonance. The wave vector $k = 2\pi/\lambda$ can now be directly measured, and we can determine the sound velocity c (see eq. (VII.1.20)):

$$c = \frac{\omega}{k}. \tag{VII.3.1}$$

Typically, the sound velocity is not constant for varying frequencies, especially close to the timescales of a relaxation process.

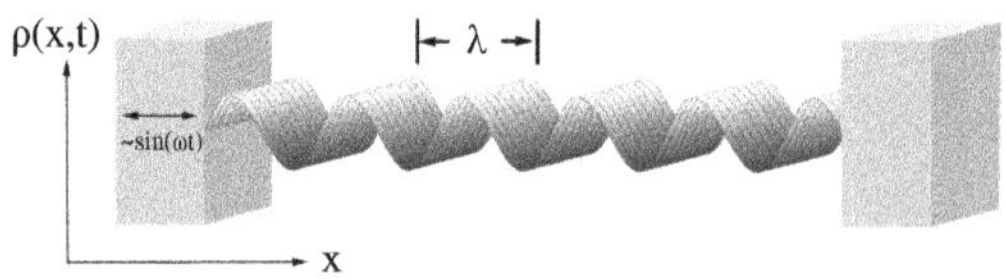

Figure VII.3.1 Standing wave with wavelength λ in a resonator. A slight periodic motion of one of the walls determines the frequency. Standing waves emerge when the length of the reservoir is a multiple of the wavelength. The wavelength can be determined by moving a wall searching for resonances as a function of position. From wavelength and frequency, the sound velocity can be determined by using eq. (VII.3.1).

Physics of Nerves and Excitatory Membranes, First Edition. Thomas Heimburg.

VII.3.1 An Approximate Dispersion Relation

Let us expand the square of the sound velocity around zero frequency:

$$c^2 = c_0^2 + \left(\frac{\partial c^2}{\partial \omega}\right)_0 \omega + \frac{1}{2}\left(\frac{\partial^2 c^2}{\partial \omega^2}\right)_0 \omega^2 + \ldots \tag{VII.3.2}$$

Since the solutions of the wave equation should be invariant against replacing ω by $-\omega$ (which is equivalent to replacing the time t by $-t$), we omit the linear term and obtain

$$c^2 \equiv c_0^2 + h_0\omega^2 + \ldots, \qquad \text{dispersion relation} \tag{VII.3.3}$$

where $h_0 = (\partial^2 c^2/\partial\omega^2)_0$ is a positive constant, and c_0 is the sound velocity in the limit of very small frequencies. The aforementioned equation implies that the sound velocity increases upon increasing the frequency. This can be understood by considering relaxation processes in soft matter. Each perturbation induced by pressure leads to a new equilibrium. This equilibrium is assumed by a relaxation process with time constant τ. If the time constant of the periodic perturbation during an oscillation ($\propto 1/\omega$) is shorter than that of the relaxation process, the material cannot equilibrate, with the effective result that the material becomes stiffer (lower compressibility). Since the sound velocity is related to the compressibility by $c = \sqrt{1/\kappa_S\rho}$, materials with lower compressibility display larger sound velocities.

The wave equation in eq. (VII.1.16) does not consider the dependence of the sound velocity on frequency. In the following, we will modify the wave equation such that the dispersion relation is taken into account. We make the Ansatz

$$\frac{\partial^2 \rho^A}{\partial t^2} = \frac{\partial}{\partial x}\left(c^2\frac{\partial \rho^A}{\partial x}\right) - h\frac{\partial^4 \rho^A}{\partial x^4} \qquad \text{wave equation with dispersion} \tag{VII.3.4}$$

Here, we ad hoc added a dispersion term to the right side of the wave equation:

$$-h\frac{\partial^4 \rho^A}{\partial x^4} \qquad \text{dispersion term}$$

The goal is to alter the wave equation in a manner that is consistent with the dispersion relation in eq. (VII.3.3). We justify the introduction of the dispersion term by the following considerations. Assuming a low-amplitude sound wave with

$$\rho^A(x,t) = \rho_0^A \sin(kx - \omega t) \tag{VII.3.5}$$

and inserting this expression into eq. (VII.3.4) yields

$$c^2 = \frac{\omega^2}{k^2} = c_0^2 + h\,k^2 = c_0^2 + \frac{h}{c^2}\omega^2 \approx c_0^2 + \frac{h}{c_0^2}\omega^2. \tag{VII.3.6}$$

When comparing eqs. (VII.3.3) and (VII.3.3), one finds identical dispersion relations with $h_0 = h/c_0^2$. This implies that eq. (VII.3.4) represents the most natural extension of the wave equation accounting for dispersion because we considered only the first non-zero coefficient of the Taylor expansion of c^2. The parameter h determines the magnitude of the frequency dependence. It has to be determined experimentally.

It should be noted that the ad hoc dispersion term in eq. (VII.3.4) is not unique. Let us consider a second dispersion term:

$$+h'\frac{\partial^4 \rho^A}{\partial x^2 \partial t^2} \qquad \text{second dispersion term}$$

It also yields a quadratic dependence on the frequency ω. The insertion of both dispersion terms into the wave equation leads to

$$\frac{\partial^2 \rho^A}{\partial t^2} = \frac{\partial}{\partial x}\left(c^2 \frac{\partial \rho^A}{\partial x}\right) - h\frac{\partial^4 \rho^A}{\partial x^4} + h'\frac{\partial^4 \rho^A}{\partial x^2 \partial t^2} \qquad \text{wave equation with two dispersion terms} \tag{VII.3.7}$$

For low-amplitude sound, we find

$$\frac{\omega^2}{k^2} = c_0^2 + h\,k^2 - h'\omega^2 \approx c_0^2 + \frac{h}{c_0^2}\omega^2 - h'\omega^2. \tag{VII.3.8}$$

This leads to the dispersion relation

$$c^2 = c_0^2 + \left(\frac{h}{c_0^2} - h'\right)\omega^2, \tag{VII.3.9}$$

which yields a quadratic dependence of the sound velocity on the frequency. The use of the second dispersion term was discussed by Engelbrecht and collaborators [8]. Presently, there exists no empirical data that justify an extra dispersion term, and therefore we will not use it for further analysis. In the following, we rely on eq. (VII.3.4).

VII.3.2 Summary: Key Ideas of Chapter VII.3

One finds in experiments that the sound velocity in membranes is frequency-dependent. This effect is called *dispersion*. The effect of dispersion can be included in the wave equation by adding a dispersion term; see eq. (VII.3.4).

- An approximate wave equation that accounts for dispersion is given by

$$\frac{\partial^2 \rho^A}{\partial t^2} = \frac{\partial}{\partial x}\left(c^2 \frac{\partial \rho^A}{\partial x}\right) - h\frac{\partial^4 \rho^A}{\partial x^4}.$$

  Here, h is a parameter that describes the magnitude of the frequency dependence of the sound velocity. This dispersion term is only correct for small frequencies because it represents the lowest order approximation of the sound velocity at low frequencies.
- The resulting dispersion relation is

$$c^2 \approx c_0^2 + \frac{h}{c_0^2}\omega_0^2 \equiv c_0^2 + h_0\omega_0^2$$

  with $h_0 = h/c_0^2$, which displays a quadratic dependence both on wave number and on frequency. Thus, the dispersion term given earlier is the most natural extension of the wave equation in the presence of dispersion.

VII.4

The Nerve Pulse as an Electromechanical Soliton

In this chapter, we focus on deriving a formalism, which describes the propagation of solitary pulses along biomembrane cylinders such as nerve axons. We assume that the pulse length is much larger than the diameter of the cylinder, which we showed in Section III.3.5 to be the case for action potentials in neurons and nerves. This allows us to employ a one-dimensional wave equation. We further assume that the membrane state is found slightly above the melting transition of the membrane. We choose these conditions because the nerve membrane is also found slightly above a melting transition under physiological conditions (see Section V.2.4). We further show that under these conditions, one finds a solitary wave (loosely called solitons) propagating in the cylinder that resembles the action potential of nerves. The soliton theory predicts an order of magnitude of thickness changes, temperature changes, contraction, pulse velocity, and length/timescale of pulses that is consistent with the experimental findings observed during the action potential in living nerves.

VII.4.1 The Nervous Impulse

In cylinders with small diameter, the nonlinear properties of the sound velocity close to melting transitions (Section VII.2.1) together with the frequency dependence of the sound velocity (dispersion, Chapter VII.3) lead to the propagation of density pulses rather than of periodic density waves. When we combine eqs. (VII.2.3) and (VII.3.4), the following modified wave equation can be obtained:

$$\frac{\partial^2 \Delta\rho^A}{\partial t^2} = \frac{\partial}{\partial x}\left[\left(c_0^2 + p\Delta\rho^A + q(\Delta\rho^A)^2\right)\frac{\partial \Delta\rho^A}{\partial x}\right] - h\frac{\partial^4 \Delta\rho}{\partial x^4}. \tag{VII.4.1}$$

We will show below that this equation with the parameters of cell membranes possesses pulse-like solutions that resemble nervous impulses. These pulses are called solitary waves, and we will loosely refer to them as *solitons*. A representative solution for the above partial differential equation using experimental values for the properties of biomembranes is displayed in Figure VII.4.1. It shows a localized pulse that travels with a velocity v along the nerve axon.

Physics of Nerves and Excitatory Membranes, First Edition. Thomas Heimburg.

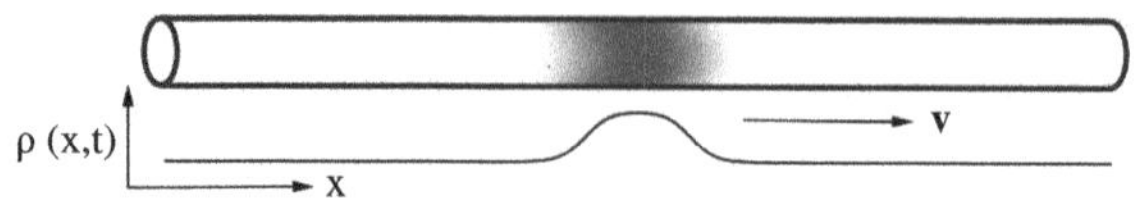

Figure VII.4.1 The pulse propagation in a membrane cylinder is a consequence of the nonlinearity of the elastic constants and the presence of dispersion. The pulse shape is the solution to the partial differential equation (eq. (VII.4.1)).

VII.4.2 Soliton Propagation

The equation governing sound propagation is given by eq. (VII.4.1). It displays solutions that are solitary pulses. We will explore the properties of these solutions by introducing a coordinate system that moves with velocity v and a coordinate z along the axon that is defined by $z = x - vt$. Here, v represents the velocity of the localized pulse. We seek solutions of the form $\Delta\rho^A(z)$.[1]

In the moving coordinate system, eq. (VII.4.1) can be rewritten as

$$v^2 \frac{\partial^2}{\partial z^2}\Delta\rho^A = \frac{\partial}{\partial z}\left[\left(c_0^2 + p\Delta\rho^A + q(\Delta\rho^A)^2\right)\frac{\partial}{\partial z}\Delta\rho^A\right] - h\frac{\partial^4}{\partial z^4}\Delta\rho^A. \quad \text{(VII.4.2)}$$

Now, we only have one single spatial variable in the moving frame. Let us further assume that localized solitary solutions exist that vanish for $|z| \to \infty$. We can perform two integrals and obtain

$$h\frac{\partial^2}{\partial z^2}\Delta\rho^A = (c_0^2 - v^2)\Delta\rho^A + \frac{1}{2}p(\Delta\rho^A)^2 + \frac{1}{3}q(\Delta\rho^A)^3. \quad \text{(VII.4.3)}$$

The integration constants are zero. Far away from the center of the soliton (as $|z| \to \infty$), the higher-order terms in $\Delta\rho^A$ can be neglected, and eq. (VII.4.3) has the form

$$h\frac{\partial^2}{\partial z^2}\Delta\rho^A = (c_0^2 - v^2)\Delta\rho^A \qquad \text{for } |z| \to \infty$$
$$\to \Delta\rho^A \sim \exp\left(-\sqrt{\frac{c_0^2 - v^2}{h}}|z|\right) \quad \text{(VII.4.4)}$$

with an exponentially decaying solution for $|z| \to \infty$. We search for a localized solution, which requires that $|v| < c_0$ because otherwise no real localized solutions exist.

Now we multiply both sides of eq. (VII.4.4) by $\partial(\Delta\rho^A)/\partial z$ and integrate once more:

$$h\left(\frac{\partial}{\partial z}\Delta\rho^A\right)^2 = (c_0^2 - v^2)(\Delta\rho^A)^2 + \frac{p}{3}(\Delta\rho^A)^3 + \frac{q}{6}(\Delta\rho^A)^4. \quad \text{(VII.4.5)}$$

The pulse profile must be symmetric around its maximum because eq. (VII.4.5) is invariant under the transformation $z \to -z$. It displays a maximum at $z = 0$ with $\partial(\Delta\rho^A)/\partial z = 0$). Since both $c_0^2 - v^2$ and q are positive (because the sound velocity displays a minimum in the transition), p must be less than zero in order to allow for a maximum with $\frac{\partial}{\partial z}\Delta\rho^A = 0$. This is in fact the case for biological membranes in their physiological state (see Section VII.2.1). We also see that the dispersion parameter does not change the solution of eq. (VII.4.5).

1 This section strongly relies on [1].

This is obvious when rescaling the z-axis by $z' \equiv \frac{z}{\sqrt{h}}$, which makes the equation independent of h. Thus, the parameter h only changes the width of the pulse but not its shape.

Equation (VII.4.5) shows us that the maximum value of $\Delta\rho^A$ at $z = 0$ is independent of h. At the maximum, the left side of eq. (VII.4.5) is zero, and the right side can be solved for $\Delta\rho^A$ to obtain the pulse amplitude:

$$\Delta\rho^A_{\max} = -\frac{p}{q} + \frac{p}{q}\sqrt{1 - \frac{6q}{p^2}(c_0^2 - v^2)}. \tag{VII.4.6}$$

When v is very close to c_0, the square root term can be Taylor-expanded (using $\sqrt{1-x} \approx 1 - 0.5x$ for $x \ll 1$) and we find that

$$\Delta\rho^A_{\max} = -\frac{3}{p}(c_0^2 - v^2) \text{ for } (c_0^2 - v^2) \to 0 \tag{VII.4.7}$$

meaning that the soliton amplitudes become small as $v \to c_0$.

If the velocity v is too small and the argument in the square-root term becomes negative, $\Delta\rho^A$ will not display a maximum, and solitons will not exist. Thus, there is a minimum velocity v_{limit} for the soliton when $1 - \frac{6q}{p^2}(c_0^2 - v^2_{\text{limit}}) = 0$ and

$$v^2_{\text{limit}} = \left(c_0^2 - \frac{p^2}{6q}\right) \tag{VII.4.8}$$

In the absence of coupling of the membrane with surface water (see Section VII.2.2), we obtain for unilamellar DPPC vesicles $c_0 = 176.6$ m/s, $p = -16.6\, c_0^2/\rho_0^A$, and $q = 79.5\, c_0^2/(\rho_0^A)^2$, with $\rho_0^A = 4.035 \times 10^{-3}$ g/m^2, assuming a bulk temperature of T = 45°C (see Chapter VII.2). In this case, the limiting velocity is $v_{\text{limit}} = 0.64985 \cdot c_0$; $c_0 \approx 115$ m/s, and $\Delta\rho^A_{\max,\text{limit}} = 0.209\ \rho_0^A$. As shown in Sections VI.5.3 and VII.2.2, the equilibrium density of the nerve can be considerably higher due to the mass of the associated water, and thus the sound velocity can be considerably slower. The aforementioned parameters are therefore expected to be valid for myelinated nerves only, when the membrane of the axon does not couple to bulk water.

The minimum velocity soliton has the largest maximum value of $\Delta\rho^A$. As mentioned, at maximum amplitude the membrane is forced partially or fully through the melting transition. Thus, there exist solitary solutions for $1 \geq v/c_0 \geq v_{\text{limit}}/c_0 = 0.64985081$ (for the experimental parameters of DPPC membranes 3°C above the melting temperature). Some of the solutions of eq. (VII.4.5) are shown in Figure VII.4.2. In order to set the scale of 10 cm for the length of a myelinated nerve, we set the dispersion parameter to $h = 2$ m^4/s^2. For each velocity v of the soliton, one finds a different amplitude and width of the pulse.

We find that the amplitude of the soliton is larger when the velocity v is slower. It reaches a maximum of $\Delta\rho^A/\rho_0^A = 0.209$. This is approximately the difference in the area density between fluid and gel membrane. Thus, the soliton consists of a gel region traveling in a fluid membrane with velocity v, as schematically shown in Figure VII.4.3.

The properties of the solitons are remarkably similar to the action potential of nerves. In fact, if the rigorous validity of thermodynamics is taken into account, the finding of solitary

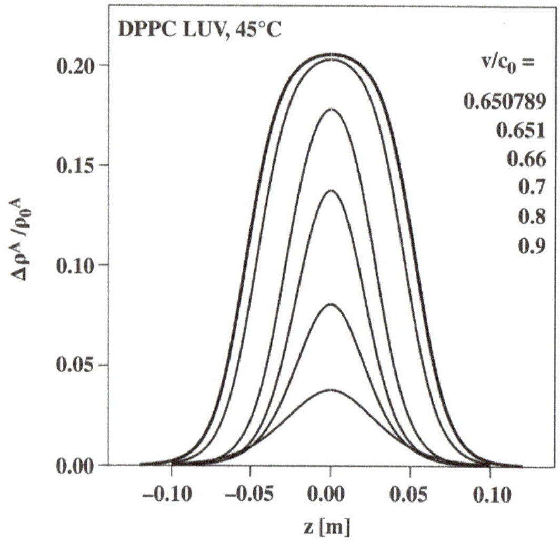

Figure VII.4.2 Solitary solutions of eq. (VII.4.5) for the parameters of a DPPC LUV membrane at 45 °C (slightly above the melting transition at about 42 °C) for different soliton velocities. The shape of the soliton depends on the pulse velocity. [1]/with permission of The National Academy of Sciences.

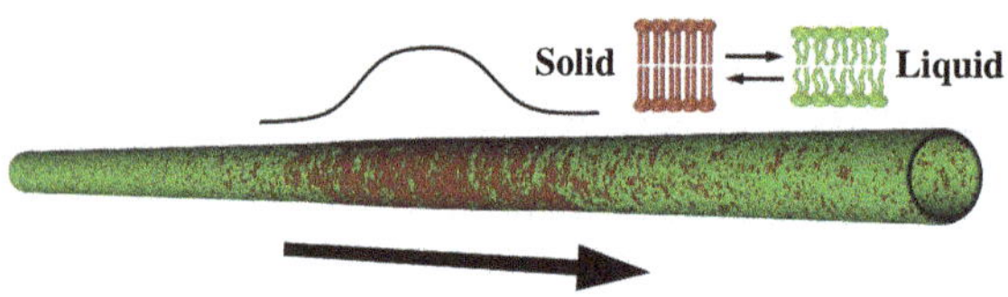

Figure VII.4.3 Schematic representation of a soliton traveling in a fluid membrane (green) and the change in lateral density. Red regions are gel-like. [9]/with permission of Elsevier.

pulses is not just an option but is absolutely unavoidable under physiological conditions – i.e., a membrane slightly above its melting temperature. We demonstrate the concept for the parameters of a unilamellar DPPC membrane slightly above its melting temperature. We find:

1. The soliton at maximum amplitude is about 115 m/s (for the parameters of DPPC membranes), similar to the nerve pulse velocity of a myelinated nerve.
2. The change in density at maximum pulse amplitude is similar to the change in area density between fluid and gel phase membranes (calculated to be 24.6% for DPPC [10]).
3. The change in membrane thickness between the fluid and gel phases of DPPC is about 0.89 nm. The thickness change during the action potential of a nerve is of the order of 1 nm [9, 11, 12]. See Section VII.4.6 for details.
4. During propagation of the pulse, the cylindrical cable contracts because the gel membrane displays a larger area density. Such a contraction has been found during the propagation of the action potential in nerves. See Chapter VII.5 for details.
5. Since the pulse consists of a gel segment within the fluid membrane, one expects a reversible heat transfer of the magnitude of the latent heat of the membrane transition. Experiments show that during the first phase of the pulse, heat is transferred

from the nerve into the environment (fluid → gel) and reabsorbed in the second phase (gel → fluid). The heat change is caused by changes in membrane entropy as proposed by [13]. See Section VII.4.7 for details.

6. There exists a minimum width of the soliton corresponding to a velocity of $v \approx 0.735 \cdot c_0$.
7. If the membrane couples to the aqueous medium as described in Section VI.5.3, pulse velocities can be much slower (in the range of 1–10 m/s), as it is found for nonmyelinated nerves. All other features of the solitons (i.e., the shape) remain similar to the case where the aqueous medium does not couple.

VII.4.3 Analytical Solution

Equation (VII.4.1) possesses analytical solutions.[2] For its mathematical form, it is convenient to introduce the following dimensionless functions:

$$\begin{aligned} u &= \frac{\Delta\rho^A}{\rho_0^A}; \quad \beta \equiv \frac{v}{c_0} \\ x' &= \frac{c_0}{\sqrt{h}} \cdot x; \quad \tau = \frac{c_0^2}{\sqrt{h}} \cdot t; \quad \xi = x' - \beta\tau \\ B_1 &= \frac{\rho_0^A}{c_0^2} \cdot p; \quad B_2 = \frac{\left(\rho_0^A\right)^2}{c_0^2} \cdot q. \end{aligned} \tag{VII.4.9}$$

With these definitions, eq. (VII.4.5) can be rewritten as

$$\left(\frac{\partial u}{\partial \xi}\right)^2 = (1-\beta^2)\,u^2 + \frac{B_1}{3}\,u^3 + \frac{B_2}{6}\,u^4. \tag{VII.4.10}$$

The maximum condition is $\partial u/\partial\xi = 0$ at $\xi = 0$. The pulse is symmetric around the maximum. Equation (VII.4.10) has two real roots, $u = a_\pm$ for $1 \le \beta \le \beta_{\text{limit}} = v_{\text{limit}}/c_0$. The roots $a_\pm$ can be written as

$$a_\pm = -\frac{B_1}{B_2}\left(1 \pm \sqrt{\frac{\beta^2 - \beta_{\text{limit}}^2}{1-\beta_{\text{limit}}^2}}\right), \tag{VII.4.11}$$

which corresponds to eq. (VII.4.6). The analytical solution of the soliton equation (eqs. (VII.4.5) and (VII.4.10)) is given by [14]

$$u(\xi) = \frac{2a_+a_-}{(a_+ + a_-) + (a_+ - a_-)cosh(\xi\sqrt{1-\beta^2})}, \tag{VII.4.12}$$

which can be translated into density and spatial coordinates (as seen in Figure VII.4.2) by letting $\Delta\rho^A = \rho_0^A \cdot u$ and $z = \xi \cdot \sqrt{h}/c_0$ (see [14] for details).

2 This section relies on [14].

VII.4.4 Dendritic Trees

As shown in the previous sections, solitons represent a one-dimensional solution of a wave equation in the presence of dispersion. Each soliton has a well-defined velocity and width. At each position, both the mass density and the energy density are given. Thus, the total energy of the soliton is also exactly specified.

If an axon branches, the shape, velocity, and amplitude remain unaffected as long as the energy density of the pulses is conserved. Thus, a condition for the unperturbed propagation of pulses in a dendritic tree is that the total sum of the circumferences of the branches is identical to the circumference of the axon such that the energy of the pulses is distributed over the same axonal membrane area. Thus, in Figure VII.4.4, the relation for the radii a_i is given by

$$a_0 = a_{11} + a_{12} \qquad \text{or more generally} \qquad a_0 = \sum_i a_{1i}, \tag{VII.4.13}$$

where

$$\begin{aligned} &a_{11} = a_{211} + a_{212} + a_{213} \quad \text{and} \quad a_{12} = a_{221} + a_{222} \\ &\text{or } a_{1i} = \sum_j a_{2ij}. \end{aligned} \tag{VII.4.14}$$

This is in fact the same rule that applies to real trees [15], where the sum of the circumferences of the branches is conserved. This conserves the flux of liquids in the tree bark. A variation of this rule for trees is known as Leonardo's principle.

Rall's model in electrochemistry (see Section IV.2.4) describes the identical problem in electrophysiology. In Rall's model, it is not the total energy that is conserved but the total current through the membranes and along the axons. Recall that we obtained $a_1^{\frac{3}{2}} = a_{11}^{\frac{3}{2}} + a_{12}^{\frac{3}{2}}$, $a_{11}^{\frac{3}{2}} = a_{211}^{\frac{3}{2}} + a_{212}^{\frac{3}{2}} + a_{213}^{\frac{3}{2}}$ and $a_{12}^{\frac{3}{2}} = a_{221}^{\frac{3}{2}} + a_{222}^{\frac{3}{2}}$. Thus, in order to be able to reduce a dendritic tree to a single cable with similar properties, one expects a slightly different change of the radii of the dendrites after branching. But both models have in common that the dendrites, after branching, display a smaller radius.

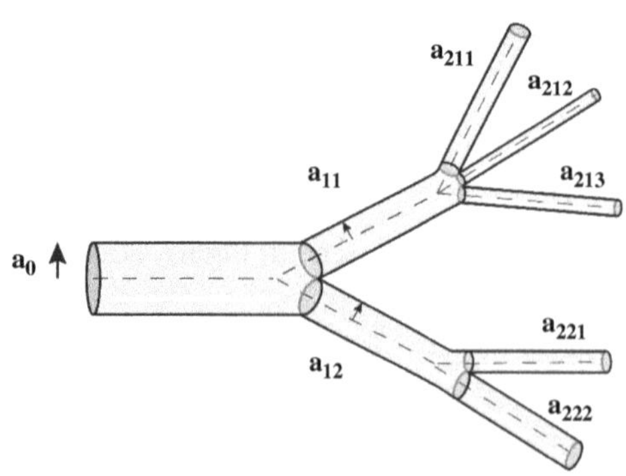

Figure VII.4.4 Representation of a dendritic tree.

VII.4.5 The Density Dependence of the Dispersion Term

The differential equation that leads to soliton propagation (eq. (VII.4.2)) displays four parameters: the equilibrium sound velocity c_0 of the membrane at physiological/experimental temperature, the parameters p and q that describe the density dependence of the phase transition in the membrane, and the dispersion parameter, h. The first three are directly accessible from experiments. The fourth parameter, h, is a free parameter that can be used to adjust the width of the pulse because its sole role consists in rescaling the z-axis along the axon ($z/\sqrt{h} \rightarrow z'$, see comment below eq. (VII.4.5)). However, it seems unlikely that the dispersion parameter is independent of the membrane state and its density. In fact, it can be shown that the dispersion parameter can be related to the relaxation times of the membrane, which can be determined from the thermodynamics of the membrane. In the melting transition of a membrane, the relaxation time is longer than outside of the transition – an effect known as "critical slowing down." However, the response of the membrane to a perturbation is different depending on whether the timescale of the perturbation is longer or shorter than the relaxation time. This also makes dispersion dependent on the relaxation time of the membrane. Since the relaxation time is maximum at the transition, the dispersion parameter is density-dependent, i.e., $h = h(\Delta\rho^A)$.

Mosgaard et al. [16] approximated the density dependence of h by employing linear response theory. They made use of the dependence of the relaxation timescale on the heat capacity, $\tau = (T^2/L)\Delta c_p$ (see Chapter VI.7). The derivation is lengthy, and we do not give the details here. They found that the dispersion parameter h can be approximated by

$$h(\rho^A) = \Delta c_p(\rho^A)\left(\frac{T^2 c_0(\rho^A)^3}{2L}\right)^2 \left[\frac{3c_1^2 + 4c_2^2}{c_2^2(c_2^2 + c_1^2)}\right], \qquad \text{(VII.4.15)}$$

where

$$\begin{aligned} c_1(\rho^A)^2 &\equiv \frac{1}{\rho^A \kappa_{T,0}^A}, \\ c_2(\rho^A)^2 &\equiv \frac{1}{\rho^A \frac{\gamma_A^2 T}{A}\Delta c_p}, \\ c_0(\rho^A)^2 &\equiv \frac{1}{(c_1(\rho^A))^{-2} + c_2(\rho^A))^{-2}}. \end{aligned} \qquad \text{(VII.4.16)}$$

Here, c_1 is the sound velocity outside of the transition range, c_2 describes the change of the sound velocity due to the melting transition in the low frequency limit, and c_0 is the total sound velocity in the low frequency limit (see Section VI.5.2). The excess heat capacity, Δc_p, is a function of the lateral density ρ^A. The compressibility $\kappa_{T,0}^A$ corresponds to the adiabatic compressibility in the low frequency limit outside of the transition regime. The coefficient L is a phenomenological coefficient that relates the heat capacity to relaxation time (see Chapter VI.7). For DMPC and DPPC LUV, it is of order $L = 0.5 \cdot 10^{12}$ [J· K/g·s]. In Figure VII.4.5 (left), the resulting function $h(\Delta\rho^A)$ is shown. It displays a maximum in the transition regime of the membrane. Thus, $h = h(rho^A)$ is no longer a free dispersion parameter, but it is rooted in the thermodynamics of the system.

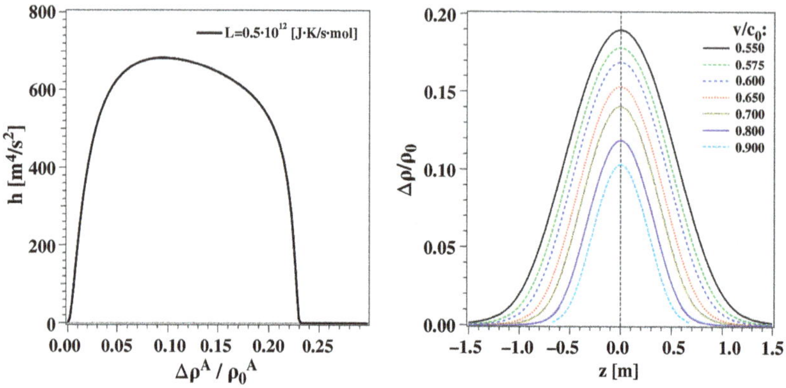

Figure VII.4.5 Left: The function $h(u)$ as a function of the lateral membrane density $u = \Delta\rho^A/\rho_0^A$ using sound velocity values from DPPC membranes and the phenomenological constant $L = 0.5 \cdot 10^{12}$ [J· K/g·s]. Adapted from [17]. Right: Resulting solitons for different velocities v for the thermodynamic data of DPPC using the dispersion function in eq. (VII.4.15) calculated for $L = 0.5 \cdot 10^{12}$ [J· K/g· s]. Adapted from [17].

When inserting eq. (VII.4.15) into eq. (VII.4.2), we see that now all four parameters c_0, p, q, and $h(\rho^A)$ can be determined from experiment, and that there is no free parameter left. Solitary pulses for the example of the lipid DPPC are shown in Figure VII.4.5 (right). The width and the timescale of the pulses are defined by the thermodynamics of the system. In the case of DPPC membranes, it is on the order of $\Delta z = 60 - 120$ cm and the duration of the pulse is $\Delta t = 4 - 12$ ms. This is a little larger than for myelinated nerves but clearly of a very similar order of magnitude (see Section III.3.1 and Table III.3.1). For myelinated motor nerves, the experimental pulse length can be around 20–50 cm and the timescale is 3–9 ms. Thus, the pulse width is determined by the relaxation behavior of the membrane.

We will take the results of this section as proof of principle. Further theoretical research on the density dependence of the dispersion terms would be desirable. For convenience, the solitons in the following chapters will be calculated using an empirical constant dispersion parameter h as used in eq. (VII.4.2), adjusted to the width of the nerve pulse found in experiments. For all practical purposes concerning excitability and the change of the thermodynamic state of the membrane during the pulse, this does not lead to different results.

An important message of this section is that timescale, velocity, and length scale of the nerve pulse can all be derived from the thermodynamics of the system.

VII.4.6 Thickness Changes of the Membrane During the Pulse

During a transition from a gel to a fluid state of the membrane in a DPPC membrane, the volume increases by 4.06% and the area increases by about 24.6% [10]. We assume that the order of magnitude is similar for other lipids and for lipid mixtures. This assumption is justified by the similar pressure dependence of the melting of both artificial membranes made of lipids with different chain lengths and of biological membranes, as described in Chapter VI.3. The aforementioned numbers imply a thickness reduction of −16.3% when

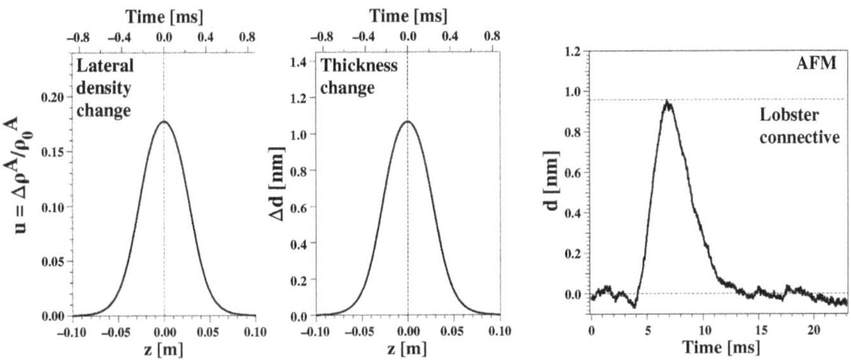

Figure VII.4.6 Left: Calculated area density and thickness change of a soliton for the parameters of DPPC with a pulse velocity of $v = 0.66c_0$. Right: Thickness change of a single axon from lobster connectives. Adapted from [9].

going from gel to fluid. This corresponds to a thickness change of 0.78 nm when going in the opposite direction from fluid to gel (as is the case at the peak of soliton), or 1.54 nm for two membranes on each side of a cylinder. A soliton that is excited in the fluid state should therefore display a thickness change on the order of up to +1.5 nm.

Figure VII.4.6 (left) shows the area density change and the calculated thickness change of two membranes during the passage of a soliton, which in this example amounts to about 1 nm. Changes of this magnitude have been found in the axons of squid, lobster connectives, and rat cortical nerves [9, 11, 12, 18]. Figure VII.4.6 (right) shows the thickness change of a giant axon from lobster connectives during an action potential. This means that the magnitude of the thickness change is very similar to that calculated from the soliton theory.

VII.4.7 Heat Exchange During the Pulse

During the solitary pulse, the membrane is shifted from the fluid to the gel state and back. This does not only change the membrane thickness, as discussed in the previous section. It also changes the enthalpy of the membrane. In order to melt the lipid membrane, one has to add the heat of melting. The latent heat of the transition is released into the environment of the membrane during the passage of the soliton. The heat is reabsorbed when the membrane goes back to the fluid phase. This happens automatically because the fluid state is the equilibrium state under physiological conditions, and the enthalpy is the same before and after the pulse. Figure VII.4.7 (top, left) shows the heat transfer rate from the membrane into its environment as calculated from the enthalpy change of the membrane during the solitary pulse. One recognizes that a phase of heat release is followed by a phase of heat reabsorption. Figure VII.4.7 (top, right) shows the integrated heat transfer rate, i.e., the total heat transferred into the environment during the pulse. We find that after the solitary pulse the total heat transferred is zero. This is a consequence of the isentropic nature of the solitary pulse, which is at the basis of the derivation of eq. (VII.4.1).

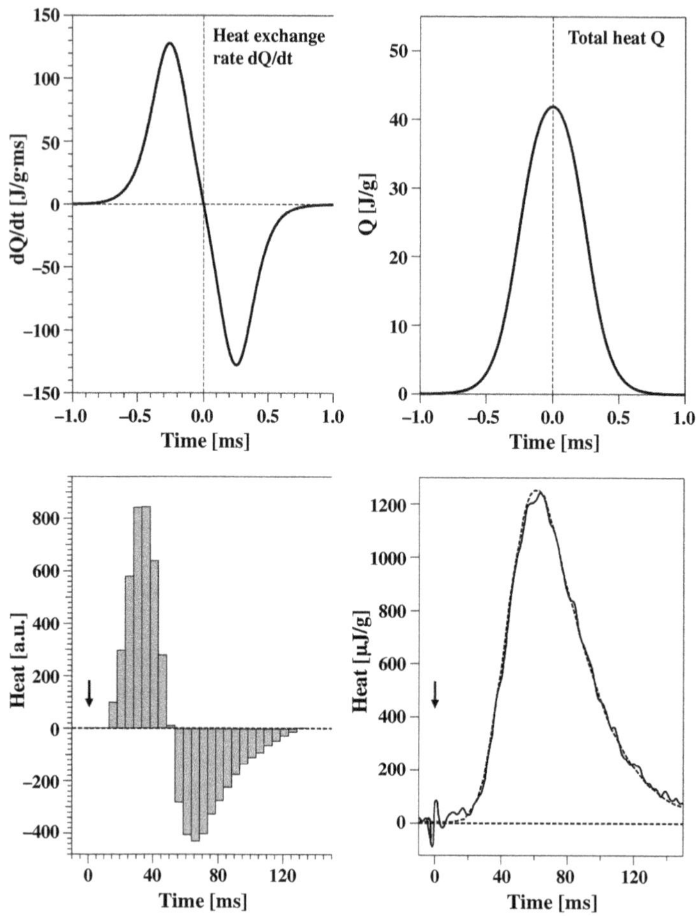

Figure VII.4.7 Top, left: Calculated exchange of heat from an axon per gram of lipid using the parameters of DPPC and $v = 0.66c_0$. Top, right: The integrated heat change. The heat at maximum approximately corresponds to the latent heat per gram of DPPC. Bottom, left: Experimental heat exchange per time in 5 ms blocks from garfish olfactory nerve. Bottom, right: Total heat. Adapted from [19].

Interestingly, this is also found during the action potential of nerves. In Chapter III.6, the heat exchange phenomenon in real nerves is described in detail. Figure VII.4.7 (bottom) shows the heat transfer rate and the total heat transfer during the action potential of nonmyelinated garfish olfactory nerves [19]. One recognizes the qualitative similarities between the theoretical and experimental behavior. Thus, the action potential in living nerves displays many properties typical of an isentropic wave or pulse. The absolute numbers for the heat transfer rate in the experiment should be considered with some care. What is measured in the experiment is rather the temperature change during the action potential. The heat is deduced from this by assuming that the temperature is uniform in the whole nerve specimen and can therefore be derived from multiplying the temperature change with the heat capacity of the aqueous medium ($Q = c_p \cdot \Delta T$). Most likely, this leads to a significant underestimation of the total heat that is transferred. It is more likely that most of the heat does not distribute equally in the medium but is transiently stored in the direct environment of the nerve membrane. Thus, the temperature change in the aqueous medium at the site of a thermocouple is probably smaller than the change close to the membrane, which is inaccessible to the thermocouples used in the experiment.

It is also notable that the heat exchanged during a nerve pulse is considerably larger than the calculated capacitive energy of the nerve membrane during the pulse. See Chapter III.6 for a detailed analysis of the heat production during the nerve pulse, and Section VII.4.8 for a comparison with the pulse's energy.

VII.4.8 The Energy Density of the Soliton

The wave equation for the soliton was calculated from a hydrodynamic perspective. One can assign an energy density and a total energy to a soliton [1, 14, 20]. It contains kinetic and potential energy terms. The potential energy density of the soliton can be written as [1, 14]

$$e = \frac{1}{2}\left[\frac{c_0^2}{\rho_0^A} \cdot \left(\Delta\rho^A\right)^2 + \frac{1}{3}\frac{p}{\rho_0^A} \cdot \left(\Delta\rho^A\right)^3 + \frac{1}{6}\frac{q}{\rho_0^A} \cdot \left(\Delta\rho^A\right)^4\right] \tag{VII.4.17}$$

with units J/m^2. The total pulse energy is given by the integral of the energy density over the nerve membrane.

$$E = \int e dA \tag{VII.4.18}$$

with units of J.

We can compare this energy with the capacitive energy. Let us assume that the nerve membrane displays a capacitance of $c_m \approx 1\mu F/cm^2$ and the capacitive energy is given by $\frac{1}{2}c_m V^2 = \frac{1}{2c_m}q^2$. In the low potential limit of the Gouy-Chapman theory (relevant for medium to high ionic strength conditions and a small fraction of charged lipids), the membrane potential is proportional to the change in charge density [21] and hence to the change in area density, $\Delta\rho^A$. We therefore calculate the capacitive energy change as

$$e_{cap} = \frac{1}{2}c_m\left(V_0\frac{\Delta\rho^A}{\rho_0^A}\right)^2. \tag{VII.4.19}$$

It is now tempting to compare the energy density of the soliton with the capacitive energy density [1]. This is shown in Figure VII.4.8. One can recognize that the capacitive energy density change is much smaller than the total energy density change during the pulse, indicating that most of the energy during the pulse does not originate from membrane electrostatics. This is in agreement with experimental findings concluding that the heat exchange with the environment during a nerve pulse is significantly larger than the capacitive energy [13, 19] (see Chapter III.6 and Section VII.4.7), even though the discrepancy is not quite as large as shown in Figure VII.4.8. Thus, the soliton theory is at least in qualitative agreement with the experiments on living nerves.

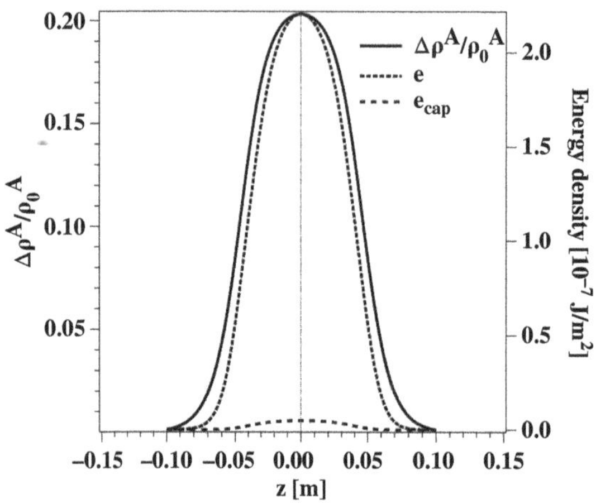

Figure VII.4.8 The relative area density change compared to the soliton energy density and the capacitive energy assuming a capacitance of 1 μF/cm^2 and a voltage at the maximum of the peak of V_0 = 100 mV. Adapted from [1].

It should be noted that the energy of the pulse calculated from hydrodynamic theory rather corresponds to a free energy in a thermodynamic treatment. One can therefore not easily compare the potential energy in this section with the transferred heat in the previous section.

VII.4.9 Soliton Genesis

Soliton genesis has been described in [14]. A soliton can be generated by perturbing the state of the membrane locally. This could be a static Gaussian perturbation (e.g., a local temperature or pressure variation) that renders the membrane solid (or gel-like). We can also use a perturbation that reflects the profile of a true soliton but with a different velocity, which we have done here. The results are shown in Figure VII.4.9. We show a moving perturbation that corresponds to the shape of a minimum width soliton with $v = 0.734761 \cdot c_0$ – however, the velocity is set to half this value (Figure VII.4.9 (top)). In Section VII.4.2, we determined a minimum soliton velocity of $v = 0.64985081 \cdot c_0$ for the parameters of DPPC membranes. Since there exists no soliton with $v = 0.367$, our perturbation will not yield a stable soliton. As a consequence, one can see in simulations that the soliton falls apart into two solitons traveling in opposite directions (Figure VII.4.9 (center)) and some small-amplitude noise that travels ahead of the solitons. Since the original perturbation traveled to the right, it falls apart into a larger soliton traveling to the right with $v = 0.799 \cdot c_0$ and a smaller soliton that travels to the left with $v = 0.948 \cdot c_0$. This is consistent with the finding that lower-amplitude solitons travel faster. The small-amplitude non-solitonic noise travels with $v \approx c_0$. In Figure VII.4.9 (bottom), the two solitons in the center panel have been subtracted. That this was possible confirms that the shapes in the center panel are indeed solitonic. According to [14], the residuals account for approximately 0.3% of the overall soliton energy.

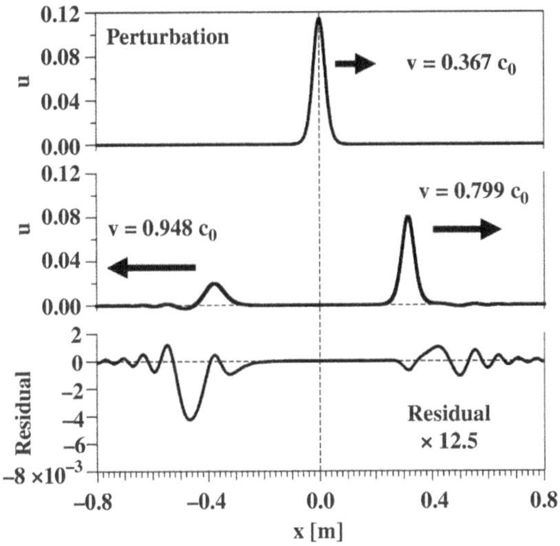

Figure VII.4.9 A single perturbation of the membrane falls apart into two solitons traveling in opposite directions with a minor contribution of small-amplitude noise traveling ahead of the solitons. Top: The original soliton with half the velocity ($v = 0.367 \cdot c_0$). Center: The original soliton falls apart into two solitons traveling in opposite directions with some residual low-amplitude noise traveling ahead of the solitons. The larger soliton traveling to the right has a velocity of $v = 0.799 \cdot c_0$, while the smaller soliton traveling toward the right displays a velocity of $v = 0.948 \cdot c_0$. Bottom: The residuals after subtracting the solitons describing the major peaks in the center panel account for less than 3% of the overall energy of the original soliton. Adapted from [14].

A local perturbation with velocity $v = 0$ will lead to the genesis of two identical solitons traveling in opposite directions. Local stimulation of a living nerve also leads to the propagation of two action potentials in opposite directions [22], a fact that is used in the so-called collision block experiments described in Section III.8.2.

VII.4.10 The Stability of Nerve Solitons in the Presence of Friction

The solitons in the previous sections were determined in the absence of friction. Lautrup et al. [14] performed a detailed analysis of solitons, which focuses on the effect that perturbations and friction have on the stability of the solitons. Friction can be included via the Navier-Stokes equations by introducing a $\nu \partial \Delta\rho^A / \partial x^2 \partial t$ term into eq. (VII.4.2) (see Section VII.1.3.1, [14]):

$$\frac{\partial^2}{\partial t^2}\Delta\rho^A = \frac{\partial}{\partial x}\left[\left(c_0^2 + p\Delta\rho^A + q(\Delta\rho^A)^2\right)\frac{\partial}{\partial x}\Delta\rho^A\right] + \nu\frac{\partial^3}{\partial x^2 \partial t}\Delta\rho^A - h\frac{\partial^4}{\partial x^4}\Delta\rho^A. \tag{VII.4.20}$$

Here, $\nu \equiv \eta/\rho_0^V$ and η is a viscosity (for derivation, see [14]) and ρ_0^V is the volume density of the membrane. The units of ν are m^2/s. In dimensionless representation as introduced in eq. (VII.4.10), this equation reads

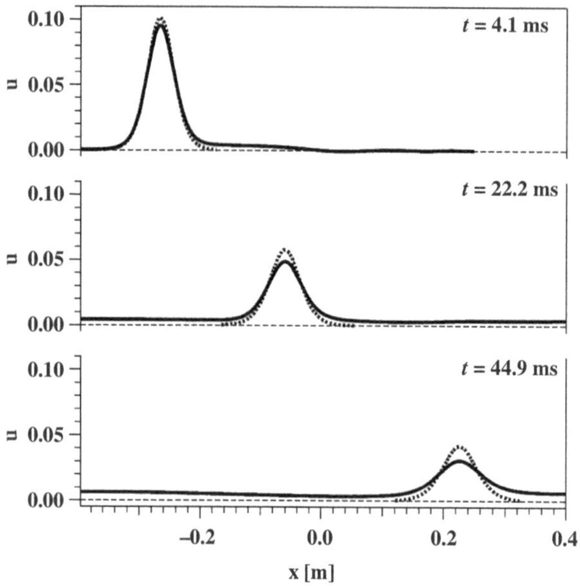

Figure VII.4.10 Decay of the soliton amplitude for a friction coefficient $\varkappa = 0.05$ (cf. eq. (VII.4.10)). The simulation was made on a periodic lattice and the solitons traveled several times the lattice size. Thus, the spatial coordinate only shows the width of the soliton correctly. This figure shows that in the presence of dissipation, the soliton remains stable. The profiles with dashed lines are the shapes of undamped solitons with the same half-width. Adapted from [14].

$$\frac{\partial^2 u}{\partial \tau^2} = \frac{\partial}{\partial x'}\left[\left(1 + B_1 u + B_2 u^2\right)\frac{\partial u}{\partial x'}\right] + \varkappa \frac{\partial^3 u}{\partial (x')^2 \partial \tau} - \frac{\partial^4 u}{\partial (x')^4}, \tag{VII.4.21}$$

where $\varkappa = \nu/\sqrt{h}$. For a value of $\varkappa = 0.05$, the propagation of the solitary wave leads to dissipation due to friction, and the amplitude of the soliton decreases as shown in Figure VII.4.10. Some of the dissipation appears as a minor low-amplitude sound contribution that travels ahead of the soliton. The timescale of passage for a single soliton is of the order of 1 ms. The soliton is shown after 4.1, 22.2, and 44.9 ms. What Figure VII.4.10 teaches us is that the solitary solutions to eq. (VII.4.1) can be stable even in the presence of friction.

VII.4.11 Summary: Key Ideas of Chapter VII.4

In this chapter, we showed that under physiological conditions the biological membrane exists in a physical state that allows for the propagation of solitary pulses. These solitons have the following properties:

- The underlying differential equation, eq. (VII.4.1), contains four parameters: the speed of sound in a membrane, c_0, two parameters (p and q) describing the shape of the melting transition in the nerve membrane, and a dispersion parameter h that defines the width of the pulse. The first three parameters can directly be obtained from the thermodynamics of the membrane, while the last parameter can be adjusted to the width of the nerve pulse.

- With linear response theory, one can show that the dispersion parameter h can be written as a function of density, $h(\rho^A)$, that is solely rooted in the thermodynamics of the membrane. When this is taken into account, the wave equation with solitary solutions does not possess any free parameter. The shape, the pulse width, and the timescale of the solitary pulse are solely determined by the relaxation behavior of the membrane, i.e., by the thermodynamics of the membrane.
- The propagation velocity of the pulse corresponds to the group velocity. It is somewhat slower than the two-dimensional speed of sound in the biomembrane but of similar order. The pulse velocity depends on the total energy of the pulse.
- In the absence of viscoelastic coupling of the nerve membrane with the aqueous environment, the pulse velocity is of the order of 100 m/s, which is the approximate velocity of action potentials in myelinated motor neurons.
- The pulse propagation velocity can be considerably slower when coupling of the motion of the membrane to the associated water volume is considered. The inertial mass of surface-associated water contributes to the mass density of the layer. One obtains velocities in the range of 1–10 m/s, similar to the velocities of nonmyelinated nerves (see Section VI.5.3).
- The solitary pulse consists of a segment of gel-like membrane in a fluid environment. As a consequence, the membrane cylinder (axon) is thicker by about 1 nm of this order in the excited region, and the axon contracts. Thickness changes and nerve contractions have been found in living nerves during the propagation of the action potential (see Chapter III.4).
- A reversible change from fluid to gel and back to fluid is consistent with a transient release from and the subsequent reabsorption of heat to the nerve membrane during the pulse. The heat measured in experiments corresponds to the latent heat of the membrane melting transition. Qualitatively similar heat transfer has been found in experiments on living nerves (see Chapter III.6).
- The overall energy of the pulse calculated from the nerve compression is larger than the electrostatic energy of the nerve membrane assuming a maximum voltage of 100 mV. This is consistent with experiments (see Chapter III.6).
- One can reduce a dendritic tree to one single axon with the same properties if the sum of the circumferences of the branched dendrites is equal to the circumference of the single axon. This is reminiscent of Rall's model in electrophysiology, which provides a slightly different relation (see Section IV.2.4).
- When the membrane is locally perturbed, two solitons are generated that travel in opposite directions.
- In the presence of friction, the solitons decay in amplitude but otherwise remain stable.

It can be concluded that the electromechanical pulses obtained in the soliton theory are consistent with many features of the action potential measured in nerve axons. Thus, we will treat them as identical phenomena.

VII.5

Nerve Contraction and Pulse Trains

The solitons described in the previous sections consist of a traveling gel-like region of nerve membrane that exists in a state of increased area density. If the overall mass is conserved, this implies that the nerve must contract if its ends are free. In experiments, this is the case when a nerve is fully removed from a body. Contraction of nerves and neurons has, in fact, been found in experiments [23, 24, 25], see Section III.4.2. In the body, nerves often have local starting and endpoints, and the overall length of the nerve is constrained. For each region of increased area density, there must be a region with a decreased region in order to conserve the overall length of the neuron. We show here that under such constraints, pulse trains emerge. We also discuss the implications that nerve contraction would have on synaptic transmission of pulses.

VII.5.1 Nerve Contraction

Let us consider an axon of length L. In Figures VII.4.2 and VII.5.2, we can see that the area density change of the soliton profile is always positive when no external constraint enforces a constant length of the axon. This results in the contraction of the nerve axon in agreement with the experimental findings shown in Section VII.5.1. Let us assume that the area density changes occur along the axon without change in its radius r. The change in length, ΔL, is then given by [26]

$$\Delta L \approx \int_0^L L_0 \left(1 - \frac{\Delta\rho^A(x)}{\rho_0}\right) dx, \tag{VII.5.1}$$

integrated over the length of the neuron, where L_0 is the length of the neuron without excitation. This is shown in Figure VII.5.1. This would render the contraction independent of nerve diameter. As one can deduce from integrating the profiles in Figure VII.4.2, an axon carrying a soliton traveling with $v = 0.651 \cdot c_0$ will contract by about $\Delta L = 2$ cm, while a pulse traveling with $v = 0.8 \cdot c_0$ will contract by about $\Delta L = 5$ mm. This is possible when there is no or only a small force acting on the nerve, as is probably the case in experiments on isolated nerves. In fact, we showed in Section III.4.2 that nerves contract during an experiment on isolated nerves (see also Refs. [23, 24, 25, 27]). Thus, the prediction from the soliton theory is qualitatively in agreement with the experiment. Experimental details are given in

Physics of Nerves and Excitatory Membranes, First Edition. Thomas Heimburg.

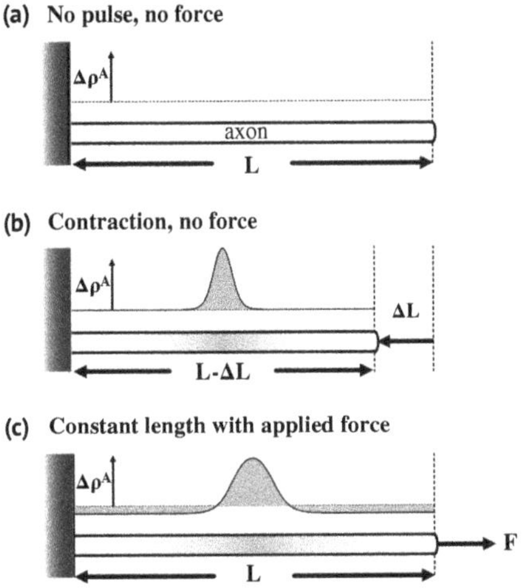

Figure VII.5.1 Contraction and the area density changes during a pulse under different constraints. (a) Schematic representation of a nerve axon fixed to a wall on one side with a total length X (A.) in the absence of a pulse. (b) The same axon in the presence of a propagating pulse without an external constraint on the length of the axon. (c) The same axon in the presence of a pulse with a constraint on the total length imposed by a constant force of the axon.

Section III.4.2. This is also important for the excitability of nerves in the presence of a force on nerves, i.e., nerve stretching. The application of stretch on nerve excitability is discussed in detail in Chapter IX.3.

VII.5.2 Pulse Trains

In the following, we show that the solitary pulse arising as a solution of eq. (VII.4.5) is a special case. In order to arrive at eq. (VII.4.5), we performed three integrations, and each integration introduces an integration constant.[1] Setting the integration constants to zero yielded exponentially localized solitons that decay to zero at long distance. We now repeat the integrations and leave the integration constants. They will allow us to define boundary conditions. One consequence of the boundary conditions is the possibility of periodic pulses and pulse trains.

In Section VII.4.2, we performed three integrations of the wave equation and let the integration constants be zero because we were searching for exponentially localized solutions that decay to zero. This is not the case anymore when we apply constraints to the length of the axon. The following derivation is similar to the one in Section VII.4.2, but now we focus on the integration constants because they contain the information on the experimental boundary condition. When we take eq. (VII.4.2) and integrate it once, we yield

$$h\frac{\partial^3}{\partial z^3}\Delta\rho^A = \left((c_0^2 - v^2) + p\Delta\rho^A + q(\Delta\rho^A)^2\right)\frac{\partial \Delta\rho^A}{\partial z}. \tag{VII.5.2}$$

1 The derivations in his section rely on Villagran-Vargas et al. (2011) [28].

Since this equation only contains first- and third-order differentials, and $\Delta\rho^A$ is a symmetric function around zero, the integration constant disappears. The next integration yields

$$h\frac{\partial^2}{\partial z^2}\Delta\rho^A = (c_0^2 - v^2)\Delta\rho^A + \frac{1}{2}p(\Delta\rho^A)^2 + \frac{1}{3}q(\Delta\rho^A)^3 + C \tag{VII.5.3}$$

with the integration constant C. Now we multiply both sides of eq. (VII.5.3) by $\partial(\Delta\rho^A)/\partial z$ and integrate once more to yield

$$h\left(\frac{\partial}{\partial z}\Delta\rho^A\right)^2 = (c_0^2 - v^2)(\Delta\rho^A)^2 + \frac{p}{3}(\Delta\rho^A)^3 + \frac{q}{6}(\Delta\rho^A)^4 + C\Delta\rho^A + V_0 \tag{VII.5.4}$$

where V_0 is a second integration constant. In this equation, there is one term that is proportional to $(\partial\Delta\rho^A/\partial z)^2$ and a second term that explicitly only depends on $\Delta\rho^A$. We call the first term a *kinetic term* and the second one a *potential term* [28]:

$$\underbrace{h\left(\frac{\partial}{\partial z}\Delta\rho^A\right)^2}_{\text{kinetic term}} \underbrace{- \left[(c_0^2 - v^2)(\Delta\rho^A)^2 + \frac{p}{3}(\Delta\rho^A)^3 + \frac{q}{6}(\Delta\rho^A)^4 + C\Delta\rho^A\right]}_{\text{potential term} = V(\Delta\rho^A)} = V_0 \tag{VII.5.5}$$

or

$$h\left(\frac{\partial}{\partial z}\Delta\rho^A\right)^2 + V(\Delta\rho^A) = V_0. \tag{VII.5.6}$$

This equation is formally equivalent to the equation of motion $\frac{1}{2}mv^2 + V(x) = \text{const.}$, where z plays the role of time and $\Delta\rho^A$ serves as a spatial coordinate. The shape of the soliton is a result of the motion in the potential $V(\Delta\rho^A)$, and all solutions of eq. (VII.5.5) can be classified according to this potential energy surface [28, 29]. This is an analogy because one of the terms depends on the square of the derivative of the density with respect to the coordinate (that contains time), while the second term is an analytic function of the density. V_0 is not relevant because it moves the potential up or down without any influence on the solutions of the differential equation. Here, we set $V_0 = 0$. The value for the integration constant C will depend on experimental constraints.

The solitons in Section VII.4.2 correspond to the case in eq. (VII.5.5), where $C = 0$. The potential representation of this case is shown in Figure VII.5.2. One can recognize a potential with two maxima and one minimum. The left-hand maximum corresponds to the unperturbed membrane with $\Delta\rho^A = 0$. Consider a ball rolling back and forth in a potential of this shape toward more positive values of the coordinate ρ^A. If the membrane is perturbed, the membrane will move in the potential until it reaches a point of larger density with the same magnitude of the potential at rest and then return to the origin. Since the slope of the potential at the origin is zero, the density pulse will approach this point infinitely slow upon return and stop at the maximum after infinite time. As a consequence, we find a singular pulse that is exponentially localized. Since during the propagation of the soliton the area density change is always positive (i.e., the membrane area of the nerve decreases), the nerve will contract as long as a pulse travels in the nerve. We discuss this in Section VII.5.1.

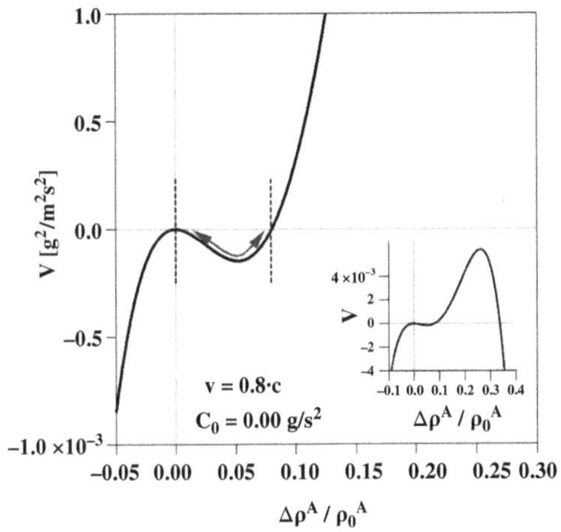

Figure VII.5.2 The potential $V(\Delta\rho^A)$ as a function of membrane density for $C = 0$ and $v = 0.8 \cdot c_0$. The constant V_0 was set to zero. The inset shows the same potential over a larger range. It displays two maxima, one of which is located at $\Delta\rho^A = 0$. In between the two maxima, a minimum is found. The motion of the soliton in the potential is indicated by the arrow. It occurs in the regime around the minimum. Adapted from [28].

VII.5.2.1 Constraint of Constant Length: Pulse Trains and Refractory Period

We showed earlier that a nerve carrying a propagating pulse contracts. However, in a physiological setting, nerves are typically within tissue and their length might not change during pulse propagation due to fixed start and end points of the neurons. The aforementioned derivations allow for treating such a situation when the integration constant C is different from zero. Figure VII.5.3 shows two examples adapted from Ref. [28]. The left side shows the potential for $C = 2.0$. There are two maxima, one at $\Delta\rho^A < 0$ and one for $\Delta\rho^A > 0$. When the membrane is close to the left-hand maximum, it will display a negative density change $\Delta\rho^A$ at baseline and a positive density change at the pulse maximum. The initial density (i.e., the baseline value of $\Delta\rho^A$) can be chosen slightly larger than density of the left-hand maximum. If there is mass conservation, the overall area underneath the baseline and above the baseline during the pulse must be equal. This generates constraints for the shape of the pulses. Since the slope of the potential is not zero at the baseline, the solution of the equation of motion becomes periodic. This is shown in Figure VII.5.4 (left) for different values of C and a fixed pulse velocity of $v = 0.8 \cdot c_0$. The initial density can be chosen such that the integral over the pulse shape is exactly zero. Under these conditions, the axon does not contract as shown in Figure VII.5.4 (right). This condition generates the pulse trains in the left-hand panel [28]. The right-hand panel of this figure shows a detail of a single pulse from a pulse train. The gray-shaded regions indicate the over- and undershoot of the density around the average density. The distance between the pulses is determined by the condition that the average change in density is zero (mass and length conservation).

Hyperpolarization and refractory period:

Figure VII.5.4 shows that in order to accommodate the condition of constant length, the nerve pulse density undershoots under the equilibrium density ρ_0^A. This corresponds to the hyperpolarization phase in the Hodgkin–Huxley model (cf., Section IV.4.2.2).

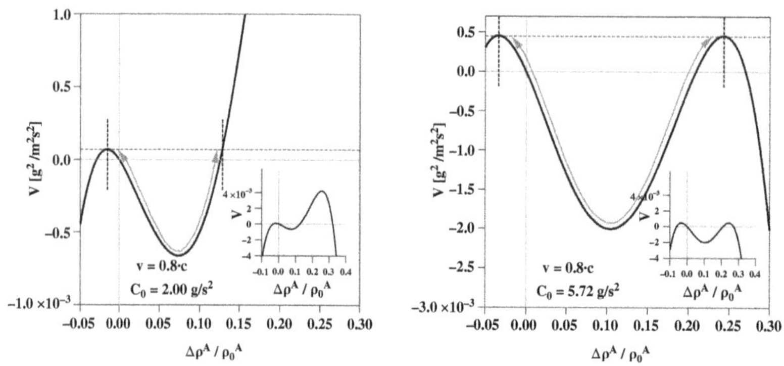

Figure VII.5.3 Influence of the integration constant C on the shape of the potential $V(\rho^A)$. Left: For $C = 2.0$ g/s^2 (left) and $v = 0.8 \cdot c_0$ g/s^2, the potential displays one maximum for $\Delta\rho^A < 0$ and a second one for $\Delta\rho^A > 0$. The second maximum displays a larger value for V. Right: For $C = 5.72$ (left) and $v = 0.8 \cdot c_0$, the potential displays one maximum for $\Delta\rho^A < 0$ and a second one for $\Delta\rho^A > 0$. Now, the two maxima display the same height. The parameters of the membrane were the same as in Figure VII.4.2. Adapted from [28].

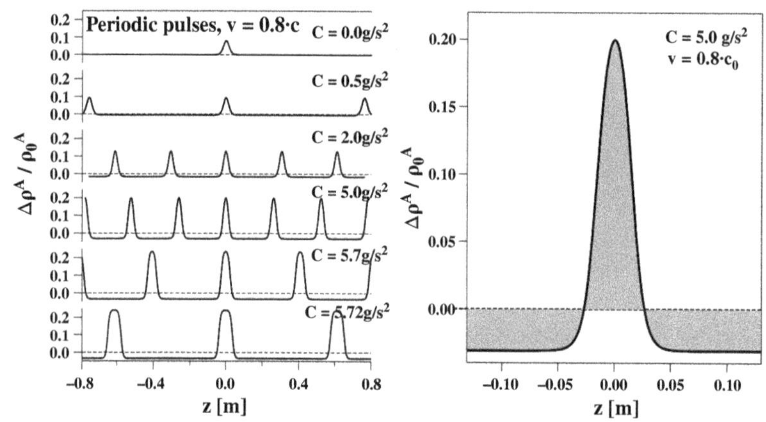

Figure VII.5.4 Left: Periodic pulses for $v = 0.8 \cdot c_0$ and different values of C. Right: Example for $C = 5.0$ g/s^2. The range of the single pulse is chosen such that the integral over the pulses is exactly zero. Under these conditions, the axon does not contract. The baseline of the pulse shape is found at a negative value for $\Delta\rho^A$. Adapted from [28].

Consecutive pulses display a minimum distance enforced by mass conservation in an axon of constant length. Pulses cannot be closer than a minimum distance that depends on C and the pulse velocity v. This is reminiscent of a refractory period (cf., Section III.2.3).

In summary, one finds the following:

1. The integration constant C depends on the experimental conditions (the length and area constraints on the nerve membrane). There exist solitary and periodic solutions for the differential equation and the constraint of constant mean density.
2. For each C and each velocity v, there is exactly one solution with a mean density change of zero. For $C = 0$ g/s^2 this is a singular pulse (soliton) and for $C > 0$ g/s^2 it is a periodic pulse.

3. There is a minimum distance between pulses dictated by the requirement that the mean density is constant, corresponding to a refractory period.
4. For each pair of pulse amplitudes and pulse distances, there is exactly one velocity.

This implies that the velocity of the pulse train depends on the amplitude and frequency of the stimulation process. It further implies that any change in the amplitude resulting from a change in the conditions (e.g., the presence of anesthetics, changes in temperature or pressure) will have a predictable consequence for the propagation velocity and the spatial distance between pulses.

VII.5.3 Thoughts About Synaptic Transmission

The finding that nerves contract during the dwell time of the action potential in a neuron (see Section III.4.2) suggests that the ends of nerves may also move when they are not constrained to a fixed position, e.g., in a synapse [26]. The synaptic gap of a chemical synapse is about 20–40 nm and that of an electrical synapse around 2–4 nm wide [30]. Let us assume that this distance does not change very much during the transmission of the pulse across the gap. The nerve pulse itself has dimensions of a few millimeters to a few centimeters in length, as can be calculated from the pulse duration and the motion in nerves (see Section III.3.5). Thus, the scale of the nerve pulse can be six or more orders of magnitude larger than the width of the gap. We have shown earlier that in the presence of solitons, nerves contract when they are not fixed at the ends. Depending on the overall length of the nerve pulse, this can lead to a contraction of hundreds of micrometers or even millimeters. Thus, the expected overall spatial motion of nerves and that of the synapse is also much larger than the width of the synaptic gap. This is schematically shown in Figure VII.5.5.

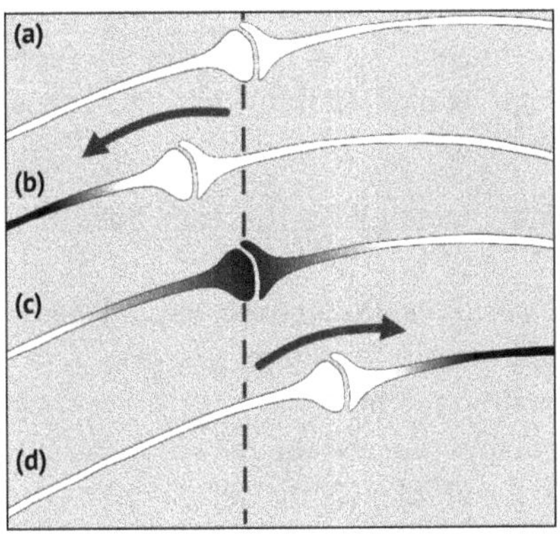

Figure VII.5.5 Schematic drawing of the motion of a soliton traveling from the left to the right side across a synapse and the resulting motion of the synapse. The dark regions represent the location of the soliton (action potential). (a) In the absence of an excitation. (b) A soliton propagates from the left side. Due to the contraction on the presynaptic side, the synapse moves to the left. (c) Motion of a soliton across the synapse. The position of the synapse is centered. (d) The soliton continues propagating to the right-hand side. Due to contraction on the postsynaptic side, the synapse moves toward the right. The overall motion is expected to be much larger than the width of the synaptic gap.

If nerves are not fixed at the end, synapses will move on scales much larger than the width of the synaptic gap. In a hydrodynamic picture of the nerve pulse, the pulse might not see a synapse at all and will cross the distance of the synaptic gap with the speed of the soliton. For a myelinated nerve with a nerve pulse velocity of 100 m/s, the distance of the chemical synaptic gap would be crossed in 0.2–0.4 ns, i.e., basically without delay. For a nonmyelinated nerve with a pulse velocity of 1–10 m/s, one would expect a passage time for the pulse maximum across the gap of 2–40 ns. The time that a gap spends in the compressed region during soliton passage corresponds to the timescale of the pulse – typically around 1–2 ms. The aforementioned considerations may suggest that synapses could be of a fast mechanical nature, i.e., that the transmission process is caused by the motion of the synapse itself.

The chemical synapse discussed in neuroscience is governed by several sequential processes, such as calcium channel open times in the millisecond range, membrane fusion processes, which should be of the order of the relaxation time of membranes (a few milliseconds at physiological temperature), and diffusion processes. Thus, synaptic transmission is expected to be relatively slow – slower than what is observed in many transmission experiments. Greengard [31] reports transmission times below 1 ms [31]. The time for a synaptic fusion event is estimated to be larger than 0.5 ms. However, synaptic calyx junctions that connect hair cells to the brain transmit signals faster than any other in the human body [32, 33], faster than can probably be understood by a chemical synapse. Therefore, a non-quantal transmission scheme for such processes was proposed [34].

It would be of high interest to test experimentally whether motion of the above kind (hundreds of micrometers to millimeters) can be observed in an active brain. It should be possible to measure this with high-resolution imaging methods.

VII.5.4 Summary: Key Ideas of Chapter VII.5

The key statement of this book is that the nerve pulse consists of an electromechanical soliton. The electromechanical solitons described here are of a region of a higher area density in which the membrane is partially moved through its melting transition. The consequences are the following:

- A nerve contracts during the dwell time of an action potential in the axon. Such contractions have been found experimentally.
- If the ends of the nerve are fixed at well-defined positions, mass conservation leads to regions of a density undershoot at some distance from the density pulse.
- To conserve the overall membrane mass, pulse trains emerge.
- This leads to phenomena reminiscent of hyperpolarization and the refractory period.
- The constant length condition might also resolve the problem of the nerve volume in a simple manner because the pulse trains would involve little or no change in the nerve volume, while contraction of nerves with loose ends also requires considerations of the change in area and volume, which imposes geometrical constraints or the formation of action potentials.
- When thinking about synaptic transmission, it seems likely that the macroscopic mechanical motion of a synapse is significantly larger than the width of the synaptic gap. It seems possible that the pulse continues on the other side of the synapse without noticing the presence of a synaptic gap. This allows for the possibility of a fast mechanical synapse.

VII.6

Excitation of Solitons

In the previous chapter, we described the emergence of solitons in cylindrical membranes, which are found in a physical state slightly above the melting transition. This chapter treats the very important problem of nerve excitability. During the propagation of the soliton, the membrane is transiently pushed through its transition. If body temperature is above the melting transition of the nerve membrane, the soliton itself consists of a gel-like region. The free energy to excite the pulse corresponds to the free energy difference between the fluid state and the gel state at physiological temperature.

VII.6.1 The Free Energy of a Solitary Pulse

We wish to estimate the free energy of soliton excitation by determining how much free energy is necessary to move the nerve membrane through its transition. We will illustrate the procedure by taking one particular soliton shown in Figure VII.6.1 as an example. This soliton was determined for the thermodynamic parameters of DPPC LUV as described in Section VII.4.2 with a pulse velocity of $v = 0.65c_0$ at a temperature of 45°C. Our goal is to estimate how much free energy must be provided to the nerve to generate this pulse. We will assume that this free energy corresponds to the free energy difference between ambient (body temperature) and the membrane state in the region of compression.

The chemical potential difference of the fluid state compared to the gel state of the membrane is given by

$$\Delta\mu_0(T) = \Delta H_0 - T\Delta S_0. \tag{VII.6.1}$$

It is linear in temperature and assumes a value of zero at the melting temperature, such that $T_m = \Delta H_0/\Delta S_0$. The Gibbs free energy difference of a membrane relative to its gel state is given by the weighted average of the chemical potential difference between fluid and gel membrane, i.e., in contrast to the chemical potential difference, it is proportional to the fraction f of fluid membrane:

$$\Delta G = f \cdot \Delta\mu_0, \tag{VII.6.2}$$

where $f(T)$ is the fluid fraction of the membrane. In Section V.2.2.3, we showed how to theoretically calculate a heat capacity profile.

Physics of Nerves and Excitatory Membranes, First Edition. Thomas Heimburg.

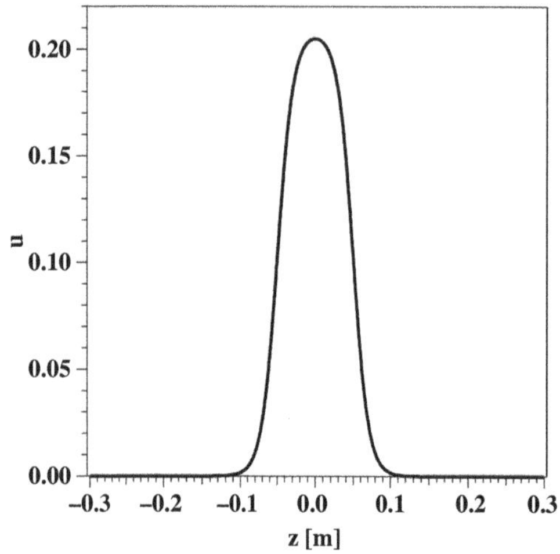

Figure VII.6.1 Soliton determined for parameters of DPPC LUV given in Section VII.4.2. For a velocity of $v = 0.65c_0$, DPPC has a transition temperature of $T_m = 41.05$ °C, and the ambient temperature is 41.05 °C, slightly above the transition temperature.

We make use of the van't Hoff description of a heat capacity profile described in section V.2.2.3. The equilibrium constant between cooperative units of fluid and gel lipids is given by

$$K = \exp\left(-n\frac{\Delta\mu(T)}{RT}\right), \tag{VII.6.3}$$

where n is the cooperative unit size. The fluid fraction of the membrane is given by $f = K/(1 + K)$. From this, we obtain the enthalpy and the heat capacity (the temperature derivative of the enthalpy) of the membrane:

$$\begin{aligned} \Delta H(T) &= \frac{K}{1+K}\cdot \Delta H_0 \\ \Delta c_p(T) &= \left(\frac{d\Delta H(T)}{dT}\right)_p = \frac{K}{(1+K)^2}\cdot n\frac{\Delta H_0^2}{RT^2}. \end{aligned} \tag{VII.6.4}$$

Further, the change of the specific area and the density of the membrane is

$$\begin{aligned} A(T) &= A_g + \frac{K}{1+K}\cdot \Delta A, \\ \rho^A(T) &= \frac{1}{A(T)}, \\ \rho_g^A &= \frac{1}{A_g}, \qquad \rho_f^A = \frac{1}{A_g+\Delta A} \equiv \rho_0^A, \\ u &\equiv \frac{\Delta\rho^A}{\rho_0^A} = \frac{1}{\rho_0^A}\left(\rho^A(T) - \rho_0^A\right), \end{aligned} \tag{VII.6.5}$$

where A_g, A_f, ρ_g, and ρ_f are the specific areas and densities of gel and fluid phases, respectively. The variable $u = \Delta\rho^A/\rho_0^A$ is the relative change in area density as a function of temperature (Figure VII.6.1).

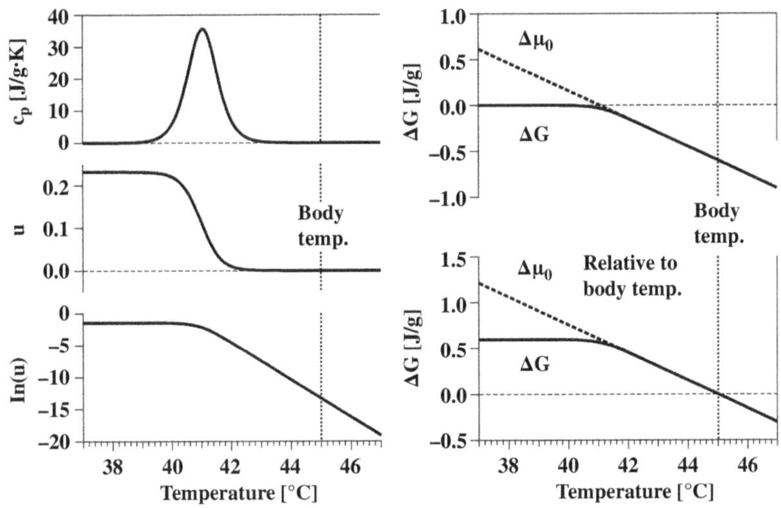

Figure VII.6.2 Left, top: The excess heat capacity calculated from a van't Hoff model with a cooperative unit size of $n = 70$. Left, center: The relative density change $\Delta u = \Delta\rho^A/\rho^A_{\text{fluid}}$. Left, bottom: The logarithm of the relative density change. Right, top: The difference in chemical potential, $\Delta\mu_0$, and Gibbs free energy density, ΔG, of the membrane lipids between fluid and gel states as a function of temperature. Right, bottom: The same graph as above. However, the differences in chemical potential and Gibbs free energy change are now relative to the value at an experimental temperature of 45 °C. The logarithm of u and the free energy change have very similar shapes and are both linear below and above the transition maximum.

Figure VII.6.2 shows the heat capacity profile calculated with the parameters of DPPC LUV (top, left). The chemical potential difference and the free energy change per gram are plotted in the top, right-hand panel. The temperature dependence of the chemical potential can be used to calculate the relative change in area density, $u = \Delta\rho^A/\rho^A_0$, as a function of temperature (Figure VII.6.2 (left, center)).

In this chapter, we wish to determine how much free energy is required in order to create a soliton. The soliton is a region of gel-like membrane that propagates in a fluid membrane at body (or experimental) temperature. Thus, we wish to determine how the density depends on the free energy difference between the solid region within the soliton and the free energy at body temperature. This is shown in Figure VII.6.2 (right, bottom). In this panel, the same free energy curve is shown as in the top panel, with the difference that the curve is shifted such that both $\Delta\mu_0$ and ΔG are zero at body temperature. This means that $\Delta\mu_0(T) \rightarrow \Delta\mu_0(T) - \Delta\mu_0(\text{body temp.})$ and $\Delta G(T) \rightarrow \Delta G(T) - \Delta G(\text{body temp.})$. In the following, we will use this definition of the free energy and the chemical potential differences.

If the free energy is defined relative to the value at body temperature, both $\Delta G(T)$ and $\Delta\mu_0(T)$ quantitatively depend on body temperature. This is shown in Figure VII.6.3 (left). The graphs are calculated for the parameters of DPPC LUV with a transition temperature of 41.05 °C. Three different values for the equilibrium temperature are shown:

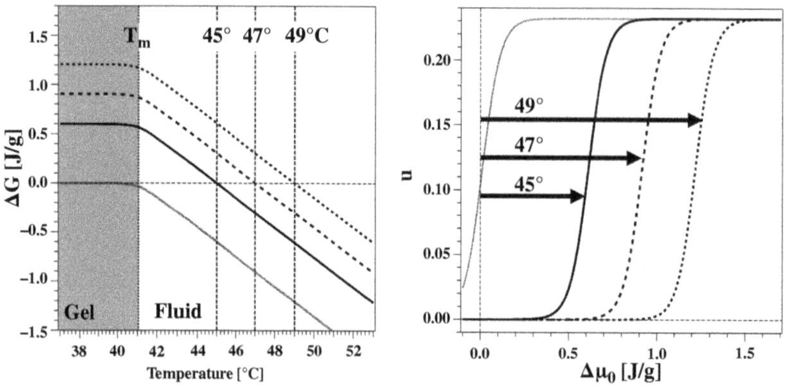

Figure VII.6.3 Left: The free energy change relative to the free energy at T_m, and three different body temperatures: 45 °C, 47 °C, and 49 °C, respectively. The higher the body temperature, the greater is the free energy needed to move the system into its gel state (shade in gray). Right: The change in density as a function of the free energy added to the system for the four different temperatures. The higher the body temperature, the more free energy has to be added to render the membrane solid (gel-like).

45 °C, 47 °C, and 49 °C, respectively.[1] One can observe that the higher the body temperature, the greater the free energy difference between that temperature and the gel membrane below T_m. This becomes even more apparent when the relative density change Δu is plotted versus the chemical potential change (Figure VII.6.3 (right)). One can recognize four sigmoidal profiles, three of which belong to body temperatures above T_m, while the fourth one corresponds to the melting temperature. The sigmoidal profiles demonstrate that in order to render a gel-like state in the membrane, a free energy threshold exists. Higher body temperature leads to a higher threshold for the excitation profile.

VII.6.2 The Threshold of Soliton Excitation

The idea of the previous section (Figure VII.6.3) is that in order to locally change the lateral density of the nerve membrane, one has to provide free energy to the membrane by perturbing the membrane with a stimulus. This stimulus could be any local change in a variable that moves the membrane into its transition range, e.g., a perturbation in voltage, local pressure, temperature, etc. The further away the membrane is from a transition, the more free energy has to be provided to excite a soliton. This is schematically shown in Figure VII.6.4, which shows how providing more free energy will lead to a threshold-like excitation pattern when the transition point is reached. This implies that changes in all intensive variables that shift the melting transition can also stimulate a nerve pulse.

It is shown in Chapter VII.8 that one finds excitation profiles of exactly this nature for pulses traveling on lipid monolayers.

---

1 The term "body temperature" makes more sense when the transitions of a biological membrane are considered. We use DPPC membranes because it is easier to demonstrate the principle. In this example, the term "body temperature" is just used in analogy and describes a temperature of the unexcited membrane above the melting temperature.

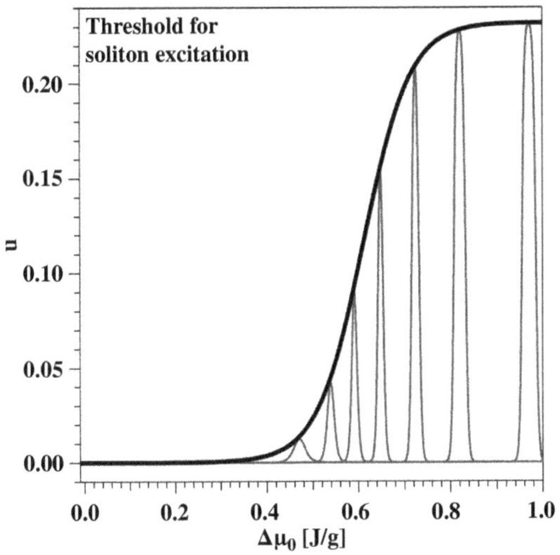

Figure VII.6.4 Schematic demonstration of the principle of an excitation threshold for solitons. The more free energy is provided, the larger the local density change (the profile is from Figure VII.6.3 at an experimental temperature of 45 °C). Solitons with corresponding maximum density were inserted into the density profile for demonstrative purposes. There is a minimum free energy that has to be provided before solitons emerge. One can define the half height of the membrane density change as the stimulation threshold.

VII.6.3 The Free Energy Density of Solitary Pulses

In Figure VII.6.2 (left, bottom), the function $\ln(u)$ is shown. It qualitatively resembles the free energy as a function of temperature. For a temperature far above the melting temperature, ($K \gg 1$), the difference of the density relative to the fluid membrane density approaches

$$\begin{aligned} u = \frac{\Delta\rho}{\rho_0} &\xrightarrow{T\gg T_m} \frac{\Delta A}{A_f}\cdot\frac{1}{K} \\ \ln(u) &\xrightarrow{T\gg T_m} const. + \frac{\Delta\mu_0}{RT}. \end{aligned} \tag{VII.6.6}$$

Likewise, for $K \gg 1$,

$$\Delta G(T) \xrightarrow{T\gg T_m} \Delta\mu_0(T). \tag{VII.6.7}$$

Thus, $\ln(u) \propto \Delta G$ above the transition. Figure VII.6.5 (left) shows that in a plot of $\ln(u)$ vs ΔG, one obtains a near-linear behavior.

Earlier, we showed that the solitons are exponentially localized (eq. (VII.4.4)). This means that away from the soliton maximum at position $z = 0$, $\ln(u) \propto |z|$, where z is the coordinate of the soliton along the axon. This is shown in Figure VII.6.5 (right). We can now use the fact that both ΔG and z are simple functions of $\ln(u)$ to calculate the free energy density as a function of the soliton coordinate. This is shown in Figure VII.6.6 for the soliton shown in Figure VII.6.1. In this diagram, the free energy is given in units of J/g. To determine the total free energy of the soliton, one has to convert this to units of J/m^2 by multiplying it with the

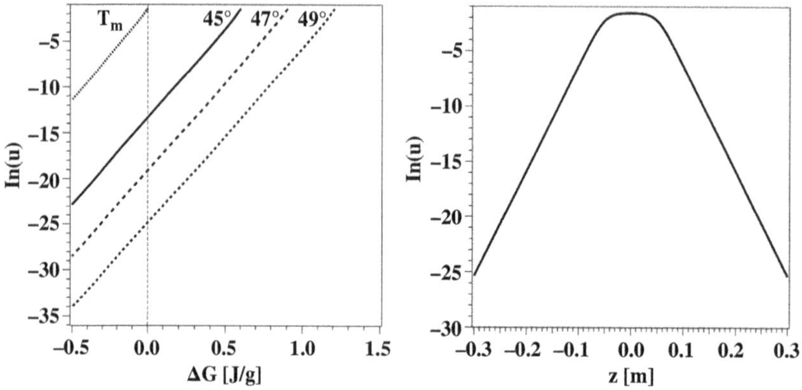

Figure VII.6.5 Left: The logarithm of the relative density change, ln(u), is roughly linear to the change in free energy, ΔG, here shown at T_m, and three different body temperatures: 45 °C, 47 °C, and 49 °C, respectively. Right: Semilogarithmic representation of the soliton profile for $v = 0.65c_0$ (Figure VII.6.1) and the parameters of DPPC. Since the solitons are exponentially localized, the semilogarithmic profiles are linear at a distance from the soliton center.

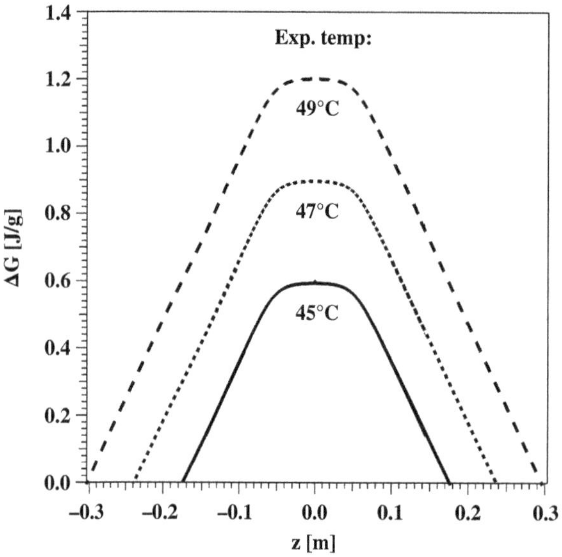

Figure VII.6.6 Since both ΔG and z are functions of ln(u) (cf. Figure VII.6.5), one can determine the free energy of the solitonic membrane. The free energy profiles for three different "body" temperatures – 45 °C, 47 °C, and 49 °C – are qualitatively similar but quantitatively different. The larger the body temperature, the higher the free energy that has to be provided to generate a soliton.

area density ($\Delta G \rightarrow \Delta G \cdot \rho_0^A$) and integrating it over the length of the soliton profile from the positions where ΔG is zero. These points correspond to the values of u at body temperature.

The message of this chapter is that the free energy necessary to excite a soliton depends on the distance of body temperature to the transition temperature, T_m. This difference can be changed by either changing body or equilibrium temperature, or by changing the transition temperature. As shown in Part V, the transition temperature depends on the other intensive variables, e.g., voltage, pH, ions, and the presence of membrane-active molecules and drugs. This implies that the excitation threshold depends on all of these variables. In Chapter IX.1, we discuss the important influence of anesthetic drugs on melting temperatures. In general, anesthetics increase the distance between body temperature and the melting temperature, and therefore the free energy of nerve excitation increases linearly with the anesthetic dose. This effect has the potential to explain the phenomenon of anesthesia quantitatively. In Chapter IX.3, we describe the influence of longitudinal tension on nerve excitability.

VII.6.4 Summary: Key Ideas of Chapter VII.6

The present chapter treats the important problem of the generation of solitons, which is directly related to nerve excitability. We show that:

- The free energy of a soliton depends on the distance between body (or equilibrium) temperature and the transition temperature of the membrane.
- The temperature enters into the expression for the free energy because the chemical potential difference between fluid and gel membranes depends linearly on the difference between temperature and transition temperature.
- The larger this temperature interval, the larger the free energy necessary to start a soliton.
- This introduces a stimulation threshold for soliton excitation.
- The excitability of the membrane can be altered by temperature changes or by everything that changes the melting temperature of the membrane. Changes in all intensive variables lead to changes in the stimulation threshold.
- This includes voltage, pressure, membrane tension, and the chemical potentials of membrane-active drugs and proteins.
- In Chapter IX.1, this is used to explain the phenomenon of general anesthesia quantitatively.

VII.7

Pulse Collisions

The soliton theory is based on hydrodynamics with conservation of momentum and energy. Likewise, solitons display inertia. Like two mechanical waves traveling in opposite directions that collide and penetrate, one might expect that solitons penetrate upon collision independent of the state of the medium during the collision. If one uses the soliton theory to describe nerves, one therefore expects that action potentials penetrate upon collision. In contrast, in the electrophysiological Hodgkin–Huxley model, one expects annihilation of action potentials due to the refractory period of the membrane in a region where a pulse has just traveled. Thus, a pulse collision experiment promises to be a good way to discriminate between the two theories.

VII.7.1 Pulses Traveling in Opposite Directions

It is generally believed that the action potential is generated in the neuron at the axon hillock [35]. Pulse propagation in the direction of the axon, the so-called orthodromic propagation, occurs from the soma toward the end of the synapses. However, in vertebrate and invertebrate nerve cells, the action potential can also be stimulated in regions remote from the axon hillock, e.g., ectopic sites located in axons or dendrites [36]. Pulse propagation in the opposite direction, called antidromic propagation, can occur. In fact, orthodromic and antidromic impulse propagation in neurons and other excitable tissues can be induced by electrical stimulation in the vicinity of the soma or in the distal part of the axon, respectively [9, 22, 37]. Thus, one can study the collision of two nerve pulses (see Figure VII.7.1). In this chapter, we compare experimental findings during pulse collisions with the prediction from the soliton theory and the Hodgkin–Huxley model.

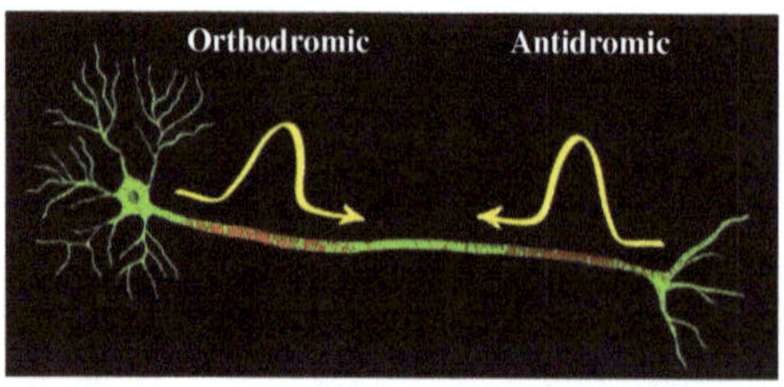

Figure VII.7.1 Schematic drawing of the orthodromic pulse traveling from the the side of the soma (from the left) and the antidromic pulse traveling from the right. The green region is the fluid nerve membrane while the red regions consist of the regions with higher area density. [37] / American Physical Society / CC BY 3.0.

Physics of Nerves and Excitatory Membranes, First Edition. Thomas Heimburg.

VII.7.2 Simulation of the Collisions of Solitons

The collision of two solitons can be determined numerically from the differential equations of the soliton theory [14]. In a linear medium, it is expected that at the position of the collision of two pulses, the amplitudes are additive in a manner that energy is conserved. The amplitude at the site of collision is higher than the amplitude of individual pulses. During the compression of the membrane in a soliton, the membrane is moved from a fluid to a gel state. However, the gel state has a rather low compressibility [10], and therefore two solitons cannot easily add their amplitude.

In Figure VII.2.1, we used a quadratic fit for the experimental c^2 versus $\Delta\rho^A$ function. In the nomenclature of Section VII.4.3, we used:

$$c^2(u) = c_0^2 \cdot \left(1 + B_1 \cdot u + B_2 \cdot u^2\right), \tag{VII.7.1}$$

where $u = \Delta\rho^A/\rho_0^A$, $B_1 = \rho_0^A/c_0^2 \cdot p$, and $B_2 = \left(\rho_0^A\right)^2/c_0^2 \cdot q$. This equation does not make any constraints on the maximum amplitudes because u values larger than the density of a gel membrane are permissible. However, as already mentioned, sound velocity and other data reveal that the membrane is nearly incompressible in the gel phase. This implies that two pulses cannot add up their amplitudes upon collision. In order to account for this near-incompressibility of the membrane in the gel phase, we modify the fit function for $c^2(u)$ to

$$c^2 = c_0^2 \cdot \left(1 + B_1 \cdot u + B_2 \cdot u^2\right) \cdot (1 + \exp(\alpha(u - u_{max}))), \tag{VII.7.2}$$

where $B_1 = -16.6$, $B_2 = 79.5$, $c_0 = 176.6$ m/s (the parameters for DPPC), and empirically $\alpha = 100$ and $u_{\max} = 0.26$. This is the dashed curve in Figure VII.7.2. It can be seen that this function fits the c^2 profile much better than the simple quadratic form. Further, the soft barrier at $u = 0.26$ prevents membrane densities that are larger than that of the gel phase membrane, which has a density of about $u = 0.246$ [10].

When simulations of colliding solitons are made using the above fit of c^2, we find the results shown in Figure VII.7.3. Pulses penetrate with the dissipation of minor amounts of energy in the form of low-amplitude sound waves. The dissipation is very small for pulse amplitudes that are not close to the maximum amplitude of the solitons (corresponding to a minimum propagation velocity of $v_{\min} = 0.649851$ for the parameters of DPPC membranes) and somewhat higher at the maximum amplitude. The loss in energy as a function of the soliton velocity (and therefore of the amplitude) is shown in Figure VII.7.4. Even if the solitons are very close to the minimum velocity, the dissipation is only a few percent and negligible for pulse velocities slightly higher than the minimum velocity. This demonstrates that the hydrodynamics of the membrane close to transitions allows for nearly complete penetration of solitons.[1]

If the action potential in nerves should in fact correspond to the electromechanical solitons described here, nerve pulses in real nerves should penetrate upon collision. Such experiments were performed by Gonzalez-Perez et al. [9, 22, 37], who in fact found that the penetration of nerve pulses is possible in living neurons and nerves. See Chapter III.8 for a description of collision experiments and technical details.

---

1 The fact that there is residual dissipation upon collision means that we describe solitary pulses rather than true solitons that penetrate without dissipation.

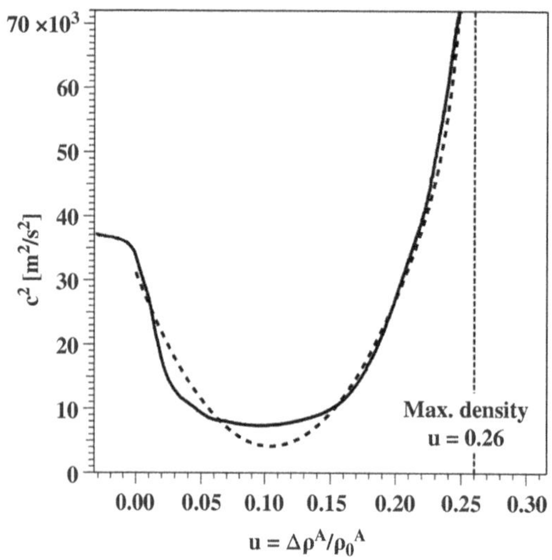

Figure VII.7.2 The modified fit to the sound velocity profile according to eq. (VII.7.2) describes the experimental profile better at higher density than the original quadratic fit and does not allow for densities large than $u = \approx 0.26$ due to a soft barrier. Adapted from [14].

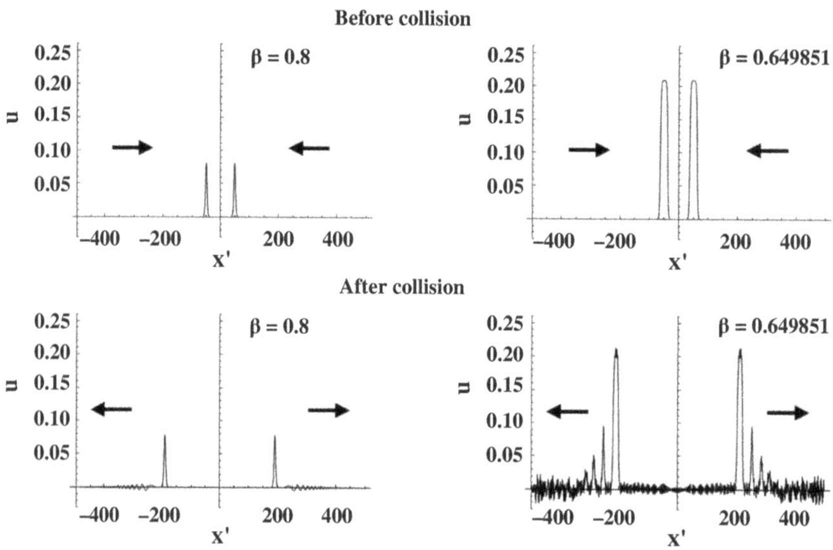

Figure VII.7.3 Colliding solitons before (top) and after the collision (bottom) for two different soliton amplitudes and velocities. In this figure, $u = \Delta\rho^A/\rho_0^A$, $x' = c_0/\sqrt{h} \cdot x$ x, and $\beta = v/c_0$. Adapted from [14].

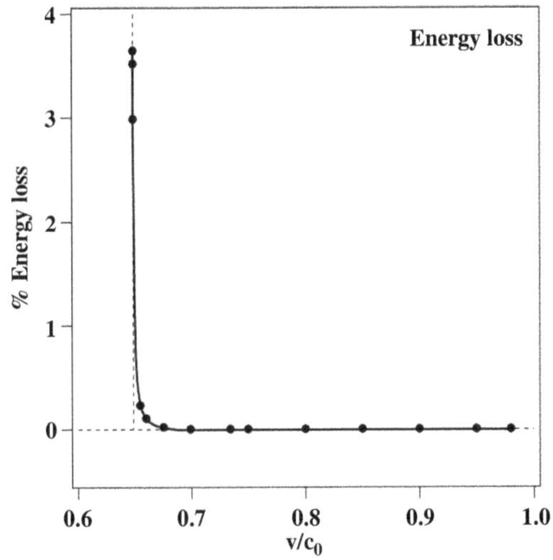

Figure VII.7.4 Energy loss after pulse collisions for solitons of different relative velocities v/c_0, where c_0 is the 2D sound velocity in the membrane at equilibrium. Only for minimum velocity soliton, some minor dissipation was found. Adapted from [14].

VII.7.3 Summary: Key Ideas of Chapter VII.7

In this chapter, we discuss the collision of nerve pulses that travel in opposite direction in a neuron. The soliton theory and the Hodgkin–Huxley model lead to different predictions about the outcome of a collision experiment.

- The soliton theory is based on hydrodynamics and considers the pulse as an isentropic phenomenon. Due to energy and momentum conservation, it is expected the two pulses penetrate each other upon collision.
- Simulations on membrane cylinders using the differential equations of the soliton theory in fact show that the solitons pass through each other with only very minor dissipation.
- The Hodgkin–Huxley model for the action potential rather expects that colliding nerve pulses annihilate due to the refractory period following the passage of a pulse. According to this model, the nerve is not excitable for some time period after the pulse, and thus, pulses cannot penetrate.
- Experiments on living nerves (earthworm and lobster connectives) demonstrate that penetration of nerve pulses is in fact possible (Chapter III.8).

VII.8

Pulses on Monolayers

The soliton theory has a lot of potential for understanding the phenomenology of events during nerve pulse propagation. It reproduces the magnitude of the thickness change (Section VII.4.6) and the pulse velocity (Section VII.4.2); it predicts nerve contraction (Section VII.5.1), the reversible heat changes (Section VII.4.7), and has the potential to explain the phenomenon of anesthesia as outlined in Chapter IX.1. All properties emerge from the thermodynamics of the membrane and its fluctuations. The theory does not require special molecular properties. All phenomena are represented in the macroscopic properties of the nerve membrane. This is possible since the nerve pulse itself is macroscopic (Section III.3.5). Therefore, everything derived in the previous chapters will also be applicable to membrane cylinders without proteins as long as a phase transition in the membrane is present slightly below equilibrium temperature. It would be a proof of principle if nerve-like phenomena could be investigated in protein-free membranes. A model system could be the membrane tethers. Such tethers are cylindrical membranes that display similar dimensions to nerves. These tethers are found in the melting regime of a DPPC:DLPC lipid mixture. Different brightnesses represent fluid and gel domains that coexist. Thus, these tethers support phase behavior and phase coexistence. Since the membrane is in its transition regime and at equilibrium, the gel-like domains do not propagate. A propagating signal is expected when a tether in the fluid regime is locally excited (see Chapter VII.6).

In order to be able to observe solitons, the tether must be significantly longer than the pulse itself. Nerve pulses, however, have the dimensions of several millimeters to a few centimeters (see Section III.3.5). This is 2–4 orders of magnitude larger than the field of view in a microscope. Presently, no method is known to generate membrane tethers that display a length of several centimeters, i.e., tethers that are longer than a soliton. Further, the field of view in a microscope is too small for soliton detection. Thus, presently, it is not feasible to measure soliton propagation in artificial membrane cylinders. However, one can generate pulses on lipid monolayers. Monolayer troughs have dimensions of several tens of centimeters and can easily be enlarged to meter dimensions. We show in the following that solitons (or pulse-like density excitations) exist in such systems that obey the physical principles outlined in Section VII.4.2.

Physics of Nerves and Excitatory Membranes, First Edition. Thomas Heimburg.

VII.8.1 Pressure and Voltage Pulses

Electromechanical pulse propagation on monolayers was first demonstrated by Griesbauer and Schneider [40, 41]. Their experiment is shown in Figure VII.8.1. On monolayer troughs, one does not adjust temperature to fix a particular membrane state (such as in calorimetry) but rather the lateral pressure Π given in [mN/m]. The pressure of the lipid layer is adjusted by a teflon barrier such that one is in the liquid-expanded (LE) phase close to the transition. The LE phase corresponds to the fluid state in a membrane. A pulse is exited by adding a little droplet of solvent at the indicated position. The pulse travels along the dotted line around a barrier (indicated in the figure). By inserting more barriers into the trough, one could, in principle, create very long pulse trajectories. The overall monolayer pressure is constant. However, the pressure can vary locally by a small amount when a pulse passes by a pressure sensor. In Figure VII.8.1, the pulse first passes by pressure sensor 2 and after some time delay dictated by the pulse velocity it arrives at pressure sensor 1. In between the two pressure sensors, a Kelvin probe is placed that measures the trans-monolayer voltage via a vibrating capacitor. The region where the pulse is recorded can be considered a monolayer cable that is analogous to a bilayer cylinder. Thus, it provides a model for the nerve. Of course, it is a synthetic system, and one has to adjust the conditions to see pulse-like behavior. But we have argued in previous chapters that the biological nerve is in a similar state as a monolayer in the liquid-expanded regime. Thus, the monolayer experiment can be used as a proof-of-principle.

The results in Figure VII.8.2 show that pulses can propagate in monolayers. They do not significantly change amplitude and shape when traveling twice the distance. The distance between the two sensors is 0.145 m, and the time difference between the signals recorded at

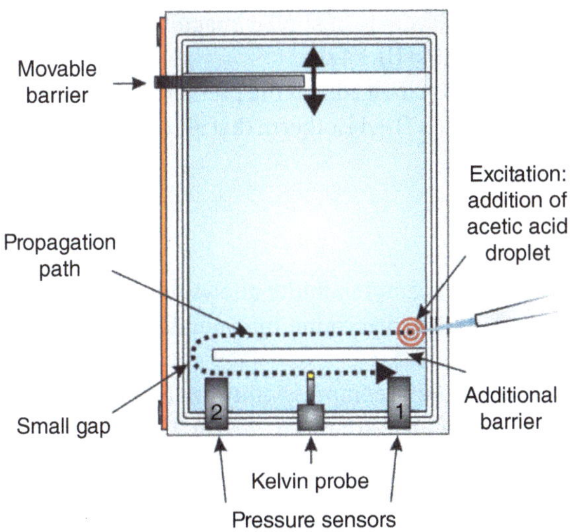

Figure VII.8.1 Monolayer trough used by Griesbauer and Schneider [40, 41]. On the top, one can see a movable barrier that is used to adjust a well-defined pressure in the lipid film such that it is found close to a transition. A pulse is exited with a small solvent droplet on the lower right side. The pulses travels along the dotted line. The local pressure is recorded by two pressure sensors at the two indicated positions. This allows to determine the pulse velocity. Between the two pressure sensors, a voltage sensor (Kelvin probe) is placed. From [41].

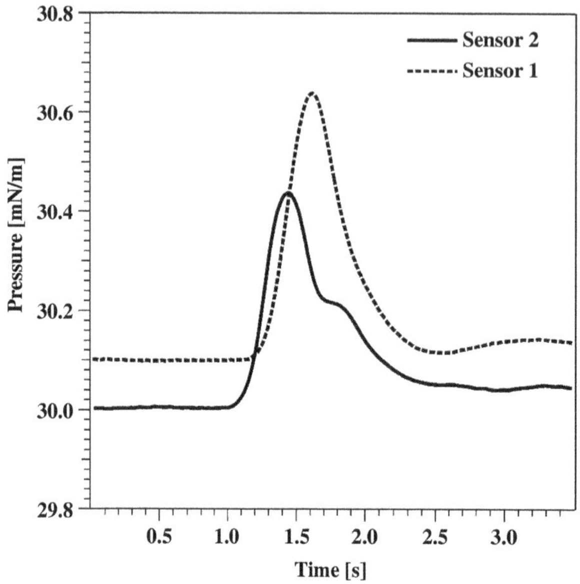

Figure VII.8.2 The pulse in the monolayer recorded at the two positions is shown. One recognizes a delay between the two signals of 0.18 s. Otherwise, the amplitude and shape are approximately conserved. For the positions of the pressure sensors, see Figure VII.8.1. This experiment demonstrates that soliton-like pulses can be propagated in protein-free lipid layers when the conditions are chosen accordingly. Adapted from [40].

the positions of the two sensors is about 0.178 s. This yields a pulse velocity of $v = 0.81$ m/s. The half-width of the pulse at sensor 1 is about 0.34 s corresponding to a pulse length of 28 cm. Thus, the propagation velocity on monolayers is of similar magnitude as in some nonmyelinated nerves (see Section III.3.1 and Table III.3.1).

Upon compression of a monolayer, the area A is changed and the pressure Π is recorded. Upon continuously changing the area, one obtains a Π–A isotherm that allows to determine the lateral isothermal compressibility of the film

$$\kappa_T^A(\Pi) = -\frac{1}{A}\left(\frac{\partial A}{\partial \Pi}\right)_T, \qquad \text{(VII.8.1)}$$

which is shown in Figure VII.8.3 (left, top). This compressibility allows one to determine the heat capacity in the same ways as it allowed to determine the lateral compressibility from the heat capacity for membranes (Chapter VI.4).

Figure VII.8.3 (left, top) shows the isothermal lateral compressibility as a function of pressure for a monolayer of DPPC. It displays a maximum at around 10 mN/m. This corresponds to the transition maximum. In Figure VII.8.3 (right, top), the pressure recording of the pulse is shown. The pulse velocity as a function of the overall monolayer pressure is shown in Figure VII.8.3 (left, bottom). The pulse velocity is of the order of 0.4–1 m/s, i.e., of similar order to the velocity in the femoral nerves of grasshoppers (0.7 m/s, [42]). The low value of the pulse velocity indicates that for single-lipid layers, the interfacial water couples to the membrane as described in Section VII.2.2. Consistent with lipid bilayers (cf. Figure VI.5.3), the pulse velocity displays a shallow minimum in the transition range. Figure VII.8.3 (right,

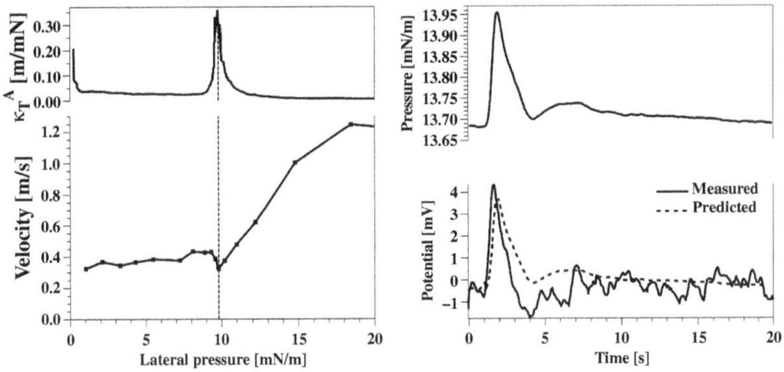

Figure VII.8.3 Left, top: The isothermal compressibility of a DPPC monolayer as a function of lateral pressure. Left, bottom: The pulse propagation velocity as a function of pressure displays a minimum at the transition pressure. Right, top: A propagating pulse monitored by lateral pressure changes. Right, bottom: Voltage changes during the pulse. The dotted line indicates the voltage change predicted from the lateral pressure change. Adapted from [41].

bottom) shows the voltage recording during the pulse, which is roughly proportional to the pressure recording. For comparison, the dashed line indicates the hypothetical voltage response if it were proportional to the pressure. The experimental result of the voltage sensor displays a similar shape to the mechanical pulse. This experiment shows that a mechanical pulse on a monolayer is correlated with an electrical pulse of similar shape. Thus, pulse propagation on monolayers is of electromechanical nature.

It is clear from the results in this chapter that electromechanical pulses can travel in linear monolayer arrays in the complete absence of proteins under similar conditions than required for solitons in bilayers. The results are in very good agreement with the soliton theory.

VII.8.2 Stimulation Thresholds

In the previous section, it was shown that soliton-like electromechanical pulses can travel on monolayers. In Chapter VII.6 (Sections VII.6.1 and VII.6.2), we showed that the excitability of the membrane depends on the distance of the membrane from a transition point. We show here that these predictions of the soliton theory also hold for linear arrays of monolayer films.

Shrivastava and Schneider [43] performed a beautiful experiment similar to the one described in the previous section. However, in this experiment, the pulse was excited mechanically by a moving razor blade, and it was recorded by Förster resonance energy transfer (FRET) between two fluorescent lipid markers (NBD-PE and Texas-Red DHPE) to monitor changes in density and lipid orientation during a propagating pulse. The authors recorded the fluorescence intensity at two wavelengths, 535 nm and 605 nm, respectively. The FRET parameter θ is defined as

$$\theta = \frac{I_{535}}{I_{605}}. \tag{VII.8.2}$$

As a brief calculation shows, the relative change in this parameter, $\Delta\theta/\theta$, is given by

$$\frac{\Delta\theta}{\theta} = \frac{\Delta I_{535}}{I_{535}} - \frac{\Delta I_{605}}{I_{605}}. \qquad \text{(VII.8.3)}$$

The value of this function was used to monitor the changes in lipid state of the monolayer. Pulses were stimulated by a moving razor blade driven by a piezoelectric device. The stimulation amplitude A is indicated on an arbitrary scale between 0 and 1, representing the strength of excitation generated by the moving blade. Figure VII.8.4 (left) shows three equilibrium conditions at which the monolayer is excited. One is in the LE (fluid) regime of the monolayer, one is close to the transition, and one is in the LC (gel) region of the film. The monolayer was excited by a constant stimulus. The right-hand side of Figure VII.8.4 shows that pulses could be excited close to the transition but not when the monolayer state was far away from the transition in the LE regime, or when it was in the LC regime. This is expected from the soliton theory; see the considerations about the excitation threshold in Chapter VII.6. Figure VII.6.3 showed that in order to obtain a situation where the membrane can be moved from fluid to solid and back, the distance of the membrane state from the transition point is essential. In the soliton theory, it is not expected that a pulse can be excited by membrane compression if the membrane is in its gel state because the membrane cannot be compressed further. This behavior is reflected in regimes (i), (ii), and (iii) in Figure VII.8.4. Note that the stimulus is constant.

If the monolayer is in state (i) (cf. Figure VII.8.4), it is possible to excite a pulse if the stimulus is increased. This is shown schematically in Figure VII.8.5 (left). A small stimulus may not be sufficient to move the monolayer into its transition regime, while a larger stimulus may succeed in doing so. The right-hand panel experimentally shows that indeed the

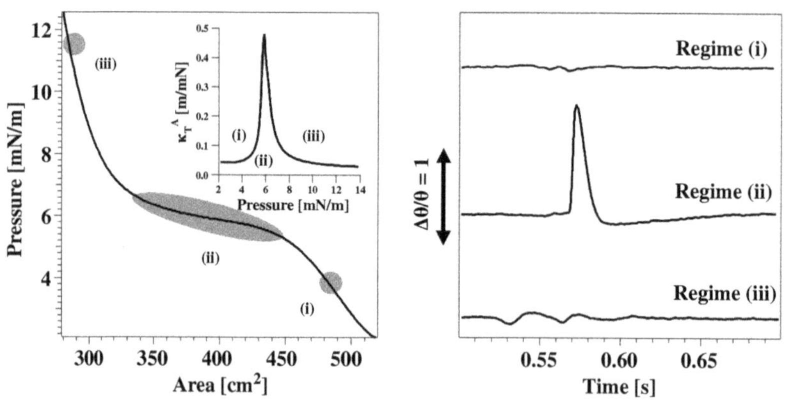

Figure VII.8.4 Pulse generation in a DPPC monolayer for constant stimulus, but different lateral pressures demonstrate that pulse excitability is a function of the monolayer state. Left: Three regimes on the pressure-area isotherm have been marked by gray shades: In the LE phase (regime i), close to the transition (regime ii), and in the liquid condensed region (regime iii). The inset shows the compressibility of the monolayer as a function of pressure. Right: FRET signal after stimulation of the monolayer. Only close to the transition a pulse could be generated, as expected from the soliton theory. Note how remarkably similar the pulse shape is to an action potential in the giant squid axon (cf., Figure IV.1.1). Far away from the transition, the monolayer cannot be excited. Adapted from [43].

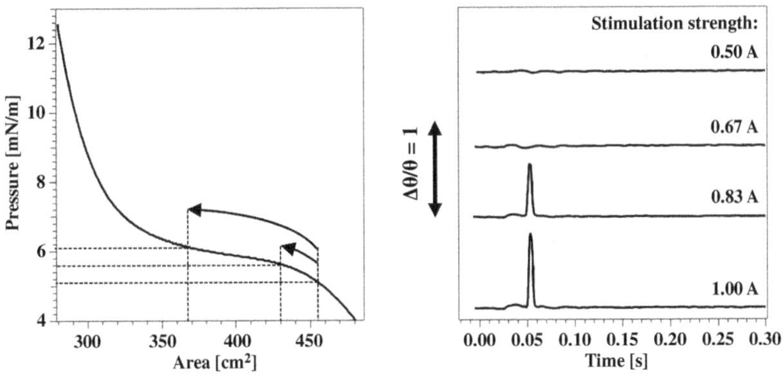

Figure VII.8.5 The same experiment as described in Figure VII.8.4, but with a constant overall monolayer pressure slightly above the transition point. What is varied is the mechanical stimulation amplitude. Left: If the membrane pressure is slightly above the transition, a small perturbation does not move the monolayer into the transition regime, resulting in a very small response to a perturbation. However, a larger perturbation can move the system through its transition, resulting in a larger response. Right: The relative change in FRET intensity is given as a function of time for different stimulation amplitudes *A*. At small stimulations, no pulse was generated. At higher stimulation amplitudes, a propagating pulse was excited. Adapted from [43].

monolayer supports pulses when the amplitude of the stimulus is increased (Figure VII.8.5). This is exactly the behavior expected from the soliton theory, which is shown in Figure VII.6.4.

Summarizing, the findings on monolayers are wonderfully consistent with the predictions from the soliton theory. One finds a regime where pulses can be excited, and the existence of a stimulation threshold. The monolayer experiments further support the notion that the pulses are electromechanical because the pulse can be observed by both pressure and voltage sensors. The monolayer experiments also show that the phenomenology of experiments on action potentials in nerves can be obtained in experimental systems that do not contain proteins.

VII.8.3 Summary: Key Ideas of Chapter VII.8

In this chapter, we showed that in a monolayer free of proteins that is found slightly above the transition in the LE phase, propagating pulses can be excited by perturbations of the lipid film. These pulses display all the properties expected for solitons in a membrane cylinder, as outlined in Chapters VII.4 and VII.6. It can be demonstrated that:

- Monolayers support mechanical pulses when being in the LE phase close to the LE-LC transition.
- The pulses have an electrical component roughly proportional to the mechanical changes due to the different polarization of lipid head groups in the two main monolayer phases. Thus, the pulses are electromechanical.
- The pulse velocity is of the order of 0.4–1 m/s, similar to the velocity of the femoral nerve of grasshoppers (0.7 m/s).
- Pulses can easily be excited close to the LE-LC transition in the monolayer.

- Further away from the transition, a larger stimulus is needed in order to excite the pulse in the monolayer.
- This provides a stimulation threshold analogous to the threshold behavior described for solitons in Section VII.6.4 and for living nerves in Section III.2.2.
- The monolayers provide a beautiful model system for the propagation of action potential-like pulses in the complete absence of proteins. The monolayer system provides an experimental proof-of-principle for all aspects of soliton theory.

It has to be stated unambiguously that electromechanical pulses can propagate in systems that are in a fluid state similar to that of a nerve membrane under physiological conditions. The properties of the pulses are exclusively dictated by the macroscopic thermodynamics of the system. Changes in the position of the transitions in the system change the excitability of the monolayer pulses. The aforementioned statements are independent of the molecular detail of the lipid layer. This means that the above statements are still valid when the lipid layers display a complex lipid composition and when they contain proteins. Thus, they will also be applicable to protein-containing biological membranes slightly above their melting transition.

Part VIII

Channels

The topic of nerve pulse transduction is very closely linked to the occurrence of currents across the nerve membrane. The Hodgkin–Huxley model requires these currents to explain the nerve pulse. In single membrane experiments, the currents often appear as quantized on-off events. Conventionally, these events are attributed to integral membrane proteins called "ion channels." There have been many books on channel proteins (e.g., [1]), and the topic has been widely covered. An important element of the electrophysiological rationale is that there exist channel proteins that are susceptible to intensive thermodynamic variables, e.g., voltage-gated channels, mechanoreceptors, and temperature-sensing proteins. Such, protein channels are described in Chapter IV.5.

However, there exists another class of membrane channels that is intrinsically sensitive to the intensive thermodynamic variables. These channels are pores in the lipid membrane (see Figure VIII.1). In this part of the book, we focus on the emergence of channels as a thermodynamic property of the lipid membrane. We will see that every change of a variable that leads to soliton excitation in nerves (see Part VII) will also lead to the appearance of channel-like conduction events in the lipid membrane. These channels are controlled by temperature, pressure, voltage, and other variables – and they are therefore temperature-sensing, mechanosensitive, and voltage-gated.

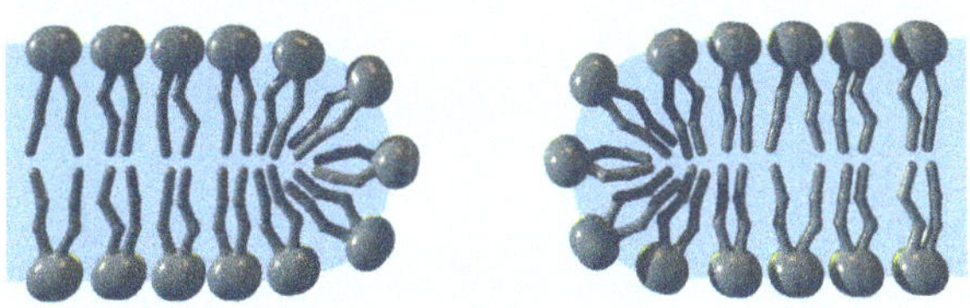

Figure VIII.1 Schematic drawing of a pore in a lipid membrane.

The membrane channels fit perfectly into an overall thermodynamic picture of excitable membranes. The lipid membrane channels behave in many aspects very similarly to protein channels, and their appearance is very similar or even indistinguishable. At the end of this part, we discuss the possibility that proteins can act as catalysts for lipid membrane channels.

VIII.1

The Permeability of Lipid Membranes

The interpretation of electrophysiological measurements of biomembranes relies on assumptions. Channel events such as those shown in Figure IV.5.2 are mostly measured by pipettes used in patch-clamp experiment. A pipette tip is schematically shown in Figure VIII.1.1. Such pipettes typically display apertures at the tip with diameters on the order of 0.5–1 μm. The diameter of a channel protein is of the order of 5 nm. Thus, about 10,000–40,000 proteins fit into the area of the patch membrane. Calculated pore diameters in the center of the protein are below 1 nm, i.e., a millionth of the membrane patch of the pipette. If one single current trace with opening and closing is observed and its conductance is determined, one tacitly makes the assumption that 99.9999% of the membrane does not significantly contribute to the signal. In particular, it is assumed that the properties of the lipid membrane are those of an insulator that does not display any signature resembling that of protein channels. We will show below that this is not correct. Lipid membranes by themselves can display a large conductance if the membrane is close to a transition. Further, they display channel-like conduction events very similar to, or even indistinguishable from, those attributed to channel proteins.

VIII.1.1 Permeability of Membranes Close to Transitions

We first show that the lipid membrane is not an inert insulator. In Section VI.4.3, we suggested that close to melting transitions, pore formation is facilitated, which will have an influence on the membrane permeability to ions and small molecules. This will be derived in more detail below in Section VIII.1.2.

There are various early measurements demonstrating significant changes in lipid membrane conductance close to melting transitions. To our knowledge, the first one is by Papahadjopoulos and collaborators [2]. They measured the conductance of an artificial membrane made of DPPG for the sodium isotope $^{22}Na^+$ (Figure VIII.1.2 (left)). It was found that the permeability maximum was close to the melting temperature of DPPG around 41 °C. Many scientists were interested in these findings, and there exists quite some theory attributing these changes to pore formation in the membrane. This will be briefly reviewed in the next section. There were several more experimental confirmations of the maximum in membrane permeability at the transition temperature, e.g., by Sabra and Mouritsen [3] who measured the permeability of DMPC membranes ($T_m \approx 24\,°C$) for Co^{2+} ions

Physics of Nerves and Excitatory Membranes, First Edition. Thomas Heimburg.

Figure VIII.1.1 Schematic drawing of a patch pipette, the patched membrane, and a single-channel protein. The area of the protein is much smaller than the area of the membrane patch.

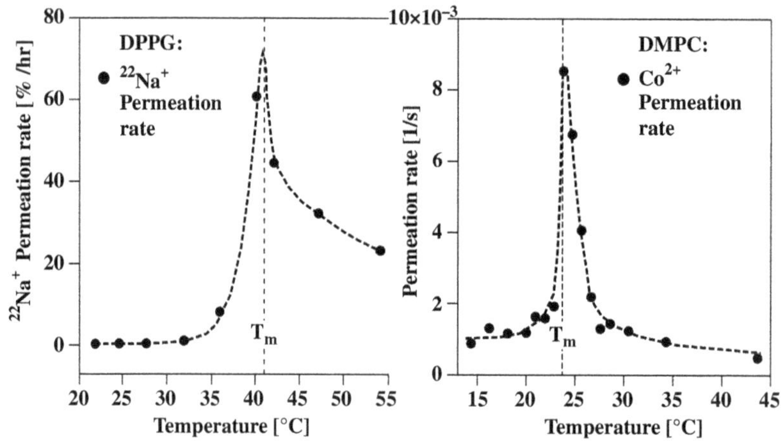

Figure VIII.1.2 Lipid membrane permeability close to transitions. Left: Permeability of a DPPG membrane (T_m = 41 °C) for $^{22}Na^+$. Adapted from [2]. Right: Time constant for Co^{2+} permeation through DMPC membranes with a melting temperature of 24 °C. Adapted from [3].

(Figure VIII.1.2 (right)). The authors further showed that drugs influencing the melting behavior also influence the temperature dependence of the membrane permeability (shown in Section VIII.5.1).

Figure VIII.1.3 shows another experiment where the leakage of the fluorescence marker rhodamine 6G (R6G) through DPPC:DPPG = 95:5 membranes is plotted versus the temperature [4]. It is compared to the excess heat capacity profile. The empirical finding is that the rate constant for the release of the fluorophore is proportional to the heat capacity within the experimental accuracy. Thus, the heat capacity maximum coincides with the permeability maximum. The diameter of an R6G molecule is of the order of 1 nm. The experiment indicates that there is a finite probability that large membrane pores are formed that are big enough to be permeable to a molecule of the size of a fluorophore.

Figure VIII.1.4 shows the conductivity of a black lipid membrane (BLM) made of a D15PC/DOPC (95:5) mixture (adapted from [5]). It is compared to the heat capacity profile

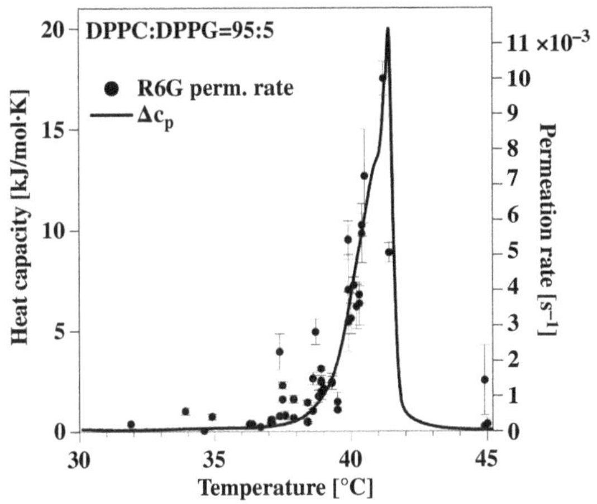

Figure VIII.1.3 Permeability of a DPPG:DPPG = 95:5 membrane for the fluorescence marker R6G. Within experimental error, the permeation rate is proportional to the excess heat capacity. Right: The same lipid mixture without and with the anesthetic octanol. Adapted from [4]. Slightly below the transition maximum, octanol increases the permeation rate of R6G. Adapted from [4].

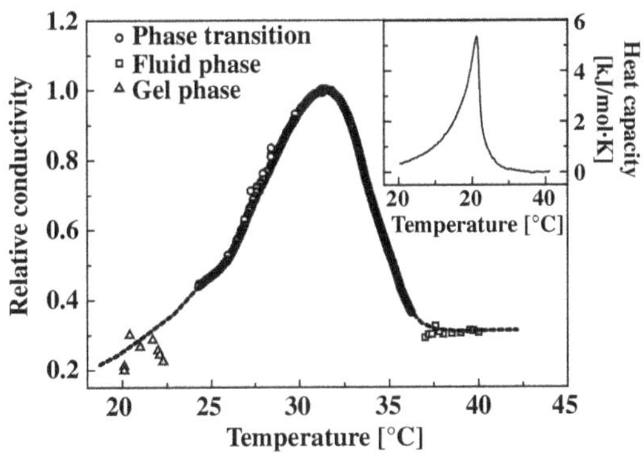

Figure VIII.1.4 Conductivity of a black lipid membrane made of D15PC/DOPC (95:5) mixture in 200 mM NaCl at 400 mV. Adapted from [5].

of the same mixture (see inset). Again, the conductivity maximum and heat capacity maximum coincide at a temperature of around 31 °C.

We can conclude from these experiments that

- lipid membranes are not generally insulators
- close to transitions, they display a conductance for ions and even for larger molecules such as fluorescence markers
- as a consequence, not all conductance of biomembranes can be due to single-molecule channels if one is close to transitions.

VIII.1.2 Theoretical Considerations

In order to create a pore of area ΔA in the lipid membrane, lipids have to be rearranged due to thermal fluctuations. The increase of the membrane permeability in the transition regime is therefore the consequence of the large fluctuations associated with the transition. Nagle and Scott [6] expanded the permeability $\mathcal{P}$ as a Taylor series

$$\mathcal{P} = \mathcal{P}_0 + C_1 \langle \Delta A \rangle + C_2 \langle \Delta A^2 \rangle + \ldots = \mathcal{P}_0 + C_2 \langle \Delta A^2 \rangle + \ldots \,, \qquad \text{(VIII.1.1)}$$

where $\langle \Delta A \rangle$ is the mean deviation of the area from the equilibrium value. Since positive and negative area fluctuations ΔA display equal likelihood, i.e., $\langle \Delta A \rangle = 0$, only the quadratic term remains. From eq. (VI.4.6), it follows that the lateral compressibility κ_T^A is given by

$$\kappa_T^A = \frac{\langle \Delta A^2 \rangle}{\langle A \rangle kT},$$

which is proportional to the second term in eq. (VIII.1.1). Therefore,

$$\mathcal{P} = \mathcal{P}_0 + C_2 \cdot \kappa_T^A + \ldots \,. \qquad \text{(VIII.1.2)}$$

In Section VI.4.1, it was derived that compressibility changes are proportional to the excess heat capacity. Therefore, we finally obtain:

$$\mathcal{P} = \mathcal{P}_0 + \alpha \cdot \Delta c_p + \ldots \,, \qquad \text{(VIII.1.3)}$$

where α is a constant.

As a consequence, one expects that there exist pores in the lipid membrane that lead to an increase in permeability. This change in the permeability is (roughly) proportional to the excess heat capacity. This is consistent with the experimental findings in Section VIII.1.1. Since the pore opening is considered a thermal fluctuation, one further expects that the fluctuation lifetimes and the mean channel open times are directly related to each other.

According to Glaser et al. [7], there may exist two kinds of pores in lipid membranes: the hydrophobic pore, in which the lipid chains are exposed to water, and the hydrophilic pore, in which one finds a rearrangement of the lipid head groups in a manner that avoids contact of lipid chains with water. This is shown schematically in Figure VIII.1.5 (top). The generation of such pores is assumed to be voltage-dependent. The emergence of pores in membranes induced by voltage pulses is a well-studied phenomenon in the field of membrane electroporation. It focuses on drug delivery facilitated by short voltage pulses. The voltage-induced transient pores allow for the transport of drugs into cells via the plasma membrane [8]. For example, it is well established that voltage pulses in fact facilitate incorporation of cytostatic drugs into cells [9, 10]. Such pores can also be seen in molecular dynamics simulations of a POPC membrane [11]. Figure VIII.1.5 (bottom) shows two steps in the pore formation process in POPC membranes. First, a transient hydrophobic pore is created, in which water molecules align in a file through the membrane. Ions can diffuse along this defect line. In a second step, a larger hydrophilic pore forms with a diameter on the order of 1 nm, similar to the area of a single lipid.

The hydrophilic pore was also simulated in 2003 in a DOPC membrane under similar conditions by Tieleman and collaborators ([13], see Figure VIII.1.6). Due to limitations of the

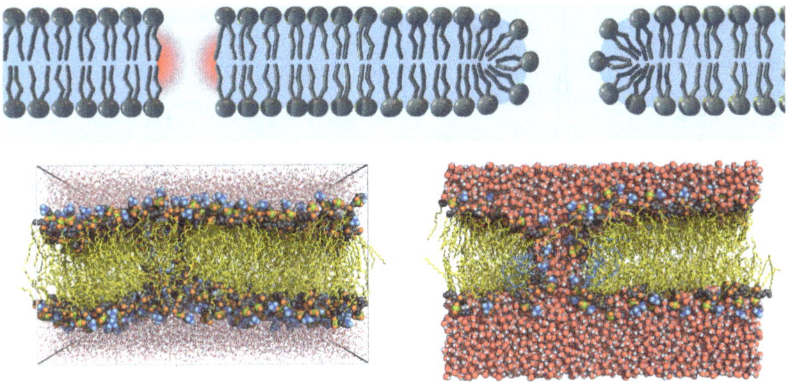

Figure VIII.1.5 Top: Schematic drawing of the emergence of hydrophobic and hydrophilic pores. [12]/with permission of American Chemical Society. Bottom: Molecular dynamics simulation of a POPC membrane at an electric field strength of E = 0.5 V/nm after 900 ps (left) and after 50 ns subsequent equilibration at a decreased field strength of E = 0.04 V/nm (right). From [11]/with permission of Elsevier.

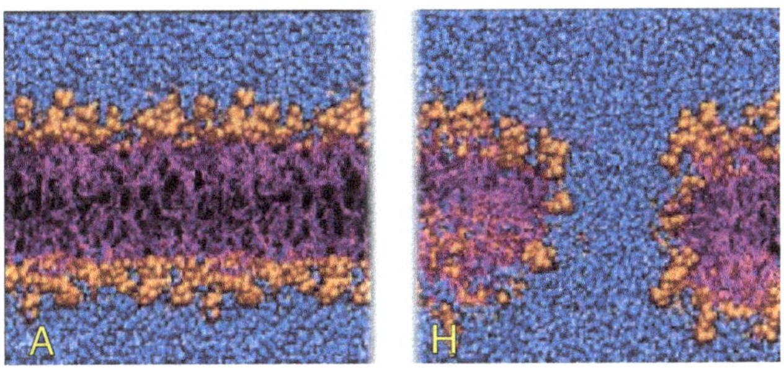

Figure VIII.1.6 Molecular dynamics simulation of the formation of a hydrophilic pore in a DOPC membrane at a field of 0.5 V/nm after a simulation time of 2.4 ns (left) and 3.55 ns (right). [13]/with permission of American Chemical Society.

MD simulation (in particular, the simulation times of only a few nanoseconds), the authors of Refs. [11, 13] used field strengths of 0.5 V/nm, which correspond to a transmembrane voltage of about 2.5 V for a complete membrane. This is about an order of magnitude higher than typical voltages in electrophysiological experiments, but of similar order than typical nerve stimulation voltages.

In Wodzinska et al. [14], a coarse-grained statistical thermodynamics model was presented to calculate the emergence of channels in a lipid matrix. In a Monte Carlo simulation, a lipid was removed from a site in the matrix and moved to the edge. Here, a pore has the size of one lipid molecule, i.e., about 0.7 nm^2 similar to the size of the pore in the MD simulations. This is also similar to the pore sizes estimated from experiments when assuming that the pore is filled with water [14]. The elementary Monte Carlo step that leads to pore formation assumes that the overall area of the membrane is unchanged upon the change in the state of the lipids and the simultaneous insertion, and that the change in the total area of fluid lipids, gel lipids, and the pore adds up to zero. This model will create pores with a

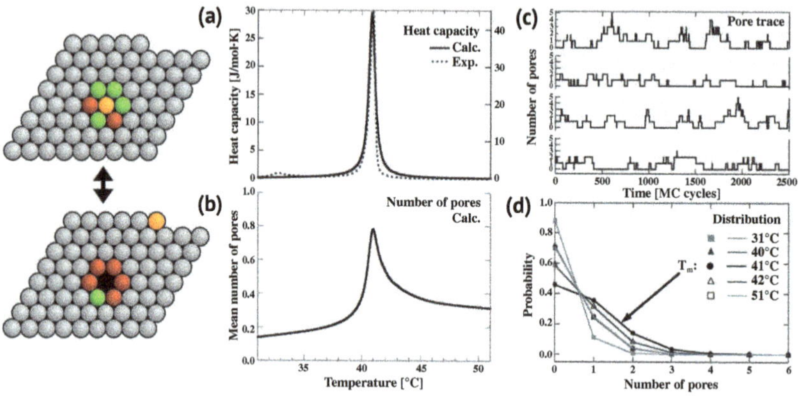

Figure VIII.1.7 Monte Carlo simulation of pore formation at the melting point of a membrane from Wodzinska et al. [14]. Left: The elementary MC step consists of a change in the number of gel and fluid lipids and the removal of a lipid in the center to form a pore with the size of a lattice site. Green circles represent fluid lipids, and red circles represent gel lipids. The yellow lipid in the center is moved to the edge of the matrix. Center: The simulated and experimental heat capacity profiles with a maximum at about 41 °C (top) and the average number of pores as a function of temperature (bottom). Right: The emergence of channels in a lipid matrix close to the melting transition maximum (top) and the distribution of pore numbers in the matrix for different temperatures following a Poisson statistics. The maximum number of simultaneously open pores is found at the melting temperature. See [14]/with permission of Royal Society of Chemistry for technical details.

higher likelihood in the transition range where the total number of gel and fluid lipids displays large fluctuations. Figure VIII.1.7 shows the elementary Monte Carlo step, the heat capacity profile, and the calculated average pore number. It also shows a trace of the number of pores as a function of MC time. Finally, histograms of the probability of finding a given number of pores are shown for different temperatures below, at, and above the transition temperature of 41 °C. We can see that the number of pores roughly corresponds to the measured permeability, and that single-channel traces seem similar in appearance to those measured in patch-clamp experiments.

- The probability of finding pores in a lipid membrane is enhanced in the melting transition.
- Close to the transition, the permeability of the membrane is higher.
- The pore lifetimes reach a maximum during the transition.

VIII.1.3 The Emergence of Lipid Ion Channels

The permeability of lipid membranes can be measured by using the BLM technique and on membrane patches using the patch-clamp technique. These two techniques are schematically described in Figure VIII.1.8 (from [15]). BLMs are generated from two monolayers over a hole in a Teflon sheet. This is known as the Montal–Müller technique [16]; see Figure VIII.1.8 (top). The holes in the Teflon sheet have characteristic diameters of 100 μm

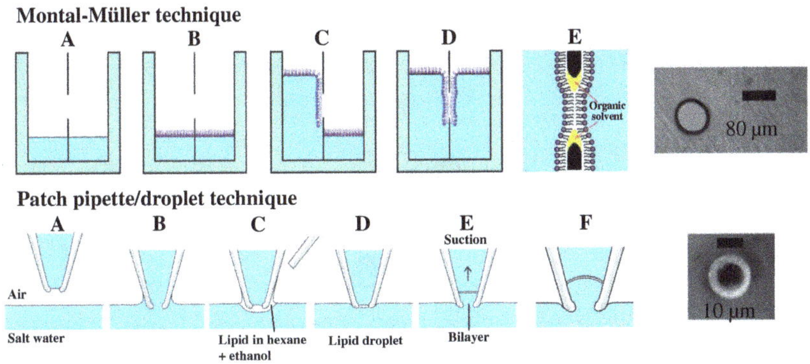

Figure VIII.1.8 Top: Schematic demonstration of the creation of a BLM using the Montal–Muller technique. The setup of a BLM experiment. A lipid membrane is created across a hole in a Teflon sheet. The size of hole in the Teflon sheet shown in the photograph displays a diameter of 80 μm. Bottom: Patch experiment on a lipid membrane created at an air–water surface. The image of the pipette tip shows an aperture of ~10 μm in diameter. Adapted from [18, 15].

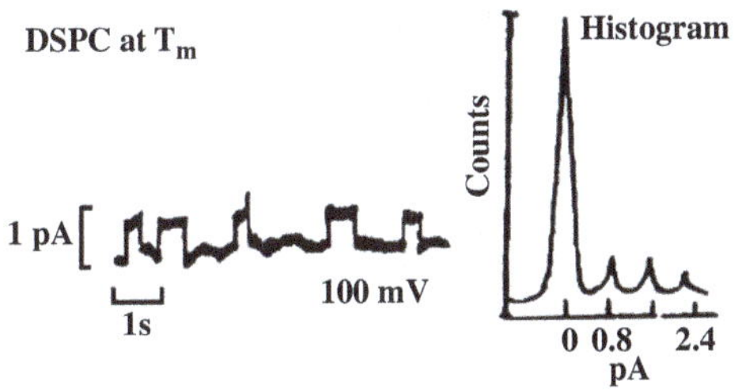

Figure VIII.1.9 First channel recordings in a BLM at the melting temperature of DSPC at 59 °C in 1 M KCl at a voltage of 100 mV. Adapted from [19].

(Figure VIII.1.8 (top, right)). Membrane patches on a pipette tip are generated with the droplet technique by Hanke and collaborators [17]; see Figure VIII.1.8 (bottom). The opening aperture of the patch pipette is of the order of 1–10 μm.

One of the first studies reporting channel-like conduction events in pure lipid membranes in the complete absence of proteins was by Antonov and collaborators [19]. Figure VIII.1.9 shows conduction events across a BLM made of the lipid DSPC at its melting temperature. At the transition temperature of 59 °C, the conductance changes drastically, and one finds pore opening and closing events very similar to those of biomembranes in the presence of proteins, with an amplitude of about 0.8 pA (conductance of 8 pS).

Another recording of channels in a BLM of a DOPC:DPPC = 2:1 mixture at the melting temperature of 19 °C (150 mM KCl, 60 mV) is shown in Figure VIII.1.10 (adapted from [4]). One can see many different conduction steps of about 8pA that are equally spaced (conductance of 133 pS). For comparison, the channels in the first paper about protein channels in the membrane of frog muscle by Neher and Sackmann [20] (Figure 2, see also [21], Figure 8) display conductances of about 32 pS, which is of a similar order to those in lipid

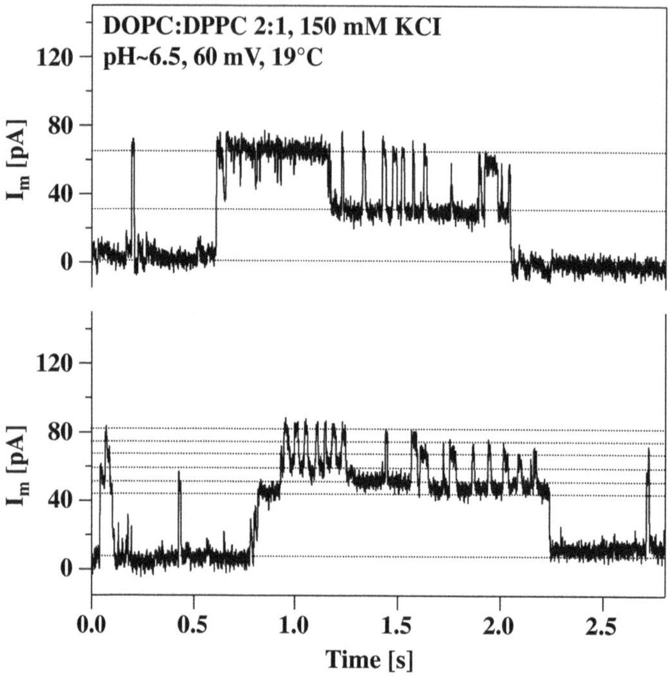

Figure VIII.1.10 Channel events emerging in BLM of a DOPC:DPPC = 2:1 lipid mixture at 60 mV. One can recognize many current steps with an amplitude of ~8 pA. This corresponds to a conductance of about 130 pS. Adapted from [4].

membranes. The conductance of the calcium-dependent potassium channel MthK in Ref. [22] is 27 pS. Also, the channel open times are on the order of a few ms to a few tens of ms in both protein channels and lipid membrane channels.

One also finds channels using the patch-clamp technique. Since the membranes on pipette tips are smaller, they can be very stable during a longer time interval. This allows for complete series of measurements where voltage, temperature, or pressure are varied. Figure VIII.1.11 shows a conduction trace of a DMPC:DLPC = 10:1 patch at 50 mV and T = 30 °C. The membrane used in this particular case was stable for more than 30 minutes [23]. One finds many open-close events with a mean amplitude of 18 pA, corresponding to a conductance of 360 pS. Such channel events are absent at some temperature distance from the transition. The influence of temperature is discussed in Chapter VIII.4.

Summarizing, one must conclude that close to melting transitions, one finds channel events that are very similar (or even indistinguishable) to those observed in biomembranes containing proteins. The typical conductances found are in the range from a few to several hundred pS, highly dependent on lipid mixture, temperature, voltage, and other experimental conditions (such as the patch pipette itself). The open timescales are of the order of milliseconds to several tens of milliseconds. Timescales are discussed in Chapter VIII.6. The lipid conduction events in artificial membranes are so similar to those reported for proteins that they were called "lipid ion channels" in Ref. [18].

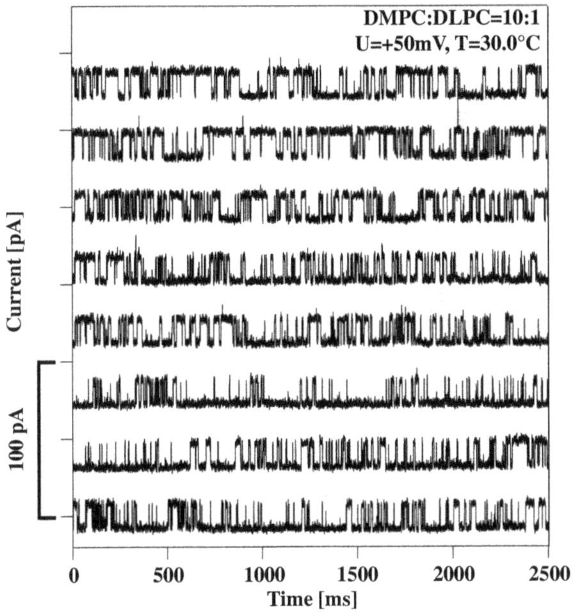

Figure VIII.1.11 Channels in a patch-clamp recording on a DMPC:DLPC = 10:1 mixture at a temperature of 30°C and a voltage of 50 mV in 150 mM KCl. Shown is a continuous trace. The offset between the segments of the trace was generated for better visibility. The step size is about 8 pA corresponding to two conductance levels with a difference of about 360 pS. From [23]/with permission of Elsevier.

VIII.1.4 Comparison of Lipid Membrane Channels and Channels from Biological Membranes

It is tempting to compare the channels in pure lipid membranes with those of real cell membranes in the presence of channel proteins. We already mentioned in the previous section that both channel conductance and channel lifetimes are very similar in preparations containing channel proteins and in pure lipid membranes close to transition. Laub et al. [23] systematically compared the conduction events of synthetic lipid membranes in the absence of any proteins with HEK293 cell membranes containing various TRP channels. All of the following experiments were performed close to room temperature (20–23 °C). We showed earlier that this temperature range is where one finds the heat capacity maximum in nerve membranes and other cell membranes (see Section V.2.4, Figure V.2.11).

Figure VIII.1.12 shows three channel traces (from Ref. [23]). The first one originates from a patch recording of an artificial membrane. It shows channel event with a conductance of 17.5 pS and open lifetimes on the order of seconds. The second trace originates from HEK293 cells containing human TRPM2 channels and H_2O_2. They display channel with lifetimes of the order of up to a second and a conductance of 5 pS. The third trace is a recording on a HEK293 cell containing human TRPM8 channels in the presence of icilin (one of the "cold"- inducing drugs). One finds typical open lifetimes of milliseconds to several tens

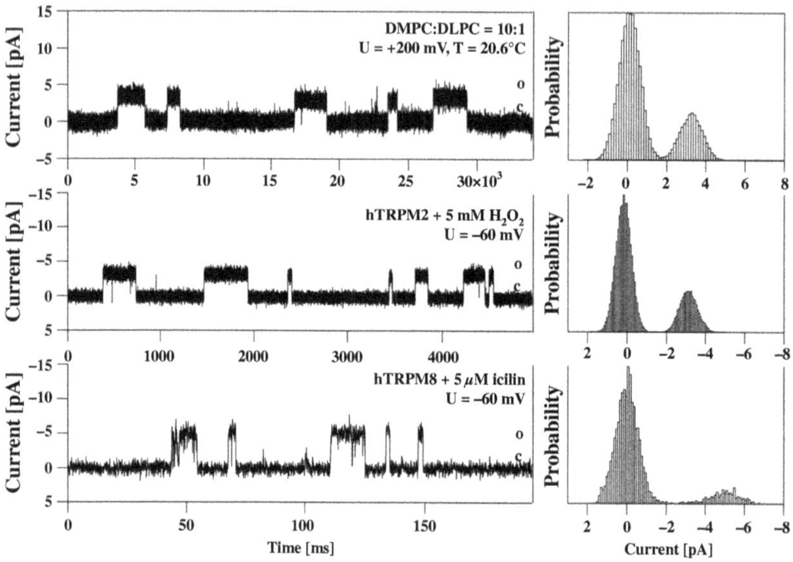

Figure VIII.1.12 Current traces of pure lipid membranes in the absence of proteins and different cell preparations with TRP channels. Top: DMPC:DLPC = 10:1 mixture in 150 mM KCl at 200 mV and T = 20.6 °C. The current amplitude is about 3.5 pA. Center: Human TRPM2 channels in HEK293 cells in the presence of 5mM H_2O_2 at −60 mV. The current amplitude of open channels is about 3 pA. Bottom: Human TRPM8 channels in HEK293 cells in the presence of 5 μM icilin at −60 mV. The current amplitude is about 5 pA. From [23]/with permission of Elsevier.

of milliseconds and a conductance of 8.3 pS. The biological preparations were measured at room temperature (21–23°C).

The appearance of channels in lipid membranes varies a lot depending on patch pipettes, voltage, temperature, and lipid mixture. One finds lipid membranes that behave very similar to experiments with cell membranes in the presence of proteins. It seems as if the biomembranes generate channels of higher reproducibility.

In Ref. [23], some other typical behaviors of membrane patches were investigated. This includes conduction bursts, flickering, multistep conduction, spiking and staircase behavior. In Figure VIII.1.13, the conduction burst behavior of a DMPC:DLPC = 10:1 membrane at 21°C and 200 mV is compared with a patch from HEK293 cells containing human TRPM8 channels in the presence of WS-12 at room temperature. The lipid membrane shows a burst lasting about 7 seconds. The current histogram shows conduction steps of 7.5 pA in amplitude (37.5 pS). The biological preparation shows a conduction burst of about 2.5 seconds. The histogram displays individual conduction steps of about 4.5 pA (7.5 pS). The appearance of both bursts, the qualitative shape of the current histogram, the amplitude of the conduction steps, and the burst lifetime are all of similar order.

Figure VIII.1.14 shows so-called flickering, which is similar to a conduction burst, but there is only one single conduction step with fast switching between an open and a closed channel or pore. The lipid preparation was a DMPC:DLPC = 10:1 mixture measured at 20.8 °C and 350 mV in 150 mM KCl. One finds a conduction step of 10.5 pA (30 pS). The duration of the flickering event was about 6.5 seconds. The biological preparation shows a

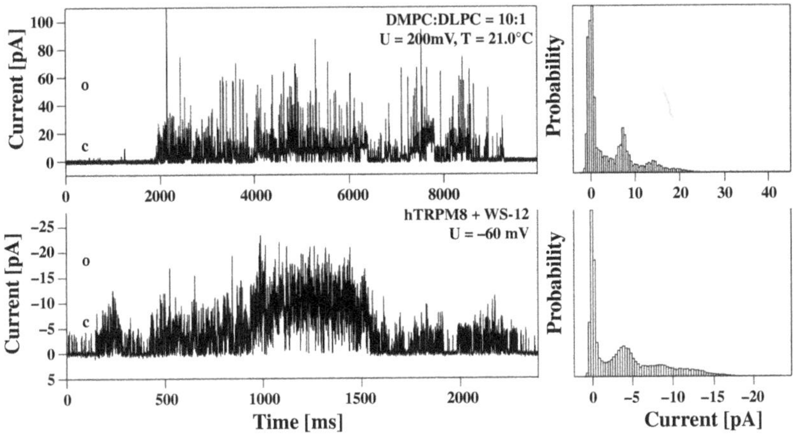

Figure VIII.1.13 Conduction bursts in a lipid membrane in the absence of proteins and in HEK293 cells containing human TRPM8 channels in the presence of WS-12. From [23]/with permission of Elsevier.

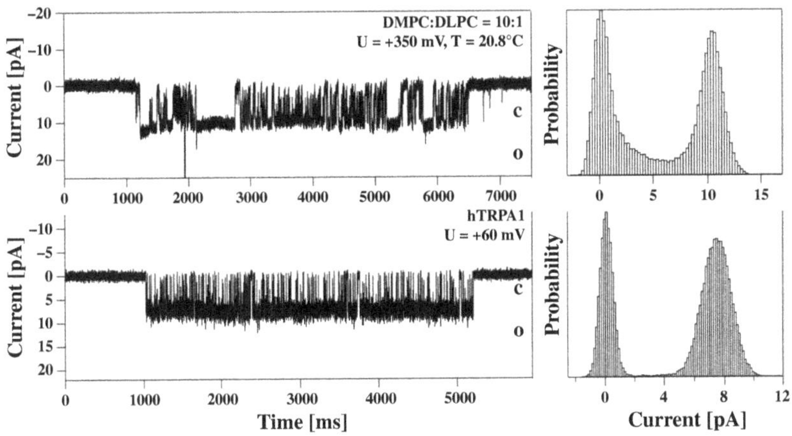

Figure VIII.1.14 Flickering in a lipid membrane (DMPC:DLPC = 10:1) in the absence of proteins and in HEK293 cells containing human TRPA1 channels. From [23]/with permission of Elsevier.

patch from HEK293 cells containing a human TRPA1 channel measured at room temperature and 60 mV. One finds a conduction step of about 7.8 pA (130 pS). The duration of the flickering event was around 4 seconds. Again, the events in the biological preparation are qualitatively and quantitatively very similar to those in the artificial membrane.

Figure VIII.1.15 shows multistep conduction events in a synthetic and a biological preparation. The synthetic membrane was again made of DMPC:DLPC = 10:1 measured at 20.6 °C, 200 mV, and 150 mM KCl. One finds at least four equally spaced conduction levels with 7 pA (35 pS) steps. The biological preparation was HEK293 cells with human TRPA1 channels in the presence of AITC (allyl isothiocyanate). It shows four conduction levels with step sizes of 1.6–2 pA (27–33 pS). Both the magnitude and the number of the

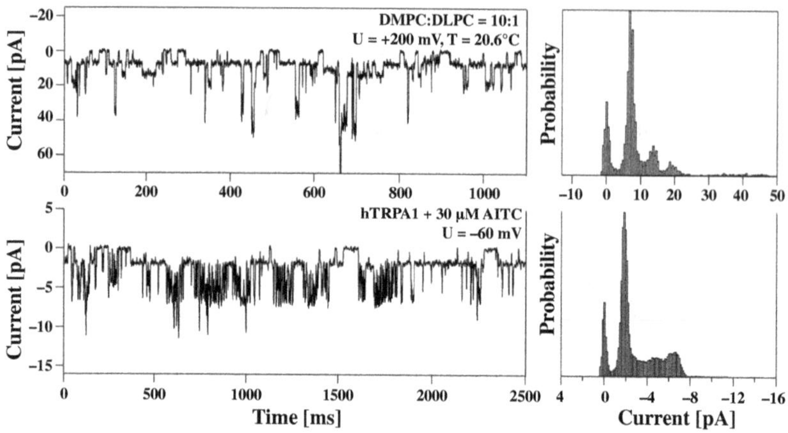

Figure VIII.1.15 Multistep conductance in a lipid membrane in the absence of proteins and in HEK293 cells containing human TRPA1 channels in the presence of AITC. From [23]/with permission of Elsevier.

conduction steps, the shape of the current histogram, and the channel lifetimes are qualitatively and quantitatively very similar in the lipid preparation and the biological patch.

The authors of Laub et al. [23] concluded that the phenomenology found in the synthetic membranes in the absence of any proteins and in the biological preparations in the presence of channel proteins is very similar. One can conclude that the traces themselves are indistinguishable in the sense that it is impossible, by visual inspection, to tell whether the trace originates from a synthetic lipid membrane or from a cell membrane. It is possible that there is additional evidence not contained in the traces that may help to identify a particular origin of the current events - but this remains to be demonstrated for each protein-containing preparation.

The similarity of conduction events in lipid membranes close to transitions and biomembrane preparations close to room temperature (with membrane transition generally being close) is surprising and has not found the attention that it deserves. In particular, these findings contradict the generally made assumption in electrophysiology that quantized conduction events in biomembranes must be due to protein channels because the lipid membrane is an insulator and cannot display channel-like events. We showed here that this assumption is incorrect. Further, while in a synthetic lipid membrane it is clear that the conduction events cannot be due to proteins, this is not evident in the biological membranes. Biomembranes display a transitional room temperature, and one cannot measure channel proteins in the absence of lipids. Therefore, it is difficult to prove without doubt that the conduction event occurs through a pore in a channel protein. Indirect evidence that one could consider would be the effect of mutations within the channel proteins, or the sensitivity to drugs such as tetrodotoxin that block the sodium channel. It is, however, safe to conclude that the conduction events are a property of the membrane patches that include both proteins and lipids.

VIII.1.5 Summary: Key Ideas of Chapter VIII.1

In this chapter, the permeability and the appearance of channel-like conduction events in lipid membranes close to melting transitions were discussed. These events were compared to those found in biological preparations. It was shown in particular that:

- Lipid membranes display a maximum in permeability in the melting regime. The change in permeability can be several orders of magnitude. The temperature dependence of the permeability (or conductivity) is, within the experimental accuracy, proportional to the excess heat capacity.
- According to Nagle and Scott [6], the proportionality of the permeability to the heat capacity can be attributed to the changes in lateral compressibility in the transition. We have shown in Chapter VI.4 that the excess compressibility is proportional to the excess heat capacity.
- Thus, the assumption in electrophysiology that lipid membranes are insulators is not correct close to room temperature, where biomembranes display a melting transition.
- Close to melting transitions, membrane patches display channel-like conduction events that have been attributed to pore formation.
- The appearance of current traces of synthetic membranes is indistinguishable from current traces of biological preparations.
- This contradicts a common assumption that is made when interpreting electrophysiological data, namely that the presence of discrete conduction steps must be due to protein activity because lipid membranes are insulators.
- Channel proteins in a biomembrane cannot be investigated in the absence of a surrounding membrane. Thus, one cannot exclude that conduction events may originate from pore formation in the lipid matrix.
- It is safe to conclude that the conduction events in biological membrane preparations are a property of the lipid-protein ensemble.

We will show in Section VIII.8.1 that proteins may alter the probability of conduction events in lipid membranes and can therefore act as catalysts for lipid membrane conductance.

VIII.2

Voltage-gated Lipid Channels

In this chapter, we show that the lipid membrane displays voltage-dependent conduction properties, and rectification (i.e., the difference in membrane conductance at positive and negative voltage typically attributed to proteins) can also be found in synthetic membranes in the complete absence of proteins. Due to the presence of membrane asymmetries or membrane polarization, current–voltage relations are not symmetric. Positive and negative applied voltage may lead to different behavior. In the following, we discuss the voltage dependence of black lipid membranes (BLMs) and membranes on patch pipettes that were made following the procedure shown in Figure VIII.1.8. These two systems differ in the diameter of the patch and the different maximum curvatures these membranes can achieve. This has consequences for the flexoelectric effect discussed in Section V.7.6.

VIII.2.1 Voltage-gated Channels in BLMs

It is a general observation that the occurrence of lipid ion channels depends on voltage. If a membrane is slightly above its melting temperature, this may be due to the voltage effect on membrane melting (see Chapter V.7, where it was shown that voltage can increase the melting temperature of membranes). Figure VIII.2.1 shows an electrophysiological experiment on BLMs made of a DOPC:DPPC = 2:1 mixture with a heat capacity maximum around 17 °C (shown in [14]) measured at 19 °C. The experimental temperature is slightly above the c_p maximum. With increasing voltage, more and more channels appear, and channel open times increase. If the mean current is plotted versus voltage, one finds a current–voltage (I–V) profile that is, within error, symmetric in voltage. While the profile is linear in the range of -150 mV $< V_m <+150$ mV, it becomes nonlinear at voltages above ±150 mV. At even higher voltages, the membranes rupture.

Figure VIII.2.2 shows another BLM experiment in which the voltage dependence of current traces of BLMs of a DMPC:DLPC:chol. = 77.3:7.7:15 BLM is shown. The conductance increases with increasing voltage and channels appear. The I–V profile is symmetric in voltage and becomes nonlinear at voltages above ±80 mV.

A symmetry of the I–V profile is a general finding in BLMs, as is the increase in the channel open probability at higher voltages.

Physics of Nerves and Excitatory Membranes, First Edition. Thomas Heimburg.

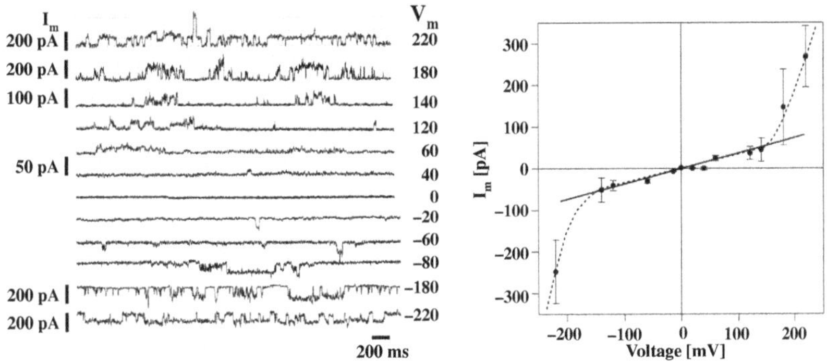

Figure VIII.2.1 Left: Voltage dependence of current traces of BLMs of a DOPC:DPPC = 2:1 mixture with a heat capacity maximum around 17 °C [14] (150 mM KCl, pH 6.5 at T=19 °C). Right: Current–voltage relation determined from the data in the left-hand panel. The profile is symmetric. At voltages above 150 mV, the I–V relation becomes nonlinear. Adapted from [14].

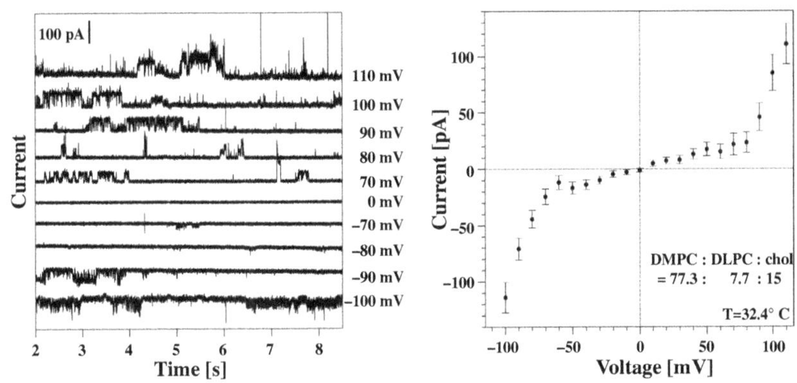

Figure VIII.2.2 Left: Voltage dependence of current traces through BLMs of a DMPC:DLPC:chol. = 77.3:7.7:15 mixture measured at 32.4 °C (150 mM KCl, 150 mM NaCl, pH of 7.6). This lipid mixture displays a broad melting profile with an upper limit at the experimental temperature [24]. Adapted from [24].

VIII.2.2 Voltage-gated Single Channels in Membrane Patches

The voltage-gated nature of channels can also be explored with patch-clamp experiments. The procedure for creating a lipid membrane patch is described in Figure VIII.1.8. In Figure VIII.2.3, single-channel events in a DMPC:DLPC = 10:1 membrane at 30 °C are shown. The left, bottom panel shows current traces at four different voltages from 40 to 70 mV. One can observe that the current becomes larger upon increasing voltage, yielding a linear current–voltage relation for the single channel (Figure VIII.2.3 (left, top)). A channel conductance of 320 pS was found. One also observes that the relative probability of finding an open channel increases with voltage because the single-channel lifetimes increase. One can see this in the current histograms shown in Figure VIII.2.3 (right, top). Thus, the relative probability of finding an open pore compared to a closed pore increases with voltage.

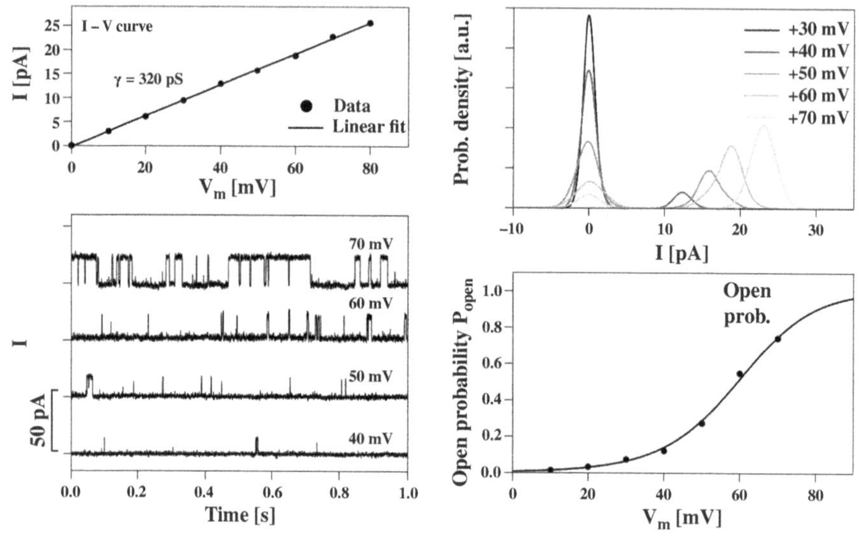

Figure VIII.2.3 Left, bottom: Current traces for a DMPC:DLPC = 10:1 mol/mol membrane (150 mM KCl, T = 30 °C) at four voltages showing an increase of single-channel conductance with voltage and an increased likelihood of channel formation. Left, top: The corresponding linear single-channel current–voltage relation indicating a single-channel conductance of γ = 320 pS. Right, top: The current histogram for the channel events in panel left, bottom at five voltages. Right, bottom: The open channel probability deduced from the current histograms indicates a sigmoidal voltage dependence of the open probability. Adapted from [25].

From the area under the two peaks in this panel, one can deduce the open probability as a function of voltage, which displays a sigmoidal shape (Figure VIII.2.3 (right, bottom)).

The behavior shown earlier is very similar to that of reported protein channels in lipid membranes. Figure VIII.2.4 shows the voltage dependence of the open probability from Figure VIII.2.3 and the open probability of the KvAP channel from Ref. [22] in a lipid membrane containing a mixture of positive and negatively charged lipids. It can be seen that both the midpoint of the sigmoidal profile (40–50 mV) and the steepness of the profile are quite similar. This indicates that the phenomenology of the voltage gating reported for protein channels is very similar to that found for pure lipid membranes.

VIII.2.3 Rectification

Figure VIII.2.5 shows the current–voltage relation of a membrane on a patch pipette tip. Experience shows that, in contrast to BLMs, the current–voltage relations on patch pipettes are asymmetric (i.e., the conductance at negative voltage is different from that at positive voltage), typically outward-rectified when using the usual convention of voltage difference between inside and outside of the pipette. Outward-rectified means that the current between inside and outside is larger for positive voltages as compared to negative voltages. Rectification is believed to be indicative of protein channels, suggesting a molecular mechanism for conduction. It is therefore surprising that one also finds it in pure lipid membranes where there is no reason to assume a molecular origin of conduction. We will see in the

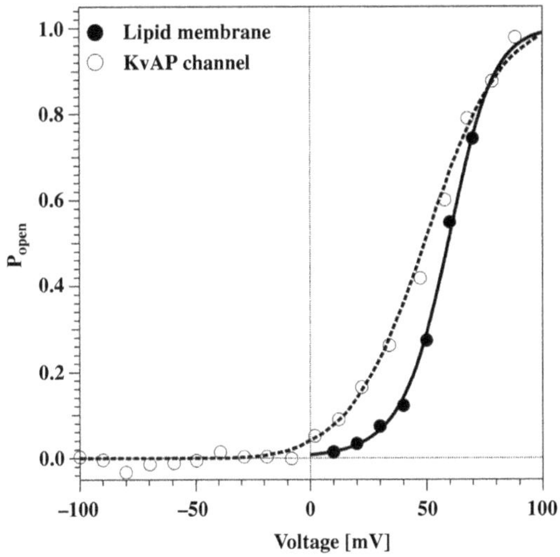

Figure VIII.2.4 The voltage dependence of the open probability of a single channel in a DMPC:DLPC = 10:1 membrane from Figure VIII.2.3 [25] (full circles and straight line) compared to the open probability of a KvAP channel in POPE:POPG vesicles (protein-to-lipid ratio 1.0 w/w) fused into bilayers of mixed EDOPC (a positively charged lipid) and POPE:POPG with percentage EDOPC being 80% (open circles and dotted line). Adapted from [22].

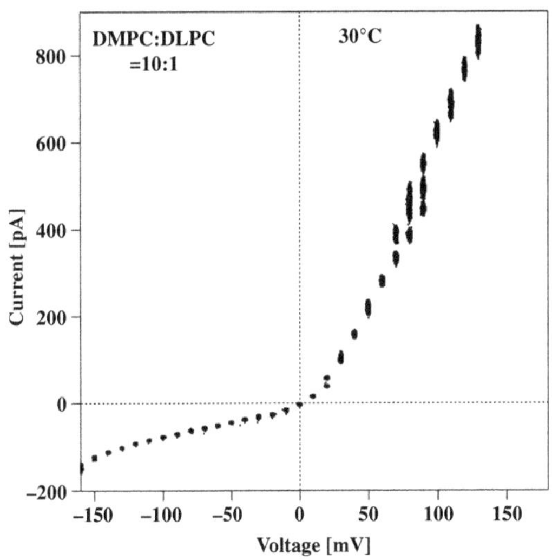

Figure VIII.2.5 Current–voltage relation of a DMPC:DLPC = 10:1 membrane at 30 °C on a patch pipette. The trace is "outward" rectified and not symmetric in voltage. Adapted from [23].

next section that rectification can arise when the membrane is polarized. Polarization does not require a molecular mechanism. It can be originated from membrane asymmetries of various natures (see Section V.7.5).

VIII.2.3.1 Theory of Rectification in Lipid Membranes

Abidor and coworkers [26] suggested that the chemical potential of a pore in a lipid membrane is given by

$$\Delta\mu_{\text{pore}} = 2\pi r \cdot \gamma - \pi r^2 \cdot \sigma - \pi r^2 \cdot \frac{1}{2} C_{m,0} V_m^2, \tag{VIII.2.1}$$

where r is the radius of the pore, γ the line tension of the edge of the pore, σ is the lateral tension in the membrane, and $C_{m,0}$ is the specific capacitance. This means that the pore energy consists of the sum of line energy, the work done to increase the area of the pore, and a contribution of the capacitive energy of a membrane segment corresponding to the size of a pore, which is quadratic in voltage. Winterhalter and Helfrich [27] suggested that the line tension of a pore also has a quadratic dependence on voltage and that voltage stabilizes a particular pore size.

Following eq. (VIII.2.1), we start our consideration by assuming that the energy of a pore with a fixed radius is given by

$$\Delta\mu_{\text{pore}} = \Delta\mu_{\text{pore},0} + \alpha \cdot V_m^2, \tag{VIII.2.2}$$

where $\mu_{\text{pore},0}$ is the chemical potential of a pore in the absence of voltage, and α is a negative coefficient. Mosgaard et al. [28] argued that the electrostatic free energy depends also on the net polarization V_0 of the membrane. The capacitive energy is given by $\pi r^2 \cdot \frac{1}{2} C_{m,0}((V_m + V_0)^2 - V_0^2)$, and the total chemical potential of the pore is now given by [24]

$$\Delta\mu_{\text{pore}} = \Delta\mu_{\text{pore},0} + \alpha \cdot \left((V_m + V_0)^2 - V_0^2\right). \tag{VIII.2.3}$$

Thus, the maximum of the chemical potential of a pore is found at $V_m = -V_0$. At this voltage, one expects the smallest likelihood to find a pore. Membrane polarization can have different origins that all involve an asymmetry of the membrane, for instance, caused by a lipid asymmetry, a protein asymmetry, or membrane curvature (see Section V.7.5).

Now we can write down a theory for pore formation. The equilibrium between an open pore and a closed one is given by

$$K = \frac{P_{\text{open}}}{P_{\text{closed}}} = \exp\left(-\frac{\Delta\mu_{\text{pore}}}{kT}\right) \tag{VIII.2.4}$$

or

$$P_{\text{open}} = \frac{K}{1+K}. \tag{VIII.2.5}$$

Thus, the total membrane current is given by

$$I = g_{\text{pore}} \cdot P_{\text{open}} \cdot (V_m - E_N), \tag{VIII.2.6}$$

where g_{pore} is the conductance of a single pore, P_{open} is the open probability of the pore, V_m is the transmembrane voltage, and E_N is the Nernst potential in the case of an asymmetry of ions on the two sides of the membrane. Such a model can in fact describe the rectified conductance of lipid membranes (and also of protein-containing membranes) very well [25].

In some cases, membranes display contributions to the conductance that are not represented by pores (e.g., the edge of the membrane at the pipette tip) and are independent of voltage. It will be described as a leak conductance, g_L [24].

$$I = (g_L + g_{pore} \cdot P_{open}) \cdot (V_m - E_N). \qquad \text{(VIII.2.7)}$$

This equation can be used to describe other I–V profiles from BLMs and from patches on pipettes.

Figure VIII.2.6 (left) shows the rectified I–V profile from Figure VIII.2.5 with an analysis according to eq. (VIII.2.7). The fit parameters include $\Delta\mu_{pore,0} = 2.3$ kJ/mol, $\alpha = -2.09 \cdot 10^3$ kJ/mol·V^2, and $V_0 = 0.15$ V. There was no leak current. The fit of the I–V profile is excellent and leads to an offset potential caused by polarization of 115 mV. This is of similar order to the resting potential in biological membranes. These parameters allow to calculate the open probability of membrane pores that can be seen as an inset in Figure VIII.2.6 (left). At a voyage of −115 mV, the pore open probability displays a minimum. Another example from a different experiment is shown in Figure VIII.2.6 (right). It also represents a DMPC:DLPC = 10:1 membrane from a different experimental series at a different temperature using a different pipette. Here, we find $\Delta\mu_{pore,0} = 3.1$ kJ/mol, $\alpha = -3.4 \cdot 10^3$ kJ/mol·V^2, and $V_0 = 0.21$ V. In order to fit this profile, a leak current was assumed. See Ref. [24] for details.

We recognize that the I–V profiles vary from experiment to experiment and are generally less reproducible than the profiles from cells, but the parameter range is always of similar order. The fits to eq. (VIII.2.7) are generally excellent.

Equation (VIII.2.7) can also be used to describe the symmetric I–V profile of the BLM shown in Figure VIII.2.2. This is shown in Figure. VII.2.7. The fit parameters include $\Delta\mu_{pore,0} = 18.7$ kJ/mol, $\alpha = -2.2 \cdot 10^4$ kJ/mol·V^2, and $V_0 = 0.005$ V. We see here that the main difference between the I–V profiles from patch pipettes and from BLMs is that the BLMs display no or just a very small offset potential, while the membranes on patch pipettes display offset potentials of the order of 50–200 mV.

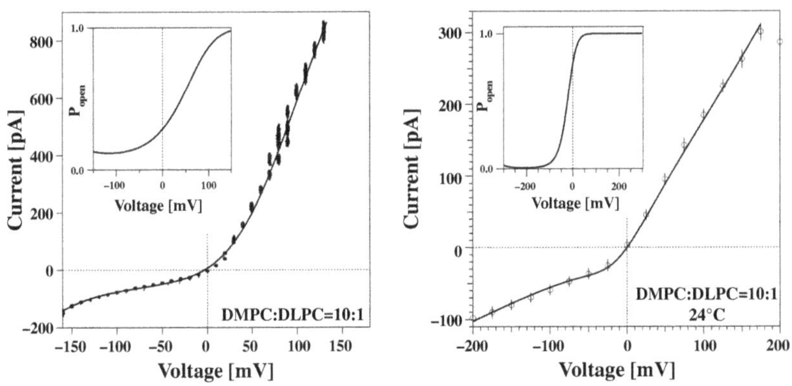

Figure VIII.2.6 Left: Fit to the membrane conductance of an artificial lipid membrane (solid line) at 30 °C in Figure VIII.2.5 and the calculated open probability of the pores (inset). Adapted from [23, 25]. Right: Another experiment on a DMPC:DLPC = 10:1 membrane made in a different series of experiments on a different pipette fitted by eq. (VIII.2.7). The offset potential is 208 mV. Adapted from [24].

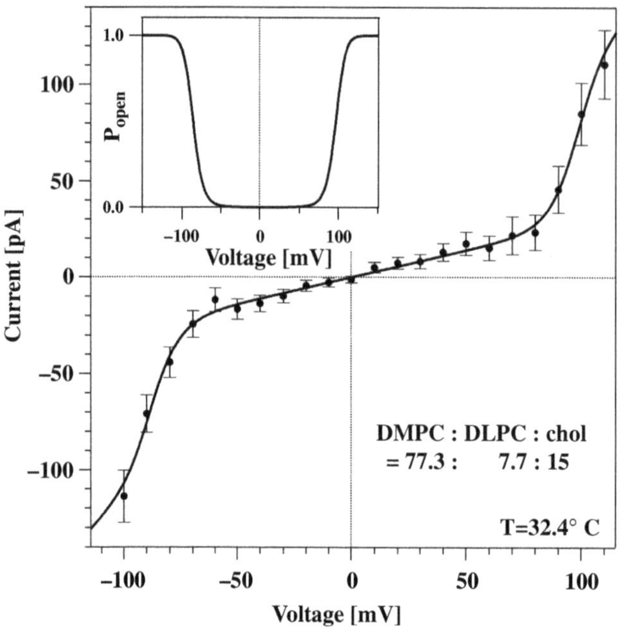

Figure VIII.2.7 Fit of the I−V profile and the calculated pore open probability in a BLM made of a DMPC:DLPC:cholesterol = 77.5:7.5:15 mixture. The offset potential V_0 is very small. Adapted from [24].

VIII.2.3.2 Why Do Lipid Patches Display a Net Polarization?

The observation is that the I–V profiles of BLMs are always symmetric in voltage, while those on patch pipettes are not symmetric but seem rectified (Figure VIII.2.7). We have seen in our analysis earlier that a likely explanation is a different membrane polarization. While the offset potential V_0 is around zero for the BLM, it is of order 100 mV in the patch pipettes. One reason for this could be the flexoelectric effect described in Section V.7.6. The flexoelectric effect (proposed by Alexander Petrov) consists of the generation of a membrane polarization by curvature [28]

$$V_m = -\beta \cdot c\,. \tag{VIII.2.8}$$

In Figure VIII.2.8, scenarios of bent BLMs on a large hole in a Teflon sheet and on a patch pipette are shown. Holes in the Teflon sheet are one to two orders of magnitude larger than the aperture of the patch pipette. This implies that the maximum possible curvature caused by pressure differences between inside and outside differs by one to two orders of magnitude, and so does the magnitude of the flexoelectric effect.

Petrov [29] found for a single lipid membrane that $f = \varepsilon_0\varepsilon\beta = 10^{\cdot 18}$ C, which yields $\beta = 2.82 \cdot 10^{-}8$ m if the dielectric constant of the membrane core is $\varepsilon = 4$ (see Ref. [28]). Therefore, the maximum voltage caused by curvature in a pipette with an aperture of 1 μm is about 56 mV, while it is less than 1 mV for a BLM on a hole of 80 μm. These numbers reflect the order of magnitude of the offset potentials found in the aforementioned experiments.

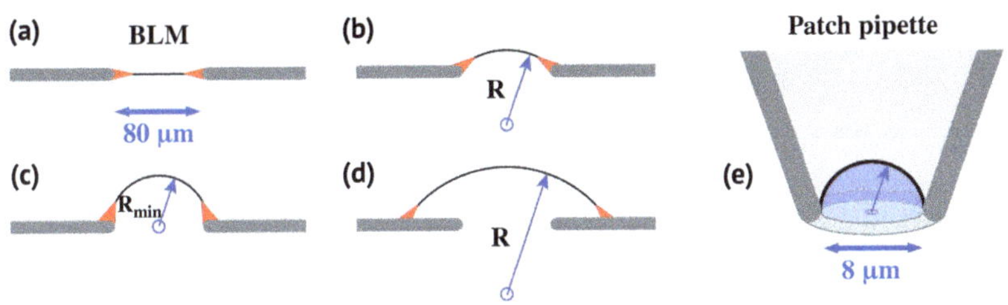

Figure VIII.2.8 Schematic drawing of curved membranes on the hole in a Teflon sheet (left) and on a patch pipette (right). The maximum curvature of a BLM over a hole of 80 μm is $c_{BLM} = 1/40$ μm. The maximum curvature over an aperture of 1 μm in a patch pipette is $c_{\text{pipette}} = 1/0.5$ μm.

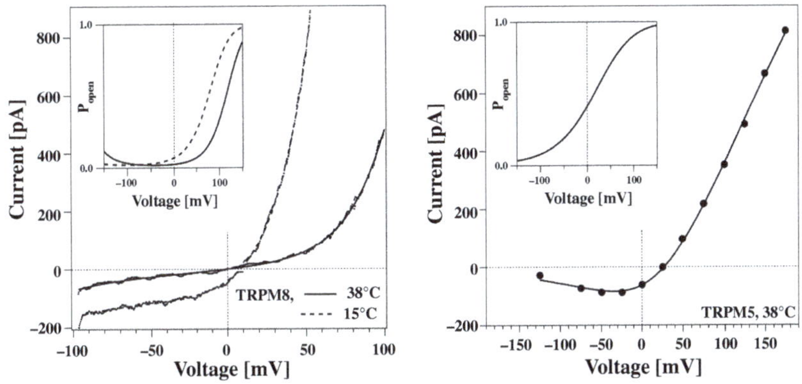

Figure VIII.2.9 Left: I–V profiles of a TRPM8 channel in HEK293 cells recorded at two different temperatures. Adapted from [30]. The fits using the theory mentioned earlier are nearly perfect and yield offset potentials of 41 mV for the profile recorded at 30 °C and 100 mV for the profile recorded at 15 °C. Right: I–V profiles of a TRPM5 channel in HEK293 cells recorded at 38 °C (Adapted from [31]) and fitted with the above theory. The inset shows again the open probability as a function of voltage.

VIII.2.4 Comparison with Proteins

The aforementioned theory also describes the I–V profiles of biological membrane preparations well. Figure VIII.2.9 (left) shows the TRPM8 channel protein in HEK293 cells at two temperatures taken from Ref. [30]. The fits are so good that they overlap the data. The curve at 30 °C was described with $dG0 = 9.1$ kJ/mol, $\alpha = -4.0 \cdot 10^4$ kJ/mol·V^2, and $V_0 = 0.041$ V. The curve recorded at 15 °C was fitted with $dG0 = 6.1$ kJ/mol, $\alpha = -2.9 \cdot 10^4$ kJ/mol·V^2, and $V_0 = 0.1$ V. The most obvious change is the doubling of the offset potential. It should be noted that Voets et al. [30] tried to describe these data with an Eyring-type model, which does not fit the data equally well (see discussion in Ref. [23]). Figure VIII.2.9 (right) shows that the I–V profile of a TRPM5 channel (data taken from [31]) can also be described with this model ($\Delta\mu_0 = 0.8$ kJ/mol, $\alpha = -0.6 \cdot 10^4$ kJ/mol·V^2, and $V_0 = 0.5$ V).

Zecchi and Heimburg analyzed channel recordings of synthetic lipid membranes and compared the open probability of pores with those found for potassium channels. Figure VIII.2.10 compares the open probability deduced for a lipid membrane (dashed line)

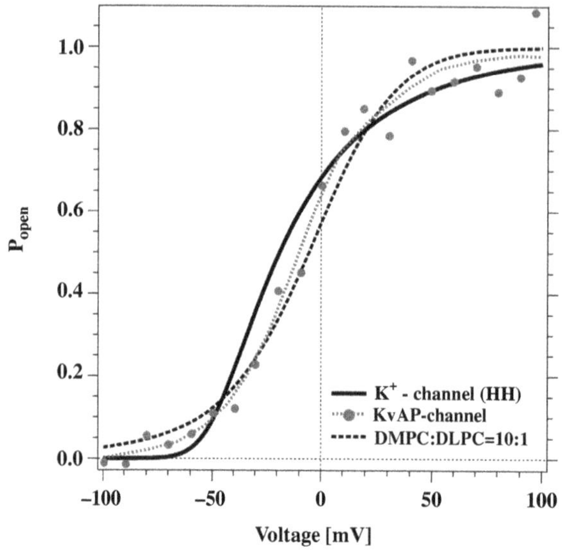

Figure VIII.2.10 Comparison of open probability of a K-channel from squid axons calculated using the parameters from the seminal paper by Hodgkin and Huxley [32], the KvAP channel [22], and a DMPC-DLPC = 10:1 lipid membrane. Figure adapted from [24].

with the calculated open probability of a potassium channel after infinite time from the Hodgkin–Huxley model for the squid axon (solid line, [32]) and the open probability of a KvAP channel reconstituted into a POPE/POPC = 3:1 membrane (symbols and dotted line). The profiles look both qualitatively and quantitatively similar and display a similar voltage at which 50% of the channels are open (around −10 to −20 mV).

Summarizing, these findings imply that the conductance properties attributed to proteins are very similar to those for lipid membranes close to transitions. Either they are similar due to a fortuitous coincidence or they actually represent the same events.

VIII.2.5 Summary: Key Ideas of Chapter VIII.2

The voltage-dependent activity of ion channel proteins is at the core of the understanding of signal transduction in molecular biology. In this chapter we showed that in electrophysiological measurements on lipid membranes, one finds channel events that display a voltage dependence that is strikingly similar to those attributed to proteins.

- BLMs display current–voltage relations that are symmetric in voltage, meaning that the conductance does depend on the magnitude of the voltage but not on its sign.
- Lipid patches on pipettes display asymmetric I–V profiles, and the profiles are rectified.
- The origin of the asymmetry is the presence of a net polarization of the membrane.
- Polarization of the membrane can arise from curvature of the membrane, which will be larger on small membrane patches (as on a pipette) than on large patches (as in a BLM experiment).
- The voltage dependence of lipid channels is nonlinear, meaning that the open probability is nonlinear.

- A simple thermodynamic theory in which the chemical potential of an open pore displays a quadratic voltage dependence can explain not only the shape of I–V profiles of lipid membranes, but also that of I–V profiles attributed to protein channels, e.g., of the TRPM8 and TRPM5 channel or potassium channels.
- The advantage of the theory presented here over a gate model as proposed by Hodgkin and Huxley [32] (see Sections IV.3.2 and IV.3.3) is that no parametrization is necessary. The voltage dependence (and also the lifetimes of the channels; see Chapter VIII.6) readily emerge from the theory of thermodynamic fluctuations.
- I–V profiles in pure lipid membranes close to transitions and of living cells are strikingly similar.

VIII.3

Mechanosensitive Lipid Channels

One of the important intensive variables is membrane tension. It shows up as a $-\Pi dA$ term in the differential of the internal energy. In Section IV.5.2, we briefly discussed the dependence of the permeability of biological membranes on lateral pressure in the presence of proteins such as Piezo1. These proteins are called mechanosensitive proteins. Here, we show that one also finds mechanosensitive channels in pure lipid membranes because the cooperative fluctuations of membranes are pressure-dependent. In Section V.4.2, it was shown that tension influences the melting temperature of membranes, and also potentially the membrane polarization via the piezoelectric and flexoelectric effects (Section V.7.6). It is therefore straightforward to hypothesize that lateral tension influences the probability of pore formation. Section IV.5.2 shows how a hydrostatic pressure difference between two sides of a membrane can lead to tension. Suction applied with the patch pipette will induce membrane tension and influence the fluctuations in the lipid membrane.

Kaufmann and collaborators [33] showed that suction on a membrane patch consisting of soybean phosphatidylcholine can lead to the emergence of channel events at constant voltage (see Figure VIII.3.1). While in the absence of suction no channel events were found,

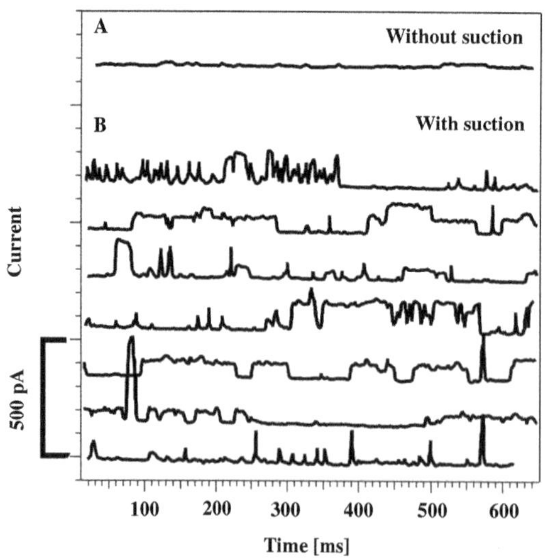

Figure VIII.3.1 Patch pipette experiment on a membrane made of soybean PC in 1 M NaCl at 500 mV. Trace A: Without suction. Traces B: With suction. Adapted from [33].

strong current fluctuations with several conductance levels were found in the presence of suction (pressure difference was not quantified in Ref. [33]).

Zecchi and Heimburg provided another study on channel generation in patch membranes. Figure VIII.3.2 shows an experiment on a single membrane of a DMPC:DLPC = 10:1 mixture at the upper end of the melting transition at two different insertion depths of a patch pipette below an aqueous surface. The experiments were performed at a depth of 4 mm and 8 mm (from Ref. [24]). Since the pipette itself contains a column of water, a deeper insertion depth lowers the pressure difference across the membrane by about 40 Pa (corresponding to about 0.3 mm Hg). In both experiments one recognizes a voltage threshold for the emergence of lipid channels. However, the threshold is lowered from about 125 mV to about 50 mV. Also, the current–voltage relations are altered without significantly changing the offset potential.

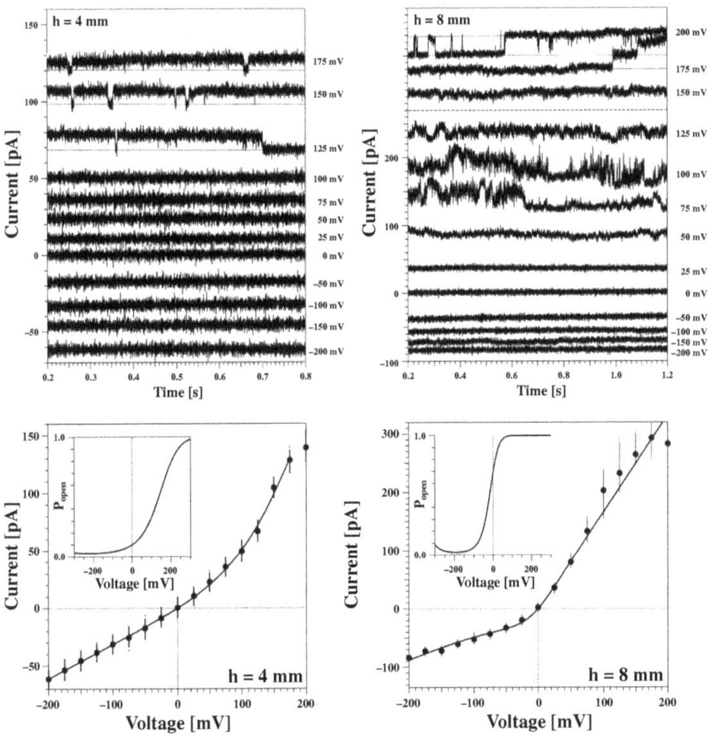

Figure VIII.3.2 Patch pipette experiment on a DMPC:DLPC membrane (150 mM KCl, 150 mM NaCl, pH 7.6, T = 24°C) at different depth *h* of the patch pipette under the surface of the buffer. Left: Insertion depth *h* = 4 mm. Current traces at different voltages and the I–V profile are shown. The solid line is the fit to eq. (VIII.2.7). Channel events become apparent above +100 mV. Left: Insertion depth *h* = 8 mm. Current traces at different voltages and the I–V profile are shown. Channel events become apparent for a voltage above +25 mV. In both experiments, no channels events could be seen for negative voltages. Adapted from [24].

VIII.3.1 Summary: Key Ideas of Chapter VIII.3

The total number of experimental studies on the mechanosensitive nature of lipid channels is very small. This phenomenon definitely deserves to be studied more extensively. The findings presented in this short chapter imply that suction on membrane patches can alter the abundance of channel events in artificial membranes. In particular

- Increasing the suction in the pipette can generate channel events.
- Increasing suction or a hydrostatic pressure difference between the two sides of the membrane patch leads to an increase in membrane tension.
- This effect is similar to the one described in Section IV.5.2 for protein channels such as the PIEZO channels.
- Increasing lateral tension leads to a change in the voltage threshold for channel generation. This implies that voltage and lateral tension are both variables that influence the channel statistics in a synergetic manner.

Summarizing, lateral tension is a variable that affects the induction of channels in lipid membranes in the complete absence of proteins.

VIII.4

Temperature sensing

Temperature is another important intensive variable. It shows up as a TdS term in the differential of the internal energy, i.e., the heat absorbed by the membrane upon temperature changes. Any change in temperature will change the phase state of the lipid membrane if one is close to a melting transition. This has a direct influence on the membrane area compressibility, the probability of forming pores, and the permeability. In Section IV.5.3, we discussed the dependence of the permeability of biological membranes on temperature and the presence of TRP channels. In Section VIII.1.1, we already found that membranes become more permeable during the transition. Here, we show that there also emerge temperature-sensing channels in pure lipid membranes in the complete absence of proteins.

Synthetic lipid membranes on pipette tips or BLMs are often not very stable and break easily. It seems that biological membranes resist perturbations better – possibly due to protein networks on the membrane surface. Channel events in lipid can consistently be seen in electrophysiological experiments, but only rarely is the membrane stable enough to survive for many minutes or even hours. For this reason, experiments that require the systematic change in conditions are not always possible. Temperature changes are slow, and we are only aware of few single-channel recording studies for a single synthetic membrane during which the temperature was changed.

It is clear that the permeability of the membrane is at maximum in the melting transition. At this temperature, one also finds the largest abundance of single-channel events in electrophysiological experiments [19]. Figure VIII.4.1 shows the melting profile of DPPC membranes and current traces above, within, and below the melting transition. One can recognize channel conduction events at the transition temperature, and one finds at least three different conduction levels with equal separation and a single-channel conductance of about 34 nS.

Another example is given in Figure VIII.4.2. DOPC:DPPC = 3:1 membranes display a heat capacity profile with a maximum at 17°C and an upper end at around 22°C (Figure VIII.4.2 (left)). The current traces at the c_p maximum display channel events with a conductance of about 88 pS. Above the melting transition at 30°C, no channel events can be recognized (Figure VIII.4.2 (right)). Thus, the membrane sensed the temperature in a similar manner as the TRPM8 channel (see Section IV.5.3).

A further example of the change in conduction properties with changes in temperature is shown in Figure VIII.4.3. The current traces of a BLM made of a D15PC:DOPC = 95:5 mixture and the corresponding current histograms are shown above the transition (33°C)

Physics of Nerves and Excitatory Membranes, First Edition. Thomas Heimburg.

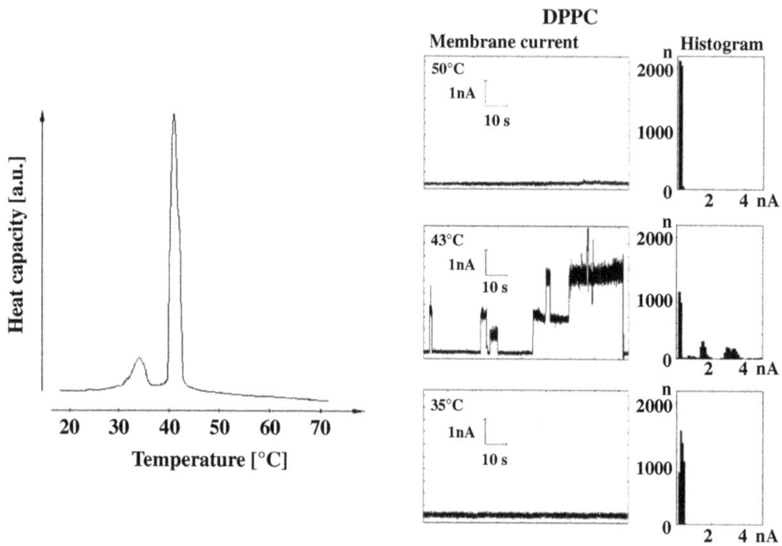

Figure VIII.4.1 Left: Melting profile of a DPPC membrane. Right: Channel activity at 50 mV. Top: At T = 50°C (above the transition). Middle: At T = 43°C (within the transition). Bottom: At T = 35°C (below the transition). The current histograms are also given. While no channel activity is seen above and below the transition, several current steps are seen for the membrane in the transition range. Adapted from [34].

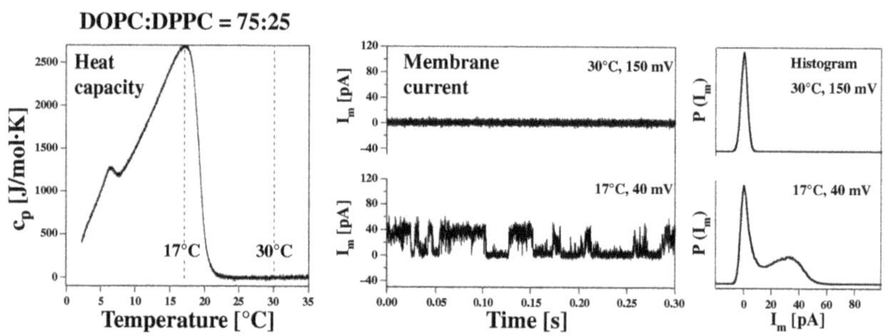

Figure VIII.4.2 Left: Melting profile of a DOPC:DPPC = 75:25 membrane. The transition peak is found at 17°C. The upper end of the transition is at 21°C. Center: At the transition temperature, the opening and closing of a pore can be seen. At 30°C, no channel events are detected. Right: Current histograms for the traces in the center panel. The conductance of the pore at the transition temperature is about 88 pS. From [4]/with permission of Elsevier.

and in the melting regime (31.5°C). For reference, see the heat capacity profile as an inset in the bottom right panel. Within the transition, one finds clearly recognizable opening and closing events, while the currents above the melting regime are better described by short-lived fluctuations. One also recognizes that the events have a much larger lifetime within the transition. Channel lifetimes are discussed in more detail in Section VIII.6.2.

Summarizing, close to transitions, the overall conductance, the emergence of channel events, and their lifetimes are closely coupled to the temperature. Thus, the lipid channels are temperature-sensing.

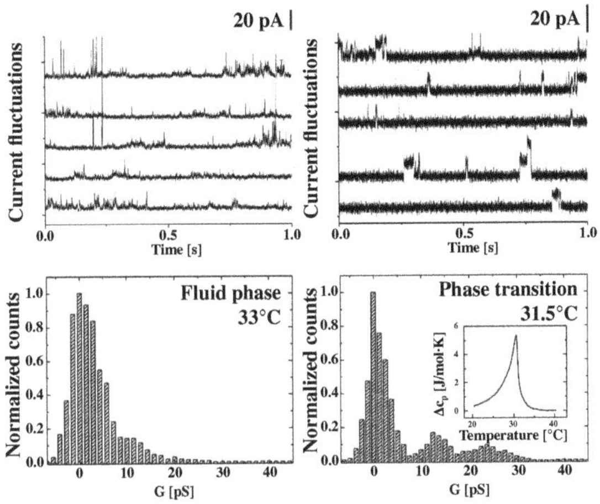

Figure VIII.4.3: BLM of a D15PC:DOPC = 95:5 mixture in the fluid phase (33°C, left) and the transition regime (31.5°C, right). The top panels show the currents as a function of time, and the bottom panel shows that current histograms. The inset in the bottom right panel shows the heat capacity profile with a maximum around 30.5°C. Adapted from [5].

VIII.4.1 Summary: Key Ideas of Chapter VIII.4

In the theory of membrane pores, the melting transition plays an important role. The heat capacity is proportional to the lateral compressibility, which in turn is proportional to the permeability of the membrane. In this chapter, we showed that temperature changes influence the probability of channel formation by moving the membrane in and out of the melting regime.

- Within the melting transition, channel events are much more likely.
- A change in temperature, therefore, can "activate" or "block" a channel.
- As shown in Section IV.5.4, proteins such as the KcsA potassium channel display a similar dependence on the heat capacity of their host membrane [35].

VIII.5

The Influence of Drugs on Membrane Permeability and Lipid Ion Channels

The chemical potentials of membrane components also contribute to the thermodynamics of membranes. This is not just true for lipids and membrane proteins, but also for membrane-soluble drugs added to membranes. These could be anesthetics described in Chapter IX.1 or other molecules such as the neurotransmitter serotonin.

VIII.5.1 Influence of Drugs on Permeability

It is well known that some drugs change the melting transition of membranes. As argued in Section VIII.1.2, the permeability of a membrane is proportional to the excess heat capacity. Thus, any drug that changes the heat capacity profile will alter the permeability. This is particularly true for drugs that are membrane-soluble. Lindane (γ-hexachlorocyclohexane) is a small molecule that acts as a neurotoxin. Figure VIII.5.1 (left) shows that increasing amounts of lindane shift the heat capacity profile of a DMPC membrane toward lower temperatures. The right-hand side of this figure indicates that this shift is reflected in permeability profiles that are also shifted toward lower temperatures. Thus, depending on the ambient temperature, a drug like lindane can either decrease the permeability (if the temperature T is above the melting temperature T_m), or it can increase it (if T is smaller than T_m).

Another example of a drug-related increase in permeability at a temperature below T_m is shown in Figure VIII.5.2. The heat capacity profile of a DPPG:DPPG = 95:5 membrane is shifted toward lower temperature when adding the general anesthetic octanol. At a temperature of 37.1 °C, which is below the heat capacity maximum, the addition of anesthetics increases the heat capacity and thus the permeability. When adding octanol, the permeation rate suddenly increases (Figure VIII.5.2 (left)). If more anesthetics are added, the effect on the permeation rate increases even further (Figure VIII.5.2 (center)). Figure VIII.5.2 (right) shows that the increase in permeability is consistent with the increase in heat capacity caused by octanol (see inset). This supports the theoretical considerations made in Section VIII.1.2, i.e., that the membrane permeability is proportional to the heat capacity, and that drugs that change the melting profile of the membranes will alter the membrane permeability due to their effect on membrane fluctuations.

Physics of Nerves and Excitatory Membranes, First Edition. Thomas Heimburg.

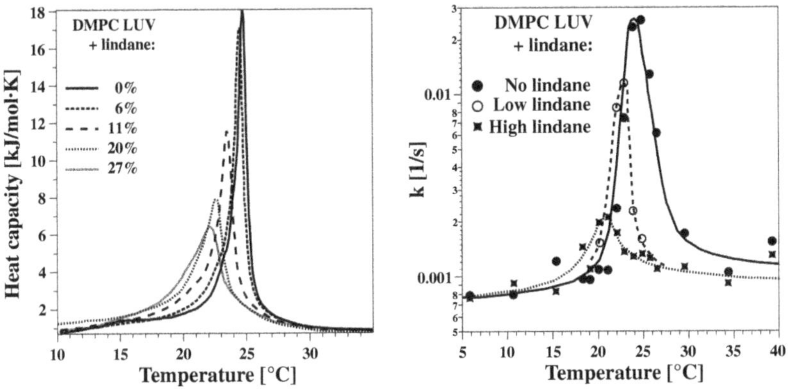

Figure VIII.5.1 DMPC large unilamellar vesicles (LUV) in the presence of different amounts of lindane. Left: Heat capacity profiles. Right: The permeability for Co^{2+}. Lindane causes both the heat capacity profiles and the permeability to shift toward lower temperature. The amplitude of both profiles decreases. Adapted from [3].

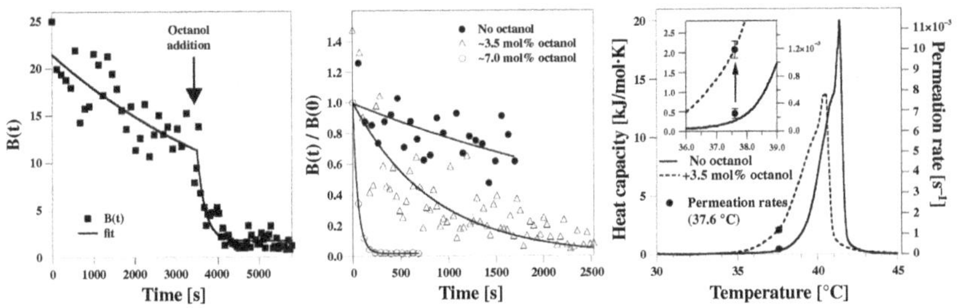

Figure VIII.5.2 Effect of the general anesthetic octanol on the permeation rate. Left: Permeation of the fluorescence marker R6G through DPPC:DPPG = 95:5 LUV at 37.1 °C after the sudden addition of 1-octanol at the time indicated by the arrow. Shown is the decrease of the marker concentration in the vesicles. The permeation rate increases drastically. Center: Permeation in the absence and presence of 3.5 mol% and 7 mol% octanol in the membrane at 37.6 °C. Right: Heat capacity profiles of the lipid LUV in the absence of octanol and in the presence of 7 mol% octanol. In the presence of octanol, the heat capacity profile is shifted toward lower temperatures. The solid circles indicate the permeation rates at 37.6 °C in the absence of octanol and in the presence of 7 mol% octanol. The increase in permeability caused by octanol correlates nicely with the increase in heat capacity at 37.6 °C. Adapted from [4].

VIII.5.2 Influence of Drugs on the Emergence of Channels

As shown in the previous chapters, the increase in permeability in the transition is accompanied by an increase in area fluctuations and consequently by an increase in the pore open probability. The emergence of pores is a consequence of the area fluctuations. Thus, one expects a higher probability of observing membrane pores in the transition. The second consequence is that the pore open lifetimes will be larger because, as shown in Chapter VI.7, the relation times in membranes and thus the timescale of the fluctuations are proportional to the heat capacity. It is likely that pore open lifetime is directly related to the fluctuation lifetime. In biological membranes at physiological temperatures, the lifetimes are in the millisecond range, which thus will be the typical timescale of pore lifetimes.

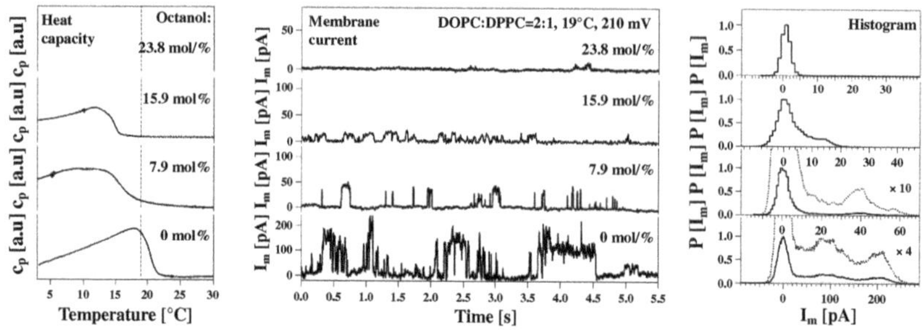

Figure VIII.5.3 The influence of the general anesthetic octanol on membranes made of a DOPC:DPPC = 2:1 mixture in 150 mM KCl. Left: Heat capacity profiles. Center: Current traces through BLMs at 19°C at 210 mV. Right: Histograms of current probabilities. Adapted from [4].

Figure VIII.5.3 shows an experiment on a BLM of a DOPC:DPPC = 2:1 membrane that displays a c_p maximum around 19°C [4]. The experimental temperature was also set to 19°C. Thus, one expects a maximum open probability and a maximum open lifetime. The addition of anesthetics shifts the transition toward lower temperature such that the heat capacity at the experimental temperature decreases. Thus, the presence of anesthetics will reduce both the open probability and the open lifetime. This is in fact found experimentally. Figure VIII.5.3 (left) shows the effect of octanol on the heat capacity, with a shift of the c_p profile toward lower temperatures upon an increase in the concentration of general anesthetic octanol. The center panel of Figure VIII.5.3 shows the corresponding current traces. In the absence of octanol (at the heat capacity maximum), one finds two current steps and open lifetimes that can be up to several hundred milliseconds or even seconds. In the presence of 7.9 mol% octanol, the open lifetimes are rather on the order of 100 ms. The amplitude of the single pore conductances becomes smaller, and the probability of finding an open channel decreases. A further increase in octanol concentration to 15.9 mol% leads to a further reduction in single-channel conductance, and the lifetimes are of an order below 100 ms (shorter than at 7.9 mol%). At 23.8 mol% octanol, channel events nearly completely disappear. One could interpret this as a blocking of pores. However, synthetic lipid membranes do not contain macromolecules that could be blocked. The disappearance of pores is a purely macroscopic thermodynamic effect that is caused by melting point depression (treated in Chapter IX.1).

Another interesting result for a membrane on a patch pipette is shown in Figure VIII.5.4. Here, three variables are varied in an experiment on a single membrane patch: voltage, anesthetics concentration, and temperature. A patch of a DMPC:DLPC = 10:1 membrane (with a heat capacity maximum at 22.1°C) measured above the melting regime at 28.9°C displays channels at 300 and 420 mV (Figure VIII.5.4 (left)). Remember that voltage increases transition temperatures (Chapter V.7). Thus, one can expect that at 300 and 430mV, the patch is close to the transition events. When a small amount of octanol is added to the membrane, the channel events disappear and do not resurface even at voltages up to 660 mV (Figure VIII.5.4 (center)). This is due to melting point depression and the shift of the membrane patch out of its transition regime into the fluid state. When the temperature is lowered by continuous cooling (temperature changes as a function of time), the overall membrane conductance increases (Figure VIII.5.4 (right)) because the membrane is moved into its

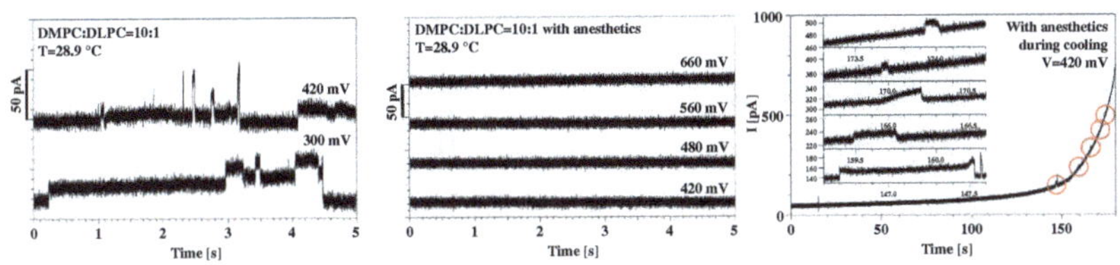

Figure VIII.5.4 Patch recordings of a DMPC:DLPC = 10:1 membrane. Left: At 300 and 420 mV, ion channel events appear. No channel events were present at lower voltages. Center: After addition of a small amount of the anesthetic octanol, channel events disappear due to lowering of the transition temperature by octanol. They do not reappear at voltages up to 660 mV. Right: Continuous cooling of the patch membrane. The overall membrane conductance increases because the membrane increasingly enters into the transition regime. Channel events reappear. The insets display channel events at the times indicated by the red circles. Adapted from [36].

transition regime again. Simultaneously, channel events resurface in the regime where the conductance increases. Thus, in this experiment, first the transition temperature increases as a function of voltage, leading to voltage-gated opening of channels. In a second step, the anesthetic octanol lowers the transition temperature again, leading to an inhibition of pore formation, and in a third step, the temperature is lowered until the membrane is in its transition regime again and channels emerge. This implies that all the three variables have the potential to enhance or reduce the probability of channel formation, and that they play hand-in-hand as is expected in a thermodynamic picture. These channels exist in a synthetic lipid membrane with a heat capacity similar to that of nerve membranes. They are voltage-gated, drug-controlled, and temperature-sensing all at the same time.

While we have only shown the effect of lindane and octanol on the permeability and the emergence of channels, it is to be expected that all molecules that alter the heat capacity profiles will affect the emergence of channel events. Notably, this includes the neurotransmitter serotonin (treated in Section IX.1.7).

In particular, one has to note that:

- the presence of channel events does not prove the activity of a macromolecular pore. It can be a consequence of large fluctuations within the lipid membrane close to transitions.
- the disappearance of channel events caused by drugs must not be caused by the blocking of a macromolecular pore. It can be caused by a macroscopic effect on the thermodynamics of the membrane as a whole (see Chapter IX.1).

VIII.5.3 Simulation of the Effect of Anesthetics

In Wodzinska et al. [14], pore formation and the effect of anesthetic on conductance were investigated both experimentally and theoretically by Monte Carlo simulations. Here, the results of the simulation are given. Several Monte Carlo steps are employed, which include the flipping of a membrane state and diffusion. The elementary Monte Carlo step for the pore formation is shown in Figure VIII.5.5 (left, top). Also shown is the number of pores as a function of simulation time for three different concentrations of anesthetics (0 mol%, 5mol%, and 10mol%) at the transition temperature of the pure lipid membrane (Figure VIII.5.5). The anesthetics shift the c_p profiles toward lower temperatures as

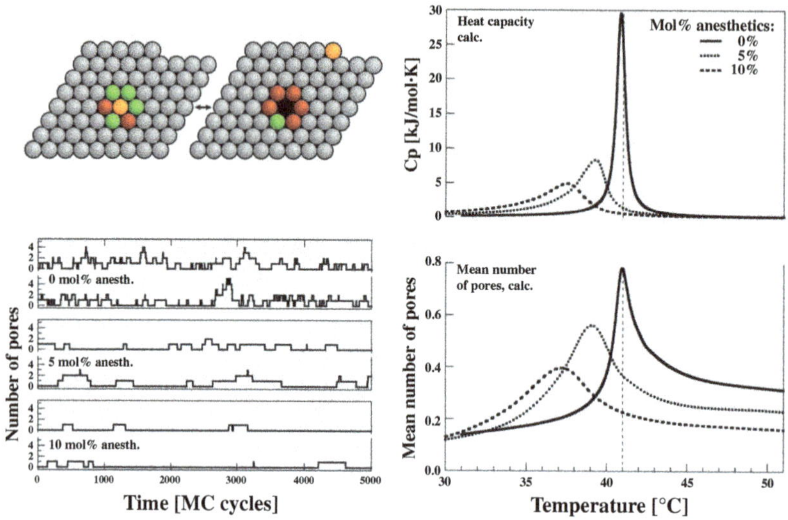

Figure VIII.5.5 Monte Carlo simulation of the effect of anesthetics on the pore formation probability. Left, top: The elementary Monte Carlo step leading to pore formation (green = fluid lipid, red = gel lipid, yellow = lipid or anesthetic molecule that is removed and placed at the edge). Left, bottom: Number of pores as a function of Monte Carlo time for three different concentrations of anesthetics in the membrane (0%, 5%, and 10%) at 41°C (=T_m). The right-hand panels show the calculated heat capacity for the three concentrations and the corresponding mean numbers of pores, which are proportional to the conductance of the membrane. Adapted from [14].

expected from the melting-point depression law (see Section IX.1.4 for a derivation, and the previous two sections for experimental verification). The mean pore number follows the behavior of the heat capacity profiles. The single-pore traces determined at $T_m = 41°C$ show that most pores appear at the melting temperature in the absence of anesthetics, but are increasingly suppressed as the anesthetic concentration increases. Since this simulation does not contain any macromolecule that can be blocked, the disappearance of channels is a consequence of alterations in the thermodynamics of the membrane caused by the ideal mixing of anesthetic molecules in the fluid phase. Note how similar the result of this simulation is to the experimental result of the interaction of the general anesthetic octanol with a DOPC:DPPC = 2:1 lipid mixture shown in Figure VIII.5.3. We can therefore assume that the pore formation process and the effect of anesthetics can be, in principle, understood by studying the thermodynamics of membranes. The phenomenon of anesthesia is discussed in depth in Chapter IX.1.

VIII.5.4 Summary: Key Ideas of Chapter VIII.5

We demonstrated here that the effect of drugs on the melting profiles of membranes can lead to a change in permeability and in the properties of membrane channels.

- In the melting transition, the permeability is enhanced. The effect of drugs (such as lindane or octanol) consists of a change of the melting profiles, which as a consequence

alters the permeability of the membrane. According to Section VIII.1.1, the permeability is roughly proportional to the excess heat capacity.
- If the experimental temperature is set to the heat capacity maximum, adding anesthetics will generally decrease the permeability of the membrane (see Chapter IX.1).
- If the experimental temperature is below the heat capacity maximum, anesthetics will increase the permeability.
- At physiological temperature, anesthetics will decrease the permeability, reduce the number of single-channel events, and reduce the open lifetime of the lipid ion channels.
- As a consequence of the above, the disappearance of channel events due to the action of drugs does not prove that a channel is blocked. In the case of lipid membranes and anesthetics, the drugs shift the heat capacity profiles, and the effect of the anesthetic drug is of a purely macroscopic thermodynamic nature. It is caused by melting point depression, which is an entropic phenomenon.

VIII.6

Channel Lifetimes

In the theory outlined in Chapter VI.7, the fluctuation lifetime (which is identical to the relaxation timescales) was found to be proportional to the heat capacity. In this chapter, we relate the probability of pore formation to the fluctuations of the membranes (see Section VIII.1.2). Thus, it is to be expected that the mean open times of channels reflect the relaxation lifetimes of the membrane itself.

VIII.6.1 Channel Lifetimes of Lipid Membranes

According to linear response theory (Chapter VI.7), the relaxation time is proportional to the heat capacity [37, 38].

$$\tau = \frac{T^2}{L}\Delta c_p. \tag{VIII.6.1}$$

Here, it is assumed that there is only one single relaxation process such that the relaxation of the variable A to a new equilibrium is given by $\Delta A(t) = \Delta A_0 \exp(-t/\tau)$. This seems a good approximation for the large enthalpy changes measured in calorimetry close to transitions described in Chapter VI.7.

While the slow enthalpy relaxation seems the dominant process, which, for lipids in melting transitions, is in the range of 1–30 s, there are faster processes down to the millisecond regime. In fact, using optical detection, Elamrani and Blume [39] found that there are fast contributions to the relaxation process in the millisecond regime. They fitted the relaxation process in membranes close to transition using a multi-exponential fit with three time constants. All time constants changed in direct correlation with the heat capacity when going through a transition. Thus, the notion that relaxation times change proportional to the heat capacity is still true, but the relaxation behavior deviates from single exponential behavior. Visual inspection of Figure VIII.5.3 indicates that the influence of anesthetics on the heat capacity of lipid membranes has a predictable effect on the lifetime. When considering lipid ion channels, the assumption is that their emergence is closely related to the fluctuation lifetimes.

One can analyze such processes by studying the current traces of channels and generating lifetime distributions of open or closed events. If we let $P(t)$ denote the probability that

Physics of Nerves and Excitatory Membranes, First Edition. Thomas Heimburg.

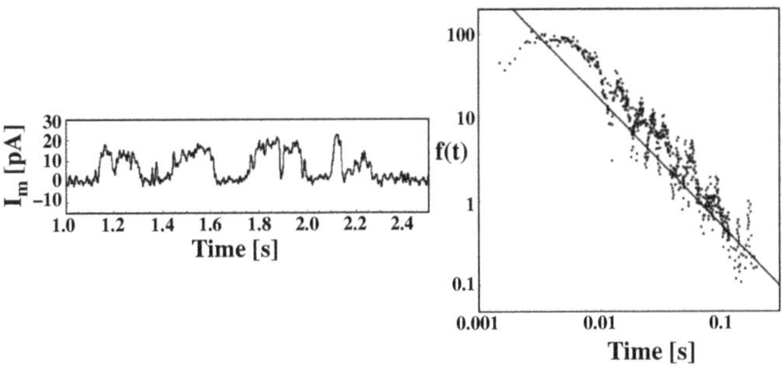

Figure VIII.6.1: Left: Current fluctuations of a DOPC:DPPC = 2:1 membrane containing 15.9 mol% octanol. Right: Lifetime distribution in a log-log representation. The slope is −3/2 for lifetimes above 3 ms. Adapted from [40].

a channel that is open at time $t = 0$ stays open until time t or longer, then the probability density function (PDF) of open times is given by $f(t) = -dP(t)/dt$. This function measures the likelihood that an open channel will close within a time interval t. Figure VIII.6.1 (left) shows the current fluctuations in a DOPC:DPPC = 2:1 membrane with 15.9 mol% octanol. The right-hand side of this figure shows the PDF of these fluctuations in a double-logarithmic plot. One can see that for lifetimes larger than 3 ms, the PDF can be fitted with a straight line with a slope close to −3/2. This corresponds to a power law $\propto t^{-3/2}$. This deviates from the single exponential behavior that was assumed in the treatment of the relaxation processes.

Gallaher et al. [40] adapted a model by Millhauser et al. [41] for protein lifetime statistics to the case of lipid pores. The kinetics of pore opening and closing was explained by a model with n states of a pore,

$$C \leftarrow O_1 \rightleftharpoons O_2 \rightleftharpoons O_2 \rightleftharpoons O_3 \rightleftharpoons \ldots \rightleftharpoons O_N, \qquad \text{(VIII.6.2)}$$

where C is a closed state surrounded by gel lipids and O_1, O_2, ... are open states with an increasing number of gel lipids surrounding them. The only transition between an open and a closed state is $O_1 \rightarrow C$. In order to reach state O_1, one needs to melt gel lipids at the pore interface. Gallaher et al. describe this model as follows: "States O_1 to O_N denote open states of the pore. We propose that going from state O_1 to state O_N corresponds to the freezing of an increasing amount of lipids in and around the pore's lining. Upon transiting to state O_2, a pore can only close again after it goes back to state O_1. States 1 to N all correspond to the same pore size and conductance. But as the state of the pore moves from state 1 to state N, the pore becomes embedded in an ever larger solid region. For the pore to close again, it is necessary that the 'frozen' surroundings melt again." Thus, the pore model in eq. (VIII.6.2) would display a strong temperature dependence close to transitions.

The probability of finding an open pore is given by the sum of the probabilities of the N open states:

$$P(t) = \sum_{1}^{N} p(i) \quad \text{and} \quad f(t) = -\frac{dP}{dt}. \qquad \text{(VIII.6.3)}$$

Gallaher et al. [40] derived the following fractal behavior using the model in eq. (VIII.6.2):

$$P(t) \approx \frac{1}{\sqrt{\pi t}} \quad \text{and} \quad f(t) \approx \frac{1}{2\sqrt{\pi t^{-3/2}}}, \tag{VIII.6.4}$$

which reproduces the experimental finding in Figure VIII.6.1, i.e., a linear behavior in a double-logarithmic plot with a slope of −3/2. Such results suggest that the lifetime kinetics of channels deviates from singe-exponential behavior.

VIII.6.2 The Control of Lifetimes by Temperature

It is clear that the heat capacity close to melting transitions is very temperature-dependent. The lifetime distribution in a lipid membrane BLM was investigated by Wunderlich et al. [5]. Figure VIII.6.2 shows a D15PC:DOPC = 95:5 membrane at 33 °C (above the melting regime) and at 31.5 °C (close to the melting regime). The corresponding current histograms were already shown in Figure VIII.4.3. While current fluctuations are fast above T_m, they display clearly visible channel events in the transition regime with a longer lifetime. Figure VIII.6.2 (right) shows the lifetime distribution for the two temperatures. Within the transition, the mean fluctuation lifetime is nearly an order of magnitude longer, which is expected from the fluctuation theorems (Chapter VI.7).

VIII.6.3 The Control of Lifetimes by Voltage

In Section VIII.2.2, we discussed the voltage gating of lipid channel events. In Section V.7.2, we showed that voltage can influence the melting behavior in both monolayers and bilayers. Thus, a likely mechanism for channel generation is the change of the melting temperature upon an increase in voltage. The emergence of channels goes along with a change

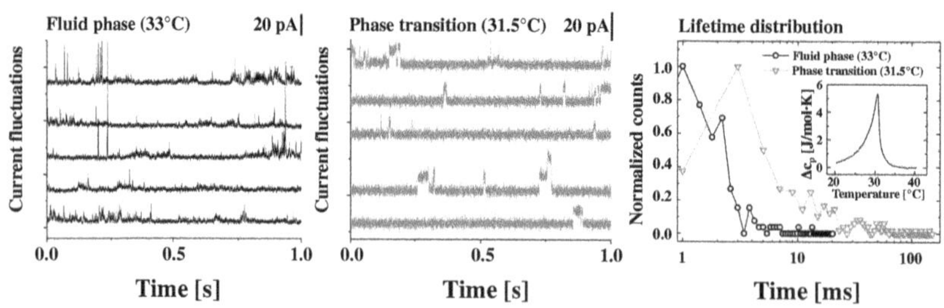

Figure VIII.6.2 Current fluctuations of a BLM of D15PC:DOPC = 95:5. The inset in the right-hand panel shows the heat capacity profiles. Left: Fluctuations at 33 °C (about 2.5 °C above the heat capacity maximum) are short-lived. Center: At 31.5 °C (about 1 °C above the heat capacity maximum), fluctuations are larger and display longer channel open events. Right: Lifetime distribution for the two temperatures. At 31.5 °C, the mean lifetime is about one order of magnitude larger. Adapted from [5].

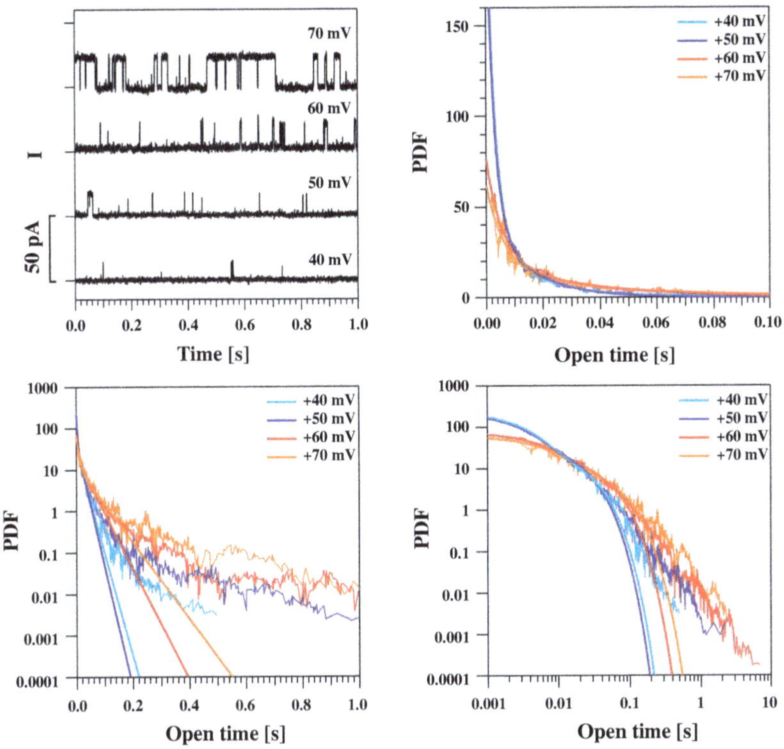

Figure VIII.6.3 The influence of voltage on lifetimes of lipid channels. Left, top: Voltage-dependent currents in a DMPC:DLPC = 10:1 mol/mol membrane (150 mM KCl, T = 30 °C) at four voltages discussed in Chapter VIII.2. Right, top: The PDF is shown as a function of voltage. A fit to a bi-exponential relation behavior (solid lines) seemingly leads to a satisfactory description of the lifetime distribution. Bottom, left: In a semilogarithmic plot of the same data, one recognizes that the bi-exponential fit (solid lines) dramatically underestimates the fraction of very long open events. Bottom, right: The double-logarithmic representation shows that the long lifetime distribution displays a linear behavior indicating that it is best described by a power law. Adapted from [25].

in their open lifetime. This phenomenon was investigated by Ref. [25]. Here, we show the lifetime distribution of the current traces shown in Figure VIII.6.3 (top, left). Displayed are the current traces at different voltages and the PDF of open lifetimes for the different voltages. Figure VIII.6.3 (top, right) shows the PDFs and the fits to bi-exponential functions, which seemingly describe the behavior very well. However, both the semilogarithmic and the double-logarithmic representations of the data (Figure VIII.6.3 (bottom, left and right)) demonstrate that bi-exponential fits underestimate the very long open times that are better described by a power law than by exponential fits. This is in agreement with the analysis in eq. (VIII.6.1) and Figure VIII.6.1. It can be seen that the lifetimes become longer when the voltage increases, which is also evident in the current traces. When membranes enter their phase transitions, lifetimes become longer due to the critical slowing down of the fluctuations. Thus, the finding that the voltage-induced opening of the lipid ion channels goes along with an increase in lifetimes is consistent with the fluctuation theorems.

VIII.6.4 Comparison to Lifetimes in Protein-containing Preparations

Membranes containing channel proteins also display lifetime distributions that are not simply single or bi-exponential. Figure VIII.6.4 shows a lifetime analysis of a single calcium-activated potassium channel (data adapted from Refs. [42, 43]). On the left-hand side, the data are shown in a linear representation. The solid line is fit to a bi-exponential curve with lifetimes of $\tau_1 = 0.053$ and $\tau_2 = 0.064$ ms. As in the case of synthetic membranes, semilogarithmic and double-logarithmic plots reveal that the probability of long lifetimes is underestimated. Korn and Horn [42] therefore tried to use three lifetimes to fit the data. Liebovitch [43] argued that the data are better fitted by a fractal behavior as we have also argued for the lipid pores earlier, and as it had been suggested by Ref. [41] for protein channels.

Figure VIII.6.5 shows the lifetime distribution for a potassium channel from mouse neurons as a function of voltage [44]. As for the synthetic lipid membrane shown in Figure VIII.6.3, increasing voltage increases the probability of long-lived pores. Thus, it seems that the lifetime distributions and their dependence on temperature and voltage are

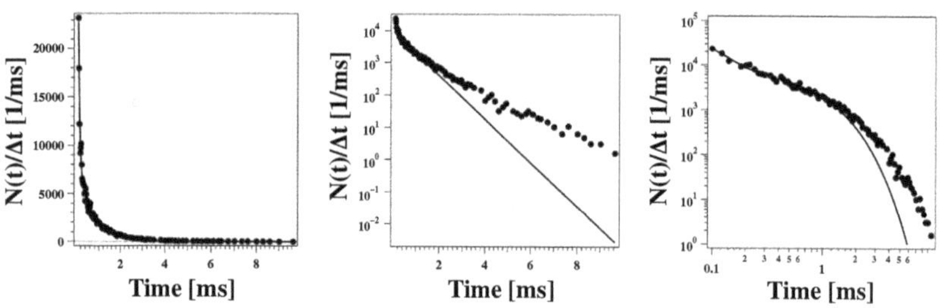

Figure VIII.6.4 Analysis of the open-time distribution of single-Ca-activated K channel currents from GH3 pituitary cells. Data adapted from [42, 43]. Left: The lifetime distribution and a seemingly good fit to two exponentials yielding two lifetimes (0.053 and 0.64 ms). Center: The semilogarithmic plot shows that a bi-exponential fit underestimates the probability of long open events. Right: This is also evident in the log-log plot.

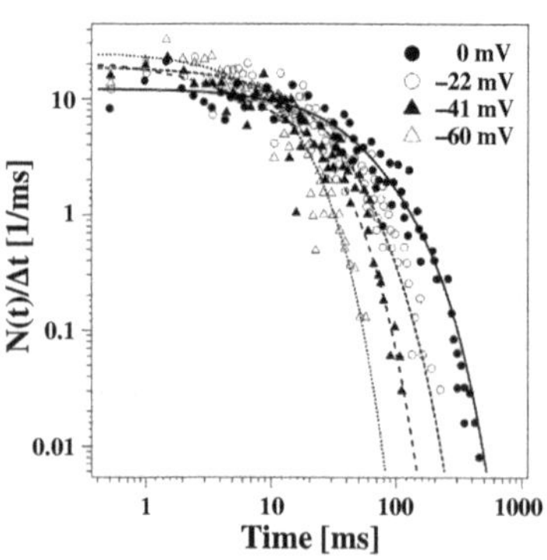

Figure VIII.6.5 Channel open-time distribution (log-log plot) for a potassium channel from mouse embryonic hippocampal neurons as a function of voltage. The lines are a guide to the eye. With increasing voltage, the open times become larger. Compare this to the voltage dependence in pure lipid membranes in Figure VIII.6.3. Adapted from [44].

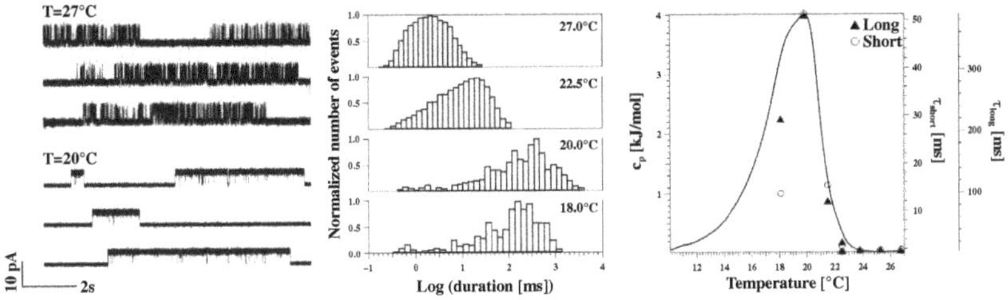

Figure VIII.6.6 Change in the lifetime distribution of KcsA potassium channels caused by changes in temperature in a synthetic membrane (DOPE:DOPG = 3:1) compared to the heat capacity profile of the host membrane. Left: Current traces at 27 °C and 20 °C. Center: Lifetime distributional four temperatures. In Seeger et al. [35], the distributions were fitted with two lifetimes corresponding to the bi-exponential fits discussed earlier. Right: The two lifetimes of the bi-exponential behavior plotted as a function of temperature. The solid line is the melting profile of the host matrix. Adapted from [35].

very similar for lipid ion channels and for protein channels. There is no indication in the data, neither quantitatively nor qualitatively, that these are different events.

An interesting experiment on the KcsA potassium channel reconstituted in a POPE: POPG = 3:1 from Seeger et al. [35] is shown in Figure VIII.6.6. We have already shown this lipid-protein system in Figure IV.5.15, where it could be seen that the conductance of the KcsA potassium channel follows the heat capacity profile of the host lipid membrane. Here, we show that the lifetimes of the potassium channel follow the heat capacity profile of the host matrix. It is an amusing fact that the electrophysiology community has, in an evolutionary process, optimized the composition of a lipid membrane for the reconstitution of channel proteins in a manner that the membrane shows melting events exactly at room (and experimental) temperature. Thus, channels are expected in the lipid membrane itself. Unfortunately, it is widely unknown in the community that this melting transition is present in their experiments.

Summarizing, the lifetimes of channels in artificial lipid membranes and in protein-containing preparations are strikingly similar. In the experiment shown in Figure VIII.6.6, one might even suspect that the protein properties were not measured at all and that one only sees the lipid membrane behavior.

VIII.6.5 Summary: Key Ideas of Chapter VIII.6

In this chapter, we showed that the open-lifetime distributions of synthetic lipid membranes close to transitions and protein-containing membranes are strikingly similar.

- Lifetimes are longer in the transition regime. This is consistent with linear response theory that predicts longer lifetimes when fluctuations are large.
- In most preparations, it seems that the lifetime distributions are not consistent with a single-exponential process. It has been proposed by various authors that the lifetimes follow a fractal or a power law behavior. This is both true in synthetic lipid membranes and in preparations containing channel proteins.
- Lipid ion channel lifetimes increase in direct relation with the heat capacity.

- Lipid ion channel lifetimes also depend on voltage – probably due to the influence of voltage on the melting behavior of lipid membranes. The same behavior was found in membranes containing channel proteins.
- Overall, there are no obvious differences in the open-time distribution of protein-containing membranes and protein-free preparations.

VIII.7

Selectivity of Lipid Channels

There exist many different channel proteins that conduct some ions more than others, e.g., sodium or potassium. To our knowledge, there exists only one experiment that addresses this phenomenon for lipid channels.

VIII.7.1 Conductance of Single Lipid Membrane Pores and the Hofmeister Series

Selectivity of channel proteins for ions is a concept originating from Hodgkin and Huxley [32, 45, 46]. The sodium channel is believed to predominantly conduct sodium ions, and the potassium channel conducts potassium ions. Since the conductances of the two proteins display different time dependences, the membrane is semipermeable in a time-dependent manner.

We have found in the previous sections that the channel events in pure lipid membranes are remarkably similar to those of proteins. Therefore, the question arises whether synthetic lipid membranes also display semipermeability. Unfortunately, this has not been well investigated. To our knowledge, the only study addressing this question is by Antonov and collaborators [34]. They studied the conductance of a black lipid membrane (BLM) made of DPPC close to its melting temperature for 1 M solutions of different alkali metal ions (see Figure VIII.7.1). These ions are lithium (Li), sodium (Na), potassium (K),rubidium (Rb), and cesium (Cs) in the order of their atomic masses. This order is also the order of the Hofmeister series that describes the lyotropic properties of ions, i.e., their ability to interact and order water molecules in their surroundings. The Hofmeister series is reviewed in Refs. [47–50]. Thus, the aforementioned order also indicates a size of the hydrated ions, with lithium being the smallest and cesium being the largest. Antonov and collaborators found that the single-channel conductance of a DPPC BLM is constant in the voltage range from −80 to 80 mV (see Figure VIII.7.1). For the alkali metals, they found that the channel conductance of large membrane pores was $g_{Li} = 30.7$ nS, $g_{Na} = 10.0$ nS, $g_K = 6.7$ nS, $g_{Rb} = 3.3$ nS, and $g_{Cs} = 2.39$ nS. Thus, the conductances follow the Hofmeister series. In Ref. [34], the pore conductance was very high, and the selectivity is small. An all-or-nothing selectivity for a particular ion could not be demonstrated. It seems unlikely that a pure lipid membrane can in fact display such a selectivity. However, more detailed experiments are lacking.

Physics of Nerves and Excitatory Membranes, First Edition. Thomas Heimburg.

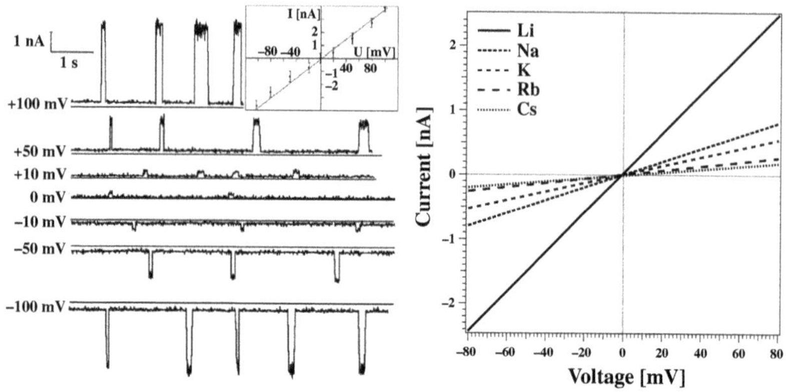

Figure VIII.7.1 Left: Channel events in DPPC membranes around 43°C in the presence of 1 M LiCl measured at different voltages indicate a linear current–voltage relation, suggesting constant single-channel conductance. Adapted from [34]. Right: The linear current–voltage relations for 1 M LiCl, NaCl, KCl, RbCl, and CsCl show that the single-channel conductance decreases in the order of how lyotropic the alkali metal ions are. Adapted from [34].

VIII.7.2 Summary: Key Ideas of Chapter VIII.7

There is only one study addressing lipid membrane specificity for the conductance of ions. It was found that

- There exists a mild selectivity for different alkali metal ions.
- The single-channel conductance follows both the atomic weight and the position in the Hofmeister series in the order $g_{Li} > g_{Na} > g_K > g_{Rb} > g_{Cs}$.
- It seems likely that this is mostly a size effect due to a different number of water molecules associated with the ions.
- There is no indication of an all-or-nothing selectivity for one particular ion.

VIII.8

Proteins as Catalysts

Membrane proteins can influence the thermodynamic properties of lipid membranes. This is true for both integral and peripheral proteins as discussed in Chapters V.3.1 and V.8 [12, 51]. Thus, the proteins possess chemical potentials and contribute to the thermodynamics of the membrane system. For instance, the presence of proteins and the variation of their chemical potentials influence the melting behavior and the magnitude of the fluctuations of the lipid membrane – just as is the case for all other intensive variables. Figure VIII.8.1 (center, top) shows the calorimetric profiles of the band 3 protein of erythrocytes and of cytochrome b_5 (Figure VIII.8.1, (center, bottom)) reconstituted into synthetic lipid membranes [52, 53]. Band 3 protein increases the temperature regime of melting in membranes, and cytochrome b_5 lowers it.

VIII.8.1 Proteins as Catalysts for Lipid Channel Formation

Since proteins influence lipid membrane fluctuations, they also affect the occurrence of lipid channels. If the hydrophobic part of the protein is more extended than the hydrocarbon core of a fluid membrane, it will match better with ordered lipids (see Section V.3.1). As a consequence, lipids tend to be more gel-like at the interface of the protein [54]. If the proteins have short hydrophobic cores (e.g., gramicidin A), it will favor fluid lipids in its vicinity (Figure VIII.8.1 (left, bottom)). The first class of proteins will shift melting events toward higher temperatures, while the second class will shift them toward lower temperatures.

If a protein that matches the gel state of the membrane is located in a fluid membrane, it tends to surround itself by a gel lipid layer, which on average becomes more fluid as a function of the distance from the protein. As a consequence, there is a regime of high fluctuations close to the protein [54, 55], meaning that both heat capacity and compressibility can be altered close to a protein. For this reason, the presence of a protein displays the potential to induce lipid pores in its proximity. Thus, proteins can facilitate the formation of lipid pores at their outer interface. Figure VIII.8.1 (right) shows Monte Carlo simulations of such a situation [12]. The simulations show a protein (black) in a membrane at temperatures below and above the melting regime. Dark red shades indicate small fluctuations, while bright yellow shades display large fluctuations. It can be seen that a protein favoring the fluid lipid state would tend to create regimes of large fluctuations in its environment at temperatures below the melting temperature (top panels). It thus seems likely

Physics of Nerves and Excitatory Membranes, First Edition. Thomas Heimburg.

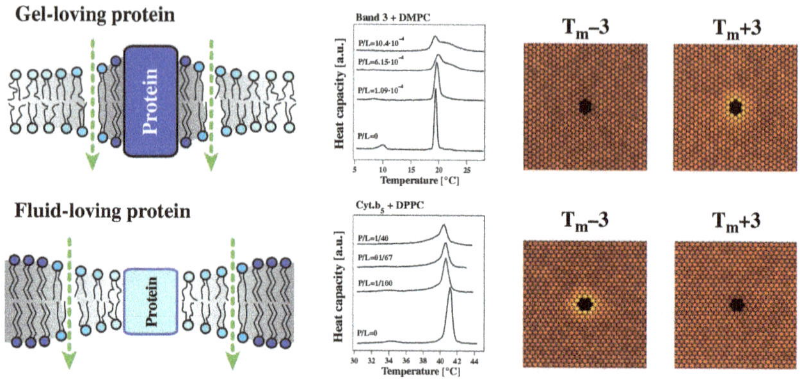

Figure VIII.8.1 Left: Interaction of proteins with lipid membranes and hydrophobic matching. The top image refers to a protein that favors the solid state, while the bottom image refers to a protein favoring the liquid state. Left: Schematic drawing of the lipid arrangement around proteins. Center: The influence of proteins on the heat capacity profile of a synthetic membrane (top: band 3 protein of erythrocytes in DMPC shifts heat capacity profiles toward higher temperatures; bottom: cytochrome b5 in DPPC shifts c_p profiles toward lower temperatures). See Section V.8.3 for details. Right: Monte Carlo simulations of the local fluctuations (yellow indicates larger fluctuations) of the lipid membrane close to the protein. Top: Fluctuations are enhanced at the protein interface above the melting temperature T_m and unaffected below. Bottom: Fluctuations are enhanced below T_m and unaffected above the transition. From [12]/with permission of American Chemical Society.

that proteins can induce channel activity and catalyze conduction without being channels themselves. This is due to the effect of the proteins on the cooperative fluctuations in the lipid membrane. It should generally be possible to estimate this effect from the influence of the protein on melting transitions. The data by Seeger et al. on the conductance of the KcsA potassium channel in lipid membranes with melting events might be an example of an integral protein acting as a catalyst for membrane fluctuations and pore generation (Section IV.5.4, see Figures IV.5.15 and VIII.6.6). Peripheral proteins can also shift transitions, for instance by shielding electrostatic charges on the surface [51]. Thus, even though peripheral proteins don't span the hydrophobic core of the membrane, it is possible that they can enhance fluctuations and thus act as catalysts for lipid channels. An interesting example was reported by Stoddart et al. [56]. They found that the peripheral truncated α-hemolysin enhances pore and channel formation in membranes.

VIII.8.2 Summary: Key Ideas of Chapter VIII.8

We demonstrate here that proteins influence membrane melting behavior. This was also discussed in Section V.8.3. This has consequences for the probability of lipid channel formation. We show in particular that:

- Fluctuations of lipid membranes at the lipid–protein interface display a different magnitude than in the rest of the membrane, depending on whether the protein prefers to dissolve in the fluid or in the gel phase.

- If an electrophysiological measurement is performed slightly above the melting transition (as typical for biological membranes at physiological temperatures), gel-loving proteins (such as the band 3 protein of erythrocytes) can induce fluctuations large enough to result in lipid channel formation. These pores emerge at the interface between lipid and protein. In this case, the protein acts as a catalyst for conduction through lipid channels.
- If the protein prefers liquid environments, the same effect would be seen below the melting temperature of the membrane.
- Peripheral proteins can also influence the melting of membranes. Thus, they can enhance or reduce membrane fluctuations and thereby promote or inhibit the formation of lipid channels. An example of such a case is truncated α-hemolysis [56].
- The striking resemblance of both the conductance and the open lifetimes of the KcsA channel reconstituted into a synthetic lipid membrane with the heat capacity profile of the host membrane (see Figures IV.5.15 and VIII.6.6) provides additional evidence for the possible role of membrane proteins as catalysts [35].

Part IX

Medical Consequences

In the previous parts, we demonstrated that solitary waves can propagate in nerve axons. The excitability of these pulses depends on the intensive thermodynamic variables, which include the chemical potential of drugs, hydrostatic pressure, tension in the axon, temperature, and pH. In the following, we show that there exist various medical conditions in which changes in these variables are relevant, e.g., in anesthesia and its pressure reversal.

Transitions in biology are the most important regulatory factors. This is because, close to transitions, the susceptibilities are high. This implies that a small change in an intensive variable leads to a large change in an extensive variable. For example, if the lateral compressibility of a membrane is high, a small change in membrane tension may lead to a large change in the area and state of a membrane. Similar statements are also true for proteins and DNA, and for this reason, it is not accidental that membranes, proteins, and DNA all exist in states close to transitions. DNA can be destabilized by changes in ionic strength, in pH, and by the binding of proteins, thus facilitating transcription processes or replication. For proteins, being closer to transitions means that conformational changes become easier, and function can be regulated by changes in the thermodynamic variables in a more subtle and sensitive manner. While this book focuses on the thermodynamics of biological membranes and its effect on nerve pulse propagation, membrane heterogeneity, and membrane permeability, one could write a similar book about the thermodynamic control of proteins and DNA.

The thermodynamic theory presented in this book considers the complete biological membrane to be a receptor for all thermodynamic variables, thereby providing a comprehensive framework for understanding the relationship between biological systems and the environment. The closer physiological temperature is to the melting transition of the membrane, the more sensitively the membrane responds to perturbations. This is due to the increase of susceptibilities such as compressibility, bending elasticity, and capacitance. The meaning of the word "susceptibility" implies that a high susceptibility makes the membrane more susceptible to small changes in the intensive variables.

We showed in the previous parts that biological membranes close to transitions support solitary electromechanical pulses that are remarkably similar to the nerve pulse. The theory of soliton excitation was described in Chapter VII.6. When discussing the medical consequences of the thermodynamics theory, we foremost focus on nerve excitability and how it is influenced by changing the physical properties of the membranes. The main idea is

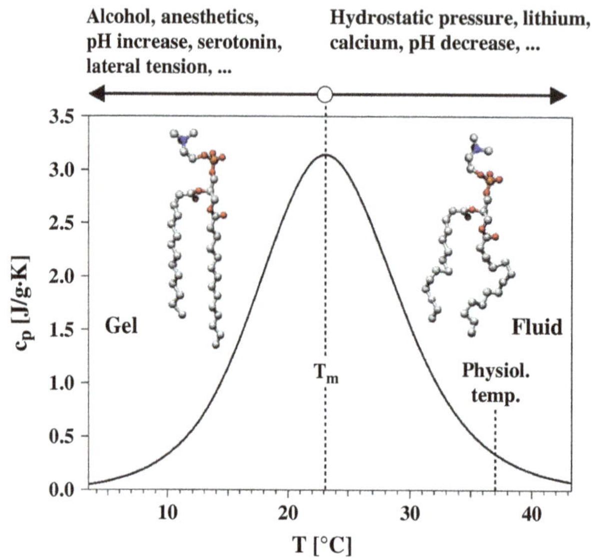

Figure IX.1 Melting transition of a biomembrane (adapted to the melting profile of porcine spine membranes). There are two important temperature: the melting temperature, T_m, and the physiological temperature. The distance between these two temperatures determines the excitation threshold.

that, to move a biomembrane through its melting transition, free energy must be provided that depends on how far the melting transition is from the physiological temperature. Everything that can influence the melting transition will thus influence the excitability of a nerve. Changes that move transitions upward will decrease the excitation threshold, and changes that lower transition temperatures will make nerves less excitable. It is easier to excite a soliton when decreasing the threshold, which is induced by

- lowering of body temperature
- increase in hydrostatic pressure
- lowering of pH, i.e., higher proton concentration
- increase in the concentration of divalent cations such as Ca^{2+} and Mg^{2+}
- increase in the concentration of lithium ions
- voltage gradients in the membrane plane

An increasing threshold is caused by

- increase in body temperature
- increase in pH, i.e., lower proton concentration
- anesthetic drugs and some neurotransmitters such as serotonin
- stretching of nerves

In general, changes in all intensive variables can change the transitions in the membrane. These changes are illustrated in Figure IX.1. This applies to all intensive variables that are conjugated to an extensive variable that changes in the transition, i.e., pressure and volume changes. Importantly, it also includes membrane proteins.

IX.1

Anesthesia

IX.1.1 Introduction to Anesthesia

Anesthesia has been a well-known phenomenon for thousands of years in the form of ethanol narcosis. In 1799, Sir Humphry Davy (1778–1829, see Figure IX.1.1 (left)) synthesized laughing gas (N_2O). He tested it on himself in a closed chamber especially designed for this purpose and found an anesthetizing effect. He tested laughing gas on other people during public "parties." The protocols of these experiments [1] – the so-called laughing gas protocols – are an exciting and funny-to-read early testimony of the euphoric effect of laughing gas.

Another anesthetic drug studied at the same time is diethyl ether (or just ether). It was already known in the Middle Ages, and its analgesic effect (as a pain reliever) was described by the Swiss physician Paracelsus (1493–1541). In 1846, the American dentist William T.G. Morton (Figure IX.1.1 (right)) demonstrated its anesthetic effect during a public surgery on a neck tumor at the Massachusetts General Hospital in Boston.

Figure IX.1.1 Left: Humphry Davy (1778–1829) at the age of 25 years (painting by Henry Howarth, 1803). Henry Howard/Wikimedia Commons/Public domain. Right: William T.G. Morton (1819–1868). Unknown author/Wikimedia Commons/Public domain.

Physics of Nerves and Excitatory Membranes, First Edition. Thomas Heimburg.

We distinguish between general anesthetics and local anesthetics. General anesthetics are used to anesthetize the whole body during surgery. General anesthesia is accompanied by a loss of conscience. The drugs exist in both gaseous and liquid forms. Well-known volatile anesthetics are desflurane, isoflurane, sevoflurane, enflurane, nitrous oxide, halothane, methoxyflurane, and the noble gas xenon. Volatile anesthetics are inhaled. Liquid anesthetics include propofol, etomidate, barbiturates (e.g., midazolam), and ketamine, which are injected. Further, the alcohols from ethanol to decanol are general anesthetics. Figure IX.1.2 shows some examples of general anesthetics. They are chemically very different and include the noble gas xenon, a single inert atom. The anesthetics distribute over the whole body. One can determine an effective dose ED_{50} *in the air or the blood at which 50% of the individuals are anesthetized. The* ED_{50} depends on the species and what is measured during the experiment. For instance, in tadpoles, one has used the dose at which 50% of the individuals sink to the ground, which may rather reflect immobilization and not be identical to a loss of consciousness. In humans, one uses other measures such as eye reflexes. Thus, while

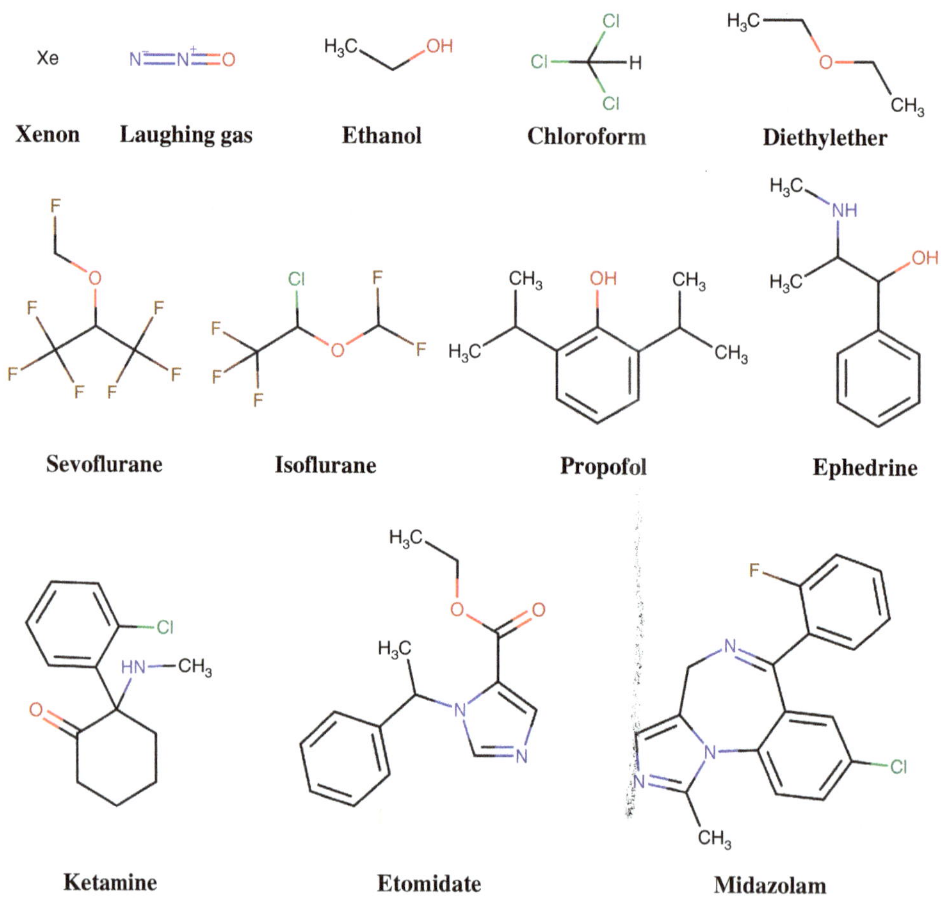

Figure IX.1.2 Selection of general anesthetics, ranging from the noble gas xenon to a large molecule such as midazolam.

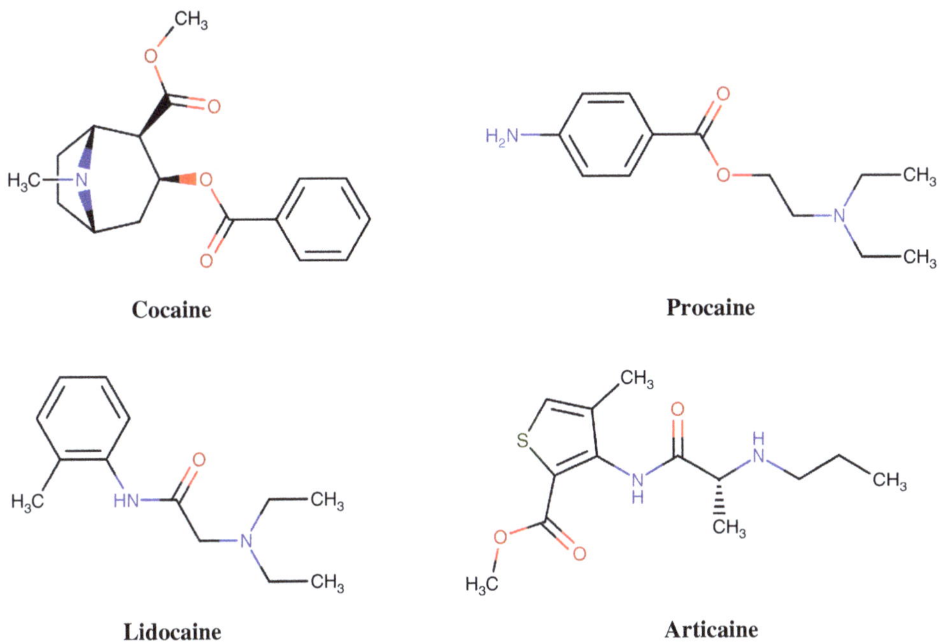

Figure IX.1.3 Cocaine and a selection of local anesthetics.

general trends are the same across different species, the absolute numbers for the critical dose may vary depending on the definition of anesthesia.

The second class of anesthetics are the local anesthetics. Local anesthetics are used to anesthetize locally without loss of consciousness. The first local anesthetic was cocaine, which, due to its addictive properties, is no longer used. Common local anesthetics are lidocaine, articaine, procaine, and bupivacaine (see Figure IX.1.3). They do not display the addictive properties of cocaine. An important feature is that they can exist in a negatively charged form at high pH and a neutral form at low pH. This feature is probably responsible for their ability to remain locally confined in the tissue into which they are injected.

IX.1.2 The Meyer–Overton Correlation

In 1901, Charles Ernest Overton (1865–1933) published a remarkable book called "Studien über die Narkose" (Studies of Narcosis) describing some universal properties of general anesthetics [2] (Figure IX.1.4) (left). The book was deemed so important that 90 years later it was translated into English [3]. In his book, Overton investigated a large range of general anesthetics and tested them on tadpoles. He defined an effective dose for the anesthetics called ED_{50} (the aqueous concentration of a general anesthetic) where 50% of the tadpoles would sink to the ground. He found the following:

- When exchanging the aqueous medium containing anesthetics for water, the effect of the tadpole anesthesia was reversed. The tadpoles woke up.

- The effective dose ED_{50} of anesthetic was found to be inversely proportional to the oil/water partition coefficient P (Overton used olive oil as the reference) with

$$log(P) = -log[ED_{50}] + \text{const.} \qquad \text{or} \tag{IX.1.1}$$

$$P \cdot [ED_{50}] = \text{const.} \qquad \text{Meyer-Overton correlation}$$

- $P{\cdot}[ED_{50}]$ corresponds to the concentration of anesthetics in the membrane. For tadpole anesthesia, it is 2.6 mol% in the membrane [4, 5].
- The effect of anesthetics is additive, i.e., 50% of the effective dose of one kind of general anesthetic plus 50% of the effective dose of another anesthetic add up to 100% of the critical dose. This implies that all general anesthetics must act by the same mechanism. Since one good anesthetic is the noble gas xenon, all anesthetics must act in a similar manner as an inert substance without specific interactions. This basically rules out any mechanism for anesthesia that involves specific binding to a receptor (Figure IX.1.4) (right).

Since Hans Meyer made similar observations for different alcohols [6] , the correlation in eq. (IX.1.1) is now known as the "Meyer–Overton" correlation. Overton drew two important conclusions from his observation. This correlation is shown in Figure IX.1.5.

Figure IX.1.4 The famous Meyer–Overton correlation is named after Charles E. Overton (1865–1933) and Hans Meyer (1853–1939). Courtesy of Dag Lundberg (Lund University).

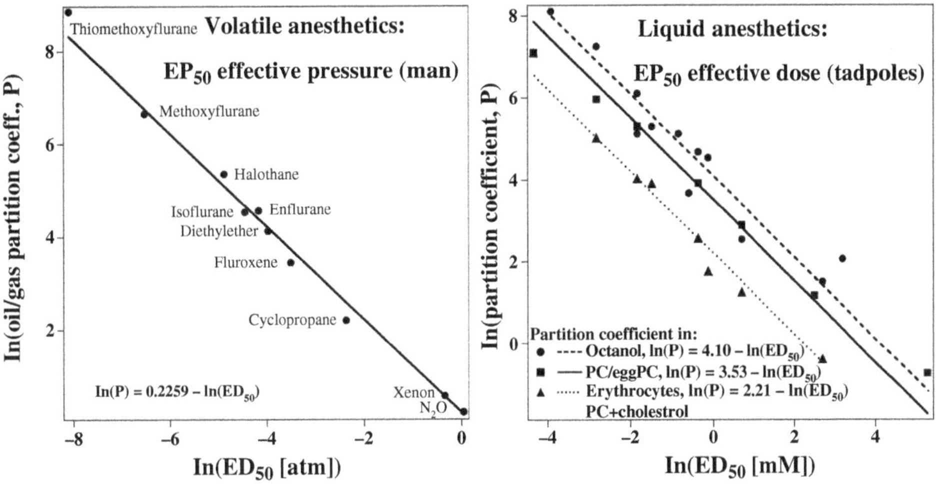

Figure IX.1.5 The Meyer–Overton correlation for volatile anesthetics (left) and for liquid anesthetics. The slope is −1, but the absolute value depends on the reference substance for the partition coefficient and the method by which the ED_{50} is measured. Data adapted from [3, 7]. Figure from [8].

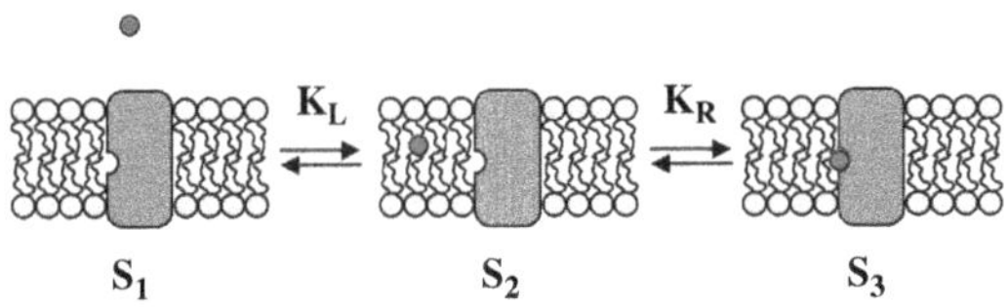

Figure IX.1.6 Hypothetical binding of an anesthetic to a receptor protein in two steps: First, the anesthetic drug (shown as a red circle) absorbs into a membrane with an association constant K_L. Second, the drug adsorbs from the membrane phase to a site on a receptor.

1. The partition coefficient of anesthetics in oil is simply related to the partition coefficient of anesthetics in the hydrophobic core of the lipid membrane.
2. The simple and unspecific inverse correlation between partition coefficient and effective dose suggests that the underlying explanation of the action of anesthetics originates from simple generic laws of physical chemistry.

IX.1.2.1 Why Is the Meyer–Overton Correlation Not Compatible with the Binding of the Drugs to a Receptor Protein?

It is assumed by some authors that anesthetic drugs act via a receptor mechanics, i.e., the action of anesthesia is caused by the binding of anesthetic drugs to receptor proteins. Nicolas P. Franks studied the effect of general anesthetics on firefly luciferase and found effects in their fluorescence properties (e.g., [9–11]). Luciferase is a protein unrelated to anesthesia but clearly demonstrates that proteins may be influenced by the presence of the drugs.

In Figure IX.1.6, a schematic diagram is shown with three states. In the first state S_1, the anesthetic drug (represented by a red circle) is dissolved in the aqueous medium. In the second state S_2, the drug is dissolved in the hydrophobic part of the lipid membrane. In the third state S_3, the drug is bound to a binding site on a receptor protein hidden in the membrane core. Since the free energy of binding is a function of state, one can describe the adsorption process to the receptor protein in two steps. State S_1 represents a membrane containing an integral receptor protein and a drug in the aqueous environment with concentrations $[A_{H_2O}]$. State S_2 is a membrane with a drug molecule dissolved in the hydrophobic core of the lipid membranes with drug concentration in the membrane $[A_m]$. State S_3 describes a membrane with the drug bound to a binding site at the membrane receptor. The total concentration of receptors shall be $[R] = [R_A] + [R_0]$, where $[R_0]$ is the concentration of receptors with empty binding sites, and R_A the concentrations of receptors with a drug bound. We will tentatively assume that the phenomenon of anesthesia is related to the concentration of $[R_A]$. The equilibrium between S_1 and S_2 shall be described by the absorption constant K_L, which is directly proportional to the partition coefficient of the anesthetic drug. The equilibrium between S_2 and S_3 is represented by the binding constant K_R of the drug to the receptor protein. Thus,

$$\begin{aligned} &\frac{[A_m]}{[A_{H_2O}]} = P \quad \rightarrow \quad [A_m] = P \cdot [A_{H_2O}] \\ &\frac{[R_A]}{[R_0][A_m]} = K_R \quad [R_A] = R_A = K_R \cdot [R_0] \cdot [A_m] = K_R \cdot [R_0] \cdot P \cdot [A_{H_2O}] \end{aligned} \tag{IX.1.2}$$

When we take into account that the total receptor concentration is $[R] = [R_0] + [R_A]$, we finally find that

$$[R_A] = [R] \cdot \frac{K_R \cdot P \cdot [A_{H_2O}]}{1 + K_R \cdot P \cdot [A_{H_2O}]} \tag{IX.1.3}$$

This equation implies two things:

1. At high concentration $[A_{H_2O}]$, the concentration of R_A becomes independent of the anesthetic concentration, i.e., the anesthetic effect saturates for high aqueous concentrations and $[R_A] = [R]$.
2. We know that the Meyer–Overton correlation implies that at critical dose, $P \cdot [ED_{50}] = C$, where C is a constant. Therefore:

$$[R_A] = [R] \cdot \frac{K_R \cdot C}{1 + K_R \cdot C}, \tag{IX.1.4}$$

   independent of the chemical nature of the anesthetic drug. We related the anesthetic action to the function of a receptor, R_A, with an associated anesthetic molecule. Its concentration should be the same for all anesthetics at critical dose. This would imply that the binding constant K_R of the drug to the receptor protein is the same for all general anesthetics, including xenon and sevoflurane. It seems highly unlikely that the binding constant of a noble gas to a receptor site can be the same for such different molecules.
3. Equation (IX.1.3) implies that the action of two different anesthetics is not additive, because $[R_A]$ is not a linear function of $[A_{H_2O}]$.

Thus, a simple receptor mechanism is inconsistent with Overton's findings. Keith Miller from Harvard Medical School noted that "..., if a protein picture was to be maintained one has to abandon a unique mechanism for anesthesia (Keith Miller, Harvard Medical School, private communication, 2006.)" This means, one would require a different protein-related mechanism for each anesthetic drug while still having to obey the Meyer–Overton correlation. This seems practically impossible.

IX.1.2.2 Deviations from the Meyer–Overton Correlation

The findings by Meyer and Overton are very convincing and are still considered to be true after more than a century. Nevertheless, the Meyer–Overton correlation is not more than a correlation and does not suggest a mechanism for anesthesia beyond the statement that simple laws are at its basis. Some authors have challenged the validity of the correlation on the grounds that there exist membrane-soluble molecules that do not cause anesthesia, most notably the lipids themselves, but also long-chain alcohols starting with more than 10 carbons, cholesterol, and other substances. Thus, they argue, the partition coefficient cannot predict the effectiveness of a drug as an anesthetic. They conclude that, for this reason, the correlation is simply not valid. We show below that a slight modification of the Meyer–Overton correlation resolves this problem.

IX.1.3 Reformulation of the Meyer–Overton Correlation

We have shown earlier that biological membranes display melting transitions slightly below physiological temperature. Let us consider a modification of the Meyer–Overton correlation, which states the following:

- Anesthetics are ideally soluble in fluid membranes, but insoluble in gel membranes.
- For the partitioning of anesthetics in the fluid phase, the Meyer–Overton correlation applies.

This does not violate the Meyer–Overton correlation because the reference substances when determining the partition coefficient P are always liquid oil and water. It adds, however, a constraint. As we will show in the following, this suggests a physical mechanism for anesthesia. It will explain the exceptions from the original correlation.

IX.1.4 Melting-point Depression

We assume now that the partition coefficient of general anesthetics in the gel (solid) membrane phase is very small, but that anesthetics are ideally soluble in the fluid membrane. The same assumption is made when calculating the freezing-point depression caused by salt on ice. In the case of ice, this is plausible because ions would create defects in a perfect crystal. This will make it unfavorable for ions to dissolve in ice. The insolubility of ions in ice leads to van't Hoff's freezing-point depression law, which turns out to be valid for anesthetics in lipid membranes. We briefly derive this law as follows.

Let us assume that the molar fraction of a molecule A (an anesthetic) in substance L (a lipid in a membrane) is given by x_A, and the molar fraction of the lipid is $x_L = 1 - x_A$. There exist two states of L, gel and fluid, and we therefore have molar factions of the lipid x_L^g

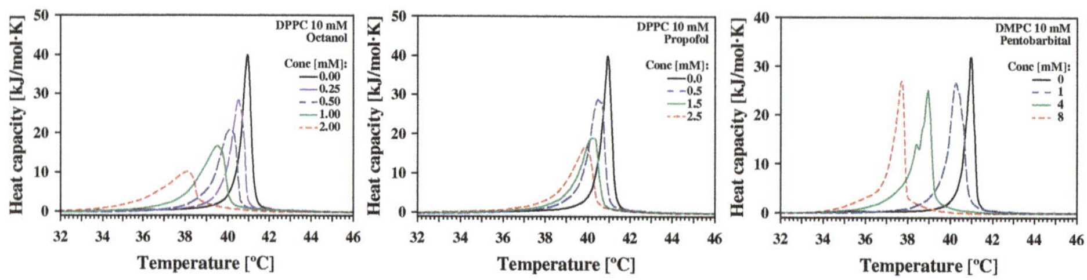

Figure IX.1.7 Heat capacity profiles of large unilamellar vesicles of DPPC in the absence and presence of three different general anesthetics. Left: Octanol, Center: Propofol, Right: Pentobarbital. All three anesthetics cause a lowering of the melting temperature.

and x_L^f in the gel and fluid, respectively. We further assume that substance A exists only in one state and is insoluble in the gel phase of the lipid. The chemical potentials of the lipid in the two states of the membrane are given by

$$\begin{aligned} \mu_L^f &= \mu_{L,0}^f + RT\ln(x_L^f) \\ \mu_L^g &= \mu_{L,0}^g. \end{aligned} \tag{IX.1.5}$$

The second equation does not contain the logarithmic term because $x_L^g = 1$. Since at the transition, $\mu_L^f = \mu_L^g$, $\Delta\mu_{L,0} = \mu_{L,0}^f - \mu_{L,0}^g = \Delta H_{L,0} - T\Delta S_{L,0}$, $\Delta S_{L,0} = \Delta H_{L,0}/T_m$, and $x_L^f = 1 - x_A$, this leads to

$$\Delta\mu_{L,0} = \Delta H_{L,0}\left(\frac{T_m - T}{T_m}\right) = RT\ln(1 - x_A) \approx -RTx_A \tag{IX.1.6}$$

for small x_A, and

$$\Delta T_m = T - T_m = -\frac{RT_m^2}{\Delta H_{L,0}}x_A. \tag{IX.1.7}$$

Here, ΔT_m is the lowering of the upper edge of the melting transition. In the derivation of the melting-point depression law, it is assumed that the interactions of the drugs with the membrane are ideal, i.e., the melting-point depression is generic. Just like the Meyer–Overton correlation, it is absolutely independent of the chemical nature of the drug [5]. In Figure IX.1.7, one can see that the lowering of the melting temperature is found in the presence of different general anesthetics (octanol, propofol, and pentobarbital).

Figure IX.1.8 shows that the calculation is consistent with experiment. It has been shown in [5] that the melting-point depression occurs. From this, the fraction of the anesthetic drugs in the membrane as a function of temperature can be calculated. Now, one can calculate both the melting-point depression and the heat capacity profiles. The details of the calculation of melting profiles are given in [12]. Calculations and experiments for the general anesthetics, octanol and pentobarbital, are given in Figure IX.1.8. A membrane–water partition coefficient of 150 for octanol and of 27 for pentobarbital was assumed.

The good agreement of theoretical and experimental profiles leads to the conclusion that the assumptions underlying the calculation were justified, i.e.,

- General anesthetics dissolve ideally in the liquid (fluid) phase of lipid membranes.
- They do not dissolve in the solid (gel) phase of the membrane.

- This is true for all general anesthetics that obey the Meyer–Overton correlation, including the noble gas xenon.
- If the Meyer–Overton correlation and the freezing-point depression law are valid, no general anesthetic can have a different affinity to a possible receptor than the noble gas xenon.

Equation (IX.1.7) implies that x_A (the fraction of anesthetics in the membrane) is proportional to $P \cdot c_A$, where P is the partition coefficient, and c_A is the concentration of anesthetics in the aqueous phase. This means that $P \cdot c_A$ is the concentration of anesthetics within the membrane, and it is proportional to the molar fraction x_A of anesthetics in the membrane. Therefore, it follows that

$$\Delta T_m \propto P \cdot c_A \quad \text{or} \quad \frac{d(\Delta T_m)}{dc_A} = P. \tag{IX.1.8}$$

The Meyer–Overton correlation implies that $P \propto 1/[\mathrm{ED}_{50}]$ (see eq. (IX.1.1)). Thus, it follows that

$$\frac{d(\Delta T_m)}{dc_A} \propto \frac{1}{[ED_{50}]} \tag{IX.1.9}$$

or

$$\ln \frac{d(\Delta T_m)}{dc_A} = \ln \frac{(\Delta T_m)}{c_A} = -\ln([ED_{50}]) + \text{const.} \tag{IX.1.10}$$

This relation is shown in Figure IX.1.9 (adapted from [4]). The melting-point depression of all alkanols from ethanol to decanol displays a linear correlation with the critical

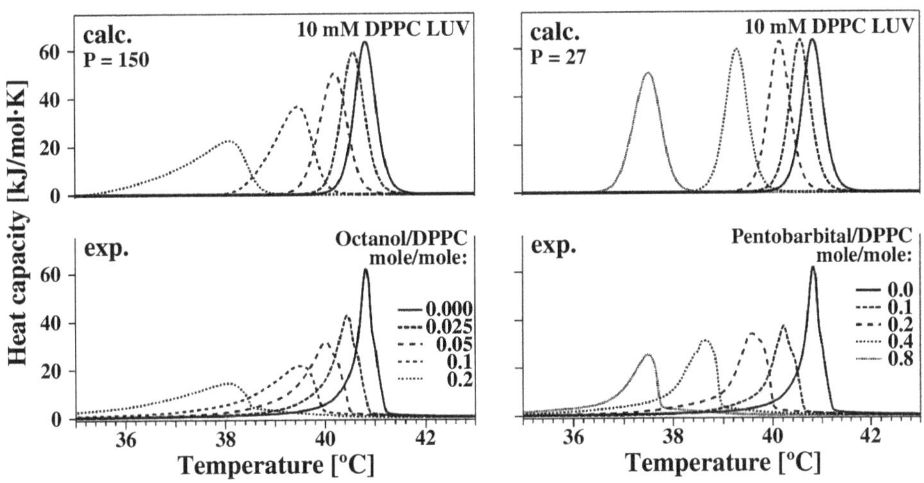

Figure IX.1.8 Theoretical and experimental melting profiles of DPPC membranes in the presence of two different general anesthetics. Left: Octanol. Right: Pentobarbital. The bottom panels show calorimetric data. The top panels show a calculation of calorimetric profiles assuming the melting-point depression law and the known partition coefficients of the two drugs. The calculations describe the experimental profiles well and are consistent with the Meyer–Overton correlation. [12]/with permission of Elsevier.

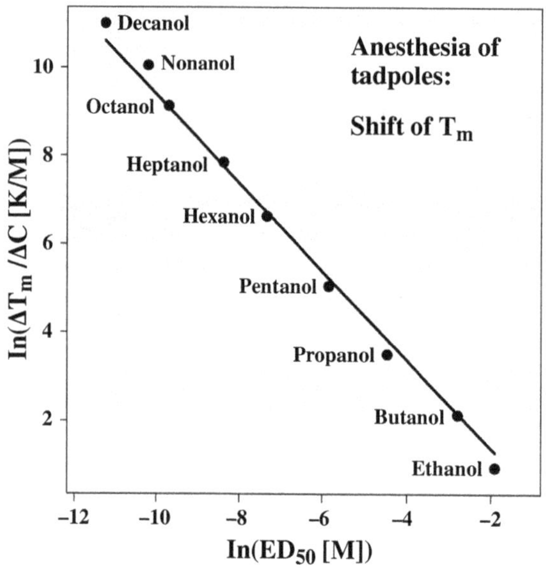

Figure IX.1.9 The change of the melting temperature upon change of anesthetics concentration versus the effective anesthetic dose for tadpoles, for the alcohols from ethanol to decanol. Adapted from [4].

anesthetic dose with a slope of −1 in a double-logarithmic plot. This plot therefore proves the validity of the aforementioned assumptions for melting-point depression and its consistency with the Meyer–Overton correlation.

Thus, we find a physical justification for the Meyer–Overton correlation when it is complemented by the assumption that anesthetics do not dissolve in solid lipid phases. This does not matter for biological membranes at physiological temperatures, which are always in the fluid state. However, it matters for temperatures below physiological temperature – the temperature regime where the melting transition takes place and where propagating solitons in membranes can be generated.

In eq. (IX.1.10), one can see that if $c_A = [ED_{50}]$, it follows that T_m is constant. The melting-point depression at critical dose is always the same, independent of the chemical nature of the anesthetic drug. For tadpole anesthesia, it is $\Delta T_m = -0.6\,°\mathrm{C}$ [5].

IX.1.4.1 Why Does Melting-point Depression Explain the Deviations from the Meyer–Overton Correlation?

The Meyer–Overton correlation has been criticized by some authors. Cantor [13] writes about exceptions such as the cutoff in potency for long n-alkanols. Further, the anomalously low potency of perfluorinated hydrocarbons and the lighter inert gases remain unexplained [14–16].

In fact, not everything that is soluble in lipid membranes causes anesthesia. This includes:

- other lipids
- cholesterol
- long-chain alcohols

For these reasons, it was proposed that anesthetics act by a binding mechanism of anesthetic drugs to receptor proteins (e.g., [11]). We have shown earlier that this is inconsistent with the Meyer–Overton correlation, because it neither explains the lack of chemical specificity in the Meyer–Overton correlation nor the additivity of the anesthetic effect.

The problem of membrane-soluble molecules that are not anesthetics was discussed in [13, 17]. For instance, cholesterol is not an anesthetic even though it dissolves in membranes. However, it also does not cause melting-point depression because it is also soluble in the gel (or liquid-ordered) phases. Cholesterol rather increases the temperature of the lipid phase transition [18]. In a similar way, although lipids are themselves soluble in membranes, they are not generally anesthetics. Many lipids display melting transitions close to the experimental temperature where one of the states is soluble in the solid phase and the other in the fluid phase. In particular, if the melting point of a secondary lipid component is higher than that of the predominant lipid species, the melting profile is usually shifted toward higher temperatures [8, 19]. Thus, they do not fulfill the requirement that they are ideally soluble in the fluid phase of the membrane only, and insoluble in the gel phase. Such considerations also apply to lipid-like molecules such as long-chain alcohols. Pringle et al. [20] report that the anesthetic potency of saturated n-alcohols increases up to dodecanol, and that anesthetic action fails at chain lengths >12–14. It is interesting to note that this correlates with the melting temperature of the pure alcohols. The melting temperature of ethanol is −114 °C, that of octanol −14.8 °C, that of decanol +6.9 °C, that of dodecanol +23.9 °C, that of tetradecanol +38.2 °C, and that of hexadecanol +49.2 °C (values taken from [21]). All alcohols with chain length 14 or longer display transitions above body temperature (37 °C). All 1-alcohols that display a general anesthetic effect display transitions well below physiological temperature and cause melting-point depression. Long-chain alcohols thus do not cause melting-point depression. Experiments from our lab show that tetradecanol fails to follow the correlation shown in Figure IX.1.9, but instead increases the transition temperature in DPPC vesicles. In a similar way, Kaminoh et al. [22] showed that 1-tridecanol and 1-tetradecanol increase the melting temperature of the synthetic lipid DPPC, whereas 1-octanol and 1-decanol decrease it. An increase in the transition temperature of DPPC in the presence of long-chain alcohols was also reported by Kamaya and Ueda [23, 24]. Thus, we must conclude that the cutoff effect of saturated n-alcohols is in agreement with the theoretical description presented here. Melting-point depression is consistent with the Meyer–Overton correlation and it explains the exceptions.

IX.1.4.2 Local Anesthetics

In the literature, local anesthetics are treated differently than general anesthetics. While it is generally admitted that general anesthesia is not well understood, it is believed that local anesthetics bind to specific receptors. Here we argue that this is not the case, and that local anesthetics work by a similar mechanism as general anesthetics.

Figure IX.1.10 (bottom) shows experimental heat capacity recordings of the two local anesthetics, lidocaine and bupivacaine, in large unilamellar vesicles of DPPC. It can be seen that in the presence of the local anesthetics, the transition midpoint is shifted toward lower

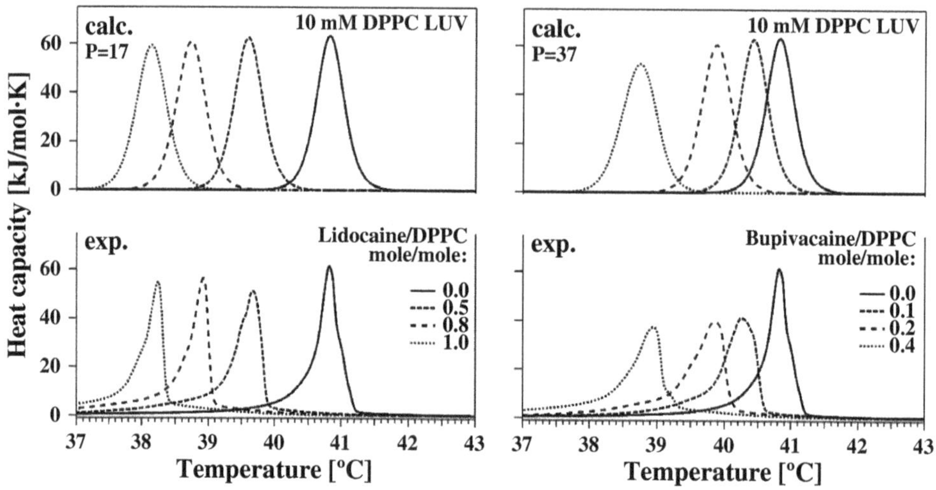

Figure IX.1.10 Melting profiles of DPPC membranes in the presence of two different local anesthetics. Left: Lidocaine. Right: Bupivacaine. The bottom panels show calorimetric data. The top panels show a calculation of calorimetric profiles assuming the melting-point depression law and the known partition coefficients of the two drugs. The calculations describe the experimental profiles equally well as in the case of general anesthetics. [12]/with permission of Elsevier.

temperatures, just as is the case for general anesthetics [12]. Let us assume that the melting-point depression law described earlier is also valid for local anesthetics. Then one can theoretically determine the melting profiles at a given concentration of the drugs (see [12] for technical details). This is shown in Figure IX.1.10 (top). One can see that the experimental profiles are qualitatively and quantitatively well described by the melting-point depression law. This suggests that local anesthetics dissolve ideally in fluid membranes and do not dissolve in gel membranes – just like general anesthetics. Therefore, local anesthetics cause melting-point depression just as general anesthetics do, and there exists no reason from a thermodynamic perspective to distinguish the effects of general and local anesthetics.

The theoretical determination of the melting profiles requires knowledge of the partition coefficient in the membrane, which in Figure IX.1.10 (top) was assumed to be $P = 17$ for lidocaine and $P = 37$ for bupivacaine. In Figure IX.1.8, we found $P = 150$ for octanol and $P = 27$ for pentobarbital. The partition coefficients of these drugs between octanol (the standard reference oil) and water are found in the literature. The partition coefficient in lipid membranes and biomembranes is generally lower by about a factor of 5, probably due to the presence of polar groups (carbonyl groups and the head group). If the values found in calorimetry are plotted versus the partition coefficients in octanol (see Figure IX.1.11), one finds a straight line with a slope of −1 in a double-logarithmic representation. Both the general and the local anesthetics are on this line (within error). This strongly suggests that the action of general and local anesthetics is similar.

Both the Meyer–Overton correlation and the law of melting-point depression are generic linear laws that depend only on the concentration of the drug in question and are entirely independent of its chemical nature. Thus, these laws explicitly predict the additivity of the

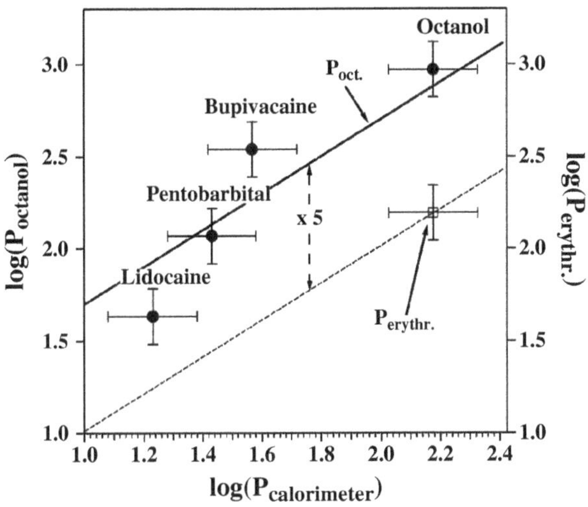

Figure IX.1.11 Octanol/water partition coefficients plotted versus the partition coefficient determined by calorimetry (solid circles). The solid line assumes that the two partition coefficients are proportional. The erythrocyte/water partition coefficient for octanol is given for comparison (open square, dashed line). The octanol/water partition coefficients differ by a factor of 5 from the erythrocyte/water coefficients (dashed vertical line). [12]/with permission of Elsevier.

anesthetic effect caused by two different anesthetics. This effect is well-documented for general anesthetics. It is not quite as obvious that the effects of general and local anesthetics are additive. There are, however, numerous publications providing evidence for the additive effects of general and local anesthesia. Himes et al. [25] reported that the critical anesthetic dose for the general anesthetic halothane in dogs can be lowered by 50% by plasma concentrations of >10 mg/mL of the local anesthetic lidocaine. In a similar way, nitrous oxide anesthesia was enhanced by lidocaine. The intramuscular administration of the local anesthetics, lidocaine and bupivacaine, increases the hypnotic effect of the general anesthetic midazolam in humans to a degree proportional to the dose of the local anesthetic [26]. The effect of bupivacaine is larger than that of lidocaine, in agreement with its larger partition coefficient in membranes. The hypnotic effect of the general anesthetic halothane is enhanced by the local anesthetics, lignocaine and bupivacaine [27]. In a similar way, Senturk et al. [28] reported that the critical dose of the general anesthetic propofol in humans was significantly lowered by intramuscular administration of both lidocaine and bupivacaine. Along the same lines, Altermatt et al. [29] reported a significantly lowered critical dose of propofol in the presence of lidocaine. Additivity of general and local anesthetic effects has also been reported for bupivacaine with the general anesthesia propofol [30]. In a similar manner, the critical dose of the general anesthetic cyclopropane is lowered by 40% in the presence of lidocaine [31]. Further, the critical dose of the general anesthetic isoflurane was lowered linearly with the administered dose of lidocaine in cats [32], with a reduction of >50% for 10 mg/mL lidocaine (in agreement with [25]). The sedative effect of lidocaine was also discussed by Szmuk et al. [33].

IX.1.4.3 "Blocking" of Lipid Channels by Anesthetics

The melting-point depression described earlier leads to the interesting phenomenon of lipid ion channel "blocking." We have argued in Part VIII that lipid ion channels can arise close to the melting transition of the membrane as a consequence of thermal fluctuations. These fluctuations are maximal at the heat capacity maximum and the mean fluctuation lifetimes are at a maximum, too. Thus, at the heat capacity maximum, one expects the largest mean conductance and the longest channel open times. For this reason, everything that influences the position of this transition will influence both channel appearance and the channel lifetimes. This was discussed in Section VIII.5.2.

The experiment shown in Figure VIII.5.3 (bottom) describes a situation where a black lipid membrane displays a heat capacity maximum at the experimental temperature. A DOPC:DPPC = 2:1 mol/mol mixture was chosen that displays a heat capacity maximum at 19 °C. One finds large current fluctuations with long lifetimes. Upon addition of the general anesthetic octanol, the transition profiles move toward lower temperatures, as also shown in Figure IX.1.7. At 7.9 mol% octanol, the mean channel open probability is reduced, and the channel lifetimes are shorter (Figure VIII.5.3, second from bottom). This trend continues with 15.9 mol% octanol (Figure VIII.5.3, second from top) and 23.8 mol% octanol (Figure VIII.5.3 (top)). In the latter case, most channel events disappear.

One could call the reduction of channel events by anesthetics the "blocking" of channels. However, the lipid channels are not molecular objects but rather fluctuations in the state of a lipid continuum. Anesthetics cause melting-point depression due to ideal solubility in the fluid membrane, and therefore, they do not bind to a particular molecular object or receptor. The change in membrane conductance and the channel lifetimes shown in Figure VIII.5.3 is in full agreement with the thermodynamics of lipid membranes and the melting-point depression law. Nothing is blocked in these experiments. The effect of anesthetics is a macroscopic thermodynamic effect.

It is tempting to compare lipid channels with protein channels. It seems that they display very similar signatures both with respect to conductance amplitudes and to mean channel open times. This is striking and raises the question of whether the physics behind these events is similar. It is remarkable that Na^+ channels and the acetylcholine receptor are affected by octanol in a manner similar to lipid channels [34–36]. The same is true for the effect of ethanol, which leads to an increase in channel lifetimes and has also been reported to have a similar effect on the acetylcholine receptor [37, 38]. In the literature, the effect of anesthetics on ion channels has led to the notion that anesthesia acts via a specific binding mechanism to channel proteins, a view that we oppose here on thermodynamic grounds. What can be concluded with certainty at the present time is that:

1. The finding of quantized currents in itself does not prove the activity of a channel protein since lipid membranes show very similar or identical events close to transitions.
2. The effect of drugs on the conduction events does not prove drug binding to a protein, because anesthetics can strongly influence conductance events in the complete absence of proteins due to their effect on the thermodynamics of the membrane.

Thus, an analysis of channel proteins exclusively based on conduction events and their alteration or blocking by drugs must be considered logically incomplete. The information is not sufficient to make such claims. To identify the action of a protein, one needs a clear-cut criterion that allows discrimination between lipid and protein ion channels.

IX.1.5 Reversal of Anesthesia by Pressure and Inflammation

IX.1.5.1 Pressure Reversal

Anesthesia can be reversed by hydrostatic pressure. The pressure needed for reversal can be very high – on the order of 50 bars or more. However, such pressures exist in the deep sea. At the bottom of the Mariana trench, the pressure is of the order of 1100 bars. The world record in apnea diving is 253 m by Herbert Nitsch (2012), and in scuba diving, it is 332 m by Ahmed Gabr (2014). However, due to problems with nitrogen narcosis at high pressures, it is difficult to estimate the effect of pressure. Therefore, many of the experiments were performed on animals without lungs that could withstand such high pressures, e.g., tadpoles and newts. Nitrogen narcosis is not an issue in such experiments.

Pressure reversal of anesthesia was first demonstrated by Johnson and Flagler [39]. They anesthetized tadpoles in 3–6 vol% ethanol. A hydrostatic pressure of 140–350 bars was found to reverse anesthesia. The pressure reversal was also studied by Miller et al. [40]. They found that in newts, pressure can reverse the anesthetic effects of laughing gas (N_2O), N_2, sulfur hexafluoride (SF_6), and carbon tetrafluoride (CF_4). Halsey and Wardley-Smith [41] also performed similar experiments in tadpoles for 12 different general anesthetics and some unconventional anesthetics. Both [40] and [41] found pressure reversal in the range of 50–200 bars, depending on the anesthetic. Pressure reversal was also studied by Mountcastle et al. [42] and Moss and Franks [9]. Thus, one has to consider the pressure reversal of anesthesia an established experimental fact.

In a previous chapter, we showed that pressure increases the melting temperature of membranes due to the change in lipid volume. This is shown in Figure IX.1.12 for large

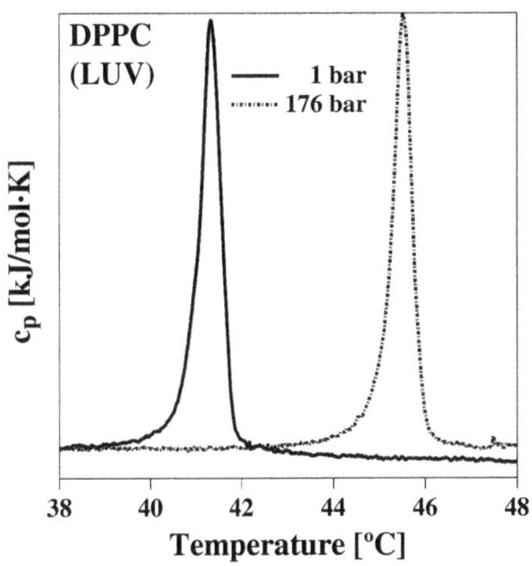

Figure IX.1.12 The shift of the melting transition of DPPC LUV is +4.3 °C upon applying a pressure of 175 bars.

unilamellar vesicles of DPPC. It was found that the shift in the transition temperature is given by

$$\Delta T_m = \gamma_V \cdot \Delta p \cdot T_m, \tag{IX.1.11}$$

where $\gamma_V = 7.8 \cdot 10^{-10}$ m^2/N and $T_m = 314.2$ K (41.05 °C) for DPPC. In Figure IX.1.12, the pressure difference of $\Delta p = 175 \cdot 10^5$ N/m^2 leads to a shift in transition temperature of 4.3° toward higher temperatures.

Following eq. (IX.1.7), we therefore find that at

$$\Delta p = \frac{1}{\gamma_V} \frac{RT_m}{\Delta H_0} x_A, \tag{IX.1.12}$$

the effect of the anesthetics is completely reversed by the pressure Δp, which for tadpole anesthesia ($x_A = 0.026$) is $p = +24.9$ bars. Note that this calculation was made for DPPC lipids where the partition coefficient is 150, which may somewhat deviate from the partition coefficients in biological membranes. The combined effect of both general and local anesthetics and pressure in shown in Figure IX.1.13.

According to Firestone et al. [43], for ethanol, the ED_{50} necessary for tadpole narcosis is 190 mM (≡1.1 vol%). This means that 3–6 times the anesthetic ethanol concentration was

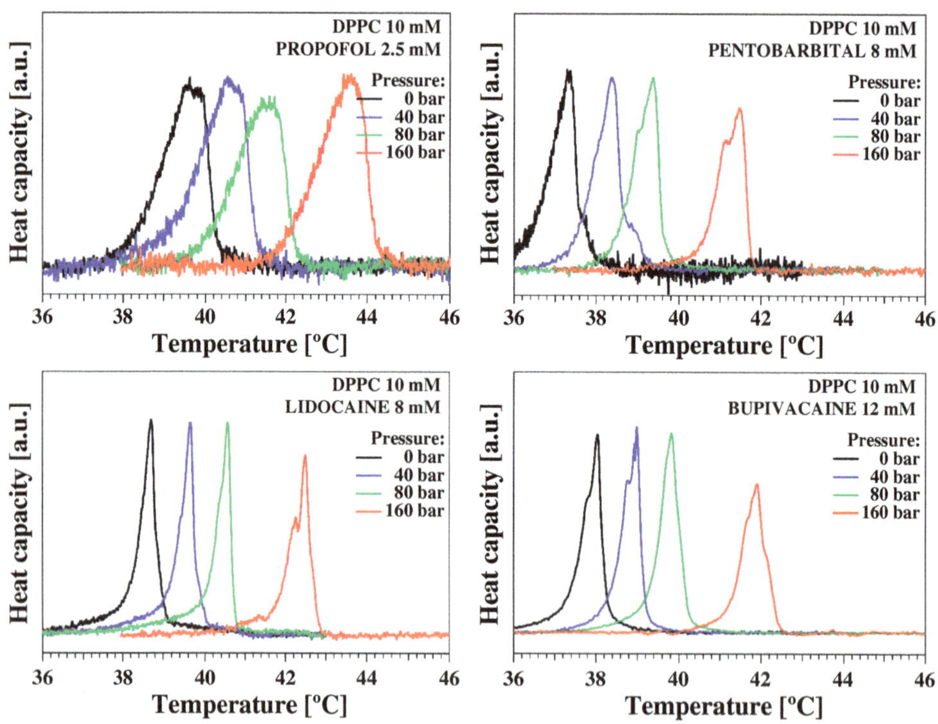

Figure IX.1.13 The pressure dependence of DPPC LUV in the presence of the general anesthetics propofol and pentobarbital, and the local anesthetics lidocaine and bupivacaine show the same pressure dependence as the lipids in the absence of anesthetics. Note that the transition temperature of DPPC is about 41 °C, and that excess hydrostatic pressure is necessary to reverse the effect of the anesthetics. [12]/with permission of Elsevier.

used in Johnson and Flagler [39]. The concentration of ethanol in the membrane in Johnson and Flagler's experiments was therefore 7.5–15 mol%. Since for the critical dose in tadpoles, the shift in transition temperature in DPPC is −0.6°, this corresponds to a transition of 3–6 times 0.6° toward lower temperatures. According to the melting-point depression law, these concentrations correspond to lowering T_m by 1.8–3.6 K. From eq. (IX.1.12), it follows that the pressure necessary to reverse this anesthetic effect is 75–150 bars. Considering the uncertainty of the partition coefficient for real biological membranes (which depends on the precise lipid mixture), this is remarkably close to the order of the values found by Johnson and Flagler. We conclude that the melting-point depression law is able to determine the pressure reversal of anesthesia quantitatively. This can also be taken as very strong evidence that anesthesia is related to the melting transition of membranes. If one were not to find a reversal pressure of similar order, the hypothesis that membrane melting is involved in anesthesia would have been falsified.

The fact that pressure increases T_m may be related to the observation that nerves fire spontaneously at high pressures [44]. Nerves fire less frequently in the presence of anesthetics. We will use this finding to explain the influence of the excitability of nerves on pressure (and other intensive variables) below.

IX.1.5.2 Reversal by Inflammation

Hydrostatic pressure is not the only way to increase melting transitions. An increase in proton concentration (lowering of pH) leads to a shift of the transition of a negatively charged membrane toward higher temperatures. Lowering the pH leads to protonation of the negatively charged lipids. This reduces repulsive electrostatic forces within the membrane plane, which stabilizes the lipid state with the lower surface area (the gel phase). Details of the calculation are given in [8, 45, 46]; see Chapter V.6 and Figure V.6.1.

Some 10% of the lipids of biological membranes are negatively charged, primarily on the inner membrane. At lower pH, some of these charges are protonated, and the electrostatic potential of the lipid membrane is reduced. Complete protonation increases the melting temperature by about 20 K [46]. While these effects depend on the precise composition of the membrane and ionic strength, they can be calculated using Debye–Hückel theory [8] or determined empirically. For example, the temperature of the melting transition in native *E. coli* membranes (in the pH range between 5 and 9) is raised by 7.3° when pH is lowered from 9 to 5 (see Figure IX.1.14), i.e., about 1.8° per pH unit. This shift corresponds approximately to that produced by 72 bars of hydrostatic pressure.

Interestingly, it is known that inflammation leads to the failure of anesthesia. The related lowering of pH in inflamed tissue, i.e., on the order of 0.5 pH units [47], is widely assumed to be responsible. According to the above, the lowering of pH from 7 to 6.5 leads to $\Delta T_m = +0.9°$, which is sufficient to reverse the action of anesthetics at the typical critical dose corresponding to $\Delta T_m = -0.6°$.

Salts can also affect the melting transition through, e.g., the binding of monovalent ions such as Li^+ [48] and divalent cations such as Mg^{2+} and Ca^{2+}. These ions shift the melting temperatures of both charged and uncharged lipids to higher temperatures; see Chapter V.6. The presence of such ions thus lowers the effectiveness of anesthetics [46, 49].

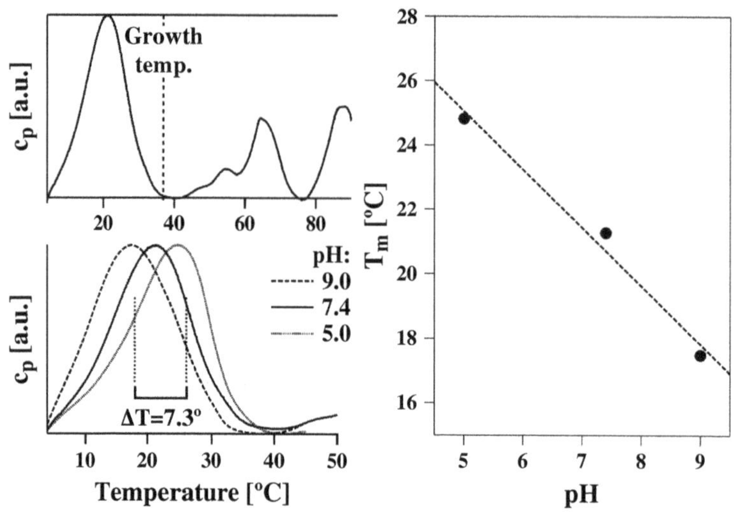

Figure IX.1.14 Shift of the transition temperature of native *E. coli* membranes at different pH values. Adapted from [5].

IX.1.6 The Influence of Anesthetics on the Nerve Excitation Threshold

In Chapter VII.6, we showed that the excitation free energy of a soliton – the nerve pulse – depends on the temperature difference between physiological temperature and the melting temperature of membranes. It is the free energy required to move the membrane from a fluid state into a gel state. This can be caused by a change in voltage, in lateral density, by a sudden drop in temperature, or other local changes in the value of intensive variables, including a local lowering of pH. In Section IX.1.4, we showed that anesthetics lower the melting temperature in membranes via a mechanism called melting-point depression – a completely unspecific effect driven by the entropy of mixing of anesthetic drugs in the fluid lipid membrane. We did not find a major difference in the action of general and local anesthetics on the thermodynamics of the membranes. However, for the local anesthetics, no critical dose is given due to the local nature of anesthesia, and one cannot determine a concentration where subjects lose conscience.

In [50], the effect of lidocaine in different bulk concentrations from 0 to 85 mM was calculated following the theory outlined in [12]. In order to do this, the melting profile of a biological membrane (porcine spine) was measured and theoretically fitted by a van't Hoff process. The fits are shown for various anesthetics concentrations in Figure IX.1.15, where 85 mM is a very large concentration that would lead to a shift in transition temperature of more than −45°. This is shown in Figure IX.1.15 (left). For comparison, the critical dose of general anesthetics leads to a shift by about −0.6 to −1°. Thus, we are talking about very high doses of lidocaine. Figure IX.1.15 (right) shows, for the doses in the left-hand panel, what change in excitation threshold of the nerve would be generated following the procedure given in Chapter VII.6. One finds a factor of up to seven times in the shift of the excitation threshold as compared to the membrane without anesthetics. This is judged from the

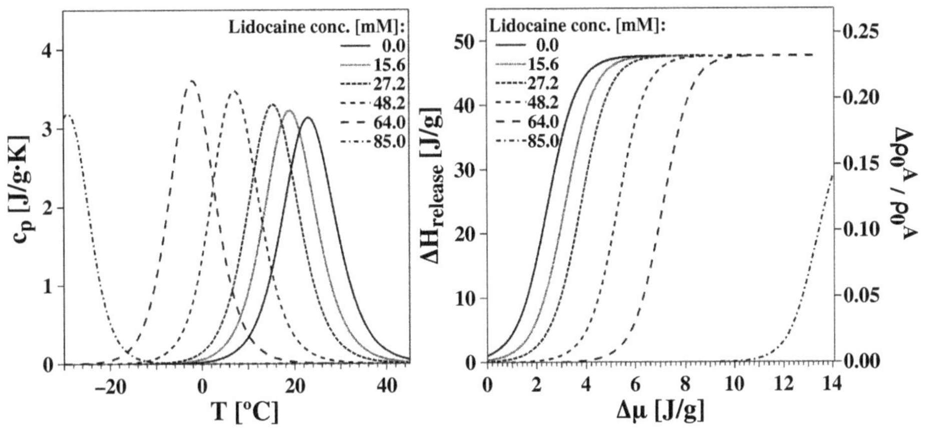

Figure IX.1.15 Left: Van't Hoff profiles of a membrane with $\Delta H_0 = 35$ kJ/mol, $T_m = 23.3$ °C (corresponding to porcine spine membranes), and a cooperative unit size of n = 5.5 in the presence of six different bulk concentrations of lidocaine. Physiological temperature is at ≈39.3 °C in pigs. Right: Excess enthalpy and relative change in area density as a function of the chemical potential difference between gel and fluid membrane relative to the chemical potential difference at body temperature. [50]/with permission of Elsevier.

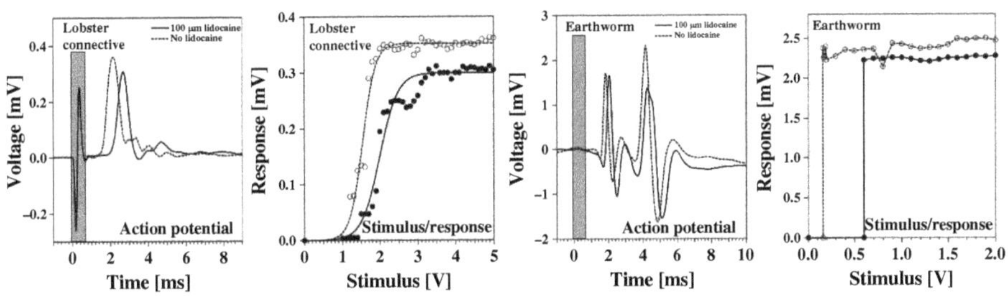

Figure IX.1.16 Action potentials of lobster connectives (left) and earthworm (right), and the associated stimulus–response curves in the presence and absence of 100 μM lidocaine. In both cases, the stimulus–response curve is shifted toward higher thresholds without changing the amplitude significantly. [50]/with permission of Elsevier.

chemical potential difference $\Delta\mu$ between the fluid and the gel membrane. In the absence of anesthetics, around 2 J/g of membrane was calculated, whereas it was around 14 J/g in the presence of 85 mM of lidocaine. The difference between physiological temperature in pigs (≈39 °C) and the melting temperature at ≈23 °C is 16°, while it is 64° in the presence of 85 mM lidocaine. Thus, the conclusion from the theoretical treatment is that anesthetics never block nerves completely, but they shift the excitation thresholds to much higher values.

The increase in threshold can be found experimentally. Figure IX.1.16 shows the changes in action potential amplitude in the absence and presence of a small concentration of lidocaine (100 μM) both for lobster connectives and for earthworm. The left-hand panels show the action potentials, and the right-hand side the amplitude of the response after a stimulus of given magnitude (stimulus–response profile). The threshold is the half-height of the sigmoidal stimulus–response curve. In the lobster connective, the threshold changes by about

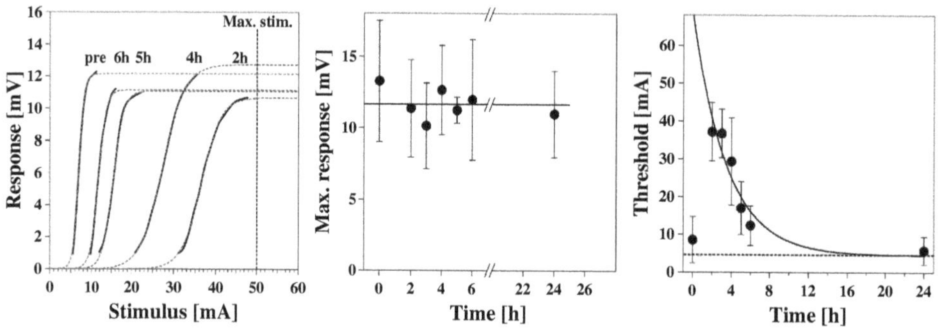

Figure IX.1.17 Anesthesia of the human median nerve [51]. Left: Stimulus–response profile for the median nerve in humans after lidocaine anesthesia, measured at different times after application of the local anesthetic (dashed profiles represent guides to the eye). Center: The mean amplitude of the response for different times after application of lidocaine. Right: Threshold for 50% response in the stimulus–response curves. The solid line represents an exponential fit to the data. The experimental data were adapted [51]/with permission of John Wiley & Sons.

30%, while it changed by about a factor of 4 in earthworm nerves. The correlation with the above theoretical results is only qualitative because both earthworm and lobster are cold-blooded animals and we have no knowledge about the exact transition temperatures of their neural membranes.

Even more convincing are data obtained at the Rigshospitalet in Copenhagen on the human median nerve [51]. In their experiments, the median nerve of healthy subjects was anesthetized at two points with 5 cm distance by injecting 5–6 mL of lidocaine (13 mg/mL, which corresponds to 55 mM). These are very high lidocaine concentrations of a similar range as discussed earlier. It should change the threshold by around 5 times if this were the concentration that the median nerve sees. The stimulus was measured as the stimulation current across two electrodes close to the elbow. The response was measured from the compound muscle action potential (CMAP) recorded from the abductor pollicis brevis (APB) at the thumb. The stimulation currents were increased in small steps up to 50 mA. Figure IX.1.17 (left) shows the stimulus–response (I–V) profiles as a function of time after the injection of the anesthetics, and before injection (indicated by "pre"). Before anesthesia, and 24 hours after anesthesia, the threshold was around 6–8 mA. In between, the threshold shifted by more than a factor of 6 such that a stimulus–response profile could not be measured in the accessible current regime. With time, the stimulus–response profile reappeared. After 2 hours, it could be completely seen with a threshold around 37 mA. The time constant of recovery was of the order of 5 hours (Figure IX.1.15 (right)). After 24 hours, any effect of lidocaine anesthesia had completely disappeared. The profiles in Figure IX.1.17 (left) qualitatively correspond to the calculations shown in Figure IX.1.15 (right).

In this experiment performed at similar concentrations as in the calculation in Figure IX.1.17, the following was found (see analysis in [50]):

1. The stimulation threshold was largely shifted by more than a factor of 6 upon the injection of lidocaine such that the threshold was above 50 mA.
2. The full amplitude of the action potential (CMAP at the APB) after sufficient anesthesia was unaffected by anesthesia.

3. There is no indication of blocking the action potential. This would have required a disappearance of signals at stimulation amplitudes below 50 mM. However, in the experiment, a continuous reappearance of the stimulus–response curves at a stimulation amplitude was found.
4. Assuming that the reappearance of the stimulus–response curves reflects the gradual decrease in the local anesthetics concentration with time, all of these findings and even the magnitude of the shift of the threshold were of similar order as predicted in the calculation shown in Figure IX.1.15, which was deduced from the thermodynamic properties of pig membranes.

In an unpublished experiment performed during the same series of experiments with the general anesthetic at critical dose, the shift of the stimulus–response profile was found to be of the order of 15% for light anesthesia ([52], unpublished data). For a shift of the transition at the critical dose of general anesthetics of about 1° compared to a position of the transition around 14° below human body temperature at 37 °C, one would expect a shift of the order of 7%. This is of the order that is expected when taking into account that the thermodynamic data for the median nerve were not known.

Summarizing, the thermodynamic analysis of the stimulus–response profile is in rough qualitative agreement with the experimental findings when measuring stimulus–response profiles in lobster connectives and earthworm, and in quantitative agreement with the detailed analysis of stimulus–response profiles in the human median nerve.

IX.1.7 Neurotransmitters

There have been speculations by Cantor [53] that some neurotransmitters may display anesthetic action. He suggested that some neurotransmitters may act as endogenous anesthetics.

> "... in addition to the rapid, saturable binding of a neurotransmitter to a receptor that results in activation, the neurotransmitter also acts indirectly on the receptor by diffusing into the postsynaptic membrane and changing its physical properties, causing a shift in receptor conformational equilibrium (desensitization). Unlike binding, this slower indirect mechanism is nonspecific: each neurotransmitter will, in principle, affect all receptors in the membrane."

This seems plausible since serotonin (Figure IX.1.18) also lowers the melting temperature of membranes [54]. Figure IX.1.19 shows the effects of both octanol and serotonin on the melting of the synthetic lipid dimyristoyl phosphatidylcholine (DMPC). Similarly, dopamine also lowers the melting temperature in a manner that is comparable to the effect of anesthetics. Thus, serotonin will exhibit an anesthetizing effect and, like other general anesthetics such as laughing gas, ethanol, and diethyl ether, cause euphoria at low doses.

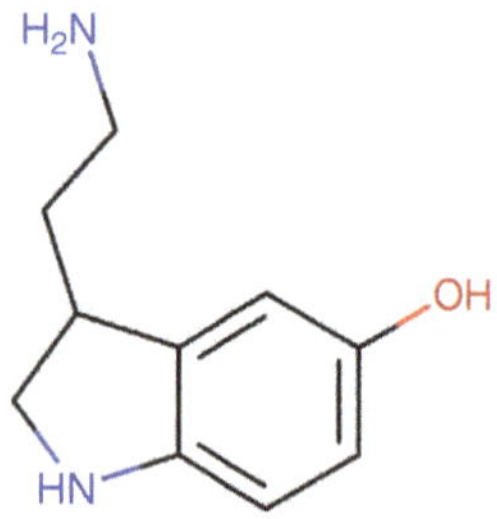

Figure IX.1.18 Chemical structure of serotonin.

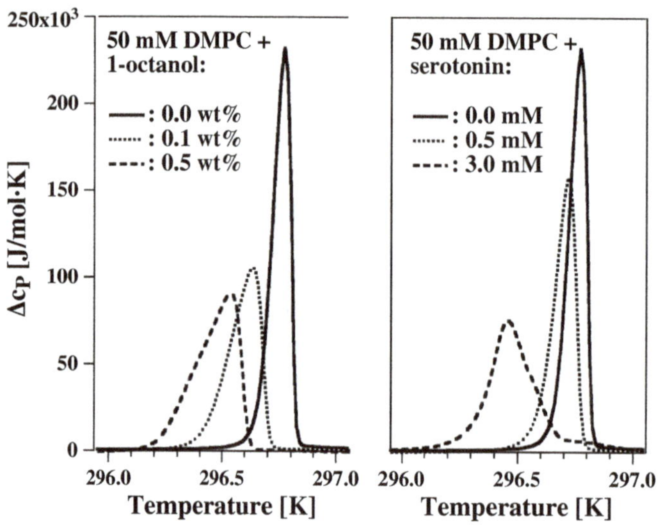

Figure IX.1.19 Heat capacity profiles of DMPC multilamellar vesicle dispersion in the presence of octanol and serotonin. Left: DMPC plus various concentrations of 1-octanol. Right: DMPC plus different concentrations of serotonin. The drugs broaden the lipid melting profile and shift it to lower temperatures. Adapted from [54].

IX.1.8 Summary: Key Ideas of Chapter IX.1

There are numerous remarkable properties of anesthesia. Here we summarize a few of the ones discussed in the present chapter.

- The Meyer–Overton correlation states that the oil–water coefficient P and the effective anesthetic dose $[ED_{50}]$, for which 50% of the individuals are anesthetized, are related via

$$P \cdot [ED_{50}] = \text{const.}, \tag{IX.1.13}$$

  meaning that the effective dose is inversely related to the solubility of anesthetics in the core of the lipid membrane. This correlation is completely independent of the chemical nature of the drug.
- The action of different general anesthetics is additive.

- At the critical concentration of anesthetics in the blood, the membrane concentration for all general anesthetics is the same. For tadpole anesthesia, it is 2.6 mol% in the membrane.
- Theories based on the binding of anesthetic drugs to receptor proteins cannot explain the Meyer–Overton correlation (i.e., the independence of anesthesia on the chemical nature of the drug), and they cannot explain the additivity of the effect of two different anesthetic drugs.
- All general anesthetics obey the aforementioned correlation. However, not every membrane-soluble molecule is an anesthetic, e.g., long-chain alcohols, cholesterol, lipids, and other membrane-soluble molecules.
- The Meyer–Overton finding is a correlation that does not explain anesthesia.
- There exists a second correlation that is rooted in thermodynamics: all general anesthetics cause melting-point depression:

$$\Delta T_m = -\frac{RT_m^2}{\Delta H} \cdot x_A = \text{const.}, \qquad \text{(IX.1.14)}$$

  where x_A is the molar fraction of anesthetics in the lipid membrane. At the effective dose, $[ED_{50}]$, all general anesthetics lower the transition temperature of membranes by exactly the same amount. For tadpole anesthesia, transition temperatures are lowered by $\Delta T_m = -0.6$ K.
- The melting-point depression law can be obtained if the Meyer–Overton correlation is slightly modified: The critical anesthetic dose $[ED_{50}]$ is inversely proportional to the solubility of the drug in the fluid membrane, while the anesthetics are insoluble in the gel membrane.
- Melting-point depression can explain the pressure reversal of anesthesia. This proves that the melting-point depression law is more than just a correlation. It hints at the true origin of anesthesia.
- Melting-point depression also provides an explanation for the reversal of anesthesia in inflamed tissue. It is caused by the decrease in pH caused by inflammation.
- Melting-point depression causes an increase in the excitation threshold of the nerve pulse, because the free energy necessary to move the membrane from fluid to gel phase increases.
- Even though the threshold increases, the maximum amplitude of the nerve pulse stays constant. The nerve is not blocked by anesthetics.
- Local anesthetics cause melting-point depression proportional to the partition coefficient, just like general anesthetics. This suggests that the mode of action of general and local anesthetics is similar. The action of general and local anesthetics is additive.
- Melting-point depression can also explain the apparent "blocking" of channels in the lipid membrane. If the melting temperature decreases, the fluctuation in the membrane at a temperature above the transition also decreases, and the likelihood of pore formation in the lipid membrane is reduced. This is consistent with numerous experiments on the change of membrane permeability within melting transitions.
- Some neurotransmitters such as serotonin lower the melting temperature of membranes. Thus, they would also increase the excitation threshold of nerves. Such neurotransmitters have been discussed as endogenous anesthetics.

IX.2

Adaptation

The condition for the excitability of biological membranes is a transition some 10–20° below physiological temperature. This is either body temperature in warm-blooded animals or ambient temperature in cold-blooded animals (such as frogs) and bacteria. The temperature difference between physiological temperature and the melting transition seems to be maintained under many conditions. In order to maintain the function of a biological membrane, its physical state has to be maintained and kept stable. However, ambient conditions can change, and the organism somehow has to adapt to new conditions. These could be changes in temperature and pressure, the ionic strength of the environment, or changes in pH and gradients in resources. A frog may hop from a sunny spot into cold water, thereby changing temperature. Cold-blooded organisms adapt to different temperatures in summer and winter. In the deep sea at the bottom of the Mariana Trench (11 km deep), one finds a pressure of 1100 bars. Such pressure leads to increases in transition temperatures by roughly 28 °C. Bacterial cells exposed to high pressure may adapt or migrate to different temperatures in a temperature gradient, i.e., the temperature requirements change. Cells may also be exposed to solvents such as ethanol after drinking a glass of wine, or to an anesthetic during surgery.

All these phenomena will change the interval between physiological temperature and the membrane transition temperature and thus the excitability of the membrane. Nature must find a way to cope with the changes in ambient conditions. It may do so by a process called adaptation, during which the composition of the lipid membranes can be changed, or another variable such as body temperature may adapt (in warm-blooded animals). Such adaptation processes are described in the following section. A number of the adaptation mechanisms described as follows were discussed in an insightful paper by D. Kharakoz [4].

IX.2.1 Adaptation of Lipid Membranes to Growth Temperature

Biological membranes adapt their membrane composition to the growth temperature. Figure IX.2.1 shows how the melting temperature of *Escherichia coli* membranes changes with growth temperature. In trout livers, the fraction of unsaturated lipids (e.g., with 22:6 fatty acids) at a water temperature of 5°C is significantly higher than at 20°C, while the concentration of saturated lipids is higher at 20°C [55]. Saturated lipids generally display higher melting temperatures in comparison to unsaturated lipids. For instance, dioleoyl phosphatidylcholine (DOPC, with two 18:1 chains) displays a melting temperature of

Physics of Nerves and Excitatory Membranes, First Edition. Thomas Heimburg.

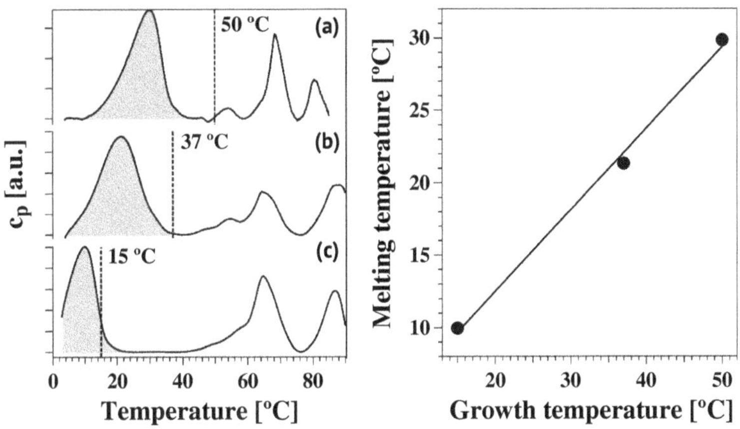

Figure IX.2.1 Adaptation changes in growth temperature for *E. coli* membranes. For lower growth temperatures, one also finds a lower melting temperature of the membrane lipids.

about −20 °C, while distearoyl phosphatidylcholine (DSPC, with two 18:0 chains) displays a melting temperature of 53 °C. Thus, one single unsaturation in each chain lowers the melting temperature by more than 70 °C. The result is a lower melting temperature of the trout liver membranes at lower water temperatures.

Adaptation of composition is probably relatively slow. The reproductive cycle in *E. coli* is about 20 minutes. Adaptation of the membranes to changes in growth temperature probably happens on this timescale. Minutes but not weeks. Another process where temperature changes is hibernation. Suri and collaborators showed that the lipid composition in the lung surfactant of squirrels adapts to body temperature (5 °C) in hibernating squirrels.

IX.2.2 Adaptation to Hyperbaric Pressure

As demonstrated in Chapter V.4, hydrostatic pressure causes a linear shift of transition temperatures toward higher values. For the lipid DPPC, that shift is about 1°/40 bars. Thus, an obvious adaptation process that helps maintain the physical state of the membrane is to increase the fraction of unsaturated lipids, which lowers the transition temperature. That this adaptation process exists was demonstrated for the deep sea bacterium CNPT3 [56]. Figure IX.2.2 shows the ratio of unsaturated to saturated lipids in the membranes of CNPT3. It displays a linear dependence on pressure.

Another possibility is the active migration to a new temperature that compensates for the pressure change. In experiments on the crustacean *Parhyale hawaiensis* in a temperature gradient, it was found that the organisms had a temperature preference that changed by about 1°C/22 bars toward higher temperature upon the application of hydrostatic pressure [57]; see Figure IX.2.3. The rough estimate of the change in transition temperature from the pressure dependence of DPPC would be about 1°C/40 bars toward higher temperature upon the application of hydrostatic pressure. Thus, the preferred temperature changes in rough agreement with the pressure dependence of the melting transition. The migration toward a different temperature is called *thermotaxis*.

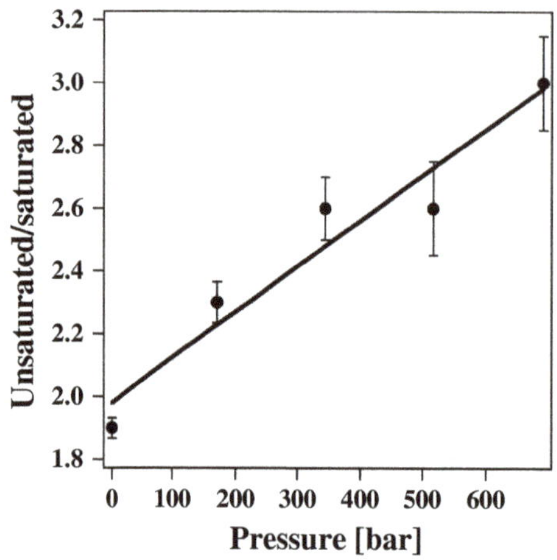

Figure IX.2.2 The ratio of unsaturated to saturated lipids in the deep sea bacterium CNPT3 is linearly dependent on growth pressure. Adapted from [8, 56].

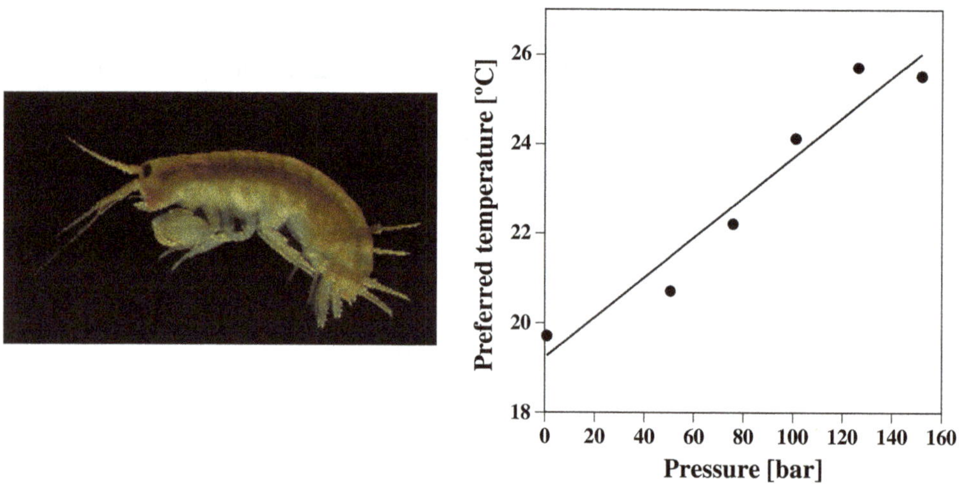

Figure IX.2.3 Left: Image of *Parhyale hawaiensis* from Yoshi Tomoyasu Kao et al., (2016)/eLife Sciences Publications Ltd/CC BY 4.0. Right: If the crustacean *Parhyale hawaiensis* is free to choose a temperature in a temperature gradient, the preferred temperature changes with external pressure. Adapted from [4, 57].

IX.2.3 Fast Adaptation by pH Changes

The fastest possible adaptation is a pH change that could, for instance, compensate for sudden temperature changes. This could be frogs jumping from a sunny spot into a cold lake. In fact, it was shown that the pH of frog muscle changes as a function of temperature. It is about 7.4 at 4 °C and about 6.9 at 37 °C, i.e., about half a pH unit between the two temperatures [58]. A pH change of −0.0147 pH units per degree in frog muscle was found by [59]. Similar temperature-dependent changes in pH were also found in mouse soleus muscle [60], in cardiac muscle [61], and in rat smooth muscle [62]. Generally, the pH is lower at higher temperatures, which implies that the transition temperature of the membranes is also higher. The lowering of pH from 9 to 5.5 in *E. coli* membranes leads to an increase in transition temperature of about 7 °C [49]. Changes of pH in hibernating animals such as newts (−0.018 pH units per degree) and salamanders (−0.041 pH units per degree, [63]) were also of similar order (reviewed in [58]). The temperature-dependent pH changes have been attributed to a Na^+-H^+ exchange mechanism [58].

IX.2.4 Adaptation to Solvents

Lipid membranes also adapt to the presence of solvents such as acetone, aniline, and anesthetics [8]. This effect was carefully studied for *E. Coli* membranes by Ingram [64]. For instance, the cardiolipin concentration changes fivefold from 1.3 to 8.4% in the presence of 48 mM acetone. In the presence of 38 mM aniline, the concentration of cardiolipin changes from 31.9 to 6.7%. In the presence of 20 mM sodium benzoate, the fraction of 12:0 chains drops from 34.1 to 5.9%. Similar changes are found for many anesthetics.

IX.2.5 Alcoholism

Bacteria adapt their lipid composition in response to permanent changes in pressure, temperature, or drug concentrations. Since ethanol is an anesthetic (as are all alcohols up to decanol), the permanent presence of ethanol will lead to an adaptation of the lipid composition. Under chronic administration of ethanol, cats adapt the lipids in their livers such that the fraction of polyunsaturated lipids decreases [65]. Likewise, chronic alcoholism in humans decreases the concentration of polyunsaturated lipids [66]. This change in lipid composition will increase the transition temperature of membranes in the absence of alcohol, but the constant presence of alcohol will compensate for this change. It seems reasonable to speculate that cell membrane function in alcoholics in the presence of alcohol is similar to that of non-alcoholics in its absence. The prediction is that, in the sudden absence of alcohol, the nerves are more excitable than usual due to the increase in the transition temperature of the membranes induced by the adaptation process. According to the soliton theory, this will cause hyperexcitability and possibly pain and tremor. Further, during an operation, a sober alcoholic is predicted to require a higher dose of anesthetics than a non-drinker. This is a known fact in clinical praxis [67, 68]. An alcoholic in the presence of his normal blood level of alcohol is predicted to require an anesthetic dose equal to that of a non-drinker in the absence of alcohol. In contrast, if a non-drinker happens to be very drunk on a particular occasion, the membrane composition has not adapted to the lipid composition, and the required anesthetic dose will therefore be smaller than normal. The connection of adaptation of membranes to alcoholism has also recently been discussed by Schneider [69] in a similar context.

IX.2.6 Hypothermia

Anesthetics lower the melting transitions of biomembranes. A possible way to adapt to the lowering of the melting temperature is a reduction in body temperature. As discussed by Kharakoz [4], anesthesia-induced hypothermia is a well-known phenomenon (see Figure IX.2.4).

He demonstrates that there is a correlation between the anesthetic potency (ED_{50}) and their potency to cause hypothermia, i.e., a lowering of the body temperature during anesthesia. Of more interest from the viewpoint of the hypothesis is the correlation between the anesthetic and hypothermic potencies. Small aquatic animals are the most appropriate objects for such analysis because their body and brain temperatures are practically equal to the temperature of the environment and, thus, can be easily controlled in experiments. O'Connor and collaborators [70, 71] showed that goldfish adapt their body temperature by swimming in a temperature gradient after administration of ethanol. Similar results were also found in mice [72] and man [73]. For details, we refer to the article by Kharakoz [4].

IX.2.7 Speculations on Fever

We have discussed earlier that inflammation leads to a lowering of pH, which goes along with an increase in membrane melting temperature. For this reason, it is difficult to anesthetize inflamed tissue. Just as hypothermia is a consequence of the presence of anesthetics, i.e., an adaptation process, fever will be appropriate to compensate for the side effects of inflammation. Thus, fever may be, just like hypothermia, an adaptation mechanism to low pH caused by inflammation.

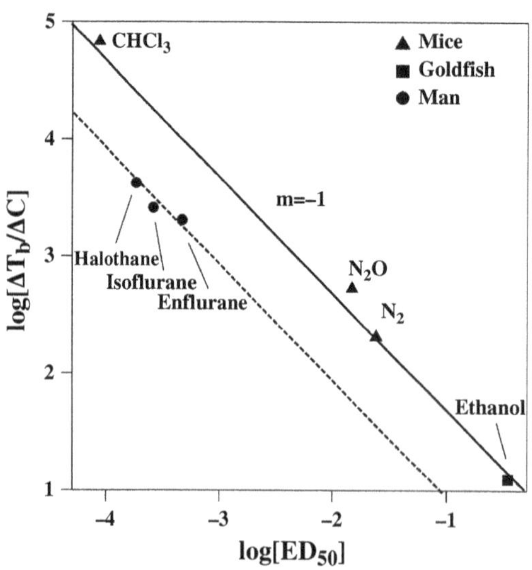

Figure IX.2.4 Hypothermia: The lowering of body temperature during anesthesia as a function of the critical dose [ED_{50}] of the anesthetic. $\Delta T_b/\Delta C$ is the dependence of the change in body temperature with anesthetic concentration. The slope of the data is m = −1, consistent with the Meyer–Overton correlation. Adapted from [4].

IX.2.8 Summary: Key Ideas of Chapter IX.2

Adaptation of the membrane composition is a common process in biology. There are various ways to adapt to a change in ambient conditions.

- The adaptation of the membrane lipid composition has been reported for changes in growth temperature, hydrostatic pressure, and the presence of solvents such as ethanol or anesthetics. This probably happens on the timescale of minutes, i.e., probably faster than the replication cycles of cells.
- Lipid compositions change in cold-blooded animals such as fish between winter and summer.
- Hypothermia: During anesthesia, the body temperature follows the shift of the melting temperature in membranes. This happens on the timescale of a surgery.
- Fever is a possible adaptation mechanism for the lowering of blood pH that occurs during inflammation.
- The fastest possible adaptation process is a change in pH. This has been observed in the muscles of cold-blooded animals, such as frogs, and in hibernating animals, such as newts and salamanders.
- Alcoholism also goes along with an adaptation of the composition of lipid membranes.
- The adaptation process could also consist of an active migration toward better ambient conditions. For example, thermotaxis is a mechanism where animals react to changes in pressure or the presence of anesthetics by choosing a different temperature environment.

IX.3

Nerve Stretching

It has been known for more than 140 years that nerves can be excited mechanically [74]. As discussed in Chapter III.4, mechanical changes in neurons have been observed in connection with the action potential [75–80]. Therefore, one might expect that the properties of nerves change in the presence of forces acting on nerves. When a body moves, nerves may be extended or stretched. As an example of the stretching of a nerves in normal motion, we show the tibial nerve in humans (Figure IX.3.1), which is stretched when the leg is extended. In this chapter, we discuss the effect of stretching on nerve excitability.

IX.3.1 Introduction to Nerve Stretching

Nerves contract upon excitation [82, 83]. Tasaki and collaborators [77] found a contraction of the garfish olfactory nerve while the action potential was traveling (Section III.4.2). This implies that during the action potential, a neuron is under increased tension when the positions of its ends are subject to a constraint. This is predicted by the soliton theory, in which nerves will contract during the time period in which the action potential travels because the pulse consists of a region of gel membrane with smaller area density and larger thickness (see Section VII.5.1). In general, stretching of nerves (the application of a force on the axon) is expected to make them less excitable, since lateral tension in a membrane leads to a shift of the phase transition toward lower temperatures [50]. The effect of the shift of the transition temperature was already discussed in Section IX.1.6 dealing with the change in excitation threshold during anesthesia. The effect of stretching is expected to go in the same direction.

Experiments show that stretching causes changes in nerve function. As early as 1935, Havasi and Rittler [84] showed that the chronaxie of motor nerve stimulation changes with stretching. The effect of stretching on reflexes has been studied extensively [85]. Guissard and collaborators [86, 87] showed that motor neuron excitability decreased during fast muscle stretch. This inhibition had been tested in the soleus muscle by recording the Hoffmann reflex (H-reflex) and the tendon reflex (T-reflex), which were both found to be reduced during fast muscle stretching. Compound action potentials decrease in amplitude during mechanical stretch in hamster sciatic nerves [88] and rat nerves [89]. The action potentials of rabbit tibial and peripheral nerves also decreased in amplitude upon stretching [90, 91]. This is in agreement with predictions from the thermodynamic theory in which

Physics of Nerves and Excitatory Membranes, First Edition. Thomas Heimburg.

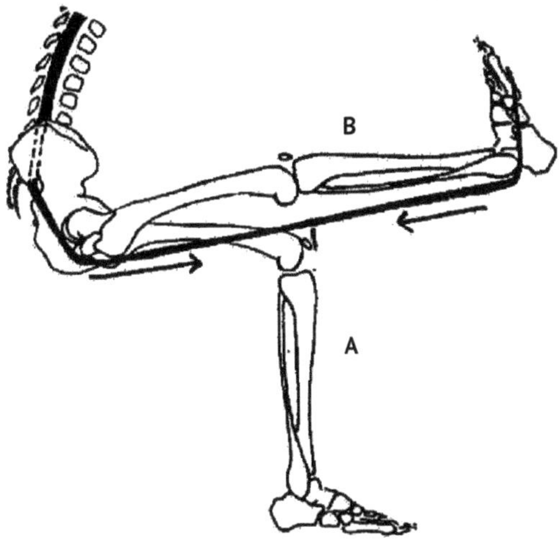

Figure IX.3.1 Stretching of the tibial nerve in the human leg upon extending the leg from position A to B. From [81]/with permission of Elsevier.

increased mechanical tension reduces excitability. Ucar et al. [92] showed that synapses react to stretching, which implies a mechanosensitivity of synapses.

In the following, we outline the effect of longitudinal tension in a nerve on its excitability and the amplitude of action potentials in the context of the thermodynamic nerve theory.

IX.3.2 Nerve Tension

We approximate the nerve axon by a cylinder filled by an incompressible fluid. Let us consider a membrane cylinder of radius r_0 and length x_0. Its volume is

$$V_0 = x_0 \cdot \pi r_0^2 \tag{IX.3.1}$$

and its membrane area is

$$A_0 = x_0 \cdot 2\pi r_0 = \sqrt{4\pi x_0 V_0}\ , \tag{IX.3.2}$$

if the cross-sectional area at the ends is not taken into account.

IX.3.2.1 Constant Volume

We assume that we are at physiological temperature above the melting transition and that the membrane permeability is small. Let us now stretch the nerve to a length $x = x_0 + \Delta x$. We assume that the aqueous medium inside the cylinder is incompressible and that, therefore, the nerve volume $V = V_0$ is constant. This implies that the membrane area increases upon increasing the length x. The membrane area is given by

$$A = A_0 \cdot \sqrt{\frac{x}{x_0}} \approx A_0 \left(1 + \frac{1}{2}\frac{\Delta x}{x_0}\right) \tag{IX.3.3}$$

for small $\Delta x/x_0$. For $x > x_0$, one finds a tension Π in the membrane. The elastic energy is given by

$$\Delta E = \frac{1}{2}K_T^A A_0 \left(\frac{\Delta A}{A_0}\right)^2, \tag{IX.3.4}$$

where K_T^A is the lateral compression modulus. The lateral pressure in the membrane is

$$\Pi = -\frac{\partial \Delta E}{\partial A} = -K_T^A \frac{\Delta A}{A_0} = -\frac{1}{2}K_T^A \frac{\Delta x}{x_0} = -\frac{1}{2\kappa_T^A}\frac{\Delta x}{x_0}, \tag{IX.3.5}$$

where $\kappa_T^A = 1/K_T^A$ is the lateral compressibility. Biological membranes exist in the fluid membrane phase. For dipalmitoyl phosphatidylcholine (DPPC) membranes at 50 °C, $\kappa_T^A = 6.9$ m/N [93]. Since DPPC is an abundant lipid in biological membranes, we will assume in the following that the lateral compressibility of fluid biological membranes at physiological temperatures is similar. Thus

$$\Pi = -\frac{1}{2 \cdot 6.9\frac{\text{m}}{\text{N}}}\frac{\Delta x}{x_0}. \tag{IX.3.6}$$

In the following, we use the term tension when the lateral pressure is negative. As an example, if $\Delta x/x_0 = 0.2$, one finds a lateral tension of $-\Pi = 0.0145$ N/m. For a neuron with a radius of $r_0 = 10$ μm, this corresponds to a force of −910 nN.

IX.3.2.2 Constant Radius

Experiments show that the change in radius of the nerve membrane during the action potential is small and reflects changes on the order of the membrane thickness. If we assume that the radius of the axon is constant, the membrane must be permeable and the volume may change. The aforementioned calculation will be slightly different under these constraints. The change in area is now proportional to the change of its length, and $A = A_0 \cdot x/x_0$. Inserting this into eq. (IX.3.5) leads to twice the tension and force for the same change in length.

IX.3.3 Changes in Melting Temperature Induced by Nerve Stretching

Figure IX.3.2 shows the heat capacity profile of native membranes from porcine spine (from [50]). One finds a pronounced peak representing lipid melting below body temperature shaded in gray. Further peaks above body temperature represent protein unfolding peaks, which are irreversible in repeated temperature scans. The body temperature of pigs is around 39 °C.

The melting temperature $T_{m,0}$ of a membrane at a lateral tension of $\Pi = 0$ is given by [8]

$$T_{m,0} = \frac{\Delta H_0}{\Delta S_0} \quad \rightarrow S_0 = \frac{\Delta H_0}{T_{m,0}}. \tag{IX.3.7}$$

For DPPC membranes, $\Delta H = 35$ kJ/mol, and $T_{m,0} = 314.2$ K (=41.05 °C). In a nerve membrane, we use the melting temperature of 23.3 °C measured in a calorimeter but we

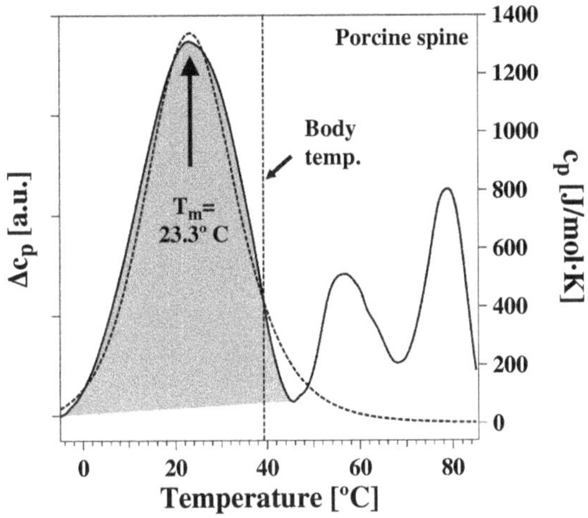

Figure IX.3.2 Melting profile of native membranes from porcine spine. The gray-shaded area represents the lipid melting peak with a maximum at 23.3 °C. The peaks above body temperature show the protein unfolding peaks. The dashed profile has been calculated using a van't Hoff profile assuming $\Delta H_0 = 35$ kJ/mol and a cooperative unit size of $n = 3.2$. Adapted from [50, 94]. See [94] for the mathematical details of the analysis.

further assume the same enthalpy of melting as for DPPC. We use this procedure because due to the large width of the transition and the exact amount of lipid in a preparation, it is difficult to determine absolute values for the enthalpy from the experiment. The enthalpy change upon changing lateral pressure is given by $\Delta H = \Delta H_0 + \Pi \Delta A_0$, where H_0 is the enthalpy in the absence of lateral pressure. Thus, the dependence of the melting temperature on lateral pressure is given by

$$T_m = \frac{\Delta H_0 + \Pi \Delta A_0}{\Delta S_0} \quad \to \Delta T_m = \frac{\Pi \Delta A_0}{\Delta H_0} \cdot T_{m,0}. \tag{IX.3.8}$$

If Π is negative (tension), the change in the transition temperature is negative, $\Delta T_m < 0$.

The molar area of a DPPC membrane in the solid state at 25 °C is $A_0^{\text{solid}} = 142728\ \text{m}^2/\text{mol}$, and the relative area change $\Delta A_0/A_0^{\text{solid}} = 0.246$ [93]. Thus, $\Delta A_0 = 35111\ \text{m}^2/\text{mol}$. Assuming that the enthalpy and area changes of a biological membrane are similar to those of DPPC and that $T_{m,0} = 23.3\,°\text{C}$, we find for a lateral tension of $\Pi = -0.02\,\text{N/m}$ that $\Delta T_m = -5.95$ K (see Figure IX.3.3 (left)). Using the aforementioned parameters, we can calculate the melting profile as a function of temperature using

$$\begin{aligned} \Delta c_p &= \frac{K}{(1+K)^2} \frac{n\Delta H}{RT^2}, \\ \text{where} \quad K &= \exp\left(-n \frac{(\Delta H - T\Delta S_0)}{RT}\right), \end{aligned} \tag{IX.3.9}$$

which was derived in [8] (see Section V.2.2.3). The cooperative unit size n indicates how many lipids on average undergo a transition simultaneously, i.e., it reflects that neighboring lipids interact and do not melt independently. It determines the width of the

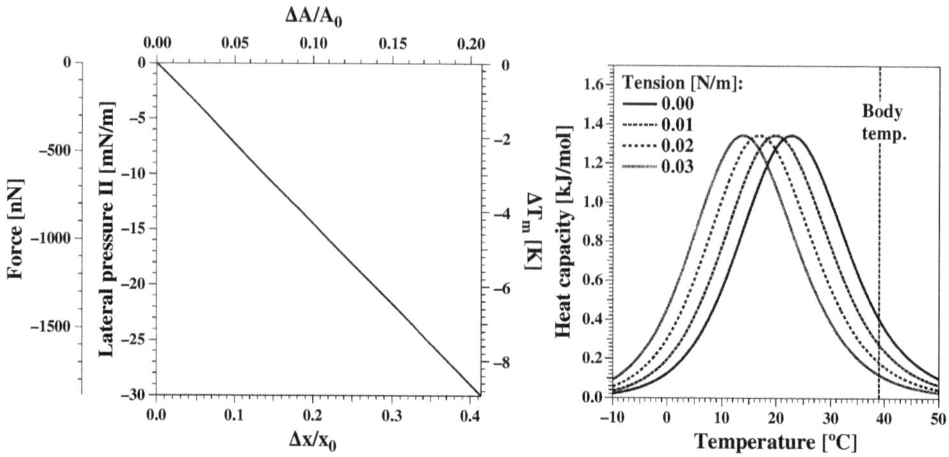

Figure IX.3.3 Left: Change in lateral tension (equal to $-\Pi$), force in an axon with a radius of $r_0 = 10$ μm, and melting temperature of a cylindrical neuron as a function of relative length changes, x/x_0. A lateral compressibility of 6.9 m/N for the liquid membrane is assumed (value for DPPC). From [93]/with permission of Elsevier. Right: Heat capacity profiles as a function of tension of a membrane with $\Delta H = 35$ kJ/mol, a relative area change of $\Delta A/A_0 = 0.246$, a melting temperature of 23.3 °C in the unperturbed axon, and a cooperative unit size of $n = 3.2$.

melting profile. The higher n, the narrower the melting profile. The dashed line in Figure IX.3.2 was calculated using the above equation for $\Delta H = 35$ kJ/mol, $T_{m,0} = 23.3$ °C, and $n = 3.2$, where n was adjusted to reproduce the width of the melting profile of pig nerves. For a more in-depth treatment of lipid membrane cooperativity, see [95]. We can now calculate the melting profiles as a function of membrane tension. The result is shown in Figure IX.3.3 (right). We see how the melting profile shifts away from physiological temperature toward lower temperatures when the membrane tension assumes larger values (more negative lateral pressures).

IX.3.4 The Effect of Tension on Nerve Excitability

The calculation of the change in threshold is exactly the same as in the case of anesthetics (see Chapter VII.6 and Section IX.1.6), and the derivation shall not be repeated here. Both anesthetics and stretching result in a lowering of the melting temperature of the membrane. Figure IX.3.4 shows the change in the density of the membrane as a function of the chemical potential difference $\Delta\mu_0$ at different values for the tension in the axon. Since tension lowers the transition temperature, the free energy necessary to change the lipid state in the membrane increases. We equate the free energy necessary to stimulate the nerve pulse with the chemical potential difference between the excited state and the resting state where $\Delta\mu_0 = 0$. If we keep the stimulus amplitude constant, i.e., we use a constant stimulus corresponding to $\Delta\mu_0$ in Figure IX.3.4 while changing the tension in the membrane, the amplitude of the excited soliton decreases [94].

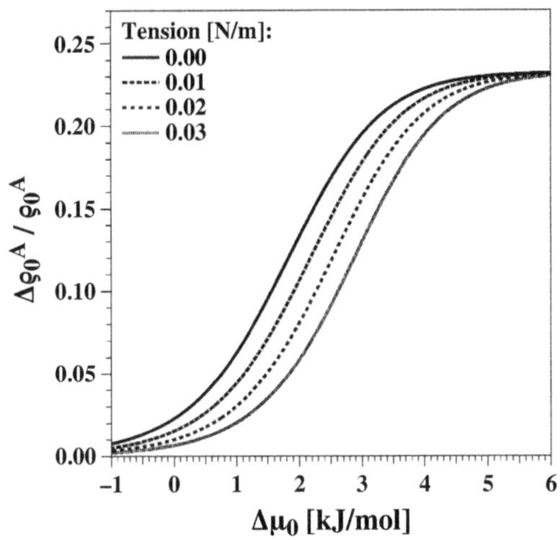

Figure IX.3.4 The change in membrane area density as a function of the chemical potential difference for different lateral tensions. $\Delta\mu_0 = 0$ corresponds to the state of the membrane at resting conditions.

IX.3.5 Examples for the Effect of Stretching

The aforementioned theoretical treatment suggests that

- The threshold for nerve stimulation increases when there is tension in the nerve. This implies that the stretched nerve is more difficult to stimulate. This can be seen in Figure IX.3.4.
- At constant stimulus, the amplitude of the action potential is lower when the nerve is stretched. This is shown schematically in Figure IX.3.5 (left).

The aforementioned predictions are in agreement with in vivo experiments on hamster sciatic nerve stretching by forces between 0 and 400 mN ([88]; Figure IX.3.5 (center)), and ex vivo stretching of guinea pig sciatic nerve by 0, 5, and 10% in length ([96]; Figure IX.3.5 (right)). In both experiments, the effect of stretching was reversible (while at larger forces, the nerve was damaged). Both in guinea pig nerves and in hamster nerves, the amplitude of the compound action potentials decreases upon stretching.

The effect of stretching on excitation threshold can also be seen in H-reflex and T-reflex. The H-reflex or finger flexor reflex is a neurological examination that can be used to determine the presence or absence of problems in the corticospinal tract. The T-reflex is the standard patella reflex or knee-cap response. Stretch reflex tests are used to determine the integrity of the spinal cord and peripheral nervous system and can be used to detect neuromuscular disease. Lengthening movements and static stretching of the human triceps have been investigated over many decades. Budini and Tilp [85] find that the H-reflex decreases immediately as static stretching is applied and in proportion to the stretch degree. Within 2 s after a single passive dorsiflexion movement, H-reflex is strongly inhibited. This implies that stretching leads to a decrease of the H-reflex, and that this is an effect with a fast onset that is reversible. Stretching over longer timescales or repetitive stretching leads to slow

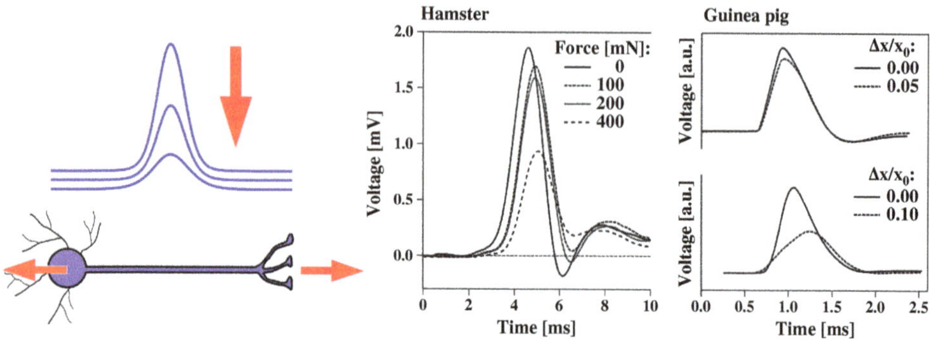

Figure IX.3.5 Left: Schematic drawing of a nerve under stretch and the resulting reduction in soliton amplitude for a constant stimulus. Center: Reduction in action potential amplitude upon stretch of hamster sciatic nerve at four different forces, 0, 100, 200, and 400 mN. Adapted from [88]. Right: Reduction of action potential amplitude of guinea pig sciatic nerve at three different elongations: $\Delta x/x_0 = 0, 0.05$, and 0.10. Adapted from [96].

recovery, possibly due to the damage of nerves. Guissard and collaborators [86, 87] showed that motor neuron excitability decreased during fast muscle stretch. This inhibition had been tested in the soleus muscle by recording the H-reflex and T-reflex, which were both found to be reduced during fast muscle stretching.

IX.3.6 Summary: Key Ideas of Chapter IX.3

In this chapter, we discuss the effect of nerve stretching on the action potentials in nerves. We found that

- The effect of stretching resembles that of the application of anesthetics. In both cases, the melting transition in the nerve membrane decreases.
- As a consequence, stretching of nerves increases the stimulation threshold.
- There exist medical examples for this prediction. Both H-reflex and T-reflex in the hand and the knee demonstrate that stretching reduces the nerve activity by increasing the stimulation threshold.
- At constant stimulation intensity, the amplitude of nerve pulses decreases upon an increase in the tension in the nerves.
- Reductions in the amplitude of compound action potentials in guinea pig and hamster sciatic nerves were found experimentally. This effect is reversible, proving that the reason for the reduction in amplitude was not caused by damage to the nerves.

IX.4

Tremor and Bipolar Disorder

There may exist many more conditions that affect melting transitions and therefore may be of medical interest. The following is a suggestion of where thermodynamic effects may play a role. While the statements on tremor, the effect of lithium, and how they are counteracted by anesthetics may not suffice as a proof of the effect and are of a rather speculative nature, there exist striking correlations that are worthwhile noting.

IX.4.1 Essential Tremor

Essential tremor (ET) is among the most common movement disorders. It consists of tremor of the upper limbs; for reviews, see [97, 98]. ET is often diagnosed when other origins of tremor can be ruled out [99].

Let us tentatively assume that ET is related to an increased excitability of nerves. We also recall that anything increasing the melting temperature in biomembranes will render the membrane more excitable and thereby favor the spontaneous firing of nerves. For instance, it is known that the spontaneous firing in nerves increases upon application of hydrostatic pressure (which increases the transition temperature and brings it closer to physiological temperature) and decreases upon application of anesthetic drugs (which lower the transition temperature). In the thermodynamic picture, the presence of tremor suggests that the membrane melting temperature in the nerve membranes is higher than usual. Then, mechanisms that lower the melting temperature should tend to suppress tremor; those that raise the melting temperature should cause tremor. Anesthetics lower the melting transition due to the melting-point depression law and therefore should reduce tremor. Interestingly, it has been known for some time that alcohol suppresses ET [100]. This has been confirmed by many studies [101–104]. Alcohol sensitivity is a secondary criterion for the diagnosis of ET [105]. This is in agreement with the prediction from thermodynamics.

Examples of other variables that will increase the likelihood of tremor are a lowering in body temperature (shivering from cold), inflammation leading to a decrease in body pH, and the application of lithium (see next section). Tremor may in fact be an indicator of inflammation.

IX.4.2 Lithium and Bipolar Disorder

Lithium is a commonly used drug in the treatment of bipolar disorder. It was introduced into clinical practice by J.F. Cade and especially by M. Schou [106–109]. About 25% of bipolar patients are susceptible to lithium treatment. The physical approaches to lithium therapy were nicely reviewed by D. Kamp [110]. Therapeutic serum levels are on the order of 0.6–2 mM [111] (0.6–0.8 mM is most common). The drug is taken over a period of at least 30 days before it becomes effective, which suggests that this is the time required to reach a steady-state concentration where intake and urination are balanced. The high dose speaks against any specific binding to a receptor because specific binding would go along with much lower critical doses. Concentrations of lithium above 1.5 mM are often toxic. The working mechanism of lithium is yet unknown. Lithium, sodium, and potassium are all alkali metals, and their ions carry a single positive change. It seems mysterious why a simple ion such as lithium should have such a profound medical effect while sodium and potassium do not affect moods.

It has been shown that, in contrast to other monovalent cations such as sodium and potassium, lithium increases phase transitions in charged lipid membranes quite significantly due to its binding to charged groups [48, 112]. Figure IX.4.1 shows the effect of K^+, Na^+, and Li^+ on the melting temperature of the negatively charged lipid DMPS as a function of ion concentration. It can be seen that the effect of lithium is 1–2 orders of magnitude larger than that of sodium and potassium Since about 10–40% of the lipids in biological membranes are charged [8], lithium may have a profound effect on membrane transitions similar to that of calcium [113].

The effect of lithium, sodium, and potassium follows the Hofmeister series (which roughly describes the interaction of the ions with water and on precipitation of proteins).

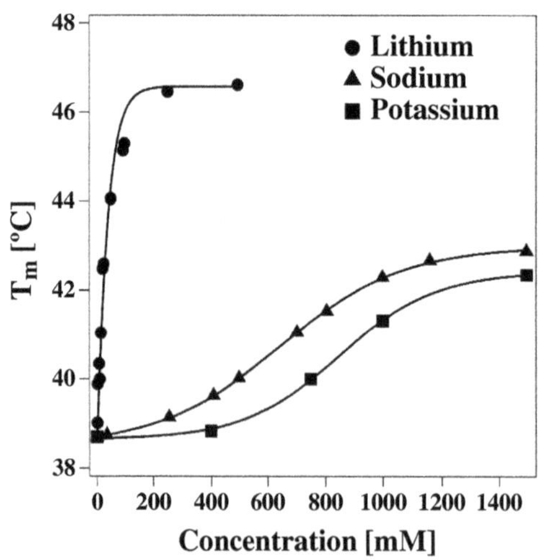

Figure IX.4.1 The influence of lithium, sodium, and potassium on the transition of the lipid DMPS. It can be seen that lithium affects phase transition much more than sodium and potassium. Adapted from [48, 112].

The Hofmeister series is: $K^+ > Na^+ > Li^+ > Mg^{2+} > Ca^{2+}$ [114–117]. Since protons (H^+) are even smaller than lithium and bind even stronger to negatively charged groups than lithium, they probably also deserve an entry in the Hofmeister series. Protons also shift transitions of negatively charged membranes dramatically [46].

IX.4.2.1 Lithium and Tremor

In the soliton theory, lithium renders the nerve membrane more excitable on first intake because it increases the transition temperature. Assuming again that tremor is related to an overexcitability of the nerves, the prediction is that overdoses of lithium may cause tremor. There is some clinical support for this expectation. Tremor is a side effect of lithium administration [118]. It has been argued that lithium-induced tremor is a subtype of ET [119, 120]. Further, it has been shown that lipid compositions changes with the chronic administration of lithium, and membranes become more fluid by adapting their unsaturated lipid composition [121]. This further supports the notion that membranes adapt to permanent changes in the thermodynamic variables. Thus, short-term effects, i.e., overdoes, and long-term effects caused by adaptation of membranes after long administration of lithium may differ.

IX.4.2.2 Lithium and Alcohol

Since lithium increases the transition temperature, it should tend to counteract the effects of alcohol. It has in fact been found that alcohol and lithium have opposing actions in bipolar patients, and chronic alcoholism increases bipolar symptoms [122]. The anesthetic ketamine is often used to treat depression, while lithium is used to attenuate mania [118]. Thus, anesthetics and lithium seem to have opposing effects.

IX.4.2.3 Depression, Body Temperature, and Lithium

In a recent study on more than 20,000 subjects, it was found that there exists a correlation between body temperature and symptoms of depression. Higher body temperatures when awake were associated with greater depression symptom severity [123]. A correlation between body temperature and depression was already proposed by [124] in a study with eight bipolar patients. Interestingly, the treatment of bipolar disorder with lithium also leads to changes in body temperature [125]. It seems that there are many couplings between depression, temperature, and lithium. Therefore, it is possible and maybe even likely that depression will also be coupled to fever, inflammation, and anesthesia.

IX.4.3 Summary: Key Ideas of Part IX.4

In this short chapter, we discussed the relation between tremor, bipolar disorder, depression, and anesthesia.

- ET consists of tremor of the upper limbs. It is often diagnosed when other origins of tremor can be ruled out.

- We assume that overexcitability of nerves is related to tremor. Tremor is found when the transition in the nerve membranes occurs at temperatures that are too close to body temperature.
- As a consequence, ET can be counteracted by the administration of ethanol (and anesthetics in general), which have an opposite effect on the transition.
- Lithium administration at typical blood concentrations of 0.6–0.8 mM is a common treatment for bipolar disorder.
- Lithium is an ion with a single positive charge. It binds to negatively charged groups much stronger than sodium and potassium. Lithium strongly affects melting temperatures in charged membranes, shifting the melting transition to higher temperatures.
- Lithium in overdose can lead to ET, in agreement with its effect on transition temperatures.
- Lithium treatment leads to changes in body temperature.
- The effect of lithium on bipolar patients is counteracted by ethanol.
- Overall, there exists evidence that neurological disorders such as tremor and mood disorders are related to transition temperatures.
- Depression is related to changes in body temperature.

Summarizing, there are many correlations between seemingly unrelated phenomena such as body temperature, lithium concentration, and anesthesia that might help in understanding bipolar disorder, depression, and tremor.

IX.5

Ultrasound Neurostimulation

Ultrasound is sound with frequencies above the range of human hearing, i.e., above about 20 kHz. It has been shown that brains can be stimulated by focused pulses of ultrasound. In medical studies, frequencies from around 250 kHz to about 7 MHz have been used. Focused ultrasound neurostimulation of the brain is an emerging area of clinical interest [126–130]. It has been found that focused ultrasound can stimulate nerves from a distance, i.e., it is a noninvasive technique. It has high spatial resolution and the ability to reach deep brain targets [131]. The spatial resolution of such an experiment is in the mm^3 range. A simulated intensity profile of ultrasound applied to a human brain is shown in Figure IX.5.1.

The setup for such an experiment includes an ultrasound transducer that is brought in direct contact with the skull of a patient or animal. This is shown in Figure IX.5.2. Focused ultrasound can stimulate motor responses by locally stimulating regions in the

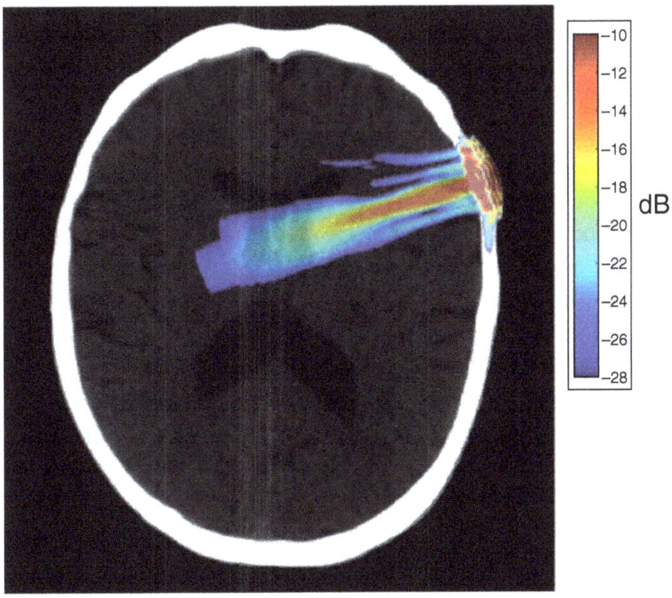

Figure IX.5.1 Focused ultrasound neurostimulation of a human brain. Simulated profile of the sound intensity close to the transducer (top, right). From [131]/Frontiers Media S.A./CC BY 4.0.

Physics of Nerves and Excitatory Membranes, First Edition. Thomas Heimburg.

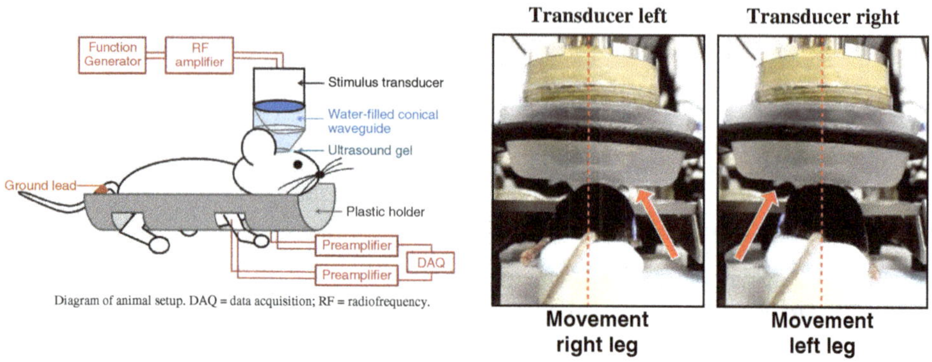

Figure IX.5.2 Left: Schematic diagram of an ultrasound stimulation experiment on a mouse brain. From [133]/with permission of Elsevier. Right: A focused ultrasound experiment as described in [132]. An ultrasonic transducer focuses sound on a part of a mouse brain. Depending on its exact position, the right or left hind leg of the mouse contracts. Other positions of the transducer can make the tail move. The images have been extracted from a supplementary video in [132]/with permission of John Wiley & Sons.

brain. Kamimura and coworkers [132] showed how one can stimulate selected brain regions in mice by focusing ultrasound on different regions of a mouse brain such that they contract the left or the right leg, or the tail (Figure IX.5.2).

It is not understood how ultrasound causes nerve stimulation, but several studies have raised the possibility that the soliton theory (described in Part VII) offers an explanation for the mechanism [126, 128, 134, 135]. This would indeed make sense since the nerve pulse possesses a mechanical component (see Chapter III.4), and the soliton theory is based on the theory of sound. Therefore, it would be interesting to compare the typical frequencies and wavelengths of pulse ultrasound neurostimulation with the length and timescale of nerve pulses. The timescale of the nerve pulse is of the order of 1 ms, corresponding to the period of a carrier frequency of $\nu = 1$ kHz. This timescale corresponds to the relaxation timescale of biomembranes above their melting transition. When ultrasound is applied to the brain, it propagates either in the aqueous cytosol or in the nerve membranes. The sound velocity in water is 1500 mm/s, and in a motor neuron, it is of the order $c = 100$ m/s [136]. The wavelength is given by $\lambda = c/\nu$. The wavelength in water and fluid membranes for different frequencies is given in Table IX.5.1.

Thus, the wavelength of sound in water in the frequency range from 100 kHz–1MHz is between 15 and 1.5 mm, much larger than cellular structures but of similar magnitude to the length of the nerve pulse. An action potential in a nonmyelinated nerve displays pulse lengths of a few millimeters or centimeters (see Sections III.3.1 and III.3.5). In in-vivo experiments, the stimulation frequency is rather between 250 kHz and 7 MHz (see tables in [126, 133]). The ultrasound is often pulsed with pulse frequencies in the range of 1–3 kHz (e.g., [126, 132]), which is close to the characteristic frequency of the nerve pulse with T = 1 ms and $\nu = 1$ kHz. Summarizing, the wavelength in water of the typical ultrasound frequencies used in in-vivo experiments is of similar order as the length of the nerve pulses in nonmyelinated nerves, and the period of the repetition frequency of pulsed ultrasound is similar to the characteristic frequency of nerve pulses. It remains to be investigated whether

Table IX.5.1 Wavelength of ultrasound in water (assuming a sound velocity of 1500 m/s) and membranes (assuming a 2D sound velocity of 100 m/s) at three different frequencies. These wavelengths are of the same order of magnitude as the length of the nerve pulse given in Table III.3.1.

Frequency	Wavelength in water	Wavelength in nerve membrane
1 kHz	1.5 m	10 cm
100 kHz	15 mm	1 mm
1 MHz	1.5 mm	0.1 mm

these similarities are meaningful for focused ultrasound neurostimulation or whether they are just fortuitous. However, experiments have evolved to favor these frequencies. It seems plausible to assume that the optimal stimulation for the brain would be a signal with the carrier frequency of the nerve signal (1 kHz) and a wavelength that corresponds to the length of the nerve pulse in space – i.e., a signal with lower frequency than normally used in neurostimulation. In fact, [133] note that focused ultrasound neurostimulation becomes more effective at lower frequencies.

IX.5.1 Summary: Key Ideas of Part IX.5

Brains can be stimulated with focused ultrasound. This is presently a growing field and hot topic. However, since this discipline is in its infancy, there are many open questions concerning the mechanism:

- It is not understood why ultrasound can stimulate nerves. This does not fit into electrical theories of nerve activity.
- It has been proposed in several studies that the soliton theory may provide a mechanism because it intrinsically contains sound as the major mechanism.
- The frequency of ultrasound in clinical studies is between a few 250 kHz and 7 MHz. This corresponds to wavelengths in aqueous media of 0.2–8 mm. This is somewhat smaller but of similar order to the typical length of a nerve pulse in nonmyelinated nerves.
- In focused ultrasound neurostimulation studies, the sound is pulsed with pulse frequencies of 1–3 kHz, corresponding to a timescale of 0.3–1 ms. This is roughly the timescale of the nerve pulse.

References

Preface

1 von Helmholtz, H. (1896). *Die Lehre von den Tonempfindungen als physiologische Grundlage für die Theorie der Musik*, Vieweg and Sohn, Braunschweig, 5th ed.
2 Denworth, L. (2023). Brain waves synchronize when people interact. *Sci. Am.* 329, 50.
3 Schrödinger, E. (1944). *What Is Life*? Cambridge University Press.
4 Einstein, A. (1910). Theorie der Opaleszenz von homogenen Flüssigkeiten und Flüssigkeitsgemischen in der Nähe des kritischen Zustandes. *Ann. Phys. (Leipzig)*, 33, 1275–1298.
5 Einstein, A. (1949). Autobiographical notes. In *Albert Einstein: Philosopher-Scientist, The Library of Living Philosophers.*, vol. VII (ed. P.A. Schilpp), *The Library of Living Philosophers*, pp. 3–95.
6 Kaufmann, K., Hanke, W., and Corcia, A. (1989). Ion Channels, Caruaru.
7 Kaufmann, K. (1989). Lipid Membrane, Caruaru.

Part I

1 Riskin, J. (2010). Machines in the garden. *Republics of Letters*, 1, 16–43.
2 Cobb, M. (2002). Timeline - exorcising the animal spirits: Jan Swammerdam on nerve function. *Nat. Rev. Neurosci.*, 3, 395–400.
3 Swammerdam, J. (1758). *The Book of Nature II.* Seyffert, London.
4 Wallace, W. (2003). The vibrating nerve impulse in Newton, Willis and Gassendi: first steps in a mechanical theory of communication. *Brain. Cogn.*, 51, 66–94.
5 Geddes, L.A. and Hoff, H.E. (1971). The discovery of bioelectricity and current electricity the Galvani-Volta controversy. *IEEE Spectr.*, 8, 38–46.
6 Galvani, L. and Green, R.M. (1953). *Commentary on the Effect of Electricity on Muscular Motion*. E. Licht, Cambridge, MA.
7 Düweke, P. (2001). *Kleine geschichte der hirnforschung*. C.H. Beck, Munich.
8 von Helmholtz, H. (1850). Messungen über den zeitlichen Verlauf der Zuckung animalischer Muskeln und die Fortpflanzungsgeschwindigkeit der Reizung in den Nerven. *Joh. Müllers Arch. Anatom. Phys.*, 276–364.

Physics of Nerves and Excitatory Membranes, First Edition. Thomas Heimburg.

9 von Helmholtz, H. (1852). Messungen über Fortpflanzungsgeschwindigkeit der Reizung in den Nerven: Zweite Reihe. *Joh. Müllers Arch. Anat. Physiol.*, 199–216.

10 von Helmholtz, H. (1848). Ueber die Wärmeentwicklung bei der Muskelaction. *Joh. Müller's Arch. Anatom. Phys. Phys.*, 144–164.

11 von Helmholtz, H. (1847). *Ueber die Erhaltung der Kraft.* G. Reimer, Berlin, Germany.

12 von Helmholtz, H. (1896). *Handbuch der physiologischen Optik.* Verlag Leopold Voss, Hamburg and Leipzig, 9th ed.

13 von Helmholtz, H. (1896). *Die Lehre von den Tonempfindungen als physiologische Grundlage für die Theorie der Musik.* Vieweg and Sohn, Braunschweig, 5th ed.

14 Bernstein, J. (1868). Ueber den zeitlichen verlauf der negativen schwankung des nervenstroms. *Pflügers. Arch.*, 1, 173–207.

15 Bernstein, J. (1912). *Elektrobiologie. Die Lehre von den elektrischen Vorgängen im Organismus auf moderner Grundlage dargestellt.* Vieweg and Sohn.

16 Bernstein, J. (1902). Untersuchungen zur Thermodynamik der bioelektrischen Ströme. 1 Theil. *Pflügers Arch.*, 92, 521–562.

17 Cole, K.S. and Curtis, H.J. (1939). Electrical impedance of the squid giant axon during activity. *J. Gen. Physiol.*, 220, 649–670.

18 Hodgkin, A.L., Huxley, A.F., and Katz, B. (1952). Measurement of current-voltage relations in the membrane of the giant axon of Loligo. *J. Physiol.*, 116, 424–448.

19 Hodgkin, A.L. and Huxley, A.F. (1952). Currents carried by sodium and potassium ions through the membrane of the giant axon of Loligo. *J. Physiol.*, 116, 449–472.

20 Hodgkin, A.L. and Rushton, W.A.H. (1946). The electrical constants of a crustacean nerve fibre. *Proc. Roy. Soc. Lond. B*, 133, 444–479.

21 Hodgkin, A.L. and Huxley, A.F. (1952). A quantitative description of membrane current and its application to conduction and excitation in nerve. *J. Physiol.*, 117, 500–544.

22 Andersen, S.S.L., Jackson, A.D., and Heimburg, T. (2009). Towards a thermodynamic theory of nerve pulse propagation. *Progr. Neurobiol.*, 88, 104–113.

23 Keynes, R.D. (1951) The ion movements during nervous activity. *J. Physiol.*, 114, 119–150.

24 Keynes, R.D. (1951). The leakage of radioactive potassium from stimulated nerve. *J. Physiol.*, 113, 99–114.

25 Hodgkin, A.L. and Keynes, R.D. (1955). The potassium permeability of a giant nerve fibre. *J. Physiol.*, 128, 61–88.

26 Caldwell, P.C. and Keynes, R.D. (1960). The permeability of the squid giant axon to radioactive potassium and chloride ions. *J. Physiol.*, 154, 177–189.

27 Keynes, R.D. and Ritchie, J.M. (1965). Movements of labelled ions in mammalian non-myelinated nerve fibres. *J. Physiol.*, 179, 333–367.

28 Hille, B., Armstrong, C.M., and MacKinnon, R. (1999). Ion channels: from idea to reality. *Nat. Med.*, 5, 1105–1109.

29 Hille, B. and Schwarz, W. (1978). Potassium channels as multi-ion single-file pores. *J. Gen. Physiol.*, 72, 409–442.

30 Doyle, D.A., Morais, J., Gulbis, J.M., Cohen, A.L., Chait, B.T., and MacKinnon, R. (1998). The structure of the potassium channel: molecular basis of k+ conduction and selectivity. *Science*, 280, 69.

31 Neher, E. and Sakmann, B. (1976). Single-channel currents recorded from membrane of denervated frog muscle fibres. *Nature*, 260, 779–802.

32 Hamill, O.P., Marty, A., Neher, E., Sakmann, B., and Sigworth, F.J. (1981). Improved patch-clamp techniques for high-resolution current recording from cells and cell-free membrane patches. *Pflügers. Arch.*, 391, 85–100.
33 LeMasurier, M., Heginbotham, L., and Miller., C. (2001). Kcsa: it's a potassium channel. *J. Gen. Physiol.*, 118, 303–314.
34 Santos, C.R.A., Duarte, A.C., Costa, A.R., Tomas, J., Quintela, T., and Goncalves, I. (2019). The senses of the choroid plexus. *Prog. Neurobiol.*, 182, 101680.
35 Schrödinger, E. (1944). *What Is Life*? Cambridge University Press.
36 Watson, J.D. and Crick, F.H.C. (1953). Molecular structure of deoxypentose nucleic acids. *Nature*, 171, 737–738.
37 Tigerstedt, R. (1880). *Studien über die mechanische Nervenreizung*. J. C. Frenkel and Son, Helsingfors.
38 Haller. (1766). Elementa physiologiae corporis humani, unleserlich zitiert nach Tigerstedt 1880, S.12.
39 Fontana. (1785). Beorbachtungen und Versuche über die Natur der thierischen Körper, unleserlich zitiert nach Tigerstedt 1880, S.12.
40 du Bois-Reymond, E. (1849). *Untersuchungen über die thierische Elektrizität*, vol. 2, Reimer.
41 Heidenhain, R. (1856). Neue Methode, motorische Nerven auf mechanischem Wege zu tetanisieren. *In Physiologische studien*, Verlag von August Hirschwald, Berlin, pp. 129–143.
42 Heidenhain, R. (1858). Ein mechanischer Tetanomotor für Vivisektionen. *In Untersuchungen zur Naturlehre des Menschen und der Thiere*, vol. 4 (ed. J. Moleschott), Verlag von Meidinger und Sohn & Comp, Frankfurt, Germany, pp. 124–133.
43 Bernstein, J. (1877). Ueber die ermüdung und erholung der nerven. *Pflügers. Arch.*, 15, 289–327.
44 Eckhard, C. (1858). Ueber den Einfluss des constanen galvanischen Stromes auf der Erregbarkeit des motorischen Nerven. *In Beiträge zur Anatomie und Physiologie*, vol. 1, J. Ricker'sche Buchhandlung, pp. 23–54.
45 Wundt, W. (1871). *Untersuchungen zur Mechanik der Nerven und Nervencentren,* vol. I, Verlag von Ferdinand Enke, Erlangen.
46 Wilke, E. (1912). Das problem der reizleitung im nerven vom standpunkte der wellenlehre aus betrachtet. *Pflügers. Arch.*, 144, 35–38.
47 Wilke, E. and Atzler, E. (1912). Experimentelle Beiträge zum Problem der Reizleitung im Nerven. *Pflügers. Arch.*, 146, 430–446.
48 Iwasa, K. and Tasaki, I. (1980). Mechanical changes in squid giant-axons associated with production of action potentials. *Biochem. Biophys. Research Comm.*, 95, 1328–1331.
49 Iwasa, K., Tasaki, I., and Gibbons, R.C. (1980). Swelling of nerve fibres associated with action potentials. *Science*, 210, 338–339.
50 Tasaki, I., Iwasa, K., and Gibbons, R.C. (1980). Mechanical changes in crab nerve fibers during action potentials. *Jap. J. Physiol.*, 30, 897–905.
51 Tasaki, I. and Iwasa, K. (1982). Further studies of rapid mechanical changes in squid giant axon associated with action potential production. *Jap. J. Physiol.*, 32, 505–518.
52 Tasaki, I. and Iwasa, K.H. (1982). Rapid pressure changes and surface displacements in the squid axon associated with production of action potentials. *Jap. J. Physiol.*, 32, 69–81.
53 Tasaki, I. and Iwasa, K.H. (1982). Further studies of rapid mechanical changes in squid giant axon associated with action potential production. *Jap. J. Physiol.*, 32, 505–518.

54 Tasaki, I. (1988). A macromolecular approach to excitation phenomena: mechanical and thermal changes in nerve during excitation. [review] [63 refs]. *Physiol. Chem. Phys. Med. NMR.*, 20, 251–268.

55 Tasaki, I., Kusano, K., and Byrne, M. (1989). Rapid mechanical and thermal changes in the garfish olfactory nerve associated with a propagated impulse. *Biophys. J.*, 55, 1033–1040.

56 Tasaki, I. and Byrne, M. (1990). Volume expansion of nonmyelinated nerve fibers during impulse conduction. *Biophys. J.*, 57, 633–635.

57 Tasaki, I. and Iwasa, K.H. (1980). Shortening of nerve fibers associated with propagated nerve impulse. *Curr. Top. Med. Chem.*, 94, 716–720.

58 du Bois-Reymond, E. (1848). *Untersuchungen über die thierische Elektrizität,* vol. 1, Verlag G. Reimer, Berlin.

59 Heimburg, T. (2021). The important consequences of the reversible heat production in nerves and the adiabaticity of the action potential. *Progr. Biophys. Mol. Biol.*, 162, 26–40.

60 Heidenhain, R. (1868). über die Reaktion der thätigen nerven. In *Studien des Physiologischen Institutes zu Breslau* vol. 4, Breitkopf & Härtel, Leipzig, pp. 248–250.

61 Rolleston, H.D. (1890). On the conditions of temperature in nerves (I) during activity, (II) during the process of dying. *J. Physiol.*, 11, 208–225.

62 Stewart, G.N. (1891). Notes on some applications in physiology of the "resistance" method of measuring temperature, with special reference to the question of heat production in mammalian nerves during excitation. *J. Physiol.*, 12, 409–425.

63 Cremer, M. (1897). Ueber neurothermische Versuche am marklosen Nerven. *Münch. med. Wochenschr.*, 44, 280–281.

64 Hill, A.V. (1912). The absence of temperature changes during the transmission of a nervous impulse. *J. Physiol.*, 43, 433–440.

65 Hill, A.V. (1959). The heat production of muscle and nerve, 1848–1914. *Ann. Rev. Physiol.*, 21, 1–18.

66 Bayliss, W.M. (1915). *Principles of General Physiology*, Longmans, Green and Co., London.

67 Downing, A.C., Gerard, R.W., and Hill, A.V. (1926). The heat production of nerve. *Proc. Roy. Soc. Lond. B.*, 100, 223–251.

68 Downing, A.C. and Hill, A.V. (1929). A new thermopile for the measurement of nerve heat production. *Proc. Roy. Soc. Lond. B.*, 105, 147–152.

69 Hill, A.V. (1929). The heat-production and recovery of crustacean nerve. *Proc. Roy. Soc. Lond. B.*, 105, 153–176.

70 Hill, A.V. (1932). A closer analysis of the heat production of nerve. *Proc. Roy. Soc. Lond. B.*, 111, 106–164.

71 Feng, T.P. and Hill, A.V. (1933). The steady state of heat production of nerve. *Proc. Roy. Soc. Lond. B.*, 113, 356–365.

72 Feng, T.P. and Hill, A.V. (1933). The effect of frequency of stimulation on the heat production of frog's nerve. *Proc. Roy. Soc. Lond. B.*, 113, 366–368.

73 Feng, T.P. and Hill, A.V. (1933). The relation between initial and recovery heat production in frog's nerve. *Proc. Roy. Soc. Lond. B.*, 113, 369–386.

74 Hill, A.V. (1933). The three phases of nerve heat production. *Proc. Roy. Soc. Lond. B.*, 113, 345–356.

75 Hill, A.V. (1933). The effect of veratrine on the heat production of medullated nerve. *Proc. Roy. Soc. Lond. B.*, 113, 386–393.

76 Hill, A.V. (1934). Nerve heat production as a physiological response to excitation. *Proc. Roy. Soc. Lond. B.*, 115, 200–216.

77 Abbott, B.C., Hill, A.V., and Howarth, J.V. (1958). The positive and negative heat production associated with a nerve impulse. *Proc. Roy. Soc. Lond. B.*, 148, 149–187.

78 Hill, A.V. and Howarth, J.V. (1958). The initial heat production of stimulated nerve. *Proc. Roy. Soc. Lond. B.*, 149, 167–175.

79 Hodgkin, A.L. (1964). *The Conduction of the Nervous Impulse*. Liverpool University Press, Liverpool, UK.

80 Abbott, B.C., Howarth, J.V., and Ritchie, J.M. (1965). Initial heat production associated with nerve impulse in crustacean and mammalian non-myelinated nerve fibres. *J. Physiol.*, 178, 368–383.

81 Howarth, J.V., Keynes, R., and Ritchie, J.M. (1968). The origin of the initial heat associated with a single impulse in mammalian non-myelinated nerve fibres. *J. Physiol.*, 194, 745–793.

82 Abbott, B.C. and Howarth, J.V. (1973). Heat studies in excitable tissues. *Physiol. Rev.*, 53, 120–158.

83 Ritchie, J.M. (1973). Energetic aspects of nerve conduction: the relationships between heat production, electrical activity and metabolism. *Prog. Biophys. Mol. Biol.*, 26, 147–187.

84 Howarth, J. (1975). Heat production in non-myelinated nerves. *Phil. Trans. Royal Soc. Lond.*, 270, 425–432.

85 Howarth, J.V. (1975). Heat production non-myelinated nerves. *Phil. Trans. Royal Soc. B.*, 270, 425–432.

86 Howarth, J.V., Ritchie, J.M., and Stagg, D. (1979). Initial heat production in garfish olfactory nerve-fibers. *Proc. Roy. Soc. Lond. B.*, 205, 347–367.

87 Howarth, J.V. and Ritchie, J.M. (1979). Recovery heat-production in non-myelinated garfish olfactory nerve fibers. *J. Physiol.*, 292, 167–175.

88 Ritchie, J.M. and Keynes, R.D. (1985). The production and absorption of heat associated with electrical activity in nerve and electric organ. *Quart. Rev. Biophys.*, 18, 451–476.

89 El Hady, A. and Machta, B.B. (2015). Mechanical surface waves accompany action potential propagation. *Nat. Commun.*, 6, 6697.

90 Heimburg, T. and Jackson, A.D. (2005). On soliton propagation in biomembranes and nerves. *Proc. Natl. Acad. Sci. USA.*, 102, 9790–9795.

91 Heimburg, T. (2024). The mechanical properties of nerves, the size of the action potential and consequences for the brain. *arXiv*, 1–14. https://arxiv.org/abs/2408.07615.

92 Kaufmann, K. (1989). Lipid Membrane, Caruaru.

93 Kobatake, Y., Tasaki, I., and Watanabe, A. (1971). Phase transition in membrane with reference to nerve excitation. *Adv. Biophys.*, 208, 1–31.

94 Tasaki, I. (1999). Evidence for phase transition in nerve fibers, cells and synapses. *Ferroelectrics*, 220, 305–316.

95 Gonzalez-Perez, A., Mosgaard, L.D., Budvytyte, R., Villagran Vargas, E., Jackson, A.D., and Heimburg, T. (2016). Solitary electromechanical pulses in lobster neurons. *Biophys. Chem.*, 216, 51–59.

Part II

1 Einstein, A. (1949). Autobiographical notes. In *Albert Einstein: Philosopher-Scientist, the Library of Living Philosophers*, vol. VII (ed. P.A. Schilpp), *The Library of Living Philosophers*, pp. 3–95.
2 Gay-Lussac, J.L. (1802). Recherches sur la dilatation des gaz et des vapeurs. *Ann. Chim. Phys.*, 43, 137–175.
3 Mayer, J.R. (1842). Bemerkungen über die kräfte der unbelebten natur. *Ann. Chem. Pharm.*, 42, 233–240.
4 von Helmholtz, H. (1847). *Ueber die Erhaltung der Kraft*. G. Reimer, Berlin, Germany.
5 Carnot, N.L.S. (1824). *Reflexions sur la puissance motrice du feu et sur les machines propres a developper cette puissance*. Bachelier, Paris.
6 Einstein, A. (1905). Über die von der molekularkinetischen Theorie der Wärme geforderte Bewegung von in ruhenden Flüssigkeiten suspendierten Teilchen. *Ann. Phys. (Leipzig).*, 17 (532), 549–560.
7 Einstein, A. (1905). über einen die Erzeugung und Verwandlung des Lichtes betreffenden heuristischen Gesichtspunkt. *Ann. Phys. (Leipzig).*, 17, 132–148.
8 Hill, T.L. (1960). *An Introduction to Statistical Thermodynamics*. Sinauer Associates Inc., Sunderland, Ma.
9 Einstein, A. (1910). Theorie der Opaleszenz von homogenen Flüssigkeiten und Flüssigkeitsgemischen in der Nähe des kritischen Zustandes. *Ann. Phys. (Leipzig).*, 33, 1275–1298.
10 Onsager, L. (1931). Reciprocal relations in irreversible processes. I. *Phys. Rev.*, 37, 405–426.
11 Onsager, L. (1931). Reciprocal relations in irreversible processes. II. *Phys. Rev.*, 38, 2265–2279.
12 Helfrich, W. (1973). Elastic properties of lipid bilayers: theory and possible experiments. *Z. Naturforsch.*, 28c, 693–703.

Part III

1 Hodgkin, A.L. (1964). *The Conduction of the Nervous Impulse*. Liverpool University Press, Liverpool, UK.
2 Katz, B. (1966). *Nerve, Muscle, and Synapse*. McGraw-Hill, London.
3 Johnston, D. and Wu, S.M.S. (1995). *Cellular Neurophysiology*. MIT Press, Boston.
4 Kandel, E.R., Schwartz, J.H., and Jessell, T.M. (2000). *Principles of Neural Science*. McGraw-Hill, New York, 4th ed.
5 Alberts, B., Johnson, A., Lewis, J., Raff, M., Roberts, K., and Walter, P. (2015). *Molecular Biology of the Cell*. Garland Science, New York, 6th ed.
6 Rozo, J.A., Martinez-Gallego, I., and Rodriguez-Moreno, A. (2024). Cajal, the neuronal theory and the idea of brain plasticity. *Front. Neuroanat.*, 18, 1331666.
7 Ramon y Cajal, S. (1888). Estructura de los centros nerviosos de las aves. *Rev. Trim. Histol. Normal Patol.*
8 Brown, K.M., Gillette, T., and Ascoli, G.A. (2008). Quantifying neuronal size: summing up trees and splitting the branch difference. *Semin. Cell Dev. Biol.*, 19, 485–493.

9 Alvarez, F.J., Pearson, J., Harrington, D., Dewey, D., Torbeck, L., and Fyffe, R.E.W. (1998). Distribution of 5-hydroxytryptamine-immunoreactive boutons on α-motoneurons in the lumbar spinal cord of adult cats. *J. Comp. Neurol.*, 393, 69–83.

10 Shapson-Coe, A., Januszewski, M., Berger, D.R., Pope, A., Wu, Y., Blakely, T., et al. (2024). A petavoxel fragment of human cerebral cortex reconstructed at nanoscale resolution. *Science*, 384, eadk4858.

11 Chung, K., Wallace, J., Kim, S.J., Kalyanasundaram, S., Andalman, A.S., Davidson, T.J., et al. (2013). Structural and molecular interrogation of intact biological systems. *Nat*, 497, 332–337.

12 Chung, K. and Deisseroth, K. (2013). CLARITY for mapping the nervous system. *Nat. Methods*, 10, 508–513.

13 Krarup, C. and Trojaborg., W. (1994). Compound sensory action potentials evoked by tactile and by electrical stimulation in normal median and sural nerves. *Muscle Nerve*, 17, 733–740.

14 Heimburg, T. (2024). The mechanical properties of nerves, the size of the action potential and consequences for the brain. *arXiv:2408.07615v1*, 1–14.

15 Tasaki, I. and Takeuchi, T. (1941). Der am Ranvierschen Knoten entstehende Aktionsstrom und seine Bedeutung für die Erregungsleitung. *Pflügers. Arch.*, 224, 696–711.

16 Huxley, A.F. and Staempfli, R. (1949). Evidence for saltatory conduction in peripheral myelinated nerve fibres. *J. Physiol.*, 108, 315–339.

17 King, R. (2013). Microscopy anomaly: normal structure. In *Handbook of Clinical Neurology*, vol. 115 (eds. G. Said and C. Krarup), Elsevier, chap. 2, pp. 7–27.

18 Taveggia, C. and Bolino, A. (2018). DRG neuron/Schwann cells myelination cocultures. In *Myelin: Methods and Protocols, Methods in Molecular Biology*, vol. 1791 (ed. A. Woodhoo), Springer Nature, pp. 115–130.

19 Mužić, T., Tounsi, F., Madsen, S.B., Pollakowski, D., Konrad, M., and Heimburg, T. (2019). Melting transitions in biomembranes. *Biochim. Biophys. Acta.*, 1861, 183026.

20 Hodgkin, A.L. and Huxley, A.F. (1939). Action potentials recorded from inside a nerve fibre. *Nature*, 144, 710–711.

21 Gonzalez-Perez, A., Budvytyte, R., Mosgaard, L.D., Nissen, S., and Heimburg, T. (2014). Penetration of action potentials during collision in the median and lateral giant axons of invertebrates. *Phys. Rev. X.*, 4, 031047.

22 Gonzalez-Perez, A., Mosgaard, L.D., Budvytyte, R., Villagran Vargas, E., Jackson, A.D., and Heimburg, T. (2016). Solitary electromechanical pulses in lobster neurons. *Biophys. Chem.*, 216, 51–59.

23 Tasaki, I., Kusano, K., and Byrne, M. (1989). Rapid mechanical and thermal changes in the garfish olfactory nerve associated with a propagated impulse. *Biophys. J.*, 55, 1033–1040.

24 Wang, T., Mužić, T., Jackson, A.D., and Heimburg, T. (2018). The free energy of biomembrane and nerve excitation and the role of anesthetics. *Biochim. Biophys. Acta.*, 1860, 2145–2153.

25 Moldovan, M., Lange, K.H.W., Aachmann-Andersen, N.J., Kjær, T.W., Olsen, N.V., and Krarup, C. (2014). Transient impairment of the axolemma following regional anaesthesia by lidocaine in humans. *J. Physiol.*, 592, 2735–2750.

26 Adrian, E.D. and Forbes., A. (1922). The all-or-nothing response of sensory nerve fibres. *J. Physiol.*, 56, 301–330.

27 Soubeyrand, M., Melhem, R., Protais, M., Artuso, M., and Crézé, M. (2020). Anatomy of the median nerve and its clinical applications. *Hand. Surg. Rehabil.*, 39, 2–18.

28 Tasaki, I. (1982). *Physiology and Electrochemistry of Nerve Fibers*, Academic Press, New York.
29 Gotch, F. and Burch, G.J. (1899). The electrical response of nerve to two stimuli. *J. Physiol.*, 24, 410–426.
30 Adrian, E.D. (1914). The temperature coefficient of the refractory period in nerve. *J. Physiol.*, 48, 453–464.
31 Tackmann, W. and Lehmann, H.J. (1974). Refractory period in human sensory nerve fibres. *Eur. Neurol.*, 12, 277–292.
32 Betts, R.P., Johnston, D.M., and Brown, B.H. (1976). Nerve fibre velocity and refractory period distributions in nerve trunks. *J. Neurol. Neurosurg. Psychiatry.*, 39, 694–700.
33 Borg, J. (1984). Refractory period of single motor nerve fibres in man. *J. Neurol. Neurosurg. Psychiatry.*, 47, 344–348.
34 Borg, J. (1984). Conduction velocity and refractory period of single motor nerve fibres in motor neuron disease. *J. Neurology Neurosurg. Psychiatry.*, 47, 349–353.
35 von Helmholtz, H. (1850). Messungen über den zeitlichen Verlauf der Zuckung animalischer Muskeln und die Fortpflanzungsgeschwindigkeit der Reizung in den Nerven. *Joh. Müllers Arch. Anatom. Phys.*, (276–364).
36 von Helmholtz, H. (1852). Messungen über Fortpflanzungsgeschwindigkeit der Reizung in den Nerven: Zweite Reihe. *Joh. Müllers Arch. Anat. Physiol.*, 199–216.
37 Hodgkin, A.L. and Huxley, A.F. (1952). A quantitative description of membrane current and its application to conduction and excitation in nerve. *J. Physiol.*, 117, 500–544.
38 Jones, G.A., Norris, S., and Henderson, Z. (1999). Conduction velocities and membrane properties of different classes of rat septohippocampal neurons recorded in vitro. *J. Physiol.*, 517, 867–877.
39 Villagran Vargas, E., Ludu, A., Hustert, R., Gumrich, P., Jackson, A.D., and Heimburg, T. (2011). Periodic solutions and refractory periods in the soliton theory for nerves and the locust femoral nerve. *Biophys. Chem.*, 153, 159–167.
40 von Muralt, A., Weibel, E., and Howarth, J. (1976). Optical spike-structure of olfactory nerve of pike and rapid birefringence changes during excitation. *Pflügers. Arch.*, 367, 67–76.
41 Matsumoto, G. and Tasaki, I. (1977). A study of conduction velocity in nonmyelinated nerve fibers. *Biophys. J.*, 20, 1–13.
42 Hursh, J.B. (1939). Conduction velocity and diameter of nerve fibers. *Am. J. Physiol.*, 127, 131–139.
43 Pumphrey, R.J. and Young, J.Z. (1938). The rates of conduction of nerve fibres of various diameter in cephalopods. *J. Exp. Biol.*, 15, 453–466.
44 Goldman, L. (1964). The effects of stretch on cable and spike parameters of single nerve fibres; some implications for the theory of impulse. *J. Physiol.*, 175, 425–444.
45 Wang, T., Gonzalez-Perez, A., Budvytyte, R., Jackson, A.D., and Heimburg, T. (2017). Reply to "comment on 'penetration of action potentials during collision in the median and lateral giant axons of invertebrates'". *Phys. Rev. X.*, 7, 028002.
46 Rosenthal, J.J.C. and Bezanilla, F. (2000). Seasonal variation in conduction velocity of action potentials in squid giant axons. *Biol. Bull.*, 199(2), 135–143.
47 Ritchie, J.M. and Keynes, R.D. (1985). The production and absorption of heat associated with electrical activity in nerve and electric organ. *Quart. Rev. Biophys.*, 18, 451–476.
48 Heimburg, T. (2025). The mechanical properties of nerves, the size of the action potential, and consequences for the brain. *Chem. Phys. Lipids.*, 267, 105461.

49 Hutzler, M., Lambacher, A., Eversmann, B., Jenkner, M., Thewes, R., and Fromherz, P. (2006). High-resolution multitransistor array recording of electrical field potentials in cultured brain slices. *J. Neurophysiol.*, 96, 1638–1645.

50 Tigerstedt, R. (1880). *Studien über die mechanische Nervenreizung*. J. C. Frenkel and Son, Helsingfors.

51 Wilke, E. (1912). Das Problem der Reizleitung im Nerven vom Standpunkte der Wellenlehre aus betrachtet. *Pflügers. Arch.*, 144, 35–38.

52 Wilke, E. and Atzler, E. (1912). Experimentelle Beiträge zum Problem der Reizleitung im Nerven. *Pflügers Arch.*, 146, 430–446.

53 Iwasa, K. and Tasaki, I. (1980). Mechanical changes in squid giant-axons associated with production of action potentials. *Biochem. Biophys. Res. Comm.*, 95, 1328–1331.

54 Iwasa, K., Tasaki, I., and Gibbons, R.C. (1980). Swelling of nerve fibres associated with action potentials. *Science*, 210, 338–339.

55 Ling, T., Boyle, K.C., Zuckerman, V., Flores, T., Ramakrishnan, C., Deisseroth, K., and Palanker, D. (2020). High-speed interferometric imaging reveals dynamics of neuronal deformation during the action potential. *Proc. Natl. Acad. Sci. USA.*, 117, 10278–10285.

56 Heimburg, T. (1998). Mechanical aspects of membrane thermodynamics. Estimation of the mechanical properties of lipid membranes close to the chain melting transition from calorimetry. *Biochim. Biophys. Acta.*, 1415, 147–162.

57 Kim, G.H., Kosterin, P., Obaid, A., and Salzberg, B.M. (2007). A mechanical spike accompanies the action potential in mammalian nerve terminals. *Biophys. J.*, 92, 3122–3129.

58 Tasaki, I. and Iwasa, K.H. (1980). Shortening of nerve fibers associated with propagated nerve impulse. *Curr. Top. Med. Chem.*, 94, 716–720.

59 Tasaki, I. and Byrne, P.M. (1982). Tetanic contraction of the crab nerve evoked by repetitive stimulation. *Biochem. Biophys. Research Comm.*, 106, 1435–1440.

60 Gray, J.A.B. and Ritchie, J. (1954). Effects of stretch on single myelinated nerve fibres. *J. Physiol.*, 124, 84–99.

61 Julian, F.J. and Goldman, D.E. (1962). Effects of mechanical stimulation on some electrical properties of axons. *J. Gen. Physiol.*, 46, 297–313.

62 Terakawa, S. and Watanabe, A. (1982). Electrical responses to mechanical stimulation of the membrane of squid giant axons. *Pflügers. Arch.*, 395, 59–64.

63 Kamimura, H.A.S., Wang, S., Chen, H., Wang, Q., Aurup, C., Acosta, C., et al. (2016). Focused ultrasound neuromodulation of cortical and subcortical brain structures using 1.9 mhz. *Med. Phys.*, 43, 5730–5735.

64 Ye, P.P., Brown, J., and Pauly, K. (2016). Frequency dependence of ultrasound neurostimulation in the mouse brain. *Ultrasound. Med. Biol.*, 4, 1512–1530.

65 Tufail, Y., Yoshihiro, A., Pati, S., Li, M.M., and Tyler, W.J. (2011). Ultrasonic neuromodulation by brain stimulation with transcranial ultrasound. *Nat. Protoc.* 6, 1453–1470.

66 Mueller, J.K. and Tyler, W.J. (2014). A quantitative overview of biophysical forces impinging on neural function. *Phys. Biol.*, 11, 051 001.

67 Feng, R., Sheng, H., and Lian, Y. (2024). Advances in using ultrasound to regulate the nervous system. *Neurol. Sci.*, 1, 1–10.

68 Cohen, L.B., Keynes, R.D., and Hille, B. (1968). Light scattering and birefringence changes during nerve activity. *Nature.*, 218, 438–441.

69 Tasaki, I., Watanabe, A., Sandlin, R., and Carnay, L. (1968). Changes in fluorescence, turbidity and birefringence associated with nerve excitation. *Proc. Natl. Acad. Sci. USA.*, 61, 883–888.

70 Stepnoski, R.A., LaPorta, A., Raccuia-Behling, F., Blonder, G., Slusher, R., and Kleinfeld, D. (1991). Noninvasive detection of changes in membrane potential in cultured neurons by light scattering. *Proc. Natl. Acad. Sci. USA.*, 88, 9382–9386.

71 Wei, L.Y. (1972). Dipole theory of heat production and absorption in nerve axon. *Biophys. J.*, 12, 1159–1170.

72 Vogel, V. and Möbius, D. (1988). Local surface potentials and electric dipole moments of lipid monolayers: contributions of the water/lipid and the lipid/air interfaces. *J. Coll. Interf. Sci.*, 126, 408–420.

73 Beitinger, H., Vogel, V., Möbius, D., and Rahmann, H. (1989). Surface potentials and electric dipole moments of ganglioside and phospholipid monolayers: contribution of the polar headgroup at the water/lipid interface. *Biochim. Biophys. Acta.*, 984, 293–300.

74 Brockman, H. (1994). Dipole potential of lipid membranes. *Chem. Phys. Lipids.*, 73, 57–79.

75 Lee, K.Y.C. and McConnell, H.M. (1995). Effect of electric-field gradients on lipid monolayer membranes. *Biophys. J.*, 68, 1740–1751.

76 Cohen, L.B., Hille, B., and Keynes, R. (1970). Changes in axon birefringence during action potential. *J. Physiol.*, 211, 495–515.

77 Kobatake, Y., Tasaki, I., and Watanabe, A. (1971). Phase transition in membrane with reference to nerve excitation. *Adv. Biophys.*, 208, 1–31.

78 Abbott, B.C., Hill, A.V., and Howarth, J.V. (1958). The positive and negative heat production associated with a nerve impulse. *Proc. Roy. Soc. Lond. B.*, 148, 149–187.

79 Heimburg, T. (2021). The important consequences of the reversible heat production in nerves and the adiabaticity of the action potential. *Progr. Biophys. Mol. Biol.*, 162, 26–40.

80 Howarth, J.V., Keynes, R., and Ritchie, J.M. (1968). The origin of the initial heat associated with a single impulse in mammalian non-myelinated nerve fibres. *J. Physiol.*, 194, 745–793.

81 Tasaki, I. and Iwasa, K.H. (1981). Temperature-changes associated with nerve excitation - detection by using polyvinylidene fluoride film. *Biochem. Biophys. Research Comm.*, 101, 172–176.

82 Tasaki, I. (1999). Evidence for phase transition in nerve fibers, cells and synapses. *Ferroelectrics.*, 220, 305–316.

83 Spyropoulos, C.S. (1961). Initiation and abolition of electrical response of nerve fiber by thermal and chemical means. *Am. J. Physiol.*, 200, 203–208.

84 Shapiro, M.G., Homma, K., Villareal, S., Richter, C.P., and Bezanilla, F. (2012). Infrared light excites cells by changing their electrical capacitance. *Nat. Commun.*, 3, 736.

85 Hallett, M. (2007). Transcranial magnetic stimulation: a primer. *Neuron*, 55, 187–199.

86 Ridding, M.C. and Rothwell, J.C. (2007). Is there a future for therapeutic use of transcranial magnetic stimulation? *Nat. Rev. Neurosci.*, 8, 559–567.

87 Wikswo, J.P., Barach, J.P., and Freeman, J.A. (1980). Magnetic field of a nerve impulse: first measurements. *Science*, 208, 53–55.

88 Swinney, K.R. and Wikswo, J.P. (1980). A calculation of the magnetic field of a nerve action potential. *Biophys. J.*, 32, 719–731.

89 Wikswo, J.P. (1982). Improved instrumentation for measuring the magnetic field of cellular action currents. *Rev. Sci. Instrum.*, 53, 1846–1850.

90 Roth, B.J. and Wikswo, J.P. (1985). The magnetic field of a single axon. A comparison of theory and experiment. *Biophys. J.*, 48, 93–109.
91 Clarke, J. (1980). Advances in SQUID magnetometers. *IEEE Trans. Electron. Devices.*, 27, 1896–1908.
92 Wikswo, J.P. (1995). Squid magnetometers for biomagnetism and nondestructive testing: important questions and initial answers. *IEEE Trans. Appl. Supercond.*, 5, 74–120.
93 Jensen, K., Budvytyte, R., Thomas, R.A., Wang, T., Fuchs, A.M., Balabas, M.V., et al. (2016). Non-invasive detection of animal nerve impulses with an atomic magnetometer operating near quantum limited sensitivity. *Sci. Rep.*, 6, 29638.
94 Xu, N., Jiang, F., Tian, Y., Ye, J., Shi, F., Lv, H., et al. (2016). Wavelet-based fast time-resolved magnetic sensing with electronic spins in diamond. *Phys. Rev. B.*, 93, 161117.
95 Zhang, C., Zhang, J., Widmann, M., Benke, M., Kübler, M., Dasari, D., et al. (2023). Optimizing NV magnetometry for magnetoneurography and magnetomyography applications. *Front. Neurosci.*, 16, 1034391.
96 Roth, B.J. and Wikswo, J.P. (1986). Electrically silent magnetic fields. *Biophys. J.*, 50, 739–745.
97 Tasaki, I. (1949). Collision of two nerve impulses in the nerve fiber. *Biochim. Biophys. Acta.*, 3, 494–497.
98 Berg, R.W., Stauning, M.T., Sørensen, J.B., and Jahnsen, H. (2017). Comment on 'penetration of action potentials during collision in the median and lateral giant axons of invertebrates'. *Phys. Rev. X.*, 4(3), 031047.

Part IV

1 Bernstein, J. (1902). Untersuchungen zur Thermodynamik der bioelektrischen Ströme. 1 Theil. *Pflügers. Arch.*, 92, 521–562.
2 Bernstein, J. (1912). *Elektrobiologie. Die Lehre von den elektrischen Vorgängen im Organismus auf moderner Grundlage dargestellt*, Vieweg and Sohn.
3 Jatta, G. (1896). *Cefalopodi viventi nel Golfo di Napoli, in: Fauna und Flora des Golfes von Neapel*, vol. 23, Friedländer and Sohn, Berlin.
4 Hodgkin, A.L. and Huxley, A.F. (1939). Action potentials recorded from inside a nerve fibre. *Nature.*, 144, 710–711.
5 Young, J.Z. (1936). The structure of nerve fibres in cephalopods and crustacea. *Proc. Roy. Soc. Lond. B.*, 121, 319–337.
6 Pumphrey, R.J. and Young, J.Z. (1938). The rates of conduction of nerve fibres of various diameter in cephalopods. *J. Exp. Biol.*, 15, 453–466.
7 Cole, K.S. (1968). *Membranes, Ions and Impulses. A Chapter of Classical Biophysics.* University of California Press.
8 Hodgkin, A.L. and Rushton, W.A.H. (1946). The electrical constants of a crustacean nerve fibre. *Proc. Roy. Soc. Lond. B.*, 133, 444–479.
9 Johnston, D. and Wu, S.M.S. (1995). *Cellular Neurophysiology*. MIT Press, Boston.
10 Rall, W. (1959). Branching dendritic trees and motoneuron membrane resistivity. *Exp. Neurol.*, 1, 491–527.

11 Rall, W., Burke, R.E., Holmes, W.R., Jack, J.J., Redman, S.J., and Segev, I. (1992). Matching dendritic neuron models to experimental data. *Physiol. Rev.*, 72(4 Suppl), S159–S186.

12 Cullheim, S., Fleshman, J.W., Glenn, L.L., and Burke, R.E. (1987). Three-dimensional architecture of dendritic trees in type-identified α-motoneurons. *J. Comp. Neurol.*, 255, 82–96.

13 Cullheim, S., Fleshman, J.W., Glenn, L.L., and Burke, R.E. (1987). Membrane area and dendritic structure in type-identified triceps surae alpha motoneurons. *J. Comp. Neurol.*, 255, 68–81.

14 Hodgkin, A.L., Huxley, A.F., and Katz, B. (1952). Measurement of current-voltage relations in the membrane of the giant axon of Loligo. *J. Physiol.*, 116, 424–448.

15 Hodgkin, A.L. and Huxley, A.F. (1952). Currents carried by sodium and potassium ions through the membrane of the giant axon of Loligo. *J. Physiol.*, 116, 449–472.

16 Hodgkin, A.L. and Huxley, A.F. (1952). A quantitative description of membrane current and its application to conduction and excitation in nerve. *J. Physiol.*, 117, 500–544.

17 Gonzalez-Perez, A., Mosgaard, L.D., Budvytyte, R., Villagran Vargas, E., Jackson, A.D., and Heimburg, T. (2016). Solitary electromechanical pulses in lobster neurons. *Biophys. Chem.*, 216, 51–59.

18 Noble, D. (1966). Applications of Hodgkin-Huxley equations to excitable tissues. *Physiol. Rev.*, 46, 1–50.

19 Heimburg, T. (2021). The important consequences of the reversible heat production in nerves and the adiabaticity of the action potential. *Progr. Biophys. Mol. Biol.*, 162, 26–40.

20 Bostock, H., Baker, M., and Reid, G. (1991). Changes in excitability of human motor axons underlying post-ischaemic fasciculations: evidence for two stable states. *J. Physiol.*, 441, 537–557.

21 Kiernan, M.C., Isbister, G.K., Lin, C.S.Y., Burke, D., and Bostock, H. (2005). Acute tetrodotoxin-induced neurotoxicity after ingestion of puffer fish. *Ann. Neurol.*, 57, 339–348.

22 Howells, J., Trevillion, L., Bostock, H., and Burke, D. (2012). The voltage dependence of Ih in human myelinated axons. *J. Physiol.*, 590, 1625–1640.

23 Moldovan, M., Lange, K.H.W., Aachmann-Andersen, N.J., Kjær, T.W., Olsen, N.V., and Krarup, C. (2014). Transient impairment of the axolemma following regional anaesthesia by lidocaine in humans. *J. Physiol.*, 592, 2735–2750.

24 Hille, B., Armstrong, C.M., and MacKinnon, R. (1999). Ion channels: from idea to reality. *Nat. Med.*, 5, 1105–1109.

25 Hille, B. and Schwarz, W. (1978). Potassium channels as multi-ion single-file pores. *J. Gen. Physiol.*, 72, 409–442.

26 Doyle, D.A., Morais, J., Gulbis, J.M., Cohen, A.L., Chait, B.T., and MacKinnon, R. (1998). The structure of the potassium channel: molecular basis of K+ conduction and selectivity. *Science*, 280, 69–77.

27 Neher, E. and Sakmann, B. (1976). Single-channel currents recorded from membrane of denervated frog muscle fibres. *Nature*, 260, 779–802.

28 Hamill, O.P., Marty, A., Neher, E., Sakmann, B., and Sigworth, F.J. (1981). Improved patch-clamp techniques for high-resolution current recording from cells and cell-free membrane patches. *Pflügers. Arch.*, 391, 85–100.

29 Bezanilla, F. (2005). Voltage-gated ion channels. *IEEE Transact. Nanobiosci.*, 4, 34–48.

30 Sands, Z., Grottesi, A., and Sansom, M.S.P. (2005). Voltage-gated ion channels. *Curr. Biol.*, 15, R44–R47.

31 Catterall, W.A. (2010). Ion channel voltage sensors: structure, function, and pathophysiology. *Neuron*, 67, 915–928.

32 Zhou, Y., Morais-Cabral, J., Kaufman, A., and MacKinnon, R. (2001). Chemistry of ion coordination and hydration revealed by a k^+ channel–Fab complex at 2.0 å resolution. *Nature*, 414, 43–48.

33 Xu, H., Li, T., Rohou, A., Arthur, C., Tzakoniati, F., Wong, E., et al. (2019). Structural basis of Nav1.7 inhibition by a gating-modifier spider toxin. *Cell*, 176, 702–715.

34 Schmidt, D., Jiang, Q.X., and MacKinnon, R. (2006). Phospholipids and the origin of cationic gating charges in voltage sensors. *Nature*, 444, 775–779.

35 Zecchi, K.A. and Heimburg, T. (2020). Nonlinear conductance, rectification and mechanosensitive channel formation of lipid membranes. *Front. Cell. Dev. Biol.*, 8, 592 520.

36 Llano, I., Webb, C.K., and Bezanilla, F. (1988). Potassium conductance of the squid giant axon. single-channel studies. *J. Gen. Physiol.*, 92, 179–196.

37 Catterall, W.A. (2000). From ionic currents to molecular mechanisms the structure and function of voltage-gated sodium channels. *Neuron*, 26, 13–25.

38 Coste, B., Xiao, B., Santos, J., Syeda, R., Grandl, J., Spencer, K., et al. (2012). Piezo proteins are pore-forming subunits of mechanically activated channels. *Nature*, 483, 176–181.

39 Kim, S.E., Coste, B., Chadha, A., Cook, B., and Patapoutian, A. (2012). The role of drosophila piezo in mechanical nociception. *Nature*, 483, 209–212.

40 Volkers, L., Mechioukhi, Y., and Coste., B. (2015). Piezo channels: from structure to function. *Pflügers Arch. Eur. J. Physiol.*, 467, 95–99.

41 Wu, J., Lewis, A., and Grandl, J. (2017). Touch, tension, and transduction – the function and regulation of piezo ion channels. *Trends Biochem. Sci.*, 42, 57–71.

42 Moroni, M., Servin-Vences, M.R., Fleischer, R., Sanchez-Carranza, O., and Lewin., R. (2018). Voltage-gating of mechanosensitive PIEZO channels. *Nat. Commun.*, 9, 1096.

43 Guo, Y.R. and MacKinnon, R. (2017). Structure-based membrane dome mechanism for piezo mechanosensitivity. *Elife*, 6, e33660.

44 Saotome, K., Murthy, S., Kefauver, J., Whitwam, T., Patapoutian, A., and Ward., A. (2018). Structure of the mechanically activated ion channel piezo1. *Nature*, 554, 481–486.

45 Brohawn, S.G., Wang, W., Handler, A., Campbell, E., Schwarz, J., and MacKinnon., R. (2019). The mechanosensitive ion channel TRAAK is localized to the mammalian node of Ranvier. *Elife*, 8, e50403.

46 Zhang, X. (2015). Molecular sensors and modulators of thermoreception. *Channels*, 9, 73–81.

47 Bautista, D.M., Siemens, J., Glazer, J., Tsuruda, P., Basbaum, A., Stucky, C., et al. (2007). The menthol receptor TRPM8 is the principal detector of environmental cold. *Nature*, 448, 204–208.

48 Diver, M.M., Cheng, Y., and Julius, D. (2019). Structural insights into TRPM8 inhibition and desensitization. *Science*, 365, 1434–1440.

49 Liao, M., Cao, E., Julius, D., and Cheng, Y. (2013). Structure of the TRPV1 ion channel determined by electron cryo-microscopy. *Nature*, 504, 107–112.

50 Clapham, D.E. and Miller, C. (2011). A thermodynamic framework for understanding temperature sensing by transient receptor potential (TRP) channels. *Proc. Natl. Acad. Sci. USA.*, 108, 19492–19497.

51 Voets, T., Droogmans, G., Wissenbach, U., Janssens, A., Flockerzi, V., and Nilius, B. (2004). The principle of temperature-dependent gating in cold- and heat-sensitive TRP channels. *Nature*, 430, 748–754.

52 Mosgaard, L.D. and Heimburg, T. (2013). Lipid ion channels and the role of proteins. *Acc. Chem. Res.*, 46, 2966–2976.

53 Mužić, T., Tounsi, F., Madsen, S.B., Pollakowski, D., Konrad, M., and Heimburg, T. (2019). Melting transitions in biomembranes. *Biochim. Biophys. Acta.*, 1861, 183026.

54 Jordt, S.E., Tominaga, M., and Julius., D. (2000). Acid potentiation of the capsaicin receptor determined by a key extracellular site. *Proc. Natl. Acad. Sci. USA.*, 97, 8134–8139.

55 Yao, J., Liu, B., and Qin, F. (2010). Kinetic and energetic analysis of thermally activated TRPV1 channels. *Biophys. J.*, 99, 1743–1753.

56 Zheng, H., Liu, W., Anderson, L.Y., and Jiang, Q.X. (2011). Lipid-dependent gating of a voltage-gated potassium channel. *Nat. Commun.*, 2, 250–259.

57 Seeger, H.M., Aldrovandi, L., Alessandrini, A., and Facci, P. (2010). Changes in single K^+ channel behavior induced by a lipid phase transition. *Biophys. J.*, 99, 3675–3683.

58 Seeger, H.M., Bortolotti, C.A., Alessandrini, A., and Facci, P. (2009). Phase-transition-induced protein redistribution in lipid bilayers. *J. Phys. Chem. B.*, 113, 16654–16659.

Part V

1 Heimburg, T. (2007). *Thermal Biophysics of Membranes*. Wiley VCH, Berlin, Germany.

2 Nägeli, C. and Cramer, C. (1855). *Pflanzenphysiologische untersuchungen*. Shulthess, Zürich.

3 de Vries, H. (1871). Sur la permeabilite du protoplasma des betteraves rouges. *Arch. Neerl. Sci. Exactes Nat.*, 61, 117–126.

4 de Vries, H. (1884). Eine methode zur analyse der turgorcraft. *Jahrb. Wiss. Bot.*, 14, 429–601.

5 Pfeffer, W. (1877). *Osmotische untersuchungen. Studien zur zellmechanik*. W. Engelmann, Leipzig.

6 Overton, E. (1895). Über die osmotischen eigenschaften der lebenden pflanzen und thierzelle. *Vierteljahresschr. Naturforsch. Ges. Zürich*. 40, 159–201.

7 Kleinzeller, A. (1999). Charles Ernest Overton's concept of a cell membrane. *Curr. Top. Membr.*, 48, 1–22.

8 Gorter, E. and Grendel, F. (1925). On bimolecular layers of lipoids on the chromacytes of the blood. *J. Exp. Med.*, 41, 439–443.

9 Danielli, J.F. and Davson, H. (1935). A contribution to the theory of permeability of thin films. *J. Cell Comp. Physiol.*, 5, 495–508.

10 Robertson, J.D. (1959). The ultrastructure of cell membranes and their derivatives. *Biochem. Soc. Symp.*, 16, 3–43.

11 Singer, S.J. and Nicolson, G.L. (1972). The fluid mosaic model. *Science*, 175, 720–731.

12 Singer, S.J. and Nicolson, G.L. (1972). The fluid mosaic model of the structure of cell membranes. In *Membranes and Viruses in Immunopathology*, Elsevier, pp. 7–47.

13 Mouritsen, O.G. and Bloom, M. (1984). Mattress model of lipid-protein interactions in membranes. *Biophys. J.*, 46, 141–153.

14 Kauzmann, W. (1959). Factors in interpretation of protein denaturation. *Adv. Prot. Chem.*, 14, 1–63.

15 Debye, P. and Hückel, E. (1923). Zur theorie der elektrolyte. I. Gefrierpunktserniedrigung und verwandte erscheinungen. *Physik. Z.*, 24, 185–206.

16 Doyle, D.A., Morais, J., Gulbis, J.M., Cohen, A.L., Chait, B.T., and MacKinnon, R. (1998). The structure of the potassium channel: molecular basis of k+ conduction and selectivity. *Science*, 280, 69–77.

17 Skou, J.C. (1957). The influence of some cations on an adenosine triphosphatase from peripheral nerves. *Biochim. Biophys. Acta.*, 23, 394–401.

18 Blume, A. (1983). Apparent molar heat capacities of phospholipids in aqueous dispersion. *Biochemistry*, 22, 5436–5442.

19 Lee, A.G. (1977) Lipid phase transitions and phase diagrams. II. Mixtures involving lipids. *Biochim. Biophys. Acta.*, 472, 285–344.

20 Krol, S., Ross, M., Sieber, M., Künneke, S., Galla, H.J., and Janshoff, A. (2000). Formation of three-dimensional protein-lipid aggregates in monolayer films induced by surfactant protein b. *Biophys. J.*, 79, 904–918.

21 Echaide, M., Autilio, C., Arroyo, R., and Perez-Gil, J. (2017). Restoring pulmonary surfactant membranes and films at the respiratory surface. *Biochim. Biophys. Acta.*, 1859, 1725–1739.

22 Mužić, T., Tounsi, F., Madsen, S.B., Pollakowski, D., Konrad, M., and Heimburg, T. (2019). Melting transitions in biomembranes. *Biochim. Biophys. Acta.*, 1861, 183026.

23 Bernardino de la Serna, J., Orädd, G., Bagatolli, L., Simonsen, A.C., Marsh, D., Lindblom, G., et al. (2009). Segregated phases in pulmonary surfactant membranes do not show coexistence of lipid populations with differentiated dynamic properties. *Biophys. J.*, 97, 1381–1389.

24 Färber, N. and Westerhausen, C. (2022). Broad lipid phase transitions in mammalian cell membranes measured by Laurdan fluorescence spectroscopy. *Biochim. Biophys. Acta.*, 1864, 183794.

25 Fedosejevs, C.S. and Schneider., F. (2022). Sharp, localized phase transitions in single neuronal cells. *Proc. Natl. Acad. Sci. USA.*, 119, e2117521119.

26 Højholt, K.L., Mužić, T., Jensen, S.D., Bilgin, M., Nylandsted, J., Heimburg, T., et al. (2019). Calcium electroporation and electrochemotherapy for cancer treatment: importance of cell membrane composition investigated by lipidomics, calorimetry and in vitro efficacy. *Sci. Rep.*, 9, 4758.

27 Heimburg, T. (1998). Mechanical aspects of membrane thermodynamics. Estimation of the mechanical properties of lipid membranes close to the chain melting transition from calorimetry. *Biochim. Biophys. Acta.*, 1415, 147–162.

28 Doniach, S. (1978). Thermodynamic fluctuations in phospholipid bilayers. *J. Chem. Phys.*, 68, 4912–4916.

29 Mouritsen, O.G., Boothroyd, A., Harris, R., Jan, N., Lookman, T., MacDonald, L., et al. (1983). Computer simulation of the main gel-fluid phase transition of lipid bilayers. *J. Chem. Phys.*, 79, 2027–2041.

30 Ipsen, J.H., Karlström, G., Mouritsen, O.G., Wennerström, H., and Zuckermann, M.J. (1987). Phase equilibria in the phosphatidylcholine-cholesterol system. *Biochim. Biophys. Acta.*, 905, 162–172.

31 Sugar, I.P., Biltonen, R.L., and Mitchard, N. (1994). Monte Carlo simulations of membranes: phase transition of small unilamellar dipalmitoylphosphatidylcholine vesicles. *Methods. Enzymol.*, 240, 569–593.

32 Heimburg, T. and Biltonen, R.L. (1996). A Monte Carlo simulation study of protein-induced heat capacity changes. *Biophys. J.*, 70, 84–96.

33 Hac, A., Seeger, H., Fidorra, M., and Heimburg, T. (2005). Diffusion in two-component lipid membranes—a fluorescence correlation spectroscopy and Monte Carlo simulation study. *Biophys. J.*, 88, 317–333.

34 Bernardino de la Serna, J., Perez-Gil, J., Simonsen, A.C., and Bagatolli, L.A. (2004). Cholesterol rules: direct observation of the coexistence of two fluid phases in native pulmonary surfactant membranes at physiological temperatures. *J. Biol. Chem.*, 279, 40715–40722.

35 Baumgart, T., Hess, S.T., and Webb, W.W. (2003). Imaging coexisting fluid domains in biomembrane models coupling curvature and line tension. *Nature*, 425, 821–824.

36 Nielsen, L.K., Bjørnholm, T., and Mouritsen, O.G. (2000). Critical phenomena - fluctuations caught in the act. *Nature*, 404, 352.

37 Nielsen, L.K., Vishnyakov, A., Jørgensen, K., Bjørnholm, T., and Mouritsen, O.G. (2000). Nanometre-scale structure of fluid lipid membranes. *J. Phys.: Condens. Matter.*, 12, A309–A214.

38 Simons, K. and Ikonen, E. (1997). Functional rafts in cell membranes. *Nature*, 387, 569–572.

39 Bagnat, M. and Simons, K. (2002). Lipid rafts in protein sorting and cell polarity in budding yeast saccharomyces cerevisiae. *Biol. Chem.*, 383, 1475–1480.

40 Bagatolli, L.A. (2006). To see or not to see: lateral organization of biological membranes and fluorescence microscopy. *Biochim. Biophys. Acta.*, 1758, 1541–1556.

41 Gaus, K., Gratton, E., Kable, E.P.W., Jones, A.S., Gelissen, I., Kritharides, L., et al. (2003). Visualizing lipid structure and raft domains in living cells with two-photon microscopy. *Proc. Natl. Acad. Sci. USA.*, 100, 15554–15559.

42 Ebel, H., Grabitz, P., and Heimburg, T. (2001). Enthalpy and volume changes in lipid membranes. I. The proportionality of heat and volume changes in the lipid melting transition and its implication for the elastic constants. *J. Phys. Chem. B.*, 105, 7353–7360.

43 Trudell, J.R., Payan, D.G., Chin, J.H., and Cohen, E.N. (1974). The effect of pressure on the phase diagram of mixed dipalmitoyl–dimyristoylphosphatidylcholine bilayers. *Biochim. Biophys. Acta.*, 373, 141–144.

44 Heimburg, T. and Jackson, A.D. (2007). The thermodynamics of general anesthesia. *Biophys. J.*, 92, 3159–3165.

45 Heimburg, T. (2022). The effect of stretching on nerve excitability. *Human. Mov. Sci.*, 86, 103000.

46 Evans, E. and Kwok, R. (1982). Mechanical calorimetry of large dimyristoylphosphatidylcholine vesicles in the phase transition region. *Biochemistry*, 21, 4874–4879.

47 Helfrich, W. (1973). Elastic properties of lipid bilayers: theory and possible experiments. *Z. Naturforsch.*, 28, 693–703.

48 Berndl, K., Käs, J., Lipowsky, R., Sackmann, E., and Seifert, U. (1990). Shape transformations of giant vesicles: extreme sensitivity to bilayer asymmetry. *Europhys. Lett.*, 13, 659–664.

49 Lipowsky, R. (1995). The morphology of lipid membranes. *Curr. Opin. Struct. Biol.*, 5, 531–540.

50 Kralj-Iglic, V., Pocsfalvi, G., Mesarec, L., Sustar, V., Hägerstrand, H., and Iglic, A. (2020). Minimizing isotropic and deviatoric membrane energy – A unifying formation mechanism of different cellular membrane nanovesicle types. *PLoS One*, 15, e0244796.

51 Sackmann, E. and Bruinsma, R.F. (2002). Cell adhesion as wetting transition? *Chem. Phys. Chem.*, 3, 262–269.

52 Weikl, T.R., Andelman, D., Komura, S., and Lipowsky, R. (2002). Adhesion of membranes with competing specific and generic interactions. *Eur. Phys. J. E.*, 8, 59–66.

53 Kozlovsky, Y. and Kozlov, M.M. (2002). Stalk model of membrane fusion: solution of energy crisis. *Biophys. J.*, 82, 882–895.

54 Seifert, U. (1997). Configurations of fluid membranes and vesicles. *Adv. Phys.*, 46, 13–137.

55 Dimova, R., Pouligny, B., and Dietrich, C. (2000). Pretransitional effects in dimyristoylphosphatidylcholine vesicle membranes: Optical dynamometry study. *Biophys. J.*, 79, 340–356.

56 Meleard, P., Gerbeaud, C., Pott, T., Fernandez-Puente, L., Bivas, I., Mitov, M.D., et al. (1997). Bending elasticities of model membranes. Influences of temperature and sterol content. *Biophys. J.*, 72, 2616–2629.

57 Träuble, H., Teubner, M., Woolley, P., and Eibl, H. (1976). Electrostatic interactions at charged lipid membranes. I. Effects of pH and univalent cations on membrane structure. *Biophys. Chem.*, 4, 319–342.

58 Hauser, H. and Shipley, G.G. (1983). Interactions of monovalent cations with phosphatidylserine bilayer membranes. *Biochem*, 22, 2171–2178.

59 Hauser, H. and Shipley, G. (1981). Crystallization of phosphatidylserine bilayers induced by lithium. *J. Biol. Chem.*, 256, 11377–11380.

60 Feld, K.E. (2012). *Physics of Non-equilibrium Membranes*, Ph.D. thesis, University of Copenhagen.

61 Thomas, N. and Agrawal, A. (2022). A lateral electric field inhibits gel-to-fluid transition in lipid bilayers. *Soft. Matter.*, 18, 6437–6442.

62 Grabitz, P., Ivanova, V.P., and Heimburg, T. (2002). Relaxation kinetics of lipid membranes and its relation to the heat capacity. *Biophys. J.*, 82, 299–309.

63 Seeger, H.M., Gudmundsson, M.L., and Heimburg, T. (2007). How anesthetics, neurotransmitters, and antibiotics influence the relaxation processes in lipid membranes. *J. Phys. Chem. B.*, 111, 13858–13866.

64 Riske, K.A. and Dimova, R. (2006). Electric pulses induce cylindrical deformations on giant vesicles in salt solutions. *Biophys. J.*, 91, 1778–1786.

65 Hodgkin, A.L. and Huxley, A.F. (1952). A quantitative description of membrane current and its application to conduction and excitation in nerve. *J. Physiol.*, 117, 500–544.

66 Zecchi, K.A., Mosgaard, L.D., and Heimburg., T. (2017). Mechano-capacitive properties of polarized membranes and the application to conductance measurements of lipid membrane patches. *J. Phys.: Conf. Ser.*, 780, 012001.

67 Heimburg, T. (2012). The capacitance and electromechanical coupling of lipid membranes close to transitions. the effect of electrostriction. *Biophys. J.*, 103, 918–929.

68 Wang, T., Mužić, T., Jackson, A.D., and Heimburg, T. (2018). The free energy of biomembrane and nerve excitation and the role of anesthetics. *Biochim. Biophys. Acta.*, 1860, 2145–2153.

69 Alvarez, O. and Latorre, R. (1978). Voltage-dependent capacitance in lipid bilayers made from monolayers. *Biophys. J.*, 21, 1–17.

70 Mosgaard, L.D., Zecchi, K.A., and Heimburg, T. (2015). Mechano-capacitive properties of polarized membranes. *Soft. Matter.*, 11, 7899–7910.

71 Petrov, A.G. (1999). *The lyotropic state of matter. Molecular physics and living matter physics.* Gordon and Breach Science Publishers, Amsterdam.

72 Petrov, A.G. and Usherwood, P.N.R. (1994). Mechanosensitivity of cell membranes. *Eur. Biophys. J.*, 23, 1–19.

73 Overton, C.E. (1901). *Studien über die Narkose.* Verlag Gustav Fischer., Jena, Germany. (English translation: Studies of Narcosis, Chapman and Hall, 1991, R. Lipnick, Ed.).

74 Trudell, J.R., Payan, D.G., Chin, J.H., and Cohen, E.N. (1975). The antagonistic effect of an inhalation anesthetic and high pressure on the phase diagram of mixed dipalmitoyl-dimyristoylphosphatidylcholine bilayers. *Proc. Natl. Acad. Sci. USA.*, 72, 210–213.

75 Kharakoz, D.P. (2001). Phase-transition-driven synaptic exocytosis: a hypothesis and its physiological and evolutionary implications. *Biosci. Rep.*, 210, 801–830.

76 Græsbøll, K., Sasse-Middelhoff, H., and Heimburg, T. (2014). The thermodynamics of general and local anesthesia. *Biophys. J.*, 106, 2143–2156.

77 DiFazio, C.A., Kennedy, S.S., Brown, R.E., Ball, C.G., and Heckel, C.G. (1972). Additive effects of anesthetics and theories of anesthesia. *Anesthesiol.*, 36, 57–63.

78 Johnson, F.H. and Flagler, E.A. (1950). Hydrostatic pressure reversal of narcosis in tadpoles. *Science*, 112, 91–92.

79 Johnson, S. and Miller, K.W. (1970). Antagonism of pressure and anaesthesia. *Nature*, 228, 75–76.

80 Miller, K.W., Paton, W., Smith, R., and Smith, E. (1973). The pressure reversal of general anesthesia and the critical volume hypothesis. *Mol. Pharmacol.*, 9, 131–143.

81 Halsey, M.J. and Wardley-Smith, B. (1975). Pressure reversal of narcosis produced by anaesthetics, narcotics and transquilizers. *Nature*, 257, 811–813.

82 Mountcastle, D.B., Biltonen, R.L., and Halsey, M.J. (1978). Effect of anesthetics and pressure on the thermotropic behavior of multilamellar dipalmitoylphosphatidylcholine liposomes. *Proc. Natl. Acad. Sci. USA.*, 75, 4906–4910.

83 Kamaya, H., Ueda, I., Moore, P.S., and Eyring, H. (1979). Antagonism between high pressure and anesthetics in the thermal phase transition of dipalmitoyl phosphatidylcholine bilayer. *Biochim. Biophys. Acta.*, 550, 131–137.

84 Cantor, R.S. (2003) Receptor desensitization by neurotransmitters in membranes: are neurotransmitters the endogenous anesthetics? *Biochem.*, 42, 11891–11897.

85 Oliynyk, V., Kaatze, U., and Heimburg, T. (2007). Pore formation of lytic peptides in lipid membranes and their influence on the thermodynamic properties of the pore environment. *Biochim. Biophys. Acta.*, 1768, 236–245.

86 Morrow, M.R., Davis, J.H., Sharom, F.J., and Lamb, M.P. (1986). Studies of the interaction of human erythrocyte band 3 with membrane lipids using deuterium nuclear magnetic resonance and differential scanning calorimetry. *Biochim. Biophys. Acta.*, 858, 13–20.

87 Freire, E., Markello, T., Rigell, C., and Holloway, P.W. (1983). Calorimetric and fluorescence characterization of interactions between cytochrome b_5 and phosphatidylcholine bilayers. *Biochemistry*, 28, 5634–5643.

88 Vahedi, A., Bigdelou, P., and Farnoud, A. (2020). Quantitative analysis of red blood cell membrane phospholipids and modulation of cell-macrophage interactions using cyclodextrins. *Sci. Rep.*, 10, 15111.

89 Peters, J., Marion, J., Becher, F.J., Trapp, M., Gutberlet, T., Bicout, D.J., et al. (2017). Thermodynamics of lipid multi-lamellar vesicles in presence of sterols at high hydrostatic pressure. *Sci. Rep.*, 7, 15339.

Part VI

1 Einstein, A. (1910). Theorie der Opaleszenz von homogenen Flüssigkeiten und Flüssigkeitsgemischen in der Nähe des kritischen Zustandes. *Ann. Phys. (Leipzig).*, 33, 1275–1298.

2 Greene, R.F. and Callen, H.B. (1951). On the formalism of thermodynamic fluctuation theory. *Phys. Rev.*, 83, 1231–1235.

3 Callen, H.B. and Greene, R.F. (1952). On a theorem of irreversible thermodynamics. *Phys. Rev.*, 86, 702–710.

4 Kubo, R. (1966). The fluctuation-dissipation theorem. *Rep. Prog. Phys.*, 29, 255–284.

5 Sugar, I.P., Biltonen, R.L., and Mitchard, N. (1994). Monte Carlo simulations of membranes: phase transition of small unilamellar dipalmitoylphosphatidylcholine vesicles. *Methods. Enzymol.*, 240, 569–593.

6 Heimburg, T. and Biltonen, R.L. (1996). A Monte Carlo simulation study of protein-induced heat capacity changes. *Biophys. J.*, 70, 84–96.

7 Ivanova, V.P. and Heimburg, T. (2001). A histogram method to obtain heat capacities in lipid monolayers, curved bilayers and membranes containing peptides. *Phys. Rev. E.*, 63, 041914.

8 Grabitz, P., Ivanova, V.P., and Heimburg, T. (2002). Relaxation kinetics of lipid membranes and its relation to the heat capacity. *Biophys. J.*, 82, 299–309.

9 Ebel, H., Grabitz, P., and Heimburg, T. (2001). Enthalpy and volume changes in lipid membranes. I. The proportionality of heat and volume changes in the lipid melting transition and its implication for the elastic constants. *J. Phys. Chem. B.*, 105, 7353–7360.

10 Kratky, O., Leopold, H., and Stabinger, H. (1969). Determination of density of liquids and gases to an accuracy of 10–6 g/cm^3. *Z. Angew. Phys.*, 27, 273–277.

11 Kratky, O., Leopold, H., and Stabinger, H. (1973). The determination of the partial specific volume of proteins by the mechanical oscillator technique. *Methods. Enzymol.*, 27, 98–110.

12 Heimburg, T. (1998). Mechanical aspects of membrane thermodynamics. Estimation of the mechanical properties of lipid membranes close to the chain melting transition from calorimetry. *Biochim. Biophys. Acta.*, 1415, 147–162.

13 Mužić, T., Tounsi, F., Madsen, S.B., Pollakowski, D., Konrad, M., and Heimburg, T. (2019). Melting transitions in biomembranes. *Biochim. Biophys. Acta.*, 1861, 183026.

14 Heerklotz, H. and Seelig, J. (2002). Application of pressure perturbation calorimetry to lipid bilayers. *Biophys. J.*, 82, 1445–1452.

15 Peters, J., Marion, J., Becher, F.J., Trapp, M., Gutberlet, T., Bicout, D.J., et al. (2017). Thermodynamics of lipid multi-lamellar vesicles in presence of sterols at high hydrostatic pressure. *Sci. Rep.*, 7, 15339.

16 Pedersen, U.R., Peters, G.H., Schrøder, T.B., and Dyre, J.C. (2010). Correlated volume-energy fluctuations of phospholipid membranes: a simulation study. *J. Phys. Chem. B.*, 114(6), 2124–2130.

17 Evans, E. and Kwok, R. (1982). Mechanical calorimetry of large dimyristoylphosphatidylcholine vesicles in the phase transition region. *Biochemistry*, 210, 4874–4879.

18 Dimova, R., Pouligny, B., and Dietrich, C. (2000). Pretransitional effects in dimyristoylphosphatidylcholine vesicle membranes: Optical dynamometry study. *Biophys. J.*, 79, 340–356.

19 Helfrich, W. (1973). Elastic properties of lipid bilayers: theory and possible experiments. *Z. Naturforsch. C.*, 28, 693–703.

20 Deuling, H.J. and Helfrich, W. (1976). The curvature elasticity of fluid membranes: a catalogue of vesicle shapes. *J. Phys. France.*, 37, 1335–1345.

21 Deuling, H.J. and Helfrich, W. (1976). Red blood cell shapes as explained on the basis of curvature elasticity. *Biophys. J.*, 16, 861–868.

22 Seifert, U., Berndl, K., and Lipowsky, R. (1991). Shape transformations of vesicles: phase diagram for spontaneous- curvature and bilayer-coupling models. *Phys. Rev. A.*, 44, 1182–1202.

23 Lipowsky, R. (1995). The morphology of lipid membranes. *Curr. Opin. Struct. Biol.*, 5, 531–540.

24 Bhatia, T., Christ, S., Steinkühler, J., Dimova, R., and Lipowsky, R. (2019). Simple sugars shape giant vesicles into multispheres with many membrane necks. *Soft. Matter.*, 16, 1246–1258.

25 Kralj-Iglic, V., Pocsfalvi, G., Mesarec, L., Sustar, V., Hägerstrand, H., and Iglic, A. (2020). Minimizing isotropic and deviatoric membrane energy – A unifying formation mechanism of different cellular membrane nanovesicle types. *PLoS One*, 15, e0244796.

26 Heimburg, T. (2007). *Thermal Biophysics of Membranes.* Wiley VCH, Berlin, Germany.

27 Fine, R.A. and Millero, F.J. (1973). Compressibility of water as a function of temperature and pressure. *J. Chem. Phys.*, 59, 5529–5536.

28 Mosgaard, L.D., Jackson, A.D., and Heimburg, T. (2013). Fluctuations of systems in finite heat reservoirs with applications to phase transitions in lipid membranes. *J. Chem. Phys.*, 139, 125101.

29 van Osdol, W.W., Johnson, M.L., Ye, Q., and Biltonen, R.L. (1991). Relaxation dynamics of the gel to liquid crystalline transition of phosphatidylcholine bilayers. effects of chainlength and vesicle size. *Biophys. J.*, 59, 775–785.

30 Halstenberg, S., Heimburg, T., Hianik, T., Kaatze, U., and Krivanek, R. (1998). Cholesterol-induced variations in the volume and enthalpy fluctuations of lipid bilayers. *Biophys. J.*, 75, 264–271.

31 Sackmann, E., Duwe, H.P., and Engelhardt., H. (1986). Membrane bending elasticity and its role for shape fluctuations and shape transformations of cells and vesicles. *Faraday. Discuss. Chem. Soc.*, 81, 281–290.

32 Fricke, K., Wirthensohn, K., Laxhuber, R., and Sackmann, E. (1986). Flicker spectroscopy of erythrocytes. *Eur. Biophys. J.*, 14, 67–81.

33 Peterson, M.A., Strey, H., and Sackmann, E. (1992). Theoretical and phase contrast microscopic eigenmode analysis of erythrocyte flicker: amplitudes. *J. Phys. France II.*, 2, 1273–1285.

34 Zilker, A., Ziegler, M., and Sackmann, E. (1992). Spectral analysis of erythrocyte flickering in the 0.3–4-μm-1 regime by microinterferometry combined with fast image processing. *Phys. Rev. A.*, 46, 7998–8001.

35 Döbereiner, H.G., Gompper, G., Haluska, C., Kroll, D., Petrov, P., and Riske., K. (2003). Advanced flicker spectroscopy of fluid membranes. *Phys. Rev. Lett.*, 91, 048301.

36 Betz, T. and Sykes, C. (2012). Time resolved membrane fluctuation spectroscopy. *Soft. Matter.*, 8, 5317–5326.

37 Tapia, J., Vera, N., Aguilar, J., González, M., Sánchez, S., Coelho, P., et al. (2021). Correlated flickering of erythrocytes membrane observed with dual time resolved membrane fluctuation spectroscopy under different D-glucose concentrations. *Sci. Rep.*, 11, 2429.

38 Meleard, P., Gerbeaud, C., Pott, T., Fernandez-Puente, L., Bivas, I., Mitov, M.D., et al. (1997). Bending elasticities of model membranes. Influences of temperature and sterol content. *Biophys. J.*, 72, 2616–2629.

39 Heimburg, T. (2000). Monte Carlo simulations of lipid bilayers and lipid protein interactions in the light of recent experiment. *Curr. Opin. Colloid. Interface. Sci.*, 5, 224–231.

40 Nagle, J.F. and Scott, H.L. (1978). Lateral compressibility of lipid mono- and bilayers. Theory of membrane permeability. *Biochim. Biophys. Acta.*, 513, 236–243.

41 Papahadjopoulos, D., Jacobson, K., Nir, S., and Isac, T. (1973). Phase transitions in phospholipid vesicles. Fluorescence polarization and permeability measurements concerning the effect of temperature and cholesterol. *Biochim. Biophys. Acta.*, 311, 330–340.

42 Sabra, M.C., Jørgensen, K., and Mouritsen, O.G. (1996). Lindane suppresses the lipid-bilayer permeability in the main transition region. *Biochim. Biophys. Acta.*, 1282, 85–92.

43 Blicher, A., Wodzinska, K., Fidorra, M., Winterhalter, M., and Heimburg, T. (2009). The temperature dependence of lipid membrane permeability, its quantized nature, and the influence of anesthetics. *Biophys. J.*, 96, 4581–4591.

44 Schrader, W., Ebel, H., Grabitz, P., Hanke, E., Heimburg, T., Hoeckel, M., et al. (2002). Compressibility of lipid mixtures studied by calorimetry and ultrasonic velocity measurements. *J. Phys. Chem. B.*, 106, 6581–6586.

45 Heimburg, T. and Jackson, A.D. (2005). On soliton propagation in biomembranes and nerves. *Proc. Natl. Acad. Sci. USA.*, 102, 9790–9795.

46 Kappler, J., Shrivastava, S., Schneider, M.F., and Netz., R. (2017). Nonlinear fractional waves at elastic interfaces. *Phys. Rev. Fluids.*, 2, 114804.

47 Schneider, M.F. (2021). Living systems approached from physical principles. *Prog. Biophys. Mol. Biol.*, 162, 2–25.

48 Heimburg, T. (2012). The capacitance and electromechanical coupling of lipid membranes close to transitions. The effect of electrostriction. *Biophys. J.*, 103, 918–929.

49 Vanselow, K. (1966). Untersuchungen über die elektrostatisch-mechanischen Kräfte an der Nervenmembran bei der Auslösung des Aktionspotentials. *Z. Dt. Gesellsch. Med. Biol. Elektronik.*, 11, 1–5.

50 Rabbitt, R.D., Ayliffe, H.E., Christensen, D., Pamarth, K., Durney, C., Clifford, S., et al. (2005). Evidence of piezoelectric resonance in isolated outer hair cells. *Biophys. J.*, 88, 2257–2265.

51 Sachs, F., Brownell, W.E., and Petrov, A.G. (2009). Membrane electromechanics in biology, with a focus on hearing. *MRS Bull.*, 34, 665–670.

52 Brownell, W.E., Qian, F., and Anvari, B. (2010). Cell membrane tethers generate mechanical force in response to electrical stimulation. *Biophys. J.*, 99, 845–852.

53 Onsager, L. (1931). Reciprocal relations in irreversible processes. II. *Phys. Rev.*, 38, 2265–2279.

54 Seeger, H.M., Gudmundsson, M.L., and Heimburg, T. (2007). How anesthetics, neurotransmitters, and antibiotics influence the relaxation processes in lipid membranes. *J. Phys. Chem. B.*, 111, 13858–13866.

Part VII

1 Heimburg, T. and Jackson, A.D. (2005). On soliton propagation in biomembranes and nerves. *Proc. Natl. Acad. Sci. USA.*, 102, 9790–9795.

2 Sommerfeld, A. (1992). *Mechanik der deformierbaren Medien, Vorlesungen über theoretische Physik*, vol. 2, Harri Deutsch.

3 Ebel, H., Grabitz, P., and Heimburg, T. (2001). Enthalpy and volume changes in lipid membranes. I. the proportionality of heat and volume changes in the lipid melting transition and its implication for the elastic constants. *J. Phys. Chem. B.*, 105, 7353–7360.

4 Bernardino de la Serna, J., Orädd, G., Bagatolli, L., Simonsen, A.C., Marsh, D., Lindblom, G., et al. (2009). Segregated phases in pulmonary surfactant membranes do not show coexistence of lipid populations with differentiated dynamic properties. *Biophys. J.*, 97, 1381–1389.

5 Heimburg, T. (2022). The effect of stretching on nerve excitability. *Human. Mov. Sci.*, 86, 103000.

6 Kappler, J., Shrivastava, S., Schneider, M.F., and Netz., R. (2017). Nonlinear fractional waves at elastic interfaces. *Phys. Rev. Fluids.*, 2, 114804.

7 Schneider, M.F. (2021). Living systems approached from physical principles. *Prog. Biophys. Mol. Biol.*, 162, 2–25.

8 Engelbrecht, J., Tamm, K., and Peets, T. (2015). On mathematical modelling of solitary pulses in cylindrical biomembranes. *Biomech. Model. Mechanobiol.*, 14, 159–167.

9 Gonzalez-Perez, A., Mosgaard, L.D., Budvytyte, R., Villagran Vargas, E., Jackson, A.D., and Heimburg, T. (2016). Solitary electromechanical pulses in lobster neurons. *Biophys. Chem.*, 216, 51–59.

10 Heimburg, T. (1998). Mechanical aspects of membrane thermodynamics. Estimation of the mechanical properties of lipid membranes close to the chain melting transition from calorimetry. *Biochim. Biophys. Acta.*, 1415, 147–162.

11 Iwasa, K. and Tasaki, I. (1980). Mechanical changes in squid giant-axons associated with production of action potentials. *Biochem. Biophys. Res. Comm.*, 95, 1328–1331.

12 Iwasa, K., Tasaki, I., and Gibbons, R.C. (1980). Swelling of nerve fibres associated with action potentials. *Science*, 210, 338–339.

13 Howarth, J.V., Keynes, R., and Ritchie, J.M. (1968). The origin of the initial heat associated with a single impulse in mammalian non-myelinated nerve fibres. *J. Physiol.*, 194, 745–793.

14 Lautrup, B., Appali, R., Jackson, A.D., and Heimburg, T. (2011). The stability of solitons in biomembranes and nerves. *Eur. Phys. J. E.*, 34, 57.

15 Grigoriev, S.V., Shnyrkov, O.D., Pustovoit, P.M., Iashina, E.G., and Pshenichnyi, K.A. (2022). Experimental evidence for logarithmic fractal structure of botanical trees. *Phys. Rev. E.*, 105, 044412.

16 Mosgaard, L.D., Jackson, A.D., and Heimburg, T. (2012). Low-frequency sound propagation in lipid membranes. In *Advances in Planar Lipid Bilayers and Liposomes*, vol. 16 (ed. A. Iglic), Elsevier, pp. 51–47.

17 Kamp, D.T. (2021). *Numerical Solutions to the Self-contained Soliton Model and the Potential Connection Between the Soliton Theory and Lithium Treatment in Bipolar Disorder*, Master thesis, University of Copenhagen, Copenhagen, Denmark.

18 Ling, T., Boyle, K.C., Zuckerman, V., Flores, T., Ramakrishnan, C., Deisseroth, K., et al. (2020). High-speed interferometric imaging reveals dynamics of neuronal deformation during the action potential. *Proc. Natl. Acad. Sci. USA.*, 117, 10278–10285.

19 Ritchie, J.M. and Keynes, R.D. (1985). The production and absorption of heat associated with electrical activity in nerve and electric organ. *Quart. Rev. Biophys.*, 18, 451–476.

20 Manoranjan, V.S., Ortega, T., and Sanz-Serna, J.M. (1988). Soliton and antisoliton interactions in the "good" Boussinesq equation. *J. Math. Phys.*, 294, 1964–1968.

21 Träuble, H., Teubner, M., Woolley, P., and Eibl, H. (1976). Electrostatic interactions at charged lipid membranes. I. Effects of pH and univalent cations on membrane structure. *Biophys. Chem.*, 4, 319–342.

22 Wang, T., Gonzalez-Perez, A., Budvytyte, R., Jackson, A.D., and Heimburg, T. (2017). Reply to "comment on 'penetration of action potentials during collision in the median and lateral giant axons of invertebrates'". *Phys. Rev. X.*, 7, 028002.

23 Wilke, E. (1912). Das problem der reizleitung im nerven vom standpunkte der wellenlehre aus betrachtet. *Pflügers. Arch.*, 144, 35–38.

24 Wilke, E. and Atzler, E. (1912). Experimentelle beiträge zum problem der reizleitung im nerven. *Pflügers. Arch.*, 146, 430–446.

25 Tasaki, I., Kusano, K., and Byrne, M. (1989). Rapid mechanical and thermal changes in the garfish olfactory nerve associated with a propagated impulse. *Biophys. J.*, 55, 1033–1040.

26 Heimburg, T. (2025). The mechanical properties of nerves, the size of the action potential, and consequences for the brain. *Chem. Phys. Lipids.*, 267, 105461.

27 Tasaki, I. and Byrne, P.M. (1982). Tetanic contraction of the crab nerve evoked by repetitive stimulation. *Biochem. Biophys. Res. Commun.*, 106, 1435–1440.

28 Villagran Vargas, E., Ludu, A., Hustert, R., Gumrich, P., Jackson, A.D., and Heimburg, T. (2011). Periodic solutions and refractory periods in the soliton theory for nerves and the locust femoral nerve. *Biophys. Chem.*, 153, 159–167.

29 Eichmann, U.A., Ludu, A., and Draayer, J.P. (2002). Analysis and classification of nonlinear dispersive evolution equations in the potential representation. *J. Phys. A - Math. Gen.*, 35, 6075–6090.

30 Hormuzdi, S.G., Filippov, M., Mitropoulou, G., Monyer, H., and Bruzzone, R. (2004). Electrical synapses: a dynamic signaling system that shapes the activity of neuronal networks. *Biochim. Biophys. Acta*, 1662, 113–137.

31 Greengard, P. (2001). The neurobiology of slow synaptic transmission. *Science*, 294, 1024–1030.

32 McCue, M.P. and Guinan, J.J. (1994). Acoustically responsive fibers in the vestibular nerve of the cat. *J. Neurosci.*, 14, 6058–6070.

33 McCue, M.P. and Guinan, J.J. (1994). Influence of efferent stimulation on acoustically responsive vestibular afferents in the cat. *J. Neurosci.*, 14, 6071–6083.

34 Govindaraju, A.C., Quraishi, I., Lysakowski, A., Eatock, R., and Raphael, R. (2023). Nonquantal transmission at the vestibular hair cell–calyx synapse: KLV currents modulate fast electrical and slow K+ potentials. *Proc. Natl. Acad. Sci. USA.*, 120, e2207466120.

35 Kandel, E.R., Schwartz, J.H., and Jessell, T.M. (2000). *Principles of Neural Science*. McGraw-Hill, New York, 4th ed.

36 Pinault, D. (1995). Backpropagation of action potentials generated at ectopic axonal loci: hypothesis that axon terminals integrate local environmental signals. *Brain. Res.*, 21, 42–92.

37 Gonzalez-Perez, A., Budvytyte, R., Mosgaard, L.D., Nissen, S., and Heimburg, T. (2014). Penetration of action potentials during collision in the median and lateral giant axons of invertebrates. *Phys. Rev. X.*, 4, 031047.

38 Leirer, C. (2008). *Dynamik und Struktur in der Phasenkoexistenz von Lipidmembranen*, Ph.D. thesis, University of Augsburg.

39 Gudmundsson, M.L. (2007). Phase behaviour in lipid tethers and vesicles - a thermodynamic study, Master's thesis, University of Copenhagen.

40 Griesbauer, J., Bössinger, S., Wixforth, A., and Schneider, M.F. (2012). Propagation of 2d pressure pulses in lipid monolayers and its possible implications for biology. *Phys. Rev. Lett.*, 108, 198103.

41 Griesbauer, J., Bössinger, S., Wixforth, A., and Schneider, M.F. (2012). Simultaneously propagating voltage and pressure pulses in lipid monolayers of pork brain and synthetic lipids. *Phys. Rev. E.*, 86, 061909.

42 Villagran Vargas, E., Rodriguez-Sosa, L., Hustert, R., Blicher, A., Laub, K., and Heimburg, T. (2013). Variations in interpulse interval of double action potentials during propagation in single neurons. *Synapse.*, 67, 68–78.

43 Shrivastava, S. and Schneider, M.F. (2014). Evidence for two-dimensional solitary sound waves in a lipid controlled interface and its implications for biological signalling. *J. R. Soc. Interface.*, 11, 20140098.

Part VIII

1 Hille, B. (1992). *Ionic Channels of Excitable Membranes.* Cambridge University Press, Cambridge.

2 Papahadjopoulos, D., Jacobson, K., Nir, S., and Isac, T. (1973). Phase transitions in phospholipid vesicles. Fluorescence polarization and permeability measurements concerning the effect of temperature and cholesterol. *Biochim. Biophys. Acta.*, 311, 330–340.

3 Sabra, M.C., Jørgensen, K., and Mouritsen, O.G. (1996). Lindane suppresses the lipid-bilayer permeability in the main transition region. *Biochim. Biophys. Acta.*, 1282, 85–92.

4 Blicher, A., Wodzinska, K., Fidorra, M., Winterhalter, M., and Heimburg, T. (2009). The temperature dependence of lipid membrane permeability, its quantized nature, and the influence of anesthetics. *Biophys. J.*, 96, 4581–4591.

5 Wunderlich, B., Leirer, C., Idzko, A., Keyser, U.F., Myles, V., Heimburg, T., et al. (2009). Phase state dependent current fluctuations in pure lipid membranes. *Biophys. J.*, 96, 4592–4597.

6 Nagle, J.F. and Scott, H.L. (1978). Lateral compressibility of lipid mono- and bilayers. Theory of membrane permeability. *Biochim. Biophys. Acta.*, 513, 236–243.

7 Glaser, R.W., Leikin, S.L., Chernomordik, L.V., Pastushenko, V.F., and Sokirko, A.I. (1988). Reversible breakdown of lipid bilayers: Formation and evolution of pores. *Biochim. Biophys. Acta.*, 940, 275–287.

8 Neumann, E., Kakorin, S., and Tænsing, K. (1999). Fundamentals of electroporative delivery of drugs and genes. *Bioelectrochem. Bioenerg.*, 48, 3–16.

9 Gehl, J. (2003). Electroporation: theory and methods, perspectives for drug delivery, gene therapy and research. *Acta Physiol. Scand.*, 177, 437–447.

10 Højholt, K.L., Mužić, T., Jensen, S.D., Bilgin, M., Nylandsted, J., Heimburg, T., et al. (2019). Calcium electroporation and electrochemotherapy for cancer treatment: importance of cell membrane composition investigated by lipidomics, calorimetry and in vitro efficacy. *Sci. Rep.*, 9, 4758.

11 Böckmann, R., de Groot, R., Kakorin, S., Neumann, E., and Grubmüller, H. (2008). Kinetics, statistics, and energetics of lipid membrane electroporation studied by molecular dynamics simulations. *Biophys. J.*, 95, 1837–1850.

12 Mosgaard, L.D. and Heimburg, T. (2013). Lipid ion channels and the role of proteins. *Acc. Chem. Res.*, 46, 2966–2976.

13 Tieleman, D.P., Leontiadou, H., Mark, A.E., and Marrink, S.J. (2003). Simulation of pore formation in lipid bilayers by mechanical stress and electric fields. *J. Am. Chem. Soc.*, 125, 6382–6383.

14 Wodzinska, K., Blicher, A., and Heimburg, T. (2009). The thermodynamics of lipid ion channel formation in the absence and presence of anesthetics. BLM experiments and simulations. *Soft. Matter.*, 5, 3319–3330.

15 Gutsmann, T., Heimburg, T., Keyser, U., Mahendran, K.R., and Winterhalter, M. (2015). Protein reconstitution into freestanding planar lipid membranes for electrophysiological characterization. *Nat. Protoc.*, 10, 188–198.

16 Montal, M. and Mueller, P. (1972). Formation of bimolecular membranes from lipid monolayers and a study of their electrical properties. *Proc. Natl. Acad. Sci. USA.*, 69(12), 3561–3566.

17 Hanke, W., Methfessel, C., Wilmsen, U., and Boheim, G. (1984). Ion channel reconstruction into lipid bilayer membranes on glass patch pipettes. *Bioelectrochem. Bioenerg.*, 12, 329–339.

18 Heimburg, T. (2010). Lipid ion channels. *Biophys. Chem.*, 150, 2–22.

19 Antonov, V.F., Petrov, V.V., Molnar, A.A., Predvoditelev, D.A., and Ivanov, A.S. (1980). The appearance of single-ion channels in unmodified lipid bilayer membranes at the phase transition temperature. *Nature*, 283, 585–586.

20 Neher, E. and Sakmann, B. (1976). Single-channel currents recorded from membrane of denervated frog muscle fibres. *Nature*, 260, 779–802.

21 Hamill, O.P., Marty, A., Neher, E., Sakmann, B., and Sigworth, F.J. (1981). Improved patch-clamp techniques for high-resolution current recording from cells and cell-free membrane patches. *Pflügers. Arch.*, 391, 85–100.

22 Schmidt, D., Jiang, Q.X., and MacKinnon, R. (2006). Phospholipids and the origin of cationic gating charges in voltage sensors. *Nature*, 444, 775–779.

23 Laub, K.R., Witschas, K., Blicher, A., Madsen, S.B., Lückhoff, A., and Heimburg, T. (2012). Comparing ion conductance recordings of synthetic lipid bilayers with cell membranes containing TRP channels. *Biochim. Biophys. Acta.*, 1818, 1–12.

24 Zecchi, K.A. and Heimburg, T. (2020). Nonlinear conductance, rectification and mechanosensitive channel formation of lipid membranes. *Front. Cell. Dev. Biol.*, 8, 592520.

25 Blicher, A. and Heimburg, T. (2013). Voltage-gated lipid ion channels. *PLoS One*, 8, e65707.

26 Abidor, I.G., Arakelyan, V.B., Chernomordik, L.V., Chizmadzhev, Y.A., Pastushenko, V.F., and Tarasevich, M.R. (1979). Electric breakdown of bilayer lipid-membranes. 1. Main experimental facts and their qualitative discussion. *Bioelectrochem. Bioenerg.*, 6, 37–52.

27 Winterhalter, M. and Helfrich, W. (1987). Effect of voltage on pores in membranes. *Phys. Rev. A.*, 36, 5874–5876.

28 Mosgaard, L.D., Zecchi, K.A., and Heimburg, T. (2015). Mechano-capacitive properties of polarized membranes. *Soft. Matter.*, 11, 7899–7910.

29 Petrov, A.G. (1999). *The lyotropic State of Matter. Molecular Physics and Living Matter Physics*, Gordon and Breach Science Publishers, Amsterdam.

30 Voets, T., Droogmans, G., Wissenbach, U., Janssens, A., Flockerzi, V., and Nilius, B. (2004). The principle of temperature-dependent gating in cold- and heat-sensitive TRP channels. *Nature*, 430, 748–754.

31 Talavera, K., Yasumatsu, K., Voets, T., Droogmans, G., Shigemura, N., Ninomiya, Y., et al. (2005). Heat activation of TRPM5 underlies thermal sensitivity of sweet taste. *Nature*, 438, 1022–1025.

32 Hodgkin, A.L. and Huxley, A.F. (1952). A quantitative description of membrane current and its application to conduction and excitation in nerve. *J. Physiol.*, 117, 500–544.

33 Kaufmann, K., Hanke, W., and Corcia, A. (1989). *Ion Channels*, Caruaru.

34 Antonov, V.F., Anosov, A.A., Norik, V.P., and Smirnova, E.Y. (2005). Soft perforation of planar bilayer lipid membranes of dipalmitoylphosphatidylcholine at the temperature of the phase transition from the liquid crystalline to gel state. *Eur. Biophys. J.*, 34, 155–162.

35 Seeger, H.M., Aldrovandi, L., Alessandrini, A., and Facci, P. (2010). Changes in single K^+ channel behavior induced by a lipid phase transition. *Biophys. J.*, 99, 3675–3683.

36 Wang, H. (2023). *The Control of Quantized Conduction Events in Lipid Membranes by Anesthetics*. Master's thesis, University of Copenhagen, Copenhagen, Denmark.

37 Grabitz, P., Ivanova, V.P., and Heimburg, T. (2002). Relaxation kinetics of lipid membranes and its relation to the heat capacity. *Biophys. J.*, 82, 299–309.

38 Seeger, H.M., Gudmundsson, M.L., and Heimburg, T. (2007). How anesthetics, neurotransmitters, and antibiotics influence the relaxation processes in lipid membranes. *J. Phys. Chem. B.*, 111, 13858–13866.

39 Elamrani, K. and Blume, A. (1983). Phase transition kinetics of phosphatidic acid bilayers. A pressure-jump relaxation study. *Biochemistry*, 22, 3305–3311.

40 Gallaher, J., Wodzinska, K., Heimburg, T., and Bier, M. (2010). Ion-channel-like behavior in lipid bilayer membranes at the melting transition. *Phys. Rev. E.*, 81, 061925.

41 Millhauser, G.L., Salpeter, E.E., and Oswald, R.E. (1988). Diffusion models of ion-channel gating and the origin of power-law distributions from single-channel recording. *Proc. Natl. Acad. Sci. USA.*, 85, 1503–1507.

42 Korn, S.J. and Horn, R. (1988). Statistical discrimination of fractal and Markov models of single-channel gating. *Biophys. J.*, 54, 871–877.

43 Liebovitch, L.S. (1989). Analysis of fractal ion channel gating kinetics: kinetic rates, energy levels, and activation energies. *Math. Biosci.*, 93, 97–115.

44 Liebovitch, L.S. and Sullivan, J.M. (1987). Fractal analysis of a voltage-dependent potassium channel from cultured mouse hippocampal neurons. *Biophys. J.*, 52, 979–988.

45 Hodgkin, A.L., Huxley, A.F., and Katz, B. (1952). Measurement of current-voltage relations in the membrane of the giant axon of Loligo. *J. Physiol.*, 116, 424–448.

46 Hodgkin, A.L. and Huxley, A.F. (1952). Currents carried by sodium and potassium ions through the membrane of the giant axon of Loligo. *J. Physiol.*, 116, 449–472.

47 Cacace, M.G., Landau, E.M., and Ramsden, J.J. (1997). The Hofmeister series: salt and solvent effects on interfacial phenomena. *Quart. Rev. Biophys.*, 30, 241–277.

48 Kunz, W., Nostro, P.L., and Ninham, B.W. (2004). The present state of affairs with Hofmeister effects. *Curr. Opin. Colloid. Interface. Sci.*, 9, 1–18.

49 Kunz, W., Henle, J., and Ninham, B.W. (2004). 'Zur Lehre von der Wirkung der Salze' (About the science of the effect of salts): Franz Hofmeister's historic papers. *Curr. Opin. Colloid. Interface. Sci.*, 9, 19–37.
50 Zhang, Y. and Cremer, P.S. (2006). Interactions between macromolecules and ions: the Hofmeister series. *Curr. Op. Chem. Biol.*, 10, 658–66.
51 Heimburg, T. and Marsh, D. (1996). Thermodynamics of the interaction of proteins with lipid membranes. In *Biological Membranes: A Molecular Perspective from Computation and Experiment* (ed. K.M. Merz and B. Roux), Birkhäuser, Boston, pp. 405–462.
52 Morrow, M.R., Davis, J.H., Sharom, F.J., and Lamb, M.P. (1986). Studies of the interaction of human erythrocyte band 3 with membrane lipids using deuterium nuclear magnetic resonance and differential scanning calorimetry. *Biochim. Biophys. Acta.*, 858, 13–20.
53 Freire, E., Markello, T., Rigell, C., and Holloway, P.W. (1983). Calorimetric and fluorescence characterization of interactions between cytochrome b_5 and phosphatidylcholine bilayers. *Biochemistry*, 28, 5634–5643.
54 Ivanova, V.P., Makarov, I.M., Schäffer, T.E., and Heimburg, T. (2003). Analyzing heat capacity profiles of peptide-containing membranes: cluster formation of gramicidin A. *Biophys. J.*, 84, 2427–2439.
55 Seeger, H., Fidorra, M., and Heimburg, T. (2005). Domain size and fluctuations at domain interfaces in lipid mixtures. *Macromol. Symposia.*, 219, 85–96.
56 Stoddart, D., Ayub, M., Hoefler, L., Raychaudhuri, P., Klingelhoefer, J.W., Maglia, G., et al. (2014). Functional truncated membrane pores. *Proc. Natl. Acad. Sci. USA.*, 111, 2425–2430.

Part IX

1 Davy, H. (1800). *Nitrous Oxide or Dephlogisticated Nitrous Air and Its Respiration.* Briggs and Cottle.
2 Overton, C.E. (1901). *Studien über die Narkose.* Verlag Gustav Fischer, Jena, Germany. (English translation: Studies of Narcosis, Chapman and Hall, 1991, R. Lipnick, Ed.).
3 Overton, C.E. (1991). *Studies of Narcosis.* Chapman and Hall, New York. English version of 'Studien der Narkose' from 1901.
4 Kharakoz, D.P. (2001). Phase-transition-driven synaptic exocytosis: a hypothesis and its physiological and evolutionary implications. *Biosci. Rep.*, 210, 801–830.
5 Heimburg, T. and Jackson, A.D. (2007). The thermodynamics of general anesthesia. *Biophys. J.*, 92, 3159–3165.
6 Meyer, H. (1899). Zur Theorie der Alkoholnarkose. Erste Mittheilung. Welche Eigenschaft der Anästhetica bedingt ihre narkotische Wirkung? *Arch. Exp. Pathol. Pharmakol.*, 425, 109–118.
7 Firestone, L.L., Janoff, A.S., and Miller, K. (1987). Lipid-dependent differential effects of stereoisomers of anesthetic alcohols. *Biochim. Biophys. Acta.*, 898, 90–96.
8 Heimburg, T. (2007). *Thermal Biophysics of Membranes.* Wiley VCH, Berlin, Germany.
9 Moss, G.W., Lieb, W.R., and Franks, N.P. (1991). Anesthetic inhibition of firefly luciferase, a protein model for general anesthesia, does not exhibit pressure reversal. *Biophys. J.*, 60, 1309–1314.
10 Franks, N.P. and Lieb, W.R. (1994). Molecular and cellular mechanisms of general anesthesia. *Nature*, 367, 607–614.

11 Franks, N.P. and Lieb, W.R. (2004). Seeing the light. *Anesthesiol.*, 101, 235–237.

12 Græsbøll, K., Sasse-Middelhoff, H., and Heimburg, T. (2014). The thermodynamics of general and local anesthesia. *Biophys. J.*, 106, 2143–2156.

13 Cantor, R.S. (1997). The lateral pressure profile in membranes: a physical mechanism of general anesthesia. *Biochemistry*, 36, 2339–2344.

14 Franks, N.P. and Lieb, W.R. (1985). Mapping of general anaesthetic target sites provides a molecular basis for cutoff effects. *Nature*, 316, 349–351.

15 Miller, K.W., Firestone, L., Alifimoff, J., and Streicher., P. (1989). Nonanesthetic alcohols dissolve in synaptic membranes without perturbing their lipids. *Proc. Natl. Acad. Sci. USA.*, 86, 1084–1087.

16 Liu, J., Laster, M., Koblin, D., Eger, E., Halsey, M., Taheri, S., et al. (1994). A cutoff in potency exists in the perfluoroalkanes. *Anesth. Analg.*, 79, 238–244.

17 Cantor, R.S. (1997). Lateral pressures in cell membranes: a mechanism for modulation of protein function. *J. Phys. Chem. B.*, 101, 1723–1725.

18 Halstenberg, S., Heimburg, T., Hianik, T., Kaatze, U., and Krivanek, R. (1998). Cholesterol-induced variations in the volume and enthalpy fluctuations of lipid bilayers. *Biophys. J.*, 75, 264–271.

19 Lee, A.G. (1977). Lipid phase transitions and phase diagrams. II. Mixtures involving lipids. *Biochim. Biophys. Acta.*, 472, 285–344.

20 Pringle, M.J., Brown, K.B., and Miller, K.W. (1981). Can the lipid theories of anesthesia account for the cutoff in anesthetic potency in homologous series of alcohols? *Mol. Pharmacol.*, 19, 49–55.

21 Lide, D.R. (2005). *Handbook of Chemistry and Physics*, CRC Press, Boca Raton, 86th ed.

22 Kaminoh, Y., Nishimura, S., Kamaya, H., and Ueda, I. (1992). Alcohol interaction with high entropy states of macromolecules: Critical temperature hypothesis for anesthesia cutoff. *Biochim. Biophys. Acta.*, 1106, 335–343.

23 Kamaya, H., Matubayasi, N., and Ueda, I. (1984). Biphasic effect of long-chain n-alkanols on the main-phase transition of phospholipid vesicle membranes. *J. Phys. Chem.*, 88, 797–800.

24 Kamaya, H., Ma, S.M., and Lin, S.H. (1986). Dose-dependent nonlinear response of the main phase-transition temperature of phospholipid membranes to alcohols. *J. Membr. Biol.*, 90, 157–161.

25 Himes, R.S., DiFasio, C.A., and Burney, R.G. (1977). Effects of lidocaine on the anesthetic requirements for nitrous oxide and halothane. *Anesthesiol.*, 47, 437–440.

26 Ben-Shlomo, I., Tverskoy, M., Fleyshman, G., Melnicko, V., and Katz, Y. (2003). Intramuscular administration of lidocaine or bupivacaine alters the effect of midazolam from sedation to hypnosis in a dose-dependent manner. *J. Basic. Clin. Physiol. Pharmacol.*, 14, 257–263.

27 Ben-Shlomo, I., Tverskoy, M., Fleyshman, G., and Cherniavsky, G. (1997). Hypnotic effect of i.v. propofol is enhanced by i.m. administration of either lignocaine or bupivacaine. *British J. Anaesth.*, 78, 375–377.

28 Senturk, M., Pembeci, K., Menda, F., Ozkan, T., Gucyetmez, B., Tugrul, M., et al. (2002). Effects of intramuscular administration of lidocaine or bupivacaine on induction and maintenance doses of propofol evaluated by bispectral index. *Br. J. Anaesth.*, 89, 849–852.

29 Altermatt, F.R., Bugedo, D.A., Delfino, A.E., Solari, S., Guerra, I., Munoz, H.R., et al. (2012). Evaluation of the effect of intravenous lidocaine on propofol requirements during total intravenous anaesthesia as measured by bispectral index. *Br. J. Anaesth.*, 108, 979–983.

30 Agarwal, A., Pandey, R., Dhiraaj, S., Singh, P.K., Raza, M., Pandey, C.K., et al. (2004). The effect of epidural bupivacaine on induction and maintenance doses of propofol (evaluated by bispectral index) and maintenance doses of fentanyl and vecuronium. *Anesth. Analg.*, 99, 1684–1688.

31 DiFasio, C.A., Neiderlehner, J.R., and Burney, R.G. (1976). The anesthetic potency of lidocaine in the rat. *Anesth. Analg.*, 55, 818–821.

32 Pypendop, B.H. and Ilkiw, J.E. (2005). The effects of intravenous lidocaine administration on the minimum alveolar concentration of isoflurane in cats. *Anesth. Analg.*, 100, 97–101.

33 Szmuk, P., Farrow-Gillespie, A., Sheeran, P., and Ezri, T. (2007). The sedative effect of high dose lidocaine. *Anesth. Analg.*, 104, 1613.

34 Zuo, Y., Aistrup, G.L., Marszalec, W., Gillespie, A., Chavez-Noriega, L.E., Yeh, J.Z., et al. (2001). Dual action of n-alcohols on neuronal nicotinic acetylcholine receptors. *Mol. Pharmacol.*, 60, 700–711.

35 Horishita, T. and Harris, R.A. (2008). n-alcohols inhibit voltage-gated Na+ channels expressed in xenopus oocytes. *J. Pharm. Exp. Ther.*, 326, 270–277.

36 Blicher, A., Wodzinska, K., Fidorra, M., Winterhalter, M., and Heimburg, T. (2009). The temperature dependence of lipid membrane permeability, its quantized nature, and the influence of anesthetics. *Biophys. J.*, 96, 4581–4591.

37 Forman, S.A. and Zhou, Q. (1999). Novel modulation of a nicotinic receptor channel mutant reveals that the open state is stabilized by ethanol. *Mol. Pharmacol.*, 55, 102–108.

38 Zuo, Y., Nagata, K., Yeh, J.Z., and Narahashi, T. (2004). Single-channel analyses of ethanol modulation of neuronal nicotinic acetylcholine receptors. *Alcoholism. Clin. Exp. Res.*, 28, 688–696.

39 Johnson, F.H. and Flagler, E.A. (1950). Hydrostatic pressure reversal of narcosis in tadpoles. *Science*, 112, 91–92.

40 Miller, K.W., Paton, W., Smith, R., and Smith, E. (1973). The pressure reversal of general anesthesia and the critical volume hypothesis. *Mol. Pharmacol.*, 9, 131–143.

41 Halsey, M.J. and Wardley-Smith, B. (1975). Pressure reversal of narcosis produced by anaesthetics, narcotics and transquilizers. *Nature*, 257, 811–813.

42 Mountcastle, D.B., Biltonen, R.L., and Halsey, M.J. (1978). Effect of anesthetics and pressure on the thermotropic behavior of multilamellar dipalmitoylphosphatidylcholine liposomes. *Proc. Natl. Acad. Sci. USA.*, 75, 4906–4910.

43 Firestone, L.L., Miller, J.C., and Miller, K.W. (1986). Tables of physical and pharmacological properties of anesthetics. In *Molecular and Cellular Mechanism of Anesthetics* (eds S.H. Roth and K.W. Miller), Plenum Press, pp. 455–470.

44 Kendig, J.J., Schneider, T.M., and Cohen, E.N. (1978). Pressure, temperature and repetitive impulse generation and synaptic transmission. *J. Appl. Physiol. Respir. Envir. Exercise Physiol.*, 45, 742–746.

45 Träuble, H. and Eibl, H. (1974). Electrostatic effects on lipid phase transitions: membrane structure and ionic environment. *Proc. Natl. Acad. Sci. USA.*, 71, 214–219.

46 Träuble, H., Teubner, M., Woolley, P., and Eibl, H. (1976). Electrostatic interactions at charged lipid membranes. I. Effects of pH and univalent cations on membrane structure. *Biophys. Chem.*, 4, 319–342.

47 Punnia-Moorthy, A. (1987). Evaluation of ph changes in inflammation of the subcutaneous air pouch lining in the rat, induced by carrageenan dextran and staphylococcus aureus. *J. Oral. Pathol.*, 16, 36–44.

48 Hauser, H. and Shipley, G. (1981). Crystallization of phosphatidylserine bilayers induced by lithium. *J. Biol. Chem.*, 256, 11377–11380.

49 Mužić, T., Tounsi, F., Madsen, S.B., Pollakowski, D., Konrad, M., and Heimburg, T. (2019). Melting transitions in biomembranes. *Biochim. Biophys. Acta.*, 1861, 183026.

50 Wang, T., Mužić, T., Jackson, A.D., and Heimburg, T. (2018). The free energy of biomembrane and nerve excitation and the role of anesthetics. *Biochim. Biophys. Acta.*, 1860, 2145–2153.

51 Moldovan, M., Lange, K.H.W., Aachmann-Andersen, N.J., Kjær, T.W., Olsen, N.V., and Krarup, C. (2014). Transient impairment of the axolemma following regional anaesthesia by lidocaine in humans. *J. Physiol.*, 592, 2735–2750.

52 Modovan, M. (2012). Personal communication.

53 Cantor, R.S. (2003) Receptor desensitization by neurotransmitters in membranes: are neurotransmitters the endogenous anesthetics? *Biochem.*, 42, 11891–11897.

54 Seeger, H.M., Gudmundsson, M.L., and Heimburg, T. (2007). How anesthetics, neurotransmitters, and antibiotics influence the relaxation processes in lipid membranes. *J. Phys. Chem. B.*, 111, 13858–13866.

55 Hazel, J.R. (1979). Influence of thermal acclimation on membrane lipid composition of rainbow trout liver. *Am. J. Physiol. Regulatory Integrative Comp. Physiol.*, 287, R633–R641.

56 DeLong, E.F. and Yayanos, A.A. (1985). Adaptation of the membrane lipids of a deep-sea bacterium to changes in hydrostatic pressure. *Science*, 228, 1101–1103.

57 Kinney, M., Jones, W.R., Royal, R., Brauer, R.W., and Sorrel, F.Y. (1981). A gradient tube system for the study of the effect of high hydrostatic pressures on temperature preference behavior in small aquatic animals. *Comp. Biochem. Physiol.*, 68, 501–505.

58 Marjanovic, M., Elliott, A., and Dawson, M. (1998). The temperature dependence of intracellular ph in isolated frog skeletal muscle: lessons concerning the Na^+-H^+ exchanger. *J. Membr. Biol.*, 161, 215–225.

59 Malan, A., Wilson, T., and Reeves, R. (1976). Intracellular pH in cold-blooded vertebrates as a function of body temperature. *Resp. Physiol.*, 28, 29–47.

60 Aickin, C.C. and Thomas, R.C. (1977). An investigation of the ionic mechanism of intracellular pH regulation in mouse soleus muscle fibres. *J. Physiol.*, 273, 295–316.

61 Ellis, D. and MacLeod, K. (1985). Sodium-dependent control of intracellular pH in Purkinje fibres of sheep heart. *J. Physiol.*, 359, 81–105.

62 Dawson, M.J. and Wray, S. (1985). The effects of pregnancy and parturition on phosphorus metabolites in rat uterus studied by ^{31}P nuclear magnetic resonance. *J. Physiol.*, 368, 19–31.

63 Johnson, D.C., Burt, C., Perng, W., and Hitzig, B. (1993). Effects of temperature on muscle pH_i and phosphate metabolites in newts and lungless salamanders. *Am. J. Physiol.*, 265, R1162–R1167.

64 Ingram, L.O. (1977). Changes in lipid composition of *Escherichia coli* resulting from growth with organic solvents and with food additives. *Appl. Environ. Microbiol.*, 33, 1233–1236.

65 Pawlosky, R.J. and Salem., N. (1995). Ethanol exposure causes a decrease in docosahexaenoic acid and an increase in docosapentaenoic acid in feline brains and retinas. *Am. J. Clin. Nutrition.*, 61, 1284–1289.

66 Pita, M.L., Rubio, J.M., Murillo, M.L., Carreras, O., and Delgado, M.J. (1997). Chronic alcoholism decreases polyunsaturated fatty acid levels in human plasma, erythrocytes, and platelets influence of chronic liver disease. *Thromb. Haemost.*, 78, 808–812.

67 Han, Y.H. (1969). Why do chronic alcoholics require more anesthesia? *Anesthesiol.*, 30, 341–342.

68 Fassoulaki, A., Farinotti, R., Servin, F., and Desmonts, J. (1993). Chronic alcoholism increases the induction dose of propofol in humans. *Anesth. Analg.*, 77, 553–556.

69 Schneider, M.F. (2021). Living systems approached from physical principles. *Prog. Biophys Mol. Biol.*, 162, 2–25.

70 O'Connor, C.S., Crawshaw, L.I., Bedichek, R.C., and Crabbe., C. (1988). The effect of ethanol on temperature selection in the goldfish, Carassius auratus. *Pharmacol. Biochem. Behav.*, 29, 243–248.

71 Crawshaw, L.I., Wollmuth, L.P., and O'Connor, C.S. (1989). Intracranial ethanol and ambient anoxia elicit selection of cooler water by goldfish. *Am. J. Physiol.*, 256, R133–R137.

72 O'Connor, C.S., Crawshaw, L.I., Kosobud, A., Bedichek, R.C., and Crabbe., C. (1989). The effect of ethanol on behavioral temperature regulation in mice. *Pharmacol. Biochem. Behav.*, 33, 315–319.

73 Ramachandra, V., Moore, C., Kaur, N., and Carli, F. (1989). Effect of halothane, enflurane and isoflurane on body temperature during and after surgery. *Br. J. Anaesth.*, 62, 409–414.

74 Tigerstedt, R. (1880). *Studien über die mechanische Nervenreizung*. J. C. Frenkel and Son, Helsingfors.

75 Iwasa, K. and Tasaki, I. (1980). Mechanical changes in squid giant-axons associated with production of action potentials. *Biochem. Biophys. Res. Commun.*, 95, 1328–1331.

76 Iwasa, K., Tasaki, I., and Gibbons, R.C. (1980). Swelling of nerve fibres associated with action potentials. *Science*, 210, 338–339.

77 Tasaki, I., Kusano, K., and Byrne, M. (1989). Rapid mechanical and thermal changes in the garfish olfactory nerve associated with a propagated impulse. *Biophys. J.*, 55, 1033–1040.

78 Kim, G.H., Kosterin, P., Obaid, A., and Salzberg, B.M. (2007). A mechanical spike accompanies the action potential in mammalian nerve terminals. *Biophys. J.*, 92, 3122–3129.

79 Gonzalez-Perez, A., Mosgaard, L.D., Budvytyte, R., Villagran Vargas, E., Jackson, A.D., and Heimburg, T. (2016). Solitary electromechanical pulses in lobster neurons. *Biophys. Chem.*, 216, 51–59.

80 Ling, T., Boyle, K.C., Zuckerman, V., Flores, T., Ramakrishnan, C., Deisseroth, K., et al. (2020). High-speed interferometric imaging reveals dynamics of neuronal deformation during the action potential. *Proc. Natl. Acad. Sci. USA.*, 117, 10278–10285.

81 Butler, D.S. (1989). Adverse mechanical tension in the nervous system: a model for assessment and treatment. *Aust. J. Physiother.*, 35, 227–238.

82 Wilke, E. (1912). Das problem der reizleitung im nerven vom standpunkte der wellenlehre aus betrachtet. *Pflügers. Arch.*, 144, 35–38.

83 Wilke, E. and Atzler, E. (1912). Experimentelle beiträge zum problem der reizleitung im nerven. *Pflügers. Arch.*, 146, 430–446.

84 Hayasi, K. and Rittler, G. (1935). über die abhängigkeit der zeiterregbarkeit des motorischen nerven vom dehnungszustand seines muskels. *Pflügers. Arch.*, 235, 50–52.

85 Budini, F. and Tilp, M. (2016). Changes in h-reflex amplitude to muscle stretch and lengthening in humans. *Rev. Neurosci.*, 27, 511–522.

86 Guissard, N., Duchateau, J., and Hainaut, K. (2001). Mechanisms of decreased motoneurone excitation during passive muscle stretching. *Exp. Brain Res.*, 137, 163–169.

87 Guissard, N. and Duchateau, J. (2006). Neural aspects of muscle stretching. *Exerc. Sport. Sci. R.*, 34, 154–158.

88 Stecker, M.M., Baylor, K., Wolfe, J., and Stevenson., M. (2011). Acute nerve stretch and the compound motor action potential. *J. Brachial. Plex. Peripher. Nerve. Inj.*, 6, 1–12.

89 Ochs, S., Pourmand, R., Si, K., and Friedman, R. (2000). Stretch of mammalian nerve in vitro: effect on compound action potentials. *J. Peripher. Nerv. Syst.*, 5, 227–235.

90 Wall, E.J., Massie, J.B., Kwan, M.K., Rydevik, B.L., Myers, R.R., and Garfin, S.R. (1992). Experimental stretch neuropathy. Changes in nerve conduction under tension. *J. Bone. Joint. Surg. Br.*, 74, 126–129.

91 Kwan, M.K., Wall, E., Massie, J., and Garfin, S. (2009). Strain, stress and stretch of peripheral nerve rabbit experiments in vitro and in vivo. *Acta. Orthop. Scand.*, 63, 267–272.

92 Ucar, H., Watanabe, S., Noguchi, J., Morimoto, Y., Iino, Y., Yagishita, S., et al. (2021). Mechanical actions of dendritic-spine enlargement on presynaptic exocytosis. *Nature*, 600, 686–689.

93 Heimburg, T. (1998). Mechanical aspects of membrane thermodynamics. Estimation of the mechanical properties of lipid membranes close to the chain melting transition from calorimetry. *Biochim. Biophys. Acta.*, 1415, 147–162.

94 Heimburg, T. (2022). The effect of stretching on nerve excitability. *Human. Mov. Sci.*, 86, 103000.

95 Ivanova, V.P. and Heimburg, T. (2001). A histogram method to obtain heat capacities in lipid monolayers, curved bilayers and membranes containing peptides. *Phys. Rev. E.*, 63, 041914.

96 Li, J. and Shi, R. (2007). Stretch-induced nerve conduction deficits in guinea pig ex vivo nerve. *J. Biomech.*, 40, 569–578.

97 Elble, R.J. (2013). What is essential tremor? *Curr. Neurol. Neurosci. Rep.*, 13, 353.

98 Boutin, E., Vaugoyeau, M., Eusebio, A., Azulay, J.P., and Witjas, T. (2015). News and controversies regarding essential tremor. *Rev. Neurol.*, 171, 415–425.

99 Espay, A.J., Lang, A., Erro, R., Merola, A., Fasano, A., Berardelli, A., et al. (2017). Essential pitfalls in "essential" tremor. *Mov Disord.*, 32, 325–331.

100 Growdon, J.H., Shahani, B.T., and Young, R.R. (1975). The effect of alcohol on essential tremor. *Neurol.*, 25, 259.

101 Koller, W.C. and Biary, N. (1984). Effect of alcohol on tremors: comparison with propranolol. *Neurol.*, 34, 221–222.

102 Koller, W.C., Busenbark, K., Miner, K., and Essential tremor study Group (1994). The relationship of essential tremor to other movement disorders: report on 678 patients. *Ann. Neurol.*, 35, 717–723.

103 Zeuner, K.E., Molloy, F.M., Shoge, R.O., Goldstein, S.R., Wesley, R., and Hallett, M. (2003). Effect of ethanol on the central oscillator in essential tremor. *Mov. Disord.*, 18, 1280–1285.

104 Klebe, S., Stolze, H., Grensing, K., Volkmann, J., Wenzelburger, R., and Deuschl, G. (2005). Influence of alcohol on gait in patients with essential tremor. *Neurol.*, 65, 96–101.

105 Knudsen, K., Lorenz, D., and Deuschl, G. (2011). A clinical test for alcohol sensitivity of essential tremor. *Mov Disord.*, 26, 2291–2295.

106 Cade, J.F.J. (1949). Lithium salts in the treatment of psychotic excitement. *Med. J. Australia.*, 2, 518–520.

107 Schou, M., Juel-Nielsen, N., Strömgren, E., and Voldby, H. (1954). The treatment of manic psychosis by the administration of lithium salts. *J. Neurol. Neurosurg. Psychiatry.*, 17, 250–260.

108 Schou, M. (1958). Lithium studies. 1. Toxicity. *Acta Pharmacol. Tox.*, 15, 70–84.

109 Schou, M. (2001). Lithium treatment at 52. *J. Affect. Disord.*, 67, 21–32.

110 Kamp, D. (2024). A physical perspective on lithium therapy. *Prog. Biophys. Mol. Biol.*, 194, 55–74.

111 Sproule, B. (2002). Lithium in bipolar disorder: can drug concentrations predict therapeutic effects? *Clin. Pharmacokinet.*, 41, 639–660.

112 Hauser, H. and Shipley, G.G. (1983). Interactions of monovalent cations with phosphatidylserine bilayer membranes. *Biochemistry*, 22, 2171–2178.

113 Binder, H. and Zschörnig, O. (2002). The effect of metal cations on the phase behavior and hydration characteristics of phospholipid membranes. *Chem. Phys. Lipids*, 115, 39–61.

114 Cacace, M.G., Landau, E.M., and Ramsden, J.J. (1997). The Hofmeister series: salt and solvent effects on interfacial phenomena. *Quart. Rev. Biophys.*, 30, 241–277.

115 Kunz, W., Nostro, P.L., and Ninham, B.W. (2004). The present state of affairs with Hofmeister effects. *Curr. Opin. Colloid. Interface. Sci.*, 9, 1–18.

116 Kunz, W., Henle, J., and Ninham, B.W. (2004). 'Zur Lehre von der Wirkung der Salze' (About the science of the effect of salts): Franz Hofmeister's historic papers. *Curr. Opin. Colloid. Interface. Sci.*, 9, 19–37.

117 Zhang, Y. and Cremer, P.S. (2006). Interactions between macromolecules and ions: the Hofmeister series. *Curr. Op. Chem. Biol.*, 10, 658–666.

118 Stahl, S.M. (2021). *Stahl's Essential Psychopharmacology. Nueroscientific Basis and Practical Applications*. Cambridge University Press, Cambridge, UK, 5th ed.

119 Hallett, M. (1986). Differential diagnosis of tremor. In *Handbook of Clinical Neurology* (eds P. Vinken, G. Bruyn, and H. Klawan), Elsevier, New York, pp. 583–595.

120 Baek, J.H., Kinrys, G., and Nierenberg, A. (2014). Lithium tremor revisited: pathophysiology and treatment. *Acta. Psychiat. Scand.*, 129, 17–23.

121 Lopez-Corcuera, B., Gimenez, C., and Aragon, C. (1988). Change of synaptic membrane lipid composition and fluidity by chronic administration of lithium. *Biochim. Biophys. Acta.*, 939, 467–475.

122 Nascimento, N.F., Carlson, K., Amaral, D., Logan, R., and Seggio, J. (2015). Alcohol and lithium have opposing effects on the period and phase of the behavioral free-running activity rhythm. *Alcohol*, 49, 367–376.

123 Mason, A.E., Kasl, P., Soltani, S., Green, A., Hartogensis, W., Dilchert, S., et al. (2024). Elevated body temperature is associated with depressive symptoms: results from the TemPredict study. *Sci. Rep.*, 14, 1884.

124 Souetre, E., Salvati, T., Wehr, A., Sack, D.A., Krebs, B., and Darcourt, G. (1988). Twenty-four-hour profiles of body temperature and plasma TSH in bipolar patients during depression and during remission and in normal control subjects. *Am. J. Psychiatry.*, 145, 1133–1137.

125 Mellerup, E.T., Widding, A., Wildschiødtz, G., and Rafaelsen, O.J. (1978). Lithium effect on temperature rhythm in psychiatric patients. *Acta. Pharmacol. Toxicol. (Copenh).*, 42, 125–129.

126 Tufail, Y., Yoshihiro, A., Pati, S., Li, M.M., and Tyler, W.J. (2011). Ultrasonic neuromodulation by brain stimulation with transcranial ultrasound. *Nat. Protoc.*, 6, 1453–1470.

127 Tyler, W.J. (2012). The mechanobiology of brain function. *Nat. Rev. Neurosci.*, 13, 867–878.

128 Mueller, J.K. and Tyler, W.J. (2014). A quantitative overview of biophysical forces impinging on neural function. *Phys. Biol.*, 11, 051001.

129 Rezayat, E. and Toostani., I. (2016). A review on brain stimulation using low intensity focused ultrasound. *Basic. Clin. Neurosci. J.*, 7, 187–194.

130 Tyler, W.J., Lani, S.W., and Hwang, G.M. (2018). Ultrasonic modulation of neural circuit activity. *Curr. Opin. Neurobiol.*, 50, 222–231.

131 Sanguinetti, J.L., Hameroff, S., Smith, E., Sato, T., Daft, C., Tyler, W., et al. (2020). Transcranial focused ultrasound to the right prefrontal cortex improves mood and alters functional connectivity in humans. *Front. Hum. Neurosci.*, 14, 52.

132 Kamimura, H.A.S., Wang, S., Chen, H., Wang, Q., Aurup, C., Acosta, C., et al. (2016). Focused ultrasound neuromodulation of cortical and subcortical brain structures using 1.9 mhz. *Med. Phys.*, 43, 5730–5735.

133 Ye, P.P., Brown, J., and Pauly, K. (2016). Frequency dependence of ultrasound neurostimulation in the mouse brain. *Ultrasound. Med. Biol.*, 42, 1512–1530.

134 Jerusalem, A., Al-Rekabi, Z., Chen, H., Ercole, A., Malboubi, M., Tamayo-Elizalde, M., et al. (2019). Electrophysiological-mechanical coupling in the neuronal membrane and its role in ultrasound neuromodulation and general anaesthesia. *Acta Biomater*, 97, 116–140.

135 Kamimura, H.A.S., Conti, A., Toschi, N., and Konofagou., E. (2020). Ultrasound neuromodulation: mechanisms and the potential of multimodal stimulation for neuronal function assessment. *Front. Phys.*, 8, 150.

136 Heimburg, T. and Jackson, A.D. (2005). On soliton propagation in biomembranes and nerves. *Proc. Natl. Acad. Sci. USA.*, 102, 9790–9795.

Epilogue

1 Wigner, E.P. (1960). The unreasonable effectiveness of mathematics in the natural sciences. *Commun. Pure Appl. Math.*, 13, 1–14.

2 Zuo, Y., Nagata, K., Yeh, J.Z., and Narahashi, T. (2004). Single-channel analyses of ethanol modulation of neuronal nicotinic acetylcholine receptors. *Alcohol. Clin. Exp. Res.*, 28, 688–696.

3 Zuo, Y., Yeh, J.Z., and Narahashi, T. (2004). Octanol modulation of neuronal nicotinic acetylcholine receptor single channels. *Alcohol. Clin. Exp. Res.*, 28, 1648–1656.

4 Wodzinska, K., Blicher, A., and Heimburg, T. (2009). The thermodynamics of lipid ion channel formation in the absence and presence of anesthetics. BLM experiments and simulations. *Soft. Matter.*, 5, 3319–3330.

5 Mullikin-Kilpatrick, D. and Treistman, S.N. (1994). Ethanol inhibition of L-type Ca^{2+} channels in PC_{12} cells: role of permeant ions. *Europ. J. Pharmacol.*, 270, 17–25.

6 Brodie, M.S., Scholz, A., Weiger, T.M., and Dopico, A.M. (2007). Ethanol interactions with calcium-dependent potassium channels. *Alcohol. Clin. Exp. Res.*, 31, 1625—1632.

7 Varley, T.F., Pope, M., Puxeddu, M., Faskowitz, J., and Sporns, O. (2023). Partial entropy decomposition reveals higher-order information structures in human brain activity. *Proc. Natl. Acad. Sci. USA.*, 120, e2300888120.

8 Chladni, E.F.F. (1817). Neue Beyträge zur akustik, Leipzig.

9 Tuan, P.H., Tung, J., Liang, H., Chiang, P., Huang, K., and Chen, Y. (2015). Resolving the formation of modern Chladni figures. *Eur. Phys. Lett.*, 111, 64004.

10 Tuan, P.H., Wen, C., Yu, Y., Liang, H., Huang, K., and Chen, Y. (2014). Exploring the distinction between experimental resonant modes and theoretical eigenmodes: from vibrating plates to laser cavities. *Phys. Rev. E.*, 89, 022911.

11 Cabral, J., Fernandes, F., and Shemesh, N. (2023). Intrinsic macroscale oscillatory modes driving long range functional connectivity in female rat brains detected by ultrafast fMRI. *Nat. Commun.*, 14, 375.

12 Toader, C., Tataru, C., Florian, I.A., Covache-Busuioc, R.A., Bratu, B.G., Glavan, L., et al. (2023). Cognitive crescendo: how music shapes the brain's structure and function. *Brain. Sci.*, 13, 1390.

13 Perani, D., Saccuman, M., Scifo, P., Spada, D., Andreolli, G., Rovelli, R., et al. (2010). Functional specializations for music processing in the human newborn brain. *Proc. Natl. Acad. Sci. USA.*, 107, 4758–4763.

14 Koelsch, S. (2014). Brain correlates of music-evoked emotions. *Nat. Rev. Neurosci.*, 15, 170–180.

15 Scholkmann, F., Haslbeck, F., Oba, E., Restin, T., Ostojic, D., Kleiser, S., et al. (2024). Creative music therapy in preterm infants effects cerebrovascular oxygenation and perfusion. *Sci. Rep.*, 14, 28249.

16 McDermott, J. and Hauser, M. (2007). Nonhuman primates prefer slow tempos but dislike music overall. *Cognition*, 104, 654–668.

17 Gess, A. (2007). Birds like music, too. *Science*, 317, 1864–1864.

18 Watanabe, S. and Nemoto, M. (1998). Reinforcing property of music in java sparrows (*Padda oryzivora*). *Behav. Process.*, 43, 211–218.

19 Watanabe, S. and Sato, K. (1999). Discriminative stimulus properties of music in java sparrows. *Behav. Process.*, 47, 53–57.

20 Chase, A.R. (2001). Music discriminations by carp (Cyprinus carpio). *Anim. Learn. Behav.*, 29, 336–353.

21 Shinozuka, K., Ono, H., and Watanabe, S. (2013). Reinforcing and discriminative stimulus properties of music in goldfish. *Behav. Process.*, 99, 26–33.

Epilogue

In 1960, Eugene Wigner wrote a thoughtful essay about "The unreasonable effectiveness of mathematics in the natural sciences" [1]. In it, he marvels at the remarkable fact that we can describe nature with mathematical equations. He concludes:

> *"The miracle of the appropriateness of the language of mathematics for the formulation of the laws of physics is a wonderful gift which we neither understand nor deserve. We should be grateful for it and hope that it will remain valid in future research and that it will extend, for better or for worse, to our pleasure even though perhaps also to our bafflement, to wide branches of learning."*

We place similar trust in the reliability of thermodynamics, which is the most fundamental theory in physics. Thermodynamic couplings are strictly true. They are neither hypothetical nor tentative. The laws of thermodynamics apply to all matter at all length scales, including biological matter and, of course, the brain. The thermodynamic correlations will also be strictly valid in the membranes that make up the nervous system. Thus, everything that has been deduced in this book is based on our confidence that thermodynamics applies to nerve cells and excitable membranes. The nerve impulse is macroscopic in scale, and we have applied thermodynamics to these scales. Such an approach can only deal with phenomena that occur at this scale. The theory of nerve cells and excitatory membranes is not suitable for dealing with phenomena that occur at the molecular scale, such as the binding of certain drugs to a macromolecule or the effect of a particular toxin. However, such effects may be inherent in the chemical potentials of such molecules – in the same way that changes in the molecular structure of a liquid may affect its macroscopic compressibility, although the compressibility is not a molecular but a macroscopic ensemble property.

Here we have provided a theory for nerve cells and excitatory membranes. It gives rise to solitary electromechanical impulses that represent the action potential. Their appearance depends on the presence of a phase transition in the nerve membrane, slightly below body or ambient temperature. It is a little-known fact that such transitions have been consistently found in the membranes of nerve cells, bacterial membranes, cancer cells, and lung surfactant. The intensive thermodynamic variables affect the transition temperature. The transition temperature is increased and nerve membranes become more excitable when certain variables such as pressure and the concentration of lithium and protons, increase. Others, such as the concentration of anesthetics and neurotransmitters like serotonin,

Physics of Nerves and Excitatory Membranes, First Edition. Thomas Heimburg.

as well as nerve stretching, lower the transition temperature and make the nerve less excitable. It is therefore easy to see how different thermodynamic variables relate to each other, and whether their effects add up or compensate for each other. For example, the effect of anesthetics is compensated for by an increase in hydrostatic pressure. As the soliton theory is a real physical theory with predictive power, the absence of pressure reversal of anesthesia would have falsified the whole concept.

Any change in the intensive variables that increases the phase transition has the potential to excite a nerve impulse. Thus, action potentials can be excited by voltage, mechanically, by lowering the pH or by cold needles touching the nerve. In Part VIII, we have shown that all these changes also induce channel-like conduction events in lipid membranes, which are indistinguishable in amplitude and open lifetime from conduction traces attributed to channel proteins. Thus, the complete phenomenology of nerve excitability and the occurrence of channel-like conduction falls out of a thermodynamic treatment. Drugs that lower the transition temperatures and thereby reduce the excitability of nerves also suppress channel events on purely thermodynamic grounds. Compared to protein channels, it has been shown that the effect of ethanol and octanol on lipid channels is very similar to their effect on the nicotinic acetylcholine receptor [2–4]. The same is true when the effect of ethanol on lipid membrane conductance is compared to voltage-gated calcium channels [5] and potassium channels [6]. This suggests that protein channels and lipid membrane channels have much in common and are subject to the same or very similar thermodynamic control mechanisms. This is unlikely to be a coincidence.

Membranes also adapt over time to changes in the intensive variable. For example, the membrane composition of certain deep-sea bacteria adapts to hydrostatic pressure, and *E. coli* membranes adapt to changes in growth temperature. Instead of changing membrane composition, the body can also change other variables such as pH or temperature. Inflammation lowers the pH of tissues, which can be compensated for by an increase in body temperature (fever). Anesthesia can be counteracted by lowering the body temperature, i.e., hypothermia. It can be seen that the predictable effect of certain variables on the thermodynamic state of the nerve membrane can lead to unexpected but interesting correlations. The thermodynamic concepts described in this book offer new perspectives on nerves, the brain and disease, and may even shed light on perception, as discussed below.

The focus of this book was on the transmission and the excitation of impulses in single axons. There has been little reference to the brain as a whole. The brain is currently understood as a network of pairwise connections of neurons. In contrast, Varley and collaborators [7] have recently explored the emergence of synergistic information in the brain, showing that its activity cannot be reduced to pairs of nodes between neurons. That is, they tried to explore properties of the brain that emerge at larger scales. Using information theory, they concluded that there is higher-order structure in brain activity, i.e., information is not stored locally. The information is widespread and invisible to network-based approaches.

In what follows, I take the opportunity to speculate about the implications of some of our findings for larger neural structures. The action potential has dimensions larger than many neurons and can span entire regions of the brain. Since our approach is based on the adiabatic theory of sound, let us discuss the possibility of acoustic modes in a brain. Such modes would have non-local properties that do not result from local interactions.

Consider a tuning fork. If two tuning forks are placed close together and one is struck by a hammer, the vibration can be transmitted through the air to a second tuning fork if it has the same resonant frequency (see Figure 1). This is called acoustic resonance. In a sense, the second tuning fork can "hear" the first. However, it can only hear the frequencies that correspond to the fundamental and the overtones. These frequencies are determined by the geometry of the tuning fork. You could call this "memory."

In two dimensions, a metal plate can be excited by a periodic force in the center of the plate, as shown in Figure 2. Standing wave patterns can be visualized by adding some sand to the vibrating surfaces, which arranges itself along the nodes of the vibrations. Distinct patterns, called "Chladni figures", are found for each eigenfrequency of the plate. Figure 3 shows some of these figures from the original work of Ernst Chladni from 1817 [8], and some more recent experiments from Tuan et al. [9]. Which pattern is produced depends not only on the frequency and the size of the plate. A plate can have several modes simultaneously, which can be found by principal mode analysis. Calculations of such modes

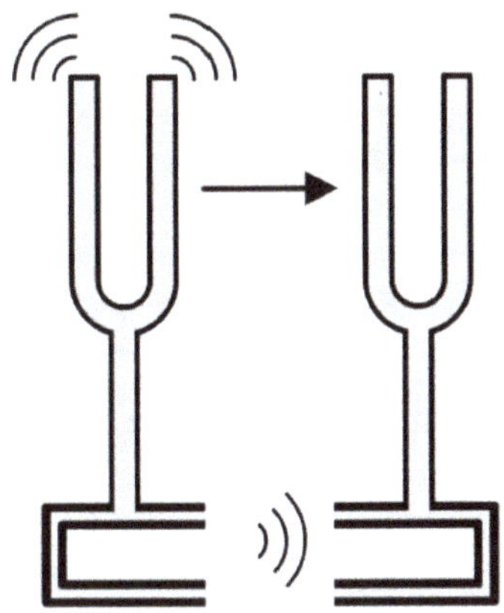

Figure 1 Acoustic resonance between two tuning forks.

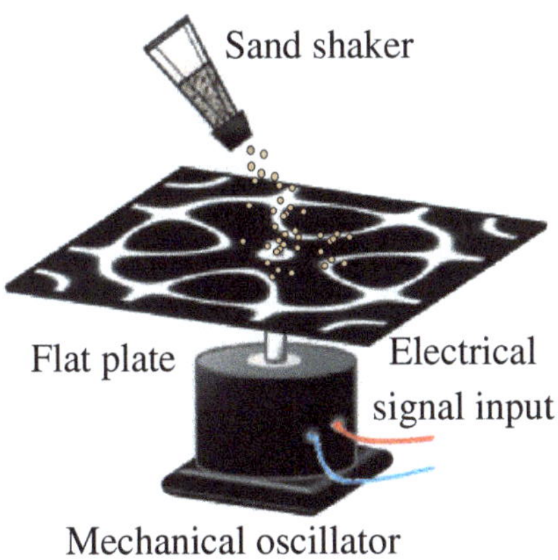

Figure 2 A quadratic metal plate is excited by a mechanical oscillator. At specific frequencies, the plate shows standing wave resonances that can be visualized by sand. The resulting patterns are called Chladni figures. [9]/with permission of IOP Publishing.

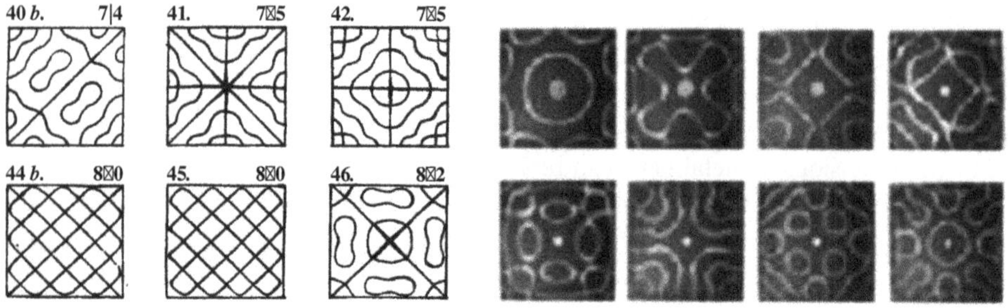

Figure 3 Left: Selected vibrational modes on rectangular plate from the original work of Chladni. [8]/Breitkopf & Härtel/Public Domain. Right: Eight Chladni figures on a rectangular plate in a frequency range between 1000 and 3000 Hz. [9]/with permission of IOP Publishing.

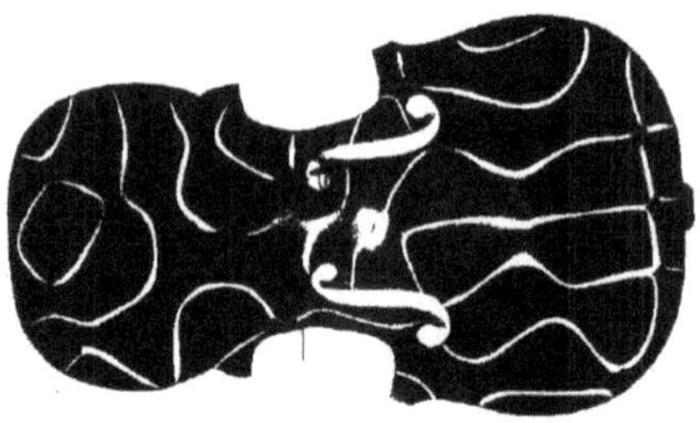

Figure 4 Chladni figures on a violin-shaped plate with a length of about 1 m at 762.4 Hz. Image by Stephen Morris. Licensed under CC BY 2.0 via Flickr (https: //tinyurl.com/h3pyayc9).

and the sum of modes are given in Ref. [10]. Modes with nodes at the position of the oscillator are not excited. Therefore, the position of the periodic excitation is also important. A Chladni figure will also form if the plate is excited by sound waves from the outside, e.g., a loudspeaker under the plate. You could say that the plate can "hear" certain frequencies that it "remembers" and that the "memory" is expressed as a standing wave pattern.

Instead of searching for resonant frequencies and suitable excitation points, one could fix the frequency and adjust the shape of the plate until it resonates. This is what is done when optimizing the shape of a violin. Figure 4 shows Chladni figures on metal in the shape of the front plate of a violin. This plate memorizes a specific sound. Adapting the shape of the plate until the sound is perfect could be called "learning." Here the memory is of a non-local geometric nature and is distributed over the entire surface of the violin. Of course, the learning process is carried out with the help of a violin maker, because the front plate of the violin cannot change shape and find resonance on its own. It also requires the trained ear of a musician. However, we note that the learning process could consist of an adaptation of the geometry.

Can the brain (or a region of the brain) be a mechanical resonator supporting resonant modes? In this book, we have treated the nerve impulse as an entropy-conserving, sound-like process. Thus, the aforementioned considerations about tuning forks and Chladni plates might also apply here. The brain consists of a complicated surface generated by billions of neurons connected in a highly nontrivial way. Because the surface is organized in three dimensions, many more principal modes are expected than on a rectangular disc. Neurons are connected by synapses. Connections can be broken and new ones formed. Therefore, the brain is a surface that can change and adapt on its own. It could learn by adapting its shape to external stimuli and create memories by forming resonances. Since excitability depends on intensive variables (as shown at various points in this book), modes in the brain could be altered by drugs such as anesthetics, by neurotransmitters, by pH, or by temperature.

A recent paper by Cabral and co-workers [11] demonstrated that the brain does indeed support macroscale resonant modes. The authors performed ultrafast functional magnetic resonance imaging (fMRI) on a rat brain. The patterns in the image oscillated in time. Principal mode analysis was used to identify coherent patterns of these oscillations – just as is done when analyzing Chladni figures [10]. Figure 5 shows the superposition of the first 10 principal modes of the analysis in one particular experiment on a rat brain. Structures can be seen on the length scale of a few millimeters, which is the scale expected for the size of nerve impulses. The timescale and appearance of the oscillations were modulated by the presence of anesthetics. Therefore, there is solid experimental support for the aforementioned speculations.

The possible existence of such resonant modes in the brain brings us back to the concept of music. In the Preface, we began with the observation that there are all sorts of analogies to music in our language to describe our perception – phrases like "being in tune," "being in harmony," and the like. These may be mere analogies and entirely coincidental. It is quite clear that the perception of harmony has a profound effect on us. There may also be a physical reason for this. If the brain were a resonator, it would automatically generate overtones and harmonies.

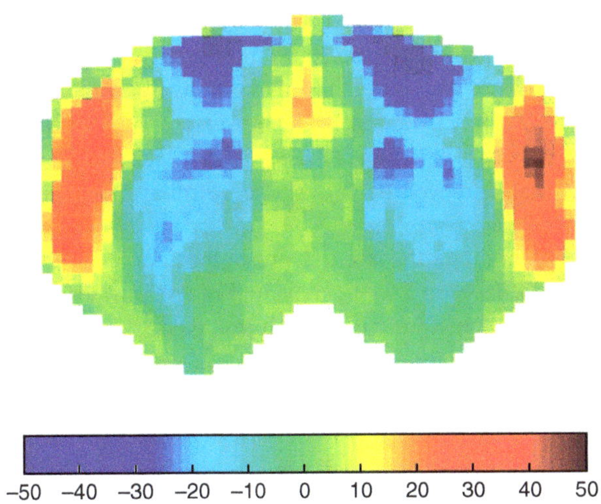

Figure 5 Standing waves in a rat brain measured by ultrafast fMRI. Shown is a superposition of the first 10 principal modes. The size of a rat brain is about 1.5–2 cm in diameter horizontally. Thus, the wavelength of the standing waves is of the order of a few millimeters. [11]/Springer Nature/CC BY 4.0.

The tone scale of many musical traditions, as well as the tuning of musical instruments such as the piano, is based on 12 semitones in an octave. An octave is a doubling of a frequency. The concert pitch of an "a" is 440 Hz. This is the lowest natural frequency (fundamental) of the respective strings of a piano. One octave higher is 880 Hz. This is the fundamental frequency of a string of half the length. An octave and a fifth is almost exactly three times the frequency.[1] This means that a fifth is 3/2 of the fundamental. If you go up the scale in fifths 12 times, you will gradually find all the keys of a piano. This can be achieved by tripling the frequency. This is why the octave and the fifth are special. Harmony results from whole-numbered frequency ratios.

So here is the question: It is obvious that a string recognizes a certain frequency. But why does the brain recognize tones and musical harmonies? This only makes sense if the body contains some sort of organ that resembles a resonator. One could argue, of course, that it is the inner ear, the cochlea, that resembles a string-like resonator, and that would not be wrong. But when frequencies are translated into electrical nerve impulses in the inner ear, the brain receives a sequence of impulses rather than music. Such signals require interpretation by the brain. Harmonies are not inherent in the signal itself. But why are harmonies recognized as something special? Why does the brain dislike dissonance and like octaves and fifths? It seems that the brain can recognize harmonies in a more direct way, without the need for learning and interpretation. The perception of music may be embedded in the geometry of the brain itself.

Music seems to be a universal phenomenon present in all cultures. Listening to music improves cognitive functions such as memory and attention span. In rehabilitation, music-based therapies have a high rate of success in treating depression and anxiety [12]. It probably does not require learning, unlike many other human abilities. Newborn babies with little or no auditory experience already process harmonies [13, 14]. Creative music therapy works as a treatment in preterm infants [15]. Even animals can recognize music. Based on the observation that non-human primates seem to be indifferent to music and prefer silence to music, McDermott and Hauser suggest that only humans like music [16]. However, Gess [17] states that this is not entirely true: "Birds like music too." Java sparrows can distinguish between Bach and Schönberg [18, 19]. In fact, they prefer Bach and Vivaldi to modern music, harmony to dissonance, and even Bach to silence. Similar reports exist for carp and goldfish [20, 21].

As described in the Preface, the starting point for the research presented here was a casual discussion with my colleague Andrew D. Jackson about the beauty of Johann Sebastian Bach's music. It is reassuring that we share this preference with koi, goldfish, and sparrows. Therefore, there is evidence that the understanding of music does not require training and human rationality, but is a property of the brain. Our perception of music may therefore provide clues to the functioning of the brain itself. The idea of the brain as an adaptive resonator is clearly speculative. But these speculations are based on undoubted thermodynamic properties of neurons that can be tested experimentally.

1 There is a slight difference, however, because $\left(\frac{3}{2}\right)^{12}/2^7 = 1.01364$, which means that after 12 fifths you are not exactly at seven octaves, but there is a slight deviation of a quarter of a key. This small deviation led to the introduction of a well-tempered tuning of the piano strings, which was corrected for this small deviation.

Index

Physics of Nerves and Excitatory Membranes, First Edition. Thomas Heimburg.

d

g

h

t

www.ingramcontent.com/pod-product-compliance
Lightning Source LLC
Chambersburg PA
CBHW081206260926
51849CB00008B/430